ACTIVITY-ORIENTED MATHEMATICS

Readings for Elementary Teachers

compiled and edited by

WILLIAM E. SCHALL

State University of New York,
College at Fredonia

D1510904

Prindle, Weber & Schmidt, Incorporated
Boston, Massachusetts

© 1976 by Prindle, Weber & Schmidt, Incorporated
20 Newbury Street, Boston, Massachusetts 02116.

Printed in the United States of America.

Library of Congress Cataloging in Publication Data

Main entry under title:

Activity-oriented mathematics.

    Includes bibliographies and indexes.
    1. Mathematics—Study and teaching (Elementary)
I. Schall, William E.
QA135.5.A28   372.7'08   75-26640
ISBN 0-87150-173-2

PREFACE

The purpose of this book of readings is practical, not theoretical. Consistent with this objective of practicality, its goal is to place at the disposal of students, teachers, and subject or curriculum specialists a collection of articles that covers a broad range of mathematical topics. These articles have been selected so that examples, ideas, and techniques can be used in the reader's classroom as soon as the opportunity presents itself.

Most of the content of this volume is designed to help the reader gain a clearer understanding of some interesting strategies and techniques in the teaching of mathematics. It is the hope that these readings of a practical nature will aid the classroom teacher to supplement or implement a mathematics program that will be relevant, exciting, challenging, and educationally sound.

This book is divided into eight parts, each of which explores a different aspect of elementary school mathematics education. A brief introduction is given before each chapter to help the reader identify with the topic, to present background information, and to stimulate the reader to pursue the issues in depth. In addition, a selected bibliography of related articles has been included after each chapter. This should provide a valuable avenue of exploration and continued research for the interested reader.

> Mathematics is fascinating to many persons because of its utility and because it presents opportunities to create and discover. It is continuously and rapidly growing because of intellectual curiosity, practical applications and the invention of new ideas.*

Many changes have and are taking place in the mathematics curriculum of the elementary school, resulting in an atmosphere in which mathematics has found acceptance in the public mind as probably never before. The time is past when teachers can be content believing they have taught mathematics if they have only taught students to compete mechanically. If more, and hopefully better, mathematics is to be taught in the elementary grades, then surely one important criterion of the effectiveness of any school's program will be the extent to which pupils *understand.* If mathematics is taught by people who do not like or understand the subject, it is highly probable that their students will not like or understand it.†

Although the selected readings cover a wide range of topics in elementary school mathematics, the compiler must freely admit that many valuable articles had to

Studies in Mathematics, Vol. VI, *Number Systems,* School Mathematics Study Group, (New Haven, Yale University Press, 1961), preface.
†*Ibid.*

be omitted because of space limitations or because they were not available. However, with these limitations in mind, the readings selected are in keeping with the main objective of this book, that of practicality. The criterion for including each article was the question, "Does this article have an idea that the reader will be able to use in class tomorrow?" It should also be noted that this book is not limited only to recent publications, but includes both current articles as well as earlier ones that are congruent with the primary objective of the book.

The compiler acknowledges and is indeed appreciative of those who granted him permission to reprint their materials. This book would not have been possible without their unselfish cooperation and assistance. The compiler is also greatly indebted to Mrs. Peggy Card and Mrs. Elaine Repasch, typists, to Mrs. Kay Riley, production editor, and last, but not least, to his family. Their support, love, and understanding patience made the difficult moments tolerable.

WILLIAM E. SCHALL

CONTENTS

vi

PART I

GOALS FOR ELEMENTARY SCHOOL MATHEMATICS

CHAPTER 1

INTRODUCTION

Mathematics is queen of the sciences and arithmetic is queen of the mathematics. She often condescends to render service to astronomy and other natural sciences, but under all circumstances the first place is her due.

Carl Frederich Gauss (1777–1855)

The objective of this chapter is to provide a strong rationale for the role that mathematics has played in the history of mankind. Even a cursory examination of the history of the world and of our society today displays a panorama of the intensive use, development, and growth of mathematics.

As Ben Sueltz indicates, the development of the Hindu-Arabic numeration system has been called the greatest single achievement in the history of the human race. Mathematics, and consequently science, invention and industry, could not have advanced as they have without the use of *this* system. Numbers and the number system provide a suitable and powerful framework not only for organizing and classifying data and information but also for the study of relationships therein. And frequently by a pure mathematical extension or venture, a whole new area of knowledge is opened.

In "Queen of the Sciences" Mr. Sueltz does an excellent job of building a powerful rationale for the importance of mathematics in the history and the future of the world. His article lays the foundation for this book of readings and for future mathematical pursuits.

QUEEN OF THE SCIENCES

Ben A. Sueltz

The Queenly state is a high estate with qualities such as beauty and charm, authority and power, and of service to her people. The popular mind of current America may think of a rose queen, a cotton queen, or a potato queen. Several years ago, however, it was common among learned men to speak of religion as the queen of the sciences. This was the era in which scholars used the pattern of logic that began with assumptions and reasoned toward conclusions and much of the reasoning was done in the area of religion. Then appeared one of the best minds of all time, Carl Frederich Gauss (1777-1855), who wrote "Mathematics is queen of the sciences and arithmetic is queen of mathematics. She often condescends to render service to astronomy and other natural sciences, but under all circumstances the first place is her due."* It is the quote from Gauss that suggests the development which follows. The status of Gauss, who was often called "The Prince of Mathematicians," is apparent from an account of Laplace, who when asked to suggest the greatest mathematician of Germany as the head of the observatory at the University of Gottingen, passed over Gauss and named Pfaff. When asked why he did not name Gauss he replied, "Oh, he is the greatest mathematician in the world."*

What is Mathematics?

The elementary student, the layman, and perhaps many teachers confuse the machinery of mathematics with the essence of mathematics. That is, their view is apt to be limited to such things as $2 + 2 = 4$, $A = \pi r^2$, sine $A = y/r$, and such daily concepts as measurement, cost, and dividends. These are mere ideas, devices, and procedures which a mathematician may use casually but they are not the sinew of mathematics which is more concerned with the study of structures and relationships and the discovery of principles and patterns. That is, mathematics involves thinking and the real mathematics is represented by a high order of logic which frequently is represented by "If A, then B." That is, if proposition A is true then it can be demonstrated that proposition B must be true. And this is not primarily concerned with the apparent truth of proposition A. It is the reasoning that is the characteristic of mathematics. Benjamin Pierce, (1809-1880), expressed it as "Mathematics is the science

---

* Gauss, Carl Frederich, quoted from Bell, "The Queen of the Sciences," The Williams and Wilkins Co., Baltimore, 1931.

Reprinted with permission from the *New York State Mathematics Teachers' Journal* (January 1964), 6–19.

that draws necessary conclusions." Later Bertrand Russell stated "Mathematics may be defined as the subject in which we never know what we are talking about, nor whether what we are saying is true." You will again recognize in this statement that mathematics is pure in that it is unconcerned with the machinery of calculation or the areas of application of the subject. Essentially then, mathematics is postulational thinking and the mathematician trusts that his proposition A may be true and that the resultant proposition B finds itself in harmony with the world of experience.

To illustrate the role of pure mathematics: in geometry, Euclid set as propositions of the class A "self evident truths" which had been established by critical observation. From these many propositions of the class B were deduced. For example; if we assume that through a point only one line can be drawn parallel to a given line, then certain conclusions are "necessary." But in 1854 Riemann discarded Euclid's straight line postulate and produced a whole new and consistent geometry which was largely for his own fancy and which he regarded as totally non-applicable to the physical world. But fifty years later Einstein used the Riemmanian geometry in his theory of relativity which became one of the cornerstones for revolutionizing the whole area we call physics. And so it was with the mathematical developments of Gibbs and thermodynamics, Clerk and Maxwell and radio and television, Cayley and the matrices of modern physics, and Einstein and others and the whole modern area of atomic fission. Yes, the pure mathematician is not concerned with the "bread and butter" values of his work. But he is not displeased when later someone puts his work to "practical" application.

Pure and Applied Mathematics

The mathematician tends to regard mathematics in the realm of what we have been calling pure mathematics while the layman thinks of applied mathematics. Let us compare these two realms with a simple geometric development.

A. A point is a location which has zero dimensions.

B. A line is a one-dimensional magnitude which may be generated by a moving point (in a straight line). We speak of *length* in one dimension.

C. A square is a two-dimensional magnitude which may be generated by a line moving at right angles to the direction of the line and for a distance equal to the line. We speak of the *area* in two dimensions.

D. A cube is a three-dimensional magnitude which may be generated by moving a square at right angles to the plane of the square and for a distance equal to the original length of the line. We speak of volume in three dimensions.

These figures and their development or generation may be pictured thus:

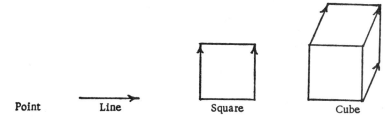

Point   Line   Square   Cube

Thus far it is easily apparent that these simple geometric figures are the basis for all measurement and calculation of distances, areas, and volumes as these are commonly used in commerce and industry. These uses of the abstract figures are called applied mathematics.

Let us extend our understanding with a little analysis of the figures:

Figure	Point	Line	Square	Cube	Tessaract
No. Dimensions	0	1	2	3	4
No. Bounding Figures		2	4	6	8
Dimensions of the Bounding Figures		0	1	2	3

The mathematician immediately asks if there isn't a 4th dimension, a 5th dimnsion, or an N-dimensional space. He postlulates such magnitudes and begins to explore propositions concerning them. Consider a simple exploration into 4th dimensional geometry which we may do by letting the cube move into a 4th dimension (direction) for a distance equal to the original length. Sketched, the figure looks like this and the layman can see that a number of cubes are formed. By an extension of the previous chart it would appear that the 4th dimensional figure is bounded by 8 cubes. We shall call this figure a tessaract. It has a number of interesting properties.

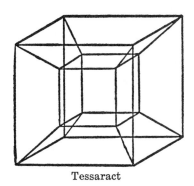

Tessaract

But let us take a non-mathematician's excursion into fancy.
1. A point-space has no dimensions. Living on this space would be like a fly pinned to a board. He has no possible motion in zero dimensions. It is a very restricted existence.

2. Not let us place the fly on a one-dimensional line. He can move back and forth the length of the line. Next consider three flies **A, B, and C on the line, thus: A B C.** They are free to move back and forth but A can never be a joining neighbor with C because B is always between them in D_1* space.
3. Now let us place our flies A, B, and C in D_2 and note their greater freedom. Now A can walk around B on the surface and be a neighbor of C. That is, a D_1 line becomes a boundary for our D_2 flies.
4. Next let us place the 3 flies in a D_3 cube and note that they are free to move back and forth, across, and up and down or in any combination of these three movements. Their limiting bounds are now made by D_2 squares. These are D_3 flies living in a D_3 space even as you and I.
5. But let us move into the tessaract of D_4. Here a D_2 wall is no longer a boundary but rather a D_3 cube becomes the restricting limit. Let us postulate a D_4 being. How does it differ from our D_3 flies that have 3 freedoms? Of course it has 4 freedoms of motion in 4 dimentions. It is a little difficult for D_3 beings like us to comprehend. But we may conclude that as a line is no boundary for a D_3 being, so a surface (wall) is no bound for a D_4 being. Are there D_2 beings? Are there D_n beings? The mathematician doesn't much care but the applied mathematician immediately seeks a reality of a mathematical proposition. To him there might be D_4 beings and he may be out looking for them. He reasons by analogy and recalls that people scoffed at microrganisms long after microscopes revealed them. Yes, there may be things invisible to us and if they are of a higher order they may be 4th dimensionally chuckling at our very crude modes of investigation of what we regard as "the secrets of the universe."

Do not abhor the work of fancy. It is a common method of our better thinkers including the pure mathematicians. The pure mathematician is an aesthetic while the applied mathematician is a pragmatist. The explorations of the one may open new vistas for the other. Both are seeking new insights into structures and relationships. It should not be assumed that the applied mathematician uses only the results of the pure mathematician. These two frequently work interchangeably to enhance each other's work and often one person embodies both aspects of mathematics.

Mathematics as a Servant

Mathematics provides the sinew and structure for many areas of intellectual endeavor. Consider in art the concepts of symmetry, perspective and projection, and proportion. The classical Greek temple design and the arrangement of Chinese public buildings exhibit a high level of symmetry while the

---

* D_1, D_2, D_3 respectively signifying the dimensionality of space.

Japanese architecture, design, and floral arrangement achieves beauty through a carefully studied asymmetry. Most analytical students of the arts are familiar with dynamic symmetry which distinguishes an incommensurate-root rectangle from one whose dimensions have a simple or static ratio. That is, a static rectangle is one in which the ratio of width to length is some simple numerical ratio such as 1 to 2, 2 to 3, or 5 to 9. In a dynamic rectangle, the ratio would involve an incommensurate value such as 1 to $\sqrt{2}$ or $\sqrt{3}$ to 5. Various scholars such as Hambidge of Yale have found, that in the analysis of classic art forms, a commensurate root rectangle suggests a static figure while an incommensurate ratio suggests mobility as for example an animal in repose versus one in motion. Similarly a curve such as an ogive with a changing ratio of curvature forces the eye to travel along the curve whereas an arc of a circle is more static. The artists Durer, Albrecht and Utrillo were masters of perspective and used this quality to give depth to their paintings.

The properties of golden section or divine proportion have long intrigued artists and mathematicians. This is the relationship where the width is to the length as the length is to the width plus the length: $(W:L = L:W + L)$. This too is an incommensurate relationship. Designers use it to achieve a pleasing relationship.

Mathematics is an organizing factor and its various number systems serve to systematize information in many areas. Consider for example the work of Charles Booth in his pioneer study in sociology in which he sought to show a numerical relationship which poverty, misery, and depravity bear to regular earnings and comparative comfort of the people of London. More recently researches in social sciences have developed interesting extensions of statistical techniques and have even borrowed from the "game theory" which might well be the handmaiden of the sophisticated gambler.

In recent years the "NeoMalthusians" have become exercised by studies of birth rates and population growth. It is so easy to take a few items of data, relate them, discover a limited relationship and then project this into the future. The man who, twenty-five years ago, "proved" that in a few decades more than half the population of this country would be living in mobile homes was a chart maker and interpreter but not a mathematician. The business analyst who ascribes to data an element of germaneness must first demonstrate this nascent characteristic.

The musical arts use mathematical scales such as the pentatonic and our own octic with majors and minors. Also the notations and the principles of harmonics are mathematical. The physics of wave motion while not understood by many composers is basic to musical structure. Frequently a composer, like a Japanese floral arranger, will seek a dissonance. No one expects or desires music to become completely mathematical and predictable. That would remove the element of artistry. It is interesting to note that it is now possible to place certain characteristics of a composer into a machine and have the machine compose music, as for example in the style of Bach, and

7

this same machine music possesses even an element of originality.

Of course such physical sciences as chemistry, physics, and astronomy depend upon mathematics for their "laws" and principles of operation. In fact, many new scientific discoveries are made mathematically instead of in the laboratory and are then verified by experiment. The same is true for new industrial processes. In fact, a whole new plant may be established because it has been demonstrated mathematically and thus the waiting period of building a pilot plant and of experiment are avoided. In any first rate college it is now expected that a student should have mastered at least a year of calculus before he begins the serious study of science.

Two events are illuminating. For a number of years it was apparent that the planet Uranus was observed not to follow its calculated orbit. The calculations were correct so the astronomer La Verrier reasoned that some body unknown and unobserved must cause the deviations and from the original data and the known deviations he calculated the existence of an additional planet (Neptune) and wrote Galle in Berlin telling him when and where to look for a new planet. This he did and thus verified Neptune which was discovered with paper and pencil. A German, Diesel, designed an engine with paper and pencil. People scoffed at him but later an engineer built such an engine just to see if it would work. It did and now the diesel engine is common property. Similar accounts abound in the fields of science and industry.

Gambling is worth a small note in the story of mathematics. It is founded upon the principles of chance and probability which a student first encounters in advanced algebra. The same principles that operate with dice and in poker also operate in insurance. It might be worthwhile for the gambler or chance taker to recall that chance has no memory. Lightning can strike twice in the same place and the theory that "bad luck must change" will only serve to drive the gambler further into debt. The idea of randomness, which in game theory is called the "random walk" can be applied to such a gamble as "A has $100, B has $200, they 'high spade' for $1 until one has all the money." The problem is to calculate how many plays before one of them has all the money. And now this same "random walk" is being applied to sociological data.

The ancient Chinese game of NIM (c.3000 B.C.) uses the base-two number system and anyone familiar with this system cannot lose. Likewise various pegboard games fall into the class $2^n - 1$ and riddle of Benares and the game of Josephus or "Turks and Christians" is based upon $2^{64} - 1$. Many of you know the story of the invention of Chess for which the Grand Vizier awarded to Shirham 1 grain of wheat for the first square on the board, 2 grains for the second square, 4 for the 3rd, etc. for the 64 squares. This amounted to $2^{64} - 1$ grains which is more wheat than our present farm surplus. In Europe about 1880 the game of 15 of Le Jeu de Taquin swept rampant. This is the little box of 16 square cells with 15 numbers which can be arranged in various sequences. The possibilities are factorial

16 (16 × 15 × 14 . .). An obscure mathematician proved that certain positions were impossible. But millions of people spent untold hours seeking to do the impossible. This is reminiscent of "proofs" of squaring the circle which has been established as impossible by the use of straightedge and compasses. As late as last spring an untutored man from California sent me a copyrighted proof and when a member of our staff wrote him pointing out his error he became irate in that a supposed mathematician would give him no credit and might even want to steal his ideas.

A Few Words on Mathematicians

The creative, imaginative mathematician is apt to be a young man. At least his more important thoughts had their incubation before he was 30 years old. He may work for many years but his genius was apparent in most cases before age 25. He has rather good intelligence. If he majored in mathematics at a good college his I.Q. was at least 125 and if he worked through the doctor's degree in mathematics he probably has an I.Q. of at least 140 and more probably 150. With this native ability he could succeed in many areas but he has other acquired characteristics. He is apt to be a "lone wolf" in his work and this may have developed because others confound him with trivia and irrelevancies. If he is a genuine creative mathematician he has a flair of genius which probably was apparent at a very early age and his I.Q. may be anywhere from 150 to 180. He is apt to have a rather fine sense of the aesthetic and an unconscious sense of beauty. He might even play chamber music and occasionally bridge or poker but he is apt not to be a country clubber or Rotarian. His recreations are creative or artistic rather than commercial.

The mathematician is imaginative. He thinks intuitively and inductively and experimentally lets his imagination work toward a conclusion via hunches, educated guesses, and analogies with simple examples. When he has a tentative conclusion he works out a most rigorous deductive proof. He states his assumptions and definitions and his conclusion must then be unique and certain. He is attuned to the pattern "If A, then B." He enjoys what he is doing and likes to work at his own pace at his own time. The chairman of the mathematics department of one of our better universities who has two respected research mathematicians on his staff when asked what they were working on said, "I don't know, they seem loath to tell me and yet I'm supposed to make an annual statement about their work." In the 19th century when Jacobi was asked why he worked in mathematics said, "Pour l'honneur de l'espirit humain." This is almost universal for a creative person.

Most lower-level mathematicians like those who teach in ordinary colleges merely travel the well marked roads in the valleys, a few reach the foothills and occasionally one has a glimpse of the peaks; but few can scale Mt. Everest and still fewer reach into the clouds beyond. But it is these who reach out who extend our knowledge.

Applied mathematics has developed many principles used in measurement in the physics world. Every elementary school pupil has learned direct measurement by applying a linear unit such as a foot or meter to a length or distance, a square unit such as the square inch to a surface or area, and a cubic unit such as the cubic foot or cubic centimeter to a prism or other volume. But soon he develops algebraic formulas for areas and volumes and uses these instead of direct measurement in two and three dimensions.

Indirect measurement is appealing to the student and the amateur mathematician because it involves an element of drama. Consider for example, finding the height of a tree or flagpole or building without measuring it directly and by using the principles of similar triangles developed by the mathematician.

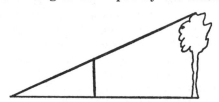

Then consider the use of trigonometric ratios in indirect measurement. Consider the surveyors who chart a tunnel through a mountain for a distance of 36 miles by boring from each side and when the two meet their lines of centers fail to coincide by a mere ½ inch. Or the Atlantic coastal survey which was based upon a single distance measurement and with thousands of angle measurements extending over hundreds of miles. When checked back the accumulated error was less than $1/_5$ inch! These are truly remarkable demonstrations. But let us extend the same method to celestial mechanics. We know the height of mountains on the moon, we know many celestial sizes and distances and orbits and now we use these procedures to chart the course of a satellite and an intercontinental missile.

Computation and Mathematics

The invention of the Hindu-Arabic numeral system has been called the greatest single achievement of the human race. Mathematics and consequently science, invention, and industry could not have advanced as they have without this system. This is attested by the fact that non-occidental cultures have been forced to adopt this system in order to advance. Note what has happened in Japan in fifty years.

Most ancient cultures required some form of counting frame such as an abacus for their computations. In fact, the very word counter as we use this in the expression "over the counter" historically signifies passing goods to the purchaser over the counting board on which reckoning was done with pebbles or calculi. Of course our word calculate literally means, "using pebbles."

Our culture has extended the system of natural whole numbers to include common and decimal fractions, negative numbers, imaginary numbers, irrational numbers, logarithmic sequences, and transfinite numbers. And in the process certain peculiar values such as the constants π, 1, e, and Planck's constant have become most important. Numbers and the number system not only provide a convenient frame work for classifying data and information but also for the study of relationships inherent therein. And frequently by a pure mathematical extension or venture a whole new area is opened as was the case with Boolean algebra and its applications to communication.

For centuries man has sought to develop aids to computation. The abacus was one, mechanical computers and cash registers are others. But the more recent electronic computer is truly a marvel of the age. It uses the same basic number system as the old Chinese game of NIM with a base two but with the principles developed for the Hindu-Arabic system which uses a base of ten.

Some 25 years ago the operator of one of our best mechanical computers was given the simple problem: "If I found a knife that when new cost one dollar but when I found it was worth 50 cents and I sold it for 25 cents, what was my per cent of gain? The operator, a rather attractive and intelligent young woman, set the problem on the machine and in the answer space the machine filled each space with a 9 and kept repeating the figure 9 in the last space. It did not stop and the operator said, "There seems to be something wrong with the machine, I'll call the mechanic." The machine was trying to show infinity but the operator did not recognize it. This is cited to demonstrate that while computers may do amazing things, may even make discoveries, they are not real thinking machines and we must depend upon programmers, operators, and readers or interpreters for these computers. A modern electronic computer alone is like an idiot. It can sort and store information and do minute things, one at a time, but it does these little things at the rate of 2 million per second and such a rapid assembly of minute results soon becomes a prodigious amount. It will not be uncommon shortly to have electronic computers with these components:

    a. a basic computer using transistors

    b. a special control unit that keeps track of the computer

    c. a third unit that feeds information to the basic computer and assembles conclusions

This new computer was partially planned by existing computers which charted the 60,000 connections and identified which wires required shielding. And when the machine breaks down it will diagnose its own difficulty, all this at a cost of some 3 million dollars.

What can you do with such a computer? All the usual computations plus such things as:

    a. It can make decisions of the nature of whether or not to launch a new product such as an automobile or a breakfast food or the proper relation of cost price to distribution costs to profit from a study of the maxima and minima from an algebraic matrix.

    b. It can apply the Monte Carlo method where many factors and variables work simultaneously and in sequence and apply this to roulette, to battle plans, and to nuclear actions.

c. It can play the Random Game repeatedly and seek a pattern and this may prove effective in the analysis of social phenomenon.

d. From coded tapes, it can make many operations semi-automatic in the solution of problems. It can similarly play checkers, dominoes, chess, and even compose music, furnished a style or pattern.

e. It can translate languages according to rule and pattern with results that are more meaningful but less picturesque, i.e., good, gooder and louse, louses.

But a computer still requires human brains to direct it. For this we pay up to $15,000 per year for programmers and $25,000 for the men with ideas.

Mathematics, A Language

A language is developed for the recording of thought and its transfer from one to another. In mathematics we use a combination of words and abstract symbols. The words may be confusing as was the case when a modern Latin teacher translated a paragraph from Gauss who wrote in a more medieval form. But the symbols have a precise meaning. Consider the common items such as $+$, $=$, $\sqrt{\,}$, and π. These are universal. They are precise in meaning and significance. But the user must be alert and intelligent and not like a colleague who requested a film rental for $600 when she intended $6.00. Again, no system can produce better results than the one who uses it. In mathematics, new symbolism is created as required but it takes a period of time for the world to agree on a symbol. Perhaps the way in which Robert Recorde described the two equal parallel lines as representative of the idea of equality was a factor in the international adoption. Frequently a symbol is copied by succeeding generations as was the case with the *fleur de lis* as the principal point of the compass.

Mathematical writing tends to parsimonious. This is perhaps an outgrowth of the adherence to what the mathematician calls "necessary and sufficient conditions" and that leads him to avoid the superfluous. For example, the British book states that the "forgetting curve is asymptotic" while an American education book would spend a paragraph or page telling what this means. The one assumes competence on the part of the reader, the other insults the intelligence of many readers and tends to make them lazy. A similar thing happened a few year ago when a prominent educator criticized a document written by this author because it omitted an important idea. Someone asked him to read a certain short paragraph which told in succinct language exactly what he wanted said. He had been lazy or incompetent in reading.

The mathematician like the scientist is apt to be a slow reader because the things he reads require a good deal of reflection and especially the average student will need to read, ponder, read, ponder, and think some more in order to grasp the ideas. This is in contrast to the situation when one of our better mathematics teachers was having dinner on the train between Detroit and New York. He was seated across from a vice president of a motor company who finally said, "Aren't you a school teacher?" and when the affirmative reply was given, the vice president said, "I thought so, you say everything three times." Think of what such a repetitious school-teacher speech has done to a generation of pupils who have been made mentally lazy.

To a mathematician working in mathematics is a pleasurable experience but it is not fun or easy in the popular sense. To attempt to "sell" mathematics by the "easy road," as so many popular educators and textbook writers have been doing during the period of "soft education," is not only a disservice to the individual but also to society. Nations, like athletes, fall when they seek the soft comforts of complacency and indolence.

Late in the 17th century the philosopher-mathematician Leibnitz proposed the development of a "universal symbolism" that would not only encompass mathematics and science but also would serve all rational thought. This should enable man to enhance and extend his thought significantly in a way similar to that in which mathematical symbolism has not only aided thinking in mathematics and science but has also internationalized it. For many years the proposition has been advanced that the "rational power" of a people is limited or determined by the linguistics of that people. Serious anthropologists have suggested that not only the "thinking power" but also the "world view" is limited by the structure of the language of a people. Is there a linguistic determinism? To test this proposition which is an outgrowth of the Leibnitz proposal a new logical language "loglan" was developed by making a mathematical analysis of the commonality of the most used languages such as English, Chinese, Hindu, Russian, Japanese, French, and German. Most important is definition of words and the structural usage of them. Words have precise rather than multiple meanings. In contrast to loglan consider our term "quarter." It can mean one-fourth of something, to cut something into four pieces, or to lodge someone. One must consider not only the definition but also the context and reference in the mind of the user when translating the precise meaning of a word. All readers of our journals know the great difficulty now found in international agreements caused by the lack of a singular meaning for words. The "loglan" experiment is now in process and we await the conclusions.

Mathematical Reasoning

The mathematician seeks truth but not an absolute truth. Rather it is a truth that surely follows an assumed proposition. We call this the "If A, then B" pattern of reasoning which follows closely the classical syllogism. Then "If A," might become a little complex when one assumes, "If A, B, and C are true," then reasons that D must follow. The brilliant mathematician may at a glance note the intermediate steps which a lesser mind such as

13

mine would require several hours to discover. Let me illustrate this mathematical reasoning with a non-mathematical situation. In "The Adventure of the Dancing Men" Sherlock Holmes and Dr. Watson had the following interchange:

H. You do not propose to invest in South African Securities.

W. How on earth did you know that?

H. It's not difficult to construct a series of inferences, each dependent upon its predecessor and each simple in itself. If, after doing so, one simple knocks out all central inferences and presents one's audience with the starting point and the conclusion, one may produce a startling, though meretorious effect . . . By an inspection of the groove between your left thumb and forefinger, I was able to feel sure that you did not propose to invest your small capital in the gold fields.

W. I see no connection.

H. Here are the missing links.
   1. You had chalk between the left forefinger and thumb when you returned from the club last night.
   2. You put chalk there when you play billiards to steady the cue.
   3. You never play billiards except with Thurston.
   4. You told me four weeks ago that Thurston had an option on some South African property which would expire within a month and he desired you to share with him.
   5. Your checkbook is locked in my drawer and you have not asked me for the key.
   6. You do not propose to invest your money in his manner.

W. How absurdly simple.

H. Quite so. Every problem is very childish when explained to you . . .

Yes, it is the in-between steps that cause the trouble for lesser minds. It is the great mind that not only readily sees the intermediate steps but also can take the step beyond the obvious.

The Modern Queen

During the past 25 years mathematics, both pure and applied, has advanced rapidly on several fronts. Some people have said that the developments of the last 50 years have been greater than those of the previous 50 centuries. That may be true but one must always remember that it is often easier to build the superstructure than to design and lay the foundation. Nevertheless, mathematics is advancing so rapidly that perhaps no one can keep abreast of all the new developments in such unifying areas as set, field, and quantum theories, in discoveries in areas such as game theory, topology, symbolic logic, and transfinite algebras, and in further extensions of mathematical analysis. The surge of new mathematics has made my own under-graduate major of some 40 semester hours look like a very mediocre achievement.

Today we have a shortage of mathematicians, especially those with a high level of training and imagination. The national defense and business and industry realize that advancement in technology depends basically upon mathematics. At long last even the layman is beginning to realize that the sciences and the arts become exact as they are founded upon a mathematical foundation. And the more sophisticated acknowledge that national survival in these critical decades may well depend upon the possession of a comfortable margin of pure and applied mathematicians. Cape Kennedy, Polaris, and still unannounced missiles are possible because of the cooperative achievement of many mathematicians and scientists.

Yes, Mathematics is queen of the sciences. She often condescends to render service to astronomy and other natural sciences, but under all circumstances the first place is her due. It is a distinct honor to be identified with "The Queen of the Sciences."

CHAPTER 2

THE CHANGING ELEMENTARY SCHOOL

Much has been written about what a contemporary elementary mathematics program should be. It is easy to become overwhelmed with articles, pamphlets, and books on modern elementary school mathematics and the changes that are taking place. Therefore, the articles in this section have been selected to give the reader insight into some specific facets of the changes in elementary school mathematics taking place in the United States and other countries.

Robert Davis presents an overview in "The New Elementary School vs The Old Curriculum." He observes that the school should be expected to build on the foundations of a child's perseverance, a child's sociability, and a child's curiosity. Unfortunately, Davis claims, schools follow the "Simon Says" approach and children learn by rote. He proposes a curriculum that has (1) a basic *minimal* core (things everybody should learn), and (2) things that go beyond this. He also gives a list of selected references which show how schools ignore (or fight) the nature and the needs of our learners, how schools often fail, and how they can get better.

In "New Trends in Teaching of Mathematics in the Primary Grades," Mary Beaton discusses three significant trends: (1) the wide use of the concept of sets, (2) introduction of modern geometry, and (3) the emphasis on certain fundamental principles of mathematics which have had a profound effect on the teaching of the basic facts. Each child is encouraged to reach generalizations on his own. When a child realizes that these fundamental principles may be powerful tools at his disposal, his interest in mathematics is claimed to increase.

Max Beberman, in "UICSM Looks at Elementary School Mathematics," suggests that the elementary school has three tasks: (1) to teach a child how to learn by himself, (2) to nourish his constantly growing capacity for thinking, and (3) to help him develop ways of expressing ideas and feelings. Beberman claims that the child must learn math via firsthand experiences with real things.

Edith Biggs, in "Trial and Experiment," extends the activity-oriented approach to teaching children mathematics. Miss Biggs, who has been a leader in this area, reports some of the work and activities being done to increase teachers' ability and expertise with the laboratory procedure. Miss Biggs summarizes by saying that "Let the people think" is a slogan for all of us concerned with teaching and learning at any level.

The articles in this chapter will challenge the reader not only to begin thinking about the changing elementary mathematics curriculum but also to recognize the new roles that teachers must assume.

THE NEW ELEMENTARY SCHOOL VS THE OLD CURRICULUM

Robert B. Davis

Curriculum is interesting in much the same way that environmental pollution is interesting: not, surely, because most of us admire empty beer cans, malodorous lakes, or piles of litter where there used to be scenery. Nor, I think, do most of us admire curriculum. What we care about are children and the ideas of mathematics (and other interesting ideas), and that is why we spend time helping children to learn these ideas.

Curriculum, like pollution, is peripheral—but each is nevertheless an inseparable part of the world we do care about: if we help children learn mathematics, can we escape from those who want to ask us which children we have helped, and what mathematics they have learned?

I think we cannot. At a quite simple-minded level we already see the problem arising at PTA meetings and parent conferences. I want to re-count one instance that could, I believe, happen to almost any teacher or school sooner or later.

Rightly or wrongly, the school in question—a K-6 elementary school—had the reputation of providing only custodial care for black children, to whom it didn't try very hard to teach very much. Graduates of this school allegedly appeared in the bottom track (out of three) in grade seven of the junior high. (The data that I've seen do not support this charge, but there is no question that many parents believed this to be the state of affairs. It is true that achievement levels in the school were below certain common norms.)

Parent pressure tended to polarize around two distinct positions. Some parents demanded that there be prepared a written document describing precisely what a "standard fifth grade child" should be able to do, that all fifth grade children be made to conform precisely to this description, and that all teachers be compelled to make each child conform. (Similar re-quirements would apply at all other grade levels.) Let me call this the "standard-grade-level" position.

Other parents were interested in multigrade classes (the so-called "family-plan" classroom seen in Bristol, England and elsewhere), in stu-dent-initiated projects, and in diversity and flexibility of various sorts.[1] For brevity, let me call this the "open" position (some authors call this, or roughly similar positions, "child-centered" or the like).

[1] Some good descriptions of this open or flexible approach to school organiza-tion are given in: R. Barth and C. Rathbone, "The Open School: A Way of Thinking about Children, Learning, Knowledge," *The Center Forum* 3:7.1–2 (July, 1969); E. Biggs and J. MacLean, *Freedom to Learn,* Addison-Wesley (1969); J. I. Goodlad et al., *Innovations in the Elementary School,* available from Information and Services Division, I.D.E.A., P.O. Box 446, Melbourne, Florida 32901; *I Do and I Understand,* Nuffield Mathematics Project, John Wiley (1967).

Reprinted with permission from the *New York State Mathematics Teachers' Journal* (April 1970), 79–88.

Most readers can probably identify certain strengths and weaknesses in each position. The open position has in its favor the promise of making school a more humane environment, of building on the varied interests of different children (and, for that matter, of different teachers), of giving children a greater variety of experiences (for example: the experience of teaching other children), of making it possible to use student projects and math labs, of relating mathematics to other subjects, and of facilitating trials of innovative programs.

In the standard-grade-level demand we seem to hear the same voice of reason that reaches us from Ralph Nader and *Consumer Reports*: the demand that he who sells a car, tire, or a sausage shall sell one that is what it appears to be and is quite safe for us to use. But I want to argue that, no matter how reasonable this "standard" position sounds at first, it is in fact not reasonable when we examine it carefully. The main difference lies in the fact that children are not automobiles—they are not mass-produced in factories, we can't throw away the rejects, and *we can't even identify the rejects*. Most of us, if we try, can find adults whom we greatly admire who were, earlier in their lives, school drop-outs. All children are highly variable from one day to the next, and extremely complex.

There are so many sides to this issue that it would be folly to try to examine them all. Consider just a few:

Do we build on the strongest foundation children offer? Informal observation of some five-year-old boys outside of school seems to reveal an attention-span often measured in *days*—e.g., I have seen a boy spend three days building an elaborate castle from kindergarten blocks, despite the frustration of frequent collapses.

Traditional school denies the child the opportunity to persevere in anything, even in his own thoughts. "Simon Says" is used to compel attention to the teacher's every word. At any point in the school day the child may be interrupted: "Put that away, now, please." "I'm sorry, but there's no more time for that now!" "Please pay attention to what I'm saying!" The child has no right to be left undisturbed. After a few years in school the boy's attention span is allegedly measured in minutes, and in precious few of those.[2] (Part of the secret of long prior-to-school attention-spans seems to lie in the fact that the activities were *selected* by the child and *initiated* by him.)

Informal observation of five-year-old girls suggests that curiosity and sociability are among their most conspicuous features. One might, then, expect school to try to build on the foundation of

(1) a child's perseverance (which would imply not interrupting his work or his thoughts too often);

(2) a child's sociability (which would imply creating an environment where he would *talk* with other children, *play* with other children, and *work* with other children);

(3) a child's curiosity (which would mean that much of the "course of study"—or whatever you call it—would arise from student initiatives and would proceed along lines of the child's interests).

Now what do we commonly find? Some nursery-school and kindergarten programs look like what we have just described, but few elementary-school programs do. Instead, we find chairs bolted to the floor in

---

[2] Boys are a severe challenge to elementary schools in many ways. See, for example, P. C. Sexton, *The Feminized Male,* Random House (1969). On attention span, at an informal level (besides some notes by the present author), cf., for example, C. Amory, "Sesame Street," *TV Guide* for February 14, 1970, p. 6.

rows in order to eliminate social contacts among children; we interrupt children according to *our* schedule, ignoring theirs; and the "course of study" was made by someone in a far away place (and often in a far away time) who did not know these children, nor the fact that today they have just seen a heavy snow-storm, nor the fact that a girl in the class who knows little English can count and can write her name in Arabic, nor the fact that black Debbie and white Claire want to talk about whether black people and white people are different, and how, and why.

One price we will usually pay for pursuing the standard-grade-level position is the loss of our most valuable foundations for building.[3]

Rote vs. resourceful: Is there a trade-off? Most of us, I suspect, if we are put in the role of a teacher who is being compelled to guarantee (as nearly as possible) that every child in class shall pass some quite specific examination, will resort to the rote device of *telling* the children exactly what to do, and advising them not to think about it too much—just do it!

I recently asked thirty adults how they learned to cope with the decimal points in such a problem in division as

$$.025 \overline{)53.4}$$

and every one of them had learned a rule of "crossing out the decimal point, moving it over, counting the number of moves, . . ." and so on. Only one recalled any effort to explain why this method worked.

These same adults gave the same sort of description of how they had learned to respond to

$$1/2 \div 3/5 = \underline{}.$$

They had all been told a rule: "invert the second fraction and multiply." Only two could explain what the expression actually means (as opposed to being able to respond by writing an "answer" without trying to understand what it means).

In industry and business, one often uses the notion of "trade-off": we design a car that is less economical to run in order to make it spew less pollutant into the atmosphere, or we build a smaller house in order to make it cheaper to heat. We trade off some economy in order to pollute the air less, or we trade off some size in order to gain economy.

In the case of school mathematics, one can wonder if there may not be some trade-off between rote and resourcefulness. We once taught a complicated algorithm for finding square roots. This algorithm worked all right if you remembered exactly how to do it — but you can conduct your own poll of how many adults still remember it, and how many can explain how (or why) it works. How many can reconstruct it when they forget it?

That old algorithm has been replaced by the "guess, divide, and average" method, which is quite easily understood by someone whose school mathematics was broad enough to include some simple analytic geometry and graphs of functions. These mathematical ideas are easy and apply quite widely (indeed, one can tackle $2^x = 5$, where the task is to deter-

---

[3] Much school procedure is a remnant of past ethical considerations, and notions of Freud, Hull, etc. (or of even more remote and less profound theorists), that suggest that a child is naturally lazy and will work only if we compel him to; a child is naturally empty of knowledge and will become full only if we stuff knowledge into him. The recently popular work of Jean Piaget could refute much of this. Children, as Piaget sees them, are eager, active, and curious, and schools (unfortunately) apply themselves to the task of making the child slow down and become vastly more submissive. But a new theory of what children are like just might lead to new ways of operating schools. See, for example, H. Ginsburg and S. Opper, *Piaget's Theory of Intellectual Development,* Prentice-Hall (1969), pp. 33–4 and elsewhere.

mine x without reference to tables); we know we can teach them to elementary-school children — but under great pressure to guarantee the highly specific result that each child shall be able to solve $\sqrt{5328}$ many of us would feel compelled to resort to rote teaching. "Don't *think;* just do it *exactly* like this. . . ."

One alternative to a rote approach is to teach more powerful and understandable general methods, such as the monotonicity method to which I have been referring (buttressed with enough guesses, graphs, and so on for the whole thing to seem exceedingly natural, not at all special or tricky or magical).

Another alternative to rote is the general problem-solving approach described by George Polya: look for similar problems; "play around" with these similar problems; then see if you can carry some workable device back to the original one. Suppose the original problem is

$$
\begin{array}{r}
1000 \\
-\ 568 \\
\hline
\end{array}
$$

This is a "messy" problem involving a binary operation where "the numbers are complicated." But this can be replaced by

$$
\begin{array}{r}
999 \\
-\ 567 \\
\hline
\end{array}
$$

yielding the same answer, and in the replacement no "borrowing" or "regrouping" is required! Or how about this one?

$$.025\,\overline{)53.4}$$

Multiplying both numbers by 10 gives

$$.25\,\overline{)534}$$

But .25 is still "messy." However, .25 is 1/4 and matters will become easy if we multiply by 4, the answer remaining unchanged:

$$1\,\overline{)2136.}$$

Can you finish this one?

I will *guess* that these "more creative" approaches

(1) will have special fascination for brighter children;
(2) will allow most children to make more significant progress mathematically;
(3) will not destroy a child's sense of his ability to "cope" the way that rote methods do (after all, if knowledge is rote, aren't we all totally dependent upon what someone has bothered to tell us?);
(4) when handled well, may reach less "academic" children who quite literally refuse to sit still for year after year of rote.

However, I will also guess that, if we are pushed for achieving a highly specific result ("Find $\sqrt{5329}$"), most of us will resort to teaching by rote. Furthermore, I will also guess that, testing being generally rather crude, it may be some time before we can measure student gains of the more generalized sort.

An obvious remedy. One obvious way to avoid this particular difficulty of teaching by rote aimed at specific test items is to make sure that neither the teacher nor anyone else around knows which particular items will appear on the test. Suppose, instead of knowing that a square-root problem will appear on the test, we knew merely that the test would contain some problem that could be solved by the "monotonicity" method.[4] That problem might be

$$\text{Find } \sqrt{5329},$$

or it might be

$$(3 \times \square) + 8 = (5 \times \square) + 13,$$

or it might be

$$2^x = 5,$$

or yet some other problem amenable to this approach.

Now we *must* teach a more general and more powerful approach, one which will leave no dangerously narrow path to a dangerously narrow goal.

It is interesting that testing for broader goals by inserting more uncertainty has not been seriously considered for elementary schools (nor even for high schools), and one is left to wonder if we don't *talk* about "flexibility" and "creativity," but actually *believe* in the memorization of specific facts and specific algorithms.

Stand up and stretch. By now even devoted readers may agree that, if economics is "the dismal science," curriculum is surely dismal if not, perhaps, a science. Every one of the mathematics problems mentioned, pedestrian as they are, could lead us into an intriguing and exciting pursuit of mathematical ideas. It seems singularly self-abnegatory to persist in our austere consideration of curriculum. Wouldn't it be more gratifying to pursue some simple but interesting mathematics?

Ah—but translate between the child's life in school and ours right here. Isn't this the whole issue we are talking about? Shall we get to do something *interesting,* or . . . ?

Limitations of the remedy. The more I think about it, the better I like the notion of generalized problems on exams (e.g. "a problem that can be solved by using monotonicity"), with specifics (such as "a problem involving square root" or "a problem involving exponents") kept a deep dark secret until the moment when a child sees his exam booklet. This would surely put a greater premium on "developing genuine mathematical sophistication," and might even put an end to the rote teaching of highly specific skills and facts. (Don't worry, though; a revolution of this magnitude is not about to happen. Still, it's worth thinking about . . .)

There are, however, some aspects to school curricula that are left unsolved even by "generalized" examination topics.

For one thing, there is the quite serious problem of the inhibition of innovations. Even though we know quite well—the matter is heavily documented—that both our urban and our suburban schools are failing in significant ways, we persist in a system where nearly all schools are alike. To imitate success is perhaps regrettable, but easy to understand. The imitation of mediocrity (and even of failure) attests to the desperation with which we cling to reeds, however frail we may know them to be. We need far more diversity and variety distinguishing one school from the

---

[4] Big words scare people, and maybe that's why they are invented in the first place. The idea behind the word "monotonicity" is not scary; in fact, it is quite simple. Cf. R. B. Davis, *Discovery in Mathematics: A Text for Teachers,* Addison-Wesley (1964), chapter 22 (pp. 136–8). A film on this topic for elementary school children has just been made by Davidson Films, in cooperation with the NCTM and with General Learning Corporation; it is entitled "Edgar's Guess," and will be released sometime in 1970.

next. I can't believe that any major problem in education can be solved so long as nearly all schools are nearly identical—so long, in David Hawkins' words, as "the independent variables are really constants."

But this, in turn, means that we need devices that will tend to make schools different, not devices that will keep them all the same.

Do YOU know what a "curriculum" really looks like? A boy was succeeding in sixth-grade mathematics—this was something of an exciting rebirth for the particular boy, attributable in part to the varied mathematics program which a creative sixth-grade teacher was using. However, a newly appointed instructional specialist had been assigned to the school, in order to individualize instruction. She administered a test, found the boy to be at the fourth-grade level, and reassigned him to grade four for mathematics, whereupon he found himself unable to do the fourth-grade work (he *had* been able to succeed with the sixth grade work), was embarrassed, developed a sense of inadequacy, and became a "behavior problem."

This story you have just read is true; the names have been changed (well, actually, omitted) in order to protect the innocent (can you decide who they are?)

Whoever said you had to be able to do long division before you could add fractions? Or add fractions before you could do long division? Do you have to study Sibelius before you study Bartok? Do you have to study Bartok before you study Sibelius?

A curriculum doesn't look like a path or a railroad track. There isn't any real, honest, necessary sequential order to things. Somebody can make a sequence that may suit *his* purposes, but it won't necessarily suit anybody else's. Which is more difficult to understand:

$$1/2 \div 3/5 = \underline{\quad} \qquad \text{or} \qquad {}^+5 + {}^-3 = \underline{\quad} \qquad ?$$

Yet the division of fractions, which I myself think is harder to understand, has traditionally come before—usually years before—the addition of signed numbers, because that is the sequence that somebody else, far away, made up. It suits neither me nor my children, but we are more or less tied to it as long as people believe that a curriculum looks like a one-track railroad line. New Haven comes before Route 128, and Greenwich comes before New Haven.

What does a curriculum look like? The answers are left as creative, open-ended exercises for the reader. (Please—just don't make it look like a single-track railroad line!)

And if we shall take a successful sixth-grader, and bring it about that he shall become an unsuccessful fourth-grader, what shall it gain us? (Perhaps a suspicion that we need to rethink a few of our basic premises?)

Enough idle talk—some action! A second-grade girl named Leslie graphed the open sentence

$$\square - 3 = \triangle$$

looked at it, and proclaimed "I don't like it; it isn't straight." Now, Leslie really meant "vertical," not straight, and she also meant business. She graphed lots of open sentences until finally she got one whose graph was vertical. She wrote a small booklet, describing all her work.[5]

Could that booklet be used as part of the description of

(1) the learning experiences Leslie encountered in school?

[5] Leslie's actual work can be seen in B. S. Cochran, R. B. Davis, and A. Barson, "Child-created Mathematics," to appear in *The Arithmetic Teacher* in March 1970.

(2) what Leslie herself was like?

(I didn't say things like "evaluation" or "achievement level" — my friends have advised me that if I talk like that I never will be able to understand schools or children.)

Now what does a "curriculum" look like? Would you have allowed a second-grade girl to go off and work on a mini-thesis like this? Was she conforming to a previously written description of exactly how second-graders have to behave?

A modest proposal. Suppose, to keep everybody happy (well, almost everybody), that we construct a curriculum in two parts:

(1) a basic minimal core: things everybody should learn. (Keep this really *minimal!* It does *not* include long division!)

(2) things that go beyond this.

This divides school learning activities into two parts, also. Leslie's mini-thesis, and lots more, fit into the second part. Student selection and student initiative shape much of the second part. The second part provides much of the breadth that converts "arithmetic" into "mathematics."

Now we have to divide our "testing" program into two parts. The first part is easy, and uses existing techniques. For the second part:

(a) we observe actual classroom lessons to see that there really is a second part;

(b) we preserve a xerox copy of Leslie's booklet as part of Leslie's "portfolio" (Why not? Art students do it!);

(c) (whatever better ideas readers may come up with).

Something else: theory. Believe it or not, there *is* a point to this essay. A brief statement of it, however, would, require more space than is available. A very brief and quite inadequate statement might go like this: Schools are not now humane places for our children to be. Also, children don't learn the things they need to learn. If the kids can't beat the rap, it's up to us adults to spring 'em. Now, there are reasons why things are the way they are, and there are methods for changing them.

In fact—and this really is the point—there is a large and impressive literature that shows quite clearly how schools ignore (or fight) the nature and needs of children, how schools fail, and how they can get better.

The trouble is that most teachers and professional educators are unfamiliar with this literature. It hasn't appeared in the journals that teachers and educators commonly read. What can be done in the present short essay is to direct teachers' attention to this important and rewarding literature. We have already mentioned a few important references, in footnote 1. *Freedom to Learn* is a good place to start. In addition, teachers tell us that the following books and articles have repaid them many times over for the effort of finding and reading them:

G. Dennison, *The Lives of Children,* Random House (1969).

P. Kael, Review of the movie *High School,* reprinted in *This Magazine is about Schools* 4:1.23–9 (Winter, 1970). (Available from *This Magazine is about Schools,* 56 Esplanade St. East, Suite 301, Toronto 215, Ontario, Canada.)

A. Barton, "The Continuum," *This Magazine is about Schools* 4:1.59–78 (Winter, 1970).

H. R. Kohl, *The Open Classroom,* Random House, Vintage Books Division (1969).

R. Hammer, "Report from a Spanish Harlem 'Fortress'," reprinted in

R. J. Havighurst, B. Neugarten, and J. Falk, *Society and Education,* Allyn and Bacon (1967), pp. 25–8.

J. Scott (with D. Cohen), "With Sticks and Rubber Bands," *The Arithmetic Teacher* 17:2.147–50 (February, 1970). (Indeed, this entire February issue of *The Arithmetic Teacher* is devoted to math labs and contains many other articles worth reading.)

Something else: actualities. The "new math" hasn't made schools more humane, partly because it didn't try very hard for that particular goal. But mathematics, taught one way, can make life more mechanistic and subhuman, and mathematics, taught another way, can add a great deal to the humane values of life. Lots of people have written about this, but— thank Heaven!—some people have acted. Robert Herse and Dan Lee, with a group of colleagues and parents, and with the cooperation of the central administration (Superintendent Roger Bardwell and Curriculum Coordinator Ann Gunning) have created the East Hill School in Ithaca, New York.[6] East Hill begins with real fundamentals: how shall we use *time?* How shall we use *space?* How shall the human beings in the building—elsewhere dichotomized as "adults" vs. "children"—relate to one another? Now—what shall the mathematics activities at East Hill look like?

Miss Jeanne Lieberman (first grade class at P.S. 169K, Brooklyn) and Miss Zina Steinberg (first grade, P.S. 42M, Manhattan) have created classrooms of a different type within New York City Public Schools; so have several teachers at P.S. 84M, Manhattan. A central part of Zina Steinberg's class is cooking real food and eating it. Reading and math are drawn from this central activity.

In Syracuse, Principal C. Felton Sayles of the Danforth School, working with several teachers, and assisted by Dr. Harry Balmer of Special Projects, has created a curriculum based on "student interest." One group of students identified athletics and physical education as their main interest, and *their entire curriculum in reading, mathematics, geography, etc., is drawn from their central work with athletic sports.* (How do you calculate a batting average? How do you calculate the odds on this week's football games? What is the trajectory of a thrown ball? What makes a football "topple" instead of sailing smoothly through the air?) Danforth, which had been described as "the second toughest elementary school in Syracuse" became, under this program, "an orderly institution where children are taking an interest in learning." [7]

How do you describe "a curriculum"?

Or perhaps what I mean is: are we in mathematics going to be helping people create new types of schools, or are we going to be making their task more difficult? As a well-known advertisement says (more-or-less) : "If we aren't part of the answer, we'll probably be part of the problem."

---

[6] See "Summerhill in Ithaca," *Newsweek* February 23, 1970, p. 65.

[7] Syracuse *Herald-Journal,* February 21, 1970, p. 2: "Danforth Pupils Honor Principal."

ADDITIONAL REFERENCES

Davis, R. B., "The War against Mediocre Schools," in G. Kinney, ed., *The Ideal School,* Kagg Press (1969).

—— and R. Greenstein, "Jennifer," this *Journal* 19:3.94–103 (June, 1969).

Detlefsen, B., "He'd Abolish Regents Exams," Syracuse *Herald-Journal,* February 20, 1970, p. 1.

Divoky, D., ed., *How Old Will You Be in 1984?,* Avon paperback (1969).

Gross, R. and B., *Radical School Reform,* Simon and Schuster (1969).

Neale, D. C., "The Role of Attitudes in Learning Mathematics," *The Arithmetic Teacher* 16:8.631–40 (December, 1969).

Schwab, J. J., "The Teaching of Science as Enquiry," in J. J. Schwab and P. Brandwein, *The Teaching of Science,* Harvard University Press (1964), pp. 1–103.

Smith, L., and W. Geoffrey, *The Complexities of an Urban Classroom,* Holt, Rinehart and Winston (1968).

New Trends in Teaching of Mathematics in the Primary Grades

Mary Beaton

THE MOST SIGNIFICANT TRENDS in the teaching of mathematics in the primary grades are (a) the wide use of the concept of sets, (b) the introduction of modern geometry, and (c) the emphasis on certain fundamental principles of mathematics which have a profound effect on the teaching of the basic facts.

Children who have the opportunity to work with the idea of sets in the elementary grades will be better prepared for junior high school mathematics. A set is an undefined term in mathematics but all of us have an intuitive idea of its meaning. It is a definite collection of objects. It must be described in such a way that it is possible to tell whether or not any given object belongs in a given set. The children attending School A form a set. The children in Room 8 in School A form a subset because every pupil in Room 8 must also be a pupil in School A.

There are two operations with sets which are appropriate for study in the early grades. The union of two sets requires that all members of the two sets be included. Actual manipulation of objects by the pupils, especially by slow learners, would help them to understand this operation. If two sets have no common members, they are called disjoint sets. The union of disjoint sets is later associated with the operation of addition.

The second operation with sets, namely the intersection of two sets, is not a required part of the primary program. However, fast learners in Grade III may learn the meaning of intersection of sets as an enrichment activity. This can be done by considering committees of children. If Committee A contains the members Jean, Bob and Carol and Committee B contains the members Florence, Bob, Jean and Jim then the intersection of the two committees would contain the children who were members of *both* committees, namely, Jean and Bob.

Ideas of *greater than, less than,* and *equal to,* are introduced by using one-to-one correspondence. At the third grade level, the child may associate a cardinal number with each set and may compare the values of these cardinal numbers. The set containing no elements is the null set, and is associated with the cardinal number, zero. In introducing the symbols $<$ and $>$, it should be emphasized that in the sentence $5 < 8$, the symbol $<$ stands for the complete verb, *is less than.* It does not mean simply, *less than.* In the primary grades, children can be given an intuitive idea of the meaning of sets and an introduction to set language.

The second significant trend is the introduction of modern geometry. One major difference between traditional geometry and modern geometry lies in the fact that the idea of sets of points is used in the definitions of modern geometry. Usually the emphasis in the primary grades is on plane geometry. A point is a geometric idea and cannot be seen or drawn but we use a dot on paper

Reprinted with permission from *The ATA Magazine,* published by the Alberta Teachers' Association (March/April 1968), 26-29.

or a chalkmark on the chalkboard to suggest a point. A curve is a set of points. A closed curve is a set of points which makes a path on which one can return to a starting point on the curve without retracing any part of the path. Chalkmarks or rope on the floor can be used to illustrate closed curves. Wool on the flannelboard can also be used. An open curve is a set of points which has a beginning and an end. Here, retracing is needed to go from the end point back to a starting point. A simple curve does not cross itself. Simple closed curves have an interior region and an exterior region. Any closed curve divides the plane into three sets of points, namely the set of points on the curve, the sets of points under and the set of points outside. A rope can be placed on the floor to represent a closed curve so that a child may stand on the curve, inside the curve, or outside the curve.

In the past one hundred years, geometers have criticized Euclid because he did not include in his geometry the idea of betweenness of points. David Hilbert stated the axioms of order in *Foundations of Geometry*. Betweenness is an undefined term but the intuitive idea involved is that in an open curve containing three given points, one of the points must be between the other two points.

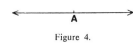

Figure 3.

ABC is a polygon made up of the segments AB, BC, AC.

Figure 4.

Any point in a line separates the line into two half-lines. Each half-line extends indefinitely in one direction.

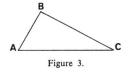

Figure 5.

A ray is made up of a point and a half-line. It is identified by the endpoint A and any other point B in the half-line.

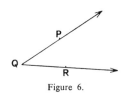

Figure 6.

An angle is the union of two rays that have a common endpoint and that are subsets of different lines. The common endpoint is the vertex and the two rays are the sides of the angle.
Angle PQR is the union of ray QP and ray QR.

Figure 1.

In Figure 1, B is between A and C because it is necessary to pass through B to reach C from A. The point D is not in the set of points for this curve so it is not between A and C.

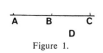

Figure 2.

There is no betweenness for closed curves. In Figure 2 there are two possible paths from A to B. C cannot be said to be between A and B because it is not necessary to pass through C in going from A to B.

The word *line* is used to refer to a special kind of curve that does not bend. It extends indefinitely in opposite directions. It is an open curve. In other words, *line* means a straight line. Frequently, we refer to a part of a line or a segment of a line. A segment is made up of two points in a line and all the points between the two given points.

If two lines have a common point, they intersect. If they do not have a common point, they do not intersect but are parallel. A polygon is a closed curve made up entirely of segments.

The pupils are encouraged to examine the angles in polygons having three or more sides.

Congruence for the primary pupil means "the same size and the same shape". Two segments are congruent if they have the same length. Pupils may use clear acetate or transparent paper to discover whether segments are congruent. Two angles are congruent if they are the same size. The tracing method with clear acetate can be used here. It must be emphasized that the length of the sides of the angles does not affect the size of the angle.

Special polygons are studied. Quadrilaterals such as the parallelogram, the rectangle, and the square are analyzed. Z. P. Dienes in *Building Up Mathematics* makes some pertinent suggestions for teaching a mathematical concept such as the parallelogram. He advises that all possible variables be changed while keeping the concept intact. Vary the shape of a parallelogram by varying the angles and by varying the lengths of the opposite sides. Vary the position of the parallelogram. To allow for individual differences, use different media for the same conceptual theme. Parallelograms may be (a) drawn on paper or on the chalkboard; (b) made of straws or sticks; (c) traced with pegs on a pegboard; (d) made with colored elastic bands on a nail board; and (e) found in physical situations in the classroom or in other parts of the child's environment.

The children learn that the common feature of all of these figures is that the opposite sides are parallel. This common feature is the mathematical concept of the parallelogram.

The most significant change in the revised program is the much greater emphasis on certain fundamental principles of mathematics. Ideas which are introduced in the first grade are treated again in greater depth in later grades. Benjamin Bloom in *Taxonomy of Educational Objectives* has emphasized that knowledge and comprehension of principles and generalizations is at a higher level than knowledge and comprehension of specifics. He has said, too, that the ability to work with generalizations indicates a still higher level of intellectual activity. Now that major generalizations of mathematics are built into the elementary mathematics program, we should make sure that these are not only understood but that students are given experience in using these at the higher levels of application, analysis and synthesis of principles.

The generalizations are not stated for the child. Each child should have the opportunity to reach the generalization on his own. The child should be allowed to state the generalization in his own way. Hopefully, he will be able to do this with progressively greater precision as he gains the necessary vocabulary.

The operations on the natural numbers which are considered in Grade I are addition and subtraction. In the primary program, the natural numbers are considered to be the set of numbers including zero and all whole numbers, i.e., $N = (0, 1, 2, 3, \ldots)$. Addition is a binary operation, in that only two numbers can be added at a time. In addition, a pair of numbers is associated with a single number. $(4, 3)$ is associated with $4 + 3$ and later with the standard name 7. The name $4 + 3$ emphasizes that things are being put together. Notice that the word *plus* is used from the beginning of the study of addition, not the word *and*. $2 + 3$ is a mathematical phrase and plus is the mathematical meaning for the symbol $+$. Similarly, the word *minus* is used to describe subtractive action or the separating process.

In the study of both of these operations, every child should have the opportunity to use objects to illustrate the actions. Piaget contends that the child's actions are most important in the formation of concepts.

The first mathematical principle that is emphasized is the commutative property of addition. The two physical situations $1 + 4$ and $4 + 1$ are different. The first could mean that one child is joined by four more, whereas the second could mean that four children are joined by one child. Similarly, $\$1,000 + \$1$ is a different physical situation from $\$1 + \$1,000$. However, after the transaction, the results can be put into one-to-one correspondence $1,000 + 1 = 1 + 1,000$. The commutative property of addition may be stated as follows:

If a and b are natural numbers, $a + b = b + a$.

Later the associative property of addition is studied. Suppose we wish to add $6 + 8 + 4$. By applying the commutative property of addition $6 + 8 + 4 = 6 + 4 + 8$. Now the associative property of addition allows us to add in either of two ways: $(6 + 4) + 8$ or $6 + (4 + 8)$.

In this case, the first procedure allows us to find the sum at a glance. We are not primarily interested in the standard name for the result but in the *idea* that the way we group the numbers does not affect the result. Here again the child should use actual objects or flannel objects to come to this generalization. The associative property of addition may be stated as follows: If a, b and c are natural numbers, $(a + b) + c = a + (b + c)$.

In Grade II the fundamental operations of multiplication and division are taught with products and dividends up to 12 although the fast learners may be able to work with products or dividends up to 18.

The use of the phrase *3 fours* emphasizes the equality of the groups. By *agreement*, we decide that the first factor in 3×4 refers to the number of groups and the second factor relates to the number of objects in each group. The purpose of this agreement is to clarify the idea that 3×4 means 3 fours whereas 4×3 means 4 threes. The children come to realize that the final result is the same, or $3 \times 4 = 4 \times 3$.

The commutative property of multiplication may be stated as follows: If a and b are any natural numbers, $a \times b = b \times a$.

Children may use objects to build arrays which illustrate this principle. An array is an orderly arrangement of objects in rows and columns. The array may be thought of as 2 rows of four blocks or as 4 columns of 2 blocks.

Multiplication is a binary operation in that only two numbers can be multiplied at one time. The associative principle of multiplication states that the way in which the numbers are grouped does not affect the product. The multiplication of $60 \times 25 \times 4$ may be done as $(60 \times 25) \times 4$ or as $60 \times (4 \times 25)$. In other words, $(60 \times 25) \times 4 = 60 \times (4 \times 25)$. Frequently, the use of this property makes our work easier. $60 \times (4 \times 25) = 60 \times 100 = 6,000$.

One very useful application of this rule is in work with the numeration system. In multiplying by 10, 100 and 1000 the meaning should be emphasized.

5×1000 means 5 thousands or 5000. The associative law of multiplication can be applied as follows: $8 \times 400 = 8 \times 4 \times 100 = (8 \times 4)$ hundreds $= 3200$.

Sufficient practice should be given in order that children may apply this knowledge in Grade IV when they study the subtractive method of division. A child is not ready to study this method of division until he has complete mastery of the multiplication of numbers by multiples of ten, one hundred and one thousand. Emphasis on the associative principle of multiplication will help to give him this mastery. This principle may be stated as follows: If a, b, and c are any natural numbers, $(a \times b) \times c = a \times (b \times c)$.

The zero facts of addition are taught by leading the pupil to an understanding of the generalization that if zero is added to any natural number, the result is the original natural number. The principle of the identity for addition may be stated as follows: If a is any natural number, $a + 0 = 0 + a = a$.

Another principle which is studied at the Grade III level is the distributive principle of multiplication over addition. One of the best ways to introduce this principle is to present the pupils with the problem of finding a product which is beyond the basic facts of multiplication already studied. For example, the problem could be to find the standard name for the product 4×7. The child

could be asked to supply another name for 7. This name could be $(5 + 2)$. The child now has $4 \times (5 + 2)$. By making arrays, the child can discover that $4 \times (5 + 2) = (4 \times 5) + (4 \times 2)$.

The distributive principle of multiplication over addition may be stated as follows: If a, b, and c are any natural numbers, $a \times (b + c) = (a \times b) + (a \times c)$.

Later, this property will make mental multiplication faster and easier. For example, $8 \times 54 = 8 \times (50 + 4) = (8 \times 50) + (8 \times 4) = 400 + 32 = 432$.

When children realize that these fundamental principles may be used as powerful tools, their interest in mathematics will increase.

This summary of mathematics for the primary grades indicates the increased emphasis on the recognition and use of undefined terms and on definitions. It also indicates that increased emphasis on the assumptions implicit in the mathematics studied, even though these are not stated for the pupil. Certainly, the teacher needs to have a mastery of these ideas in order that the program be meaningful for the student.

UICSM Looks at Elementary School Mathematics

MAX BEBERMAN*

The University of Illinois Committee on School Mathematics (UICSM) was established in 1951 to make improvements in the content and teaching of college preparatory mathematics. At the present writing, UICSM has produced a four-volume textbook series for grades 9–12 (*High School Mathematics, Courses* 1–4, D. C. Heath and Co.), a two-volume series on vector geometry (*A Vector Approach to Euclidean Geometry*, Macmillan Co.), a two-year program for underachieving seventh and eighth grade students (*Stretchers and Shrinkers* and *Motion Geometry*, Harper and Row), and a teacher-training film series on the teaching of algebra (distributed by Modern Learning Aids). UICSM has trained thousands of secondary school teachers in the use of its materials and has worked with many school systems throughout the United States in the implementation of these materials.

Although UICSM's work in the secondary school field in the 1950s had a profound effect on the 'new math' in the elementary school (UICSM invented the technique of using frames as variables, made a substantial contribution to the art of designing sequences of exercises leading to the discovery of concepts and principles, and (regrettably) stirred up a fuss about the distinction

between numbers and numerals), it had not done any formal work on the development of a mathematics curriculum for the elementary school. During the 1960s, I became increasingly disturbed by what I saw happening in elementary school mathematics. Great efforts were underway to retrain teachers and to assure parents that children would love and thrive on the new mathematics. But in many places the promised potential did not appear to be realised. Children and teachers were struggling with a new vocabulary ('one-to-one correspondence', 'union of sets', 'commutativity', 'bases of numeration') and were being drilled in the manipulation of trivial abstractions ('a number is the property of a set', 'subtracting is a renaming process', 'to divide by a non-zero number, multiply by its multiplicative inverse'). The results hardly seemed worth the effort. Children certainly did no better, and perhaps they did worse, on tests of computation than they would have under the traditional programme (and I can't recall *ever* hearing a seventh grade teacher expressing satisfaction with the mathematical equipment of his incoming students). What was even worse, children did not seem able to apply mathematics in new science programmes in the elementary and junior high school.

All of this led me to make public denunciations of the new elementary school mathematics curriculum. (For the record, the first such denunciation was my talk at the 1964 Montreal meeting of the American Association for the Advancement of Science.) But denunciations are not enough,

* Professor Beberman wrote this article in its present form for publication in the United States and was going to revise it for publication in *Mathematics Teaching*. He died before carrying out these revisions. His colleagues at the Curriculum Laboratory nevertheless feel it appropriate to continue with the publication of the article.

This article is reproduced from *Mathematics Teaching* (Summer 1971), 26–28, the Journal of the Association of Teachers of Mathematics, by permission of the author and the editor.

especially when the only available alternative – the traditional programme – was not acceptable to me. The opportunity for me and my colleagues to undertake development work in elementary school mathematics arose in the fall of 1968 when the University of Illinois entered into partnership with the Unit 4 School District in Champaign, Illinois in operating a laboratory school for children from age 5 through 11.

We set two guiding principles for this development: children would have to be treated as individuals and be allowed to learn at their own rate; and the mathematics to be learned would have to function in the lives of the children – at the very least, in their lives in school.

The first principle was based on our reaction to the almost universally (at least among people who spend their lives in elementary classrooms) acknowledged fact that mass instruction neglects the needs of a sizeable portion of the school population. The second principle reflects our adverse reaction to the teaching of mathematics for its own sake, at least in the elementary school. Although some mathematics can be taught and learned at any level as an end in itself, the attempt to cast the entire elementary school programme in this light is miseducational for it ignores the important fact that mathematics is an eminently useful subject and that children need to use mathematics to help them solve problems that arise in their study of other subjects.

In the spring of 1968 I had visited several primary (i.e., elementary) schools in Britain to examine at firsthand some of the things I had heard about the teaching of mathematics – the Nuffield Project and the active learning approach expounded by Edith Biggs and Dora Whittaker. Not only did I see this approach in action – children working singly or in small groups on real things, and learning and applying mathematics – but I encountered a philosophy of primary education in general that was appealing to me. Here were classroom teachers and head teachers (i.e., principals) who treated children as individuals and really tried to fashion a personal curriculum for each child. In subsequent visits in the spring of 1969 and the spring of 1970, I saw more schools trying to do this, and I learned that not all was perfect. Teachers with classes of 35 to 45 children sometimes did a less than adequate job of catering to individual needs. But the philosophy was constant – it was the individual child who was the concern of the teacher, not the class. This contrasts sharply with what I see in elementary schools in our country where we worry about class averages and grade-level norms.

The teachers at the Washington Elementary School in Champaign, the laboratory school I referred to above, are now experimenting with ways of organising their classrooms so that children can learn by themselves, at their own rates, and in accordance with their own interests and needs. They have been helped in this by visits from consultants with extensive experience in active learning – Edith Biggs of London, F. Frank Blackwell of East Croydon, Elwyn Richardson of Auckland, and Dora Whittaker of Nottingham – and through day-by-day contact with various members of the University research staff who have taught in British primary schools. But whatever success they have had or will have in this very difficult transition from formal to informal education should in complete fairness be attributed to the efforts of the teachers themselves. I was told that this was the case in the British experience and it is certainly the case at Washington Elementary School.

I have become so excited and intrigued by what I have seen happening in British primary schools that I elected to spend the 1970–71 school year in Bristol. Several of the University research staff in science or mathematics education are here with me. We spend our time observing or teaching in classrooms, talking with classroom teachers and head teachers, observing in teacher-training colleges, attending professional meetings, participating in inservice courses, reading, and thinking.

My views on the teaching of mathematics are changing. For one thing, as a mathematics educator I now find that I must first become interested in the total education of the elementary school child. No matter how convenient it may be for curriculum developers to pretend that a child can be 'partitioned' into subjects, in fact he cannot. The child comes to the school as an integral being; hopefully he leaves with the integrity intact. What is the elementary school supposed to do for children? All educators must first settle on an answer to this question before they can undertake any kind of curriculum development work. At the moment, I have settled on three tasks for the elementary school: it should teach a child how to learn by himself; it should nourish his constantly growing capacity for thinking; and it should help him develop ways of expressing his ideas and feelings. The process of elementary education is largely that of finding a workable match between each child and the culture. There are infinitely many things children can do in school. Some things will help him learn to learn; some will encourage him to think; some will help him find ways of expressing himself; and many will do nothing but waste his time, or worse, make him bitter about school. Mathematics as a school subject has the capacity for doing all these things, the negative ones as well as the positive ones. It is relatively easy to arrange things so that mathematics will promote only the negative outcomes. For example, you can design a curriculum in which the mathematical concepts are arranged in logical sequence and the algorithms are rationalised in terms of mathematical laws; then you parcel this out to children in small pieces, each to be mastered in its turn, but with provision for considerable remedial work to rescue the numerous children who have failed to learn. With such a program you can be sure that a child will have done plenty of memorising (he will even have memorised his 'understanding'), that he will not have learned how to learn (the authors of the instructional materials will have arranged the 'optimal' instructional sequence), and that he will not have done much thinking. He will certainly have developed feelings about mathematics (and school) but will probably have not learned how to express them except under his breath.

I now think that it is a fallacy of mathematics curriculum development *for young children* that logical organisation of the subject determines its pedagogical organisation. When a child learns mathematics via firsthand experiences with real things, the reality of the context provides him with all he may need *at that time* to make sense out of what he is learning. A six-year-old trying to build a tower 10 inches high might discover that he can do this with a $6\frac{1}{2}$-inch block and a $3\frac{1}{2}$-inch block. He does not have to be taught how to add 'mixed numbers' before he can make that discovery, nor would this be the time to teach him an algorithm for doing so. I believe that children need a protracted period in which to work with real things and discover mathematical facts. For some children, these may be isolated facts; for others, the facts may point to generalisations. It is the teacher's

responsibility to intervene in a gentle way when a child is in the neighborhood of a generalisation and needs just a bit of a push to achieve it. Needless to say, this is part of the artistry of teaching.

There does come a time when a child should bring generalisations together and see that they are linked in logical structures. It is difficult to determine when this should happen. I am convinced from my own observation and from what I know of psychological findings that, although the appropriate time will differ from child to child, we should not begin a serious search for children who are ready for structural organisation of generalisations until they have had four to five years of elementary education behind them. (The fact that one has heard of a mathematician's nephew who could cope with these abstractions when he was seven years old is not a sign that one should build a curriculum designed to bring all seven-year-olds to this level.)

I am unable to say at this time what will be the product of UICSM's involvement in the elementary school programme. I doubt that we shall produce the customary line of textbooks and manuals, for this kind of material does not lend itself readily to personalised instruction. Perhaps we shall have given a good enough account of ourselves if we merely find ways of encouraging and enabling classroom teachers to find their own ways of personalising education for their children.

Trial and experiment

EDITH BIGGS

Many variations have been tried in the pattern and timing of mathematics workshops in England to encourage more teachers to provide their pupils with opportunities to learn mathematics actively. At initial workshops, which may be of three to five days duration, one important part of the program is the preparation of topics to try out in the classroom. For this, teachers of pupils of age range five to sixteen years work together, and each group chooses a topic.

Preparation for classroom experiment

Certain general points are kept in mind in the discussion which precedes the planning.

1. For most pupils (and adults!) one discovery is not enough. It is therefore essential to include varied activities on one particular topic at different stages. These experiences should become successively more demanding.

2. Practical work can be time-consuming. Teachers should therefore have a clear mathematical purpose in mind when planning the work. Running around with a tape measure can be a waste of time. Activities such as "measure the classroom" can be just as boring as practicing too many calculations. Questions such as "Is your classroom a square? In how many ways can you find out?" or, "Is the hall twice as long as it is wide?" or, at a later stage: "Peg out a ten yard square plot by eye; measure the plot and make a scale drawing to see how near you were" could engage a pupil's interest and give him a more thought-provoking experience, especially if standard measuring instruments are hidden.

One piece of practical work might provide the starting point for several ideas in mathematics such as approximation, averages, ratio, proportion, and scale.

3. Discussion is essential if children are to clarify their ideas and if teachers are to know whether a concept has been understood or not.

4. Pupils should be given practice in calculations *after* they have devised and refined their *own* methods as, for example, in long multiplication or division.

Topics may be prepared under the following headings:

1. Stages of development (concepts).

2. Sample experiences appropriate to the various stages. Good, thought-provoking questions allowing pupils opportunities for developing their own suggestions. These are to indicate starting points only.

3. Methods of communicating the relationships discovered.

Many activities are suggested for each stage so that considerable experience is pooled. Each teacher therefore has a variety from which to choose in making his

Reprinted from *The Arithmetic Teacher* (January 1970), 26–32, by permission of the author and the publisher.

own framework. Phrasing of challenging questions is difficult. Some say that their pupils would require more directed questions if they are ever to discover relationships for themselves. But all agree that (1) open-ended questions are very hard to phrase if ambiguity is to be avoided, (2) open questions can always be closed when the pupil requires more help, but the reverse is not true. A direct question usually gives too much help at too early a stage. It is very difficult to make some questions undirected. There are a variety of methods of communicating relationships that are discovered. In the early stages communication may be oral, or by pictures or diagrams, or using three-dimensional material. Mapping, tabular forms, graphs of various kinds are useful at different stages (block graphs and connected graphs when appropriate).

Before a choice is made, the range of topics covered at the elementary stage is discussed and analyzed. The topics considered are number; the measures—length, weight, and density; time and rate; capacity and volume; area; temperature and pressure; money; and shapes. In planning the topic, certain general ideas in mathematics are of such cardinal importance that these should be kept in mind at every stage. Here is the list, not necessarily in order of importance:

1. Classification, sorting, matching
2. Comparison, at first descriptive, then quantitative (by subtraction and by division, i.e., as a difference and as a ratio); inequalities; putting in order; and conservation
3. Operations of addition, subtraction, multiplication, and division
4. Approximation (inbetweenness)
5. Ratio and proportion, similarity and scale
6. Relations, variables, graphs
7. Limits—maximum and minimum

One of the initial workshops I conducted was for principals only. In order to understand the meaning of some of the above terms, each principal took a container of some kind and, working in groups of four or five as the children would, they put the containers in order in as many ways as they could contrive. In doing this they found, for example, that two containers of very different shapes had the same volume and that two others had the same perimeter. These are examples of conservation. Questions such as (1) How much more does the largest container hold than the smallest? (2) How many small containers could be filled from the largest? were examples of subtraction and division (and of the two ways of comparing capacity).

Classroom experiments undertaken by principals

After an initial workshop, there is normally an interval before a follow-up is held to allow teachers and principals time to experiment in their own classrooms and schools. In this instance, the topic chosen was volume and capacity. One principal of an infant school (ages five to seven) found her staff (mostly teachers in their first year) so lacking in background knowledge that she made a simplified scheme on "volume and capacity" and gave the teachers themselves some practical experience. She reported steady progress but lack of confidence still.

Another principal gave an amusing account of the building operations at her school of 60 percent immigrants, age range five to seven plus. The children had been warned to keep out of the workmen's way, but they were soon imitating building operations in their play. When the principal introduced some large bricklike wooden blocks of identical size and shape, the children immediately began to build brick walls illustrating the two types of bonding used by the workmen. They also used the blocks for comparing various areas, which they covered with blocks (and counted). They used string to compare the perimeters of the shapes they made. The pattern of the mesh of the wire netting (used to keep them out) intrigued them so much that they drew the pattern over and over again in their drawings and paintings.

A third principal brought a good sample of infant work of all kinds because she found it hard to isolate one topic from the many others with which it is associated. This work was particularly useful because each class in this school is organized as a family group and comprises forty children of ages five through seven. One topic introduced was "aircraft." Each child had made his own model from trash material, covered this with newspaper pasted to fit on, and painted the whole with thick powder colors. There were some weird, futuristic models, but all had one factor in common—without exception the length was considerably greater than the wing span. The older children compared lengths, wingspan, and body perimeters using strips of paper marked in inches so that they could record the lengths. They made paper kites to find good "flying" models. Discussion of size of airplane and number of passengers carried led to another investigation of the sizes of various containers. At first they used pebbles to fill the containers but rejected these because of the variation in size. They tried marbles but were worried about the air spaces left. Finally they filled each container with sand, poured the contents into a glass container, and compared the levels.

Not one of the principals of junior schools (seven to eleven age range) had undertaken any consistent work nor had they discussed with their staffs the aims and content of our workshops. A few had worked with various classes themselves during the two or three days prior to the follow-up. The range and quality of this work were good, and it was pointed out how much of more permanent value might have been achieved if the work had started soon after the initial workshop. One principal followed the framework he had drawn up but, because of pressure of time,

said he felt he had directed the work more than he needed, had he had more time.

Some eight-year-olds compared the capacities of a collection of bottles by counting the number of small cupfuls of water it took to fill each bottle. They then filled a quart jug using, in turn, bottles which held 2 pints, 1 pint, ½ pint, ⅓ pint. They recorded their results in two ways: as a table and as a block graph using unit squares.

Capacity of bottle in pints	Number of bottles to fill quart jug
2	1
1	2
½	4
⅓	6

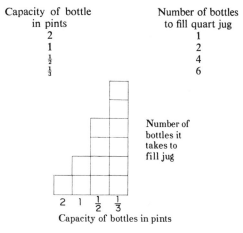

FIGURE 1

In the discussion that followed, the principals decided that the graph could be misleading unless it was carefully labelled and fully understood. (It was only later that I realized that this was an example of inverse proportion. Find the relation between each ordered pair of numbers!)

The most interesting work was undertaken by nine-year-olds who were making a fish tank. The principal suggested they investigate to find out the most suitable material to use for the floor of the tank. Would it be better to use sand or coarse gravel? In which material would the fish food be more likely to sink out of reach? In order to solve this problem the children decided to compare the volume of air space contained in equal volumes of the two materials. They were provided with glass measuring jars of 250 cc capacity. (Younger children had made their own measuring jars using a cubic inch as unit.) They filled two identical jars—one with sand and the other with coarse gravel. A jar containing 250 cc of water was then slowly poured into the jar of sand until the children judged that all air space was filled. They then repeated the experiment with the jar of gravel and measured the volumes of water used for each.

They wrote: There were 76 cc of air in 250 cc of sand and 100 cc of air in 250 cc of gravel. So the gravel was 100/76 or 1⅓ times as coarse as sand. The results of this experiment persuaded the children to use fine sand for their aquarium.

Classroom experiments on time

It was with considerable interest that on the following Saturday I visited an outer London borough where I had directed a three-day workshop for teachers and principals some three months previously. Eighty teachers and principals of pupils of ages five through sixteen attended the initial course and between forty and fifty came to the day's follow-up. This time there was a very extensive exhibition of work from pupils of all ages from five to sixteen although most of the work came from classes containing younger children. The work from one infant school was outstanding. Here the principal and her staff had concentrated on "Time." There had been much discussion, some reading about the history of time, and extensive preliminary planning concerning the materials and books to be made available to the children. Teachers found that time could be introduced into every aspect of the curriculum, and there was a wide variety of imaginative and illustrative work. Here are some examples from the range exhibited:

1. Telling the time long, long ago. The story told by the teacher had captivated the children. They had notched a stick at the time of each new moon. They observed and drew the moon at night and painted the landscape as they imagined it to be at the time of cavemen.

2. Time and growth; time and age.

(a) Some green peas were planted and a daily record was kept of the young plants by "matching" and measuring the height and by counting the number of leaves ("It's the two times table," they wrote).

(b) One group of children drew pictures of the six ages of man, illustrating and writing about each one.
Nicola wrote: "At first I was a baby."
Donna: "I learnt to walk when I was one year old."
Karen: "Now I am five years old. I am a schoolgirl.
Gillian: "One day I will be an adult like my mummy and wear earrings."
Jane: Then I will grow older like Granny. John has a great granny. She is 70. She is very old.

Six-year-old Neil became fascinated by old cars. His pictures came from a cigarette card series but the writing was his own. For example: "200 yise of time Through car Fashions. 50 yise ago cars wer tall and naro like the Peugeot. There wer more open cars in olden days. 55 yise ago the first saloon car was made."

(c) Birthdays and bedtimes were also topics that attracted these young children. The teacher tried to help the seven-year-olds to find out how many hours they were in bed by providing them with this framework for finding "How long do I sleep?" (See fig. 2.)
Although these children were accustomed to making block graphs, they found the idea of measuring time by the length of a line very difficult. The teacher had to help them shade in the "length of time" they were in bed.

Their comments included:

Afternoon					Night							Morning								
1	2	3	4	5	6	7	8	9	10	11	12	1	2	3	4	5	6	7	8	9

FIGURE 2

"We are all in bed between 8 and 7 o'clock. I sleep thirteen hours. Debra goes to bed first."

"My grandad works at night. He sleeps in the day."

"Four children go to sleep at 14 o'clock."

The last comment was one of many which revealed that although the younger children wanted to be included they had no idea what they were doing.

Older pupils, at the secondary stage, had also worked on the topic of time. They experimented with pendulums and made a series of graphs of the results obtained as they changed one variable after another (weight of bob, angle of swing, length of string, method of timing). They took several readings and gained insight into the meaning and purpose of finding an average. They discussed the graphs and used them to obtain other results. They timed themselves walking, running, and cycling and found their "average" speeds in kilometers per hour and the time to travel one kilometer.

Number patterns

At the junior stage (seven to eleven), the most outstanding piece of work was on number patterns. The principal himself had planned the program with teachers at various levels in the school. The work of the nine- and ten-year-olds was very varied, and the children showed individuality and enterprise in developing the work further. At first they used unit squares to build a sequence of squares with edges 1, 2, 3, . . . , 10 units, and cubes to build a sequence of cubes. They were excited to find that "square numbers are square and cube numbers are cubes!" They wished they could visualize a fourth power! When they decided to use the unit squares in each square in the sequence to make a block graph, they discovered the difference pattern of the square numbers. (One ten-year-old girl was fascinated by the difference patterns and continued to the fourth and fifth powers. She suggested that her peers should find the successive differences

of the sixth powers of the counting numbers 1^6, 2^6, 3^6, etc. (i.e., 1, 64, 729, etc.) to see if they agreed with her findings. They were attracted by multibase arithmetic (which the principal introduced through Dienes multibase arithmetic blocks). They used the Sieve of Eratosthenes to find the prime numbers and then turned their attention to magic squares. Here they discovered how to make such squares for themselves. When some of the class joined the ordered pairs of the following sequences by connected lines, they asked if the graphs continued beyond zero. "Count backwards and see," was the teacher's suggestion. In this way they meet the negative numbers.

(0, 0)	(0, 0)
(1, 3)	(1, 1)
(2, 6)	(2, 4)
(3, 9)	(3, 9)
(4, 12)	(4, 16)
etc.	etc.

From then on the children experimented widely with number patterns and ordered pairs. This concentration on pattern in number had a marked effect on the children's number knowledge and manipulative ability with numbers. They looked at numbers in a different way, always expecting to discover a pattern. When a discovery was made, a question was asked to give other children the chance to make the discovery for themselves.

An introduction to area

Let us look at two other pieces of work concerned with area. The first is an astonishing accomplishment for seven-year-olds and shows how children (and their teachers) sometimes excel themselves in their effort to solve a problem in which they are deeply involved. I am *not* suggesting that this piece of work should be undertaken by other children of this age—in general, the numbers used would be far beyond their comprehension—but the urge to solve the problem drove them (and their teacher) on. At this overcrowded school, there were 320 children of ages five to seven years. They complained one day that the school playground was so crowded that they could not play games without knocking each other over. The teacher asked them if they could find out how much space each one would have. They decided that they must cover the rectangular yard with paper to find out. Some measured the yard with the 33-foot tape while others began to cut out foot squares to cover the ground. Different groups recorded: "Me and Kerry measured the width of the playground. We used the 33-foot tape measure once but the second time we only used 32 feet of it, so the playground is 65 feet wide."

$$\begin{array}{r} 33 \\ +32 \\ \hline 65 \end{array}$$

Kerry and Colin measured the playground. "We used the long measure three times and the last time we only used 29 feet of it."

$$\begin{array}{r} 33 \\ 33 \\ 33 \\ +29 \\ \hline 128 \text{ feet long} \end{array}$$

The children soon realized that there was no future in covering the yard with foot squares. The teacher gave them some large sheets of centimeter-square paper, which they cut and joined together to represent the rectangular yard 128 squares by 65 squares. By now they understood the need to use a small square to represent a square foot.

```
1  2  3  ···  100    1  2  ···  28
29 30 31 ···
```

FIGURE 3

The children decided to take turns to number all the squares on the paper and to start again at 1 after each 100 (which would be numbered in red). So the top row was numbered 1 to 100 and then 1 to 28, etc. They continued to number each

row taking it in turns to write numbers in 100 squares.

"Lots of children counted in hundreds and I did, too," wrote Susan. There was an exciting moment when the twenty-fifth row was complete. "Stephen and Michael found out where to put the hundreds. They make a pattern." At this stage, no more consecutive squares were numbered; successive hundreds were marked instead, according to the pattern Stephen and Michael found. The red 100s were then counted. "There were 83 hundreds and 20 more so the playground is 8,320 square feet," wrote Trevor and Peter.

The children were now faced with the problem of dividing 8,320 square feet among 320 children. It is interesting to notice that they invented an ancient method of division.

Different children worked each stage.

	8,320 square feet for 320 children
Colin:	4,160 square feet for 160 children
Trevor:	2,080 square feet for 80 children
Kerry:	1,040 square feet for 40 children
Robert:	520 square feet for 20 children
Raymond:	260 square feet for 10 children—(Notice the opportunity missed here.)
Antony:	130 square feet for 5 children

"Raymond then used the number patterns and Julie and Theresa used the interlocking unifix blocks to find that there are 26 square feet for each child." To make quite sure that every child understood what 26 square units would look like, the children used Cuisenaire rods to make all the patterns they could think of to represent 26. (As far as I know, they did not cut out 26-foot squares to see what the space looked like in the playground.)

The second piece of work was done by

ten-year-olds. After they had weighed themselves in stones and pounds and converted this to pounds (1 stone = 14 pounds), the teacher asked the children to calculate how many pounds they carried on each square inch of their feet. They found the area of both feet by drawing round their feet on square-inch paper and counting the squares. The results (obtained by division) when put in order by the children ranged from 1.6 to 2.8 pounds per square foot. This idea attracted the children so they decided to find out a corresponding rate for an elephant. Hazel wrote: "Before Christmas I wrote to Belle Vue Zoo asking them to draw round an elephant's foot. In about another week I received the elephant's foot below."

Area of this foot is 136 square inches. Weight of the elephant is 3 tons 5 cwt. The pressure is 13.4 pounds per square inch.

Practical help for teachers

What have I learned from this experiment and many others during the past twelve years of helping teachers to introduce active learning in their classrooms?

1. Teachers must have firsthand experience of investigating mathematical problems for themselves at their level if they are to be convinced that active learning of mathematics is possible.

2. They must be helped to plan worthwhile activities for the classroom and must be encouraged to exchange ideas with colleagues in their own (and other) schools at regular intervals.

3. There must be a follow-up at which teachers from several schools (covering a *wide* age range) meet to show pupils' work and to discuss successes and failures. Teachers require sympathy and encouragement at every stage. It is not easy to organize for active learning.

Above all, those of us engaged in in-service and in preservice training of teachers must *never* be discouraged. Educational reform comes slowly and is perhaps all the better for this. In the United Kingdom we may not have achieved a great deal in twelve years—not more than 75 percent of elementary schools have carried out any classroom experiments (and many of these are still very tentative). At the secondary stage, although probably 50 percent of the schools have changed the mathematical content to some degree, less than 1 percent give opportunities for investigation by small groups before instruction "sets in." But the effect on those children and teachers who have been concerned in these changes is there for all to see. "Let the people think" is a slogan for all of us concerned with teaching and learning any subject at any stage.

SELECTED BIBLIOGRAPHY FOR CHAPTER 2
The Changing Elementary School

Biggs, Edith E. "The Role of Experience in the (Classroom?)." *The Arithmetic Teacher* (May 1971), 278-95.

Brownell, William A. "The Revolution in Arithmetic." *The Arithmetic Teacher* (February 1954), 1-5.

Court, Nathan Allishiller. "Mathematics in the History of Civilization." *The Mathematics Teacher* (February 1968), 148-56.

Davis, Robert. "New Math: Success/Failure." *Instructor* (February 1974), 53.

Glennon, Vincent. "Too-Heavy Input of Mathematicians." *Instructor* (February 1974), 45.

Grass, Benjamin A. "This Is the New Math???" *New York State Mathematics Teachers' Journal* (January 1967), 29-32.

Hartung, Maurice L. "The Role of Experience in the Learning of Mathematics." *The Arithmetic Teacher* (May 1971), 279-85.

Henderson, George L. "Tired 'Rithmetic Teacher—Hark!" *New York State Mathematics Teachers' Journal* (June 1970), 166-68.

Hilton, Peter. "The Continuing Work of the Cambridge Conference on School Mathematics (CCSM)." *The Arithmetic Teacher* (February 1966), 145-48.

Johnson, Donovan A. "Behavioral Objectives for Mathematics." *School Science and Mathematics* (February 1971), 109-15.

Kline, Morris. "Why Teach Mathematics?" *New York Mathematics Teachers' Journal* (April 1974), 55-70.

Lowry, William C. "Structure and the Algorisms of Arithmetic." *The Arithmetic Teacher* (February 1965), 146-50.

Matthews, G. and Comber, J. "Beginning Mathematics." *New York State Mathematics Teachers' Journal* (June 1971), 87-90.

Metzner, Seymour. "The Elementary Teacher and the Teaching of Arithmetic: A Study in Paradox." *School Science and Mathematics* (June 1971), 479-82.

Neufeld, K. Allen. "Structure—Key Word of the Sixties." *The Arithmetic Teacher* (December 1965), 612.

Newsom, Carroll V. "Mathematics for the Millions." *New York State Mathematics Teachers' Journal* (January 1966), 2-13.

Rising, Gerald R. "Elementary School Mathematics Curriculum Revision—The State of the Art." *New York State Mathematics Teachers' Journal* (June 1966), 90-109.

Rising, Gerald R. and Kaufman, Burt. "Some Thoughts About Behaviorists and Curriculum." *New York State Mathematics Teachers' Journal* (January 1973), 12-15.

PART II

NUMBER SYSTEMS AND OPERATIONS

CHAPTER 3

SETS, LOGIC AND NUMERATION SYSTEMS

Much has been written about the set concept and it is certainly one of the most basic ideas in mathematics. Z. P. Dienes and E. W. Golding,* when discussing why a study of sets is important, stated, "If we aim at a deeper understanding of number in the child's learning, then the preparation for number must also be at a greater depth."

Set terminology and set ideas are useful because they can provide many early experiences that prepare a child for dealing with more complex number concepts at a later time. In this chapter, B. M. Hives, in "A Question of Infinity," discusses how to illustrate the concept of infinity to "non-believing" children using such ideas as fractions approaching zero, squares, and mirrors. You too can help your learners look right along the mirror tunnel to infinity.

Evelyn Knowles uses a fun, creative, and a very rewarding concrete learning experience for each child in her article, "Fun With One-to-One Correspondence." All you need for this activity is paper, a ruler, a sharp pencil—and, of course, know how to draw a straight line.

James M. Moser, in "Grouping of Objects As a Major Idea at the Primary Level," demonstrates how a child who is given a set of objects can be taught to separate them into smaller subsets or groups, each having the same number of objects. The author contends that continuous use of the grouping process will simplify learning the concepts of division and multiplication.

Grossnickle and Reckzeh† have stated, "The basic logic essential in mathematics and a number of other disciplines can be treated effectively in elementary school situations by using set language and ideas." James J. Roberge, in "Some Elementary Concepts of Logic," demonstrates this quite well. He illustrates the logical concepts "all," "no," and "some" by using Venn diagrams. He then goes on to illustrate syllogisms and extends these basic ideas into classroom activities. This article supplies several interesting activities which an imaginative and creative elementary teacher could use to introduce some of the basic concepts of logic.

Number bases other than ten have received considerable emphasis in the past several years. The learner, through the study of nondecimal numeration systems, should gain a broader insight into the structure of a place-value system, and con-

---

Sets, Numbers and Powers (New York: Herder and Herder, 1966), p. 9.

†*Discovering Meanings in Elementary School Mathematics,* 6th ed. (New York: Holt, Rinehart and Winston, 1973), p. 56.

sequently should develop greater skill in the fundamental operations by performing these operations in different nondecimal systems. Wayne Peterson in "Numeration—A Fresh Look" and Harold F. Rahmlow in "Understanding Different Number Bases" each develop the idea of number bases other than ten through the use of a place-value chart and an abacus. In "Let's Use Our Checkers and Checkerboards to Teach Number Bases" Lucile LaGanke introduces the idea of a checkerboard as an arithmetic teaching tool—a checkerboard abacus. Be sure to read this article to find out how to use the checkerboard to build numbers, to add and subtract, and to compare how a certain number appears in different number bases.

A Question of Infinity

B. M. HIVES

I asked my eight year old,
'What is the biggest number you know?'
'A googol'.
'Who told you about a googol?'
'Philippa'. (11 year old sister)
'What if I gave you one more?'
'It would be a googol and one'.
'And one more?'
'A googol and two. (After thought) Then it would go on to a googol and a hundred, and then a thousand, and then . . . two googols'.
'And then?'
'Two googols and one. (More thought) That's a lot to count – I've only counted up to three hundred!'
'So do you think I could count up to a googol?'
'No, you couldn't – well you could, but it would take absolutely ages'.
'Anyway, where does counting stop?'
'I don't know. I can't think. I don't think it does stop. So it probably goes on. Am I right?'
'Yes'

At this point the 11 year old joined the discussion, and said firmly,
'That's infinity'.

I then introduced the idea of infinity on a line. 'Could you draw a line on and on and never come to the end of it?' (demonstrating a straight line and a spiral)
8 year old: 'No, you'd get fed up.'
11 year old: 'Yes, that's infinity again', and, as a helpful explanation to the 8 year old, 'Say it was drawn by, say, magic, straight up and up, it could go on for ever'.

Now the 13 year old joined us, and I posed the question of halving the piece of ribbon.
'Would your hands eventually touch?'
8 year old (with complete conviction): 'Of course'.
11 year old: 'Yes'.
13 year old: 'Yes'.
'But what about that last little half?'
8 year old: 'It doesn't matter. Your hands would be touching, and you'd throw the last tiny bit away'.
11 year old: 'They'd touch. They'd have to in the end'.
13 year old: 'I see what you mean'.

I then started halving fractions, a half, a quarter, . . . a hundredth, a thousandth . . . etc., and the eight year old discovered a googolth! I let it rest there.

But later on, in discussion with the 13 year old only, he saw that however many times the fractions were halved you would never reach zero.

From this fruitful discussion with the different age-groups, it appears that the 8 year old came eventually to her first appreciation of the infinity of the counting numbers, but found the infinite line unacceptable – it was too abstract for her concrete thinking. The 11 year old, on the other hand, had attained the concept of the infinity of the natural numbers, and a line, but still maintained that the hands would touch (she 'shut off' when I was halving fractions). Only the 13 year old was able to appreciate the implications of the fractions approaching zero but never quite getting there.

The 8 year old's reaction, to throw the last tiny bit away, was very interesting in view of what Sawyer says when discussing calculus in *Mathematician's Delight*. He points out that you can get ever nearer the actual speed at the point of collision by measuring the very short distances covered in very short periods of time, and still find only the average speed over the last, say, millionth of a second, but 'the engineer thinks the discussion is a waste of time. He does not mind whether the speed was 50 m.p.h. or 50.0031 m.p.h.' He is happy to 'throw the tiny bit away', as the 8 year old is. There are no absolutes to the engineer. But the pure mathematician wants to find the exact result, of course.

In order to give a child the necessary practical experience of mathematical limits – the infinitely small and infinitely large, there is a splendid ex-

This article is reproduced from *Mathematics Teaching* (Spring 1971), 6–7, the Journal of the Association of Teachers of Mathematics, by permission of the author and the editor.

ample in 'Mathematics in Primary Schools' *Curriculum Bulletin No. 1*'. Children were given a large sheet of $\frac{1}{4}$ inch graph paper and told to draw the largest possible square on it, find the mid-points of each side, join these, and continue in this way, until they realised they could go on drawing smaller and ever smaller squares. And furthermore, the original square could be expanded ad infinitum.

The experience and understanding of this particular illustration of the infinite would be of great help towards the understanding, at a later stage, of both infinite and finite transformations of the plane – and the 'discovery' of the fascinating snowflake curve. This curve is itself a very interesting one – its length is infinite, yet the area it encloses is finite!

The study of infinity is full of paradoxes – Achilles and the tortoise, the 'curve of pursuit', etc., but the paradoxical relationships of infinite numerical series is one of the most difficult to comprehend.

In the following two sets,
set of all natural numbers 1, 2, 3, 4, 5, . . .
set of all squares of all
          natural numbers 1, 4, 9, 16, 25, . . .
all the members of the second set are members of the first, but there are infinitely more members of the first set which are not members of the second. Is the first set somehow 'greater than' the other? How can it be – both sets have an infinite number of terms, and the terms of each set can be placed in one-to-one correspondence with each other. But can any set with an infinite number of terms be deemed greater than another? Cantor thought so, and it was Cantor's concept of a class of numbers with which to classify the number of numbers of infinite sets, that allowed degrees of infinitude to be defined. He called his first 'transfinite' number A_1, and as the set of all natural numbers is fundamental to the number structure, all those infinite sets which can be placed in one-to-one correspondence with the set of all natural numbers (e.g. all odd numbers, all even numbers, all square numbers, all rational numbers) were allotted this first transfinite number A_1. Cantor then looked for an infinite set whose number is greater than A_1. He had to find a set, just one member of which could not be placed in one-to-one correspondence with the set of natural numbers – and he proved that the set of all real numbers between 0 and 1 is such a set. So the transfinite number of the set of all real numbers between 0 and 1 is in fact greater than A_1. He called this new transfinite number C – not A_2 (although C may be A_2) because it could not be proved whether or not there exists a transfinite number greater than A_1 and less than C.

To quote Constance Reid in her book *From Zero to Infinity* (1956),
'The challenge of the finite set is merely physical. With the infinite set, the challenge is mental'.

My personal appreciation of this statement leads me back, subdued, to the conversation between the children still in their First School, whose mental development does not allow very much abstract thought.

If the children appealed to me in their discussion, I should take into consideration their age and ability, of course, but so long as they were at the top of the First School or at the lower end of the Middle School, and were of at least average ability, I would probably use the same method as I did with my 8 year old, to establish the difference between the very large finite and the infinite. As well as using the language of the set of natural numbers in 'conversational reasoning', we could use the classroom number line as a good visual aid. (Ours stops at a hundred, but could it not carry on round the classroom and out through the door . . . ?) We could then talk of large numbers, and they might like to see, say, a thousand pebbles in groups of a hundred – this would also help notation understanding. If they were still interested, we might see how long it would take to count a large number – they would be surprised!

In the second discussion, the argument as to whether or not the hands touch implies that one child at least thinks they won't. He would, of course, have my support, but this is a much more difficult concept to appreciate, and perhaps beyond the ken yet of the others. However, if appealed to, I must try at best to enlighten, at worst not to confuse. Obviously we must cut a ribbon in halves. This might clarify the situation for some, and would in any case lead to further useful discussion – and eventually to exasperation at the impracticability of the exercise (this in itself might bring a ray of light, the very small pieces illustrating the possibility of even smaller pieces, and taking the imagination into the heart of the matter). We could then transfer the problem to pencil and paper, but there I think I would let it rest for the time being.

But I would get out the $\frac{1}{4}$ inch graph paper, so that they could discover the expanding and contracting squares. This to me is an ideal expression of the nature of the infinity.

There is another visual example, though, which was the source of great pleasure to me. While I was in discussion with my 13 year old on the subject of infinity, I showed him, just briefly, a picture in one of my books showing a boy's image reflected an infinite number of times in two facing mirrors. He seemed unimpressed – his thoughts were on halving halves. But a few days later, hand-mirror in hand, he burst into the room,
'I've seen it' he said.
'Seen what?' I asked.
'Infinity!'
And he showed me how to stand in front of a large mirror, position the hand-mirror just below eye-level, line it up with its reflection, and look right along the tunnel to infinity.

Fun with one-to-one correspondence

EVELYN KNOWLES

A fascinating discovery for an intermediate grade student is the construction of a geometric picture using only straight lines.

This lesson extends practice in basic geometry, division, and one-to-one correspondence. The student will need to know how to hold a ruler, draw a straight line, and be accurate. The lesson also offers a review in mathematical terminology. After the few basic rules are learned, other objectives, such as balance, color, and the blending of colors, may be developed.

Any kind of paper, a ruler, and a sharp pencil are the only necessary materials. Colored pencils may be used later if desired.

Starting anywhere on the paper, draw a large angle. Fairly long lines and an angle of at least 50° are best for the beginner.

Place a dot every half-inch on line A (Fig. 1), extending or shortening the line as necessary. The dot at the vertex of the angle (point C) should be a little above the intersecting line B.

Beginning at the vertex of the angle (point C) on line A, number the dots one, two, three, etc., to the end of the line.

Measure line B. Beginning just to the right of point C, divide line B into an equal number of dots to correspond with line A. Number the dots on line BC, beginning at point C, one, two, three, etc.

The teacher should check each paper to see that the work thus far is correct, and that each child has a clear understanding of what he is doing.

Place the ruler on line A at point number 1, and connect to line B point number 1. Draw one straight line between these two points. Move the ruler to point

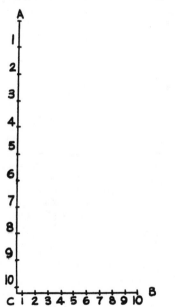

Figure 1

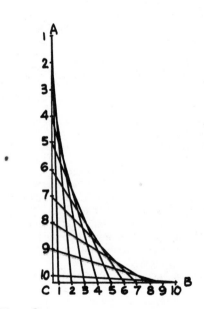

Figure 2

Reprinted from *The Arithmetic Teacher* (May 1965), 370–372, by permission of the author and the publisher.

number 2 on line *A* and connect to point number 2 on line *B*. Continue to connect corresponding numbers on each line with a straight line. As the student will soon discover the interior of the angle gives the illusion that curved lines have been used (Fig. 2).

After the student has an understanding of the principles involved, let him experiment with more than one angle and different positions of angles as in Figures 3 and 4.

This experience can be made more permanent by using tagboard. To prevent any pencil work from showing on the front of the finished picture have the student work on the back of the board, using the same principles, but reversing his work.

Thin, inexpensive crochet thread can be purchased in many brilliant colors. It is easier to use than yarn and from a distance gives a completely different effect than that of thread. The student will need to know how to thread a needle and tie a knot.

Two colors may be blended very effectively by dividing the number of points on line *A* in half, and working the first half of the numbers in one color, the second half of the numbers in another color. For example, work numbers 1–5 in blue and numbers 6–10 in yellow. The interior of the angle will produce a green shading, while one end is blue and the other yellow.

Animals, flowers, designs, and numerous other things may be made using one-to-one correspondence and angles. It's fun and creative, and a very rewarding, concrete learning experience for each child.

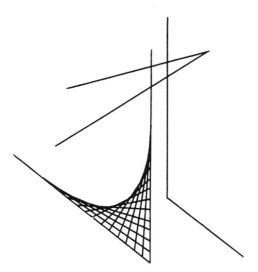

Figure 3

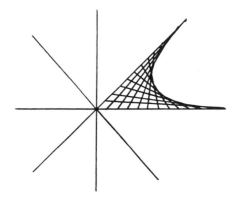

Figure 4

Grouping of objects as a major idea at the primary level

JAMES M. MOSER

The Commission on Mathematics, in its well-known report written in 1959, called for the "judicious use of unifying ideas" in mathematical instruction. While it is true that the report was written for the main purpose of improving secondary mathematics instruction, this particular recommendation has just as much importance for elementary education as it does for secondary.

One mathematical process that has a wide range of use and application in the instructional program of elementary mathematics and therefore qualifies as a unifying idea is the process of grouping and partitioning.

In the grouping process, a child is given a set of objects, and he separates the set into smaller subsets or groups, each of the same number with the possible exception of a single remainder group, which will be smaller than the other groups. For example, suppose the child is given a set of 35 objects and is told to "group by 4." He performs the grouping, finds out there are eight groups of 4 and one group of 3, and writes the following expression:

$$8(4) + 3.$$

Why is the equation

$$35 = 8(4) + 3$$

not written? The answer to this query involves one of the main applications of the process: the grouping process can be used to teach place value to young children. In the above example, it is perfectly reasonable to give 35 objects to a young first or second grader without his knowing that

group by

3   5   7   10

record what you have done on the work sheet

Fig. 1

Reprinted from *The Arithmetic Teacher* (May 1971), 301–305, by permission of the author and the publisher.

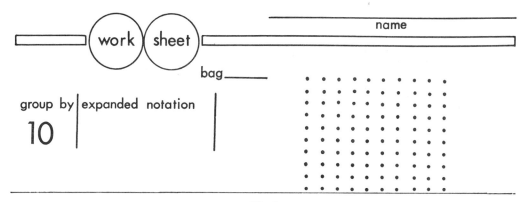

Fig. 2

there are 35 objects in the set. In fact, it is just this process that can be used to let him find out how many objects there are in the set given to him. In this application of the process, the child would be directed to "group by 10." He would respond by actually performing the grouping and writing

$$3(10) + 5.$$

The only prerequisite behaviors needed to perform such a task are the ability to count to 10 and the ability to write the numerals from 0 to 10.

At the University of Wisconsin we are currently involved in a curriculum-development effort that is producing an instructional program called Developing Mathematical Processes. It is an activity approach to learning mathematics, and in this pro-

gram the grouping process is used to help the children learn about place value. We begin by teaching the process of grouping. A child is given a sack of objects and is directed to perform certain groupings by means of an activity card, a sample of which is shown in figure 1. Results of the physical grouping can be recorded on a separate sheet, a portion of which is shown in figure 2. Notice the place for a pictorial representation of the results of the work. For the younger child who does not yet have a command of larger numbers, it is quite easy to structure the problems so that all the numbers *he* works with at the time are less than or equal to 10.

Grouping can also be performed on somewhat abstract objects as well as real objects. Figure 3 shows a picture of "creatures" that are to be grouped. We also ask

group by 5

Fig. 3

children to group the pegs on their geo-boards (see fig. 4).

Once the child has learned the grouping process, it then becomes rather easy to make the transition to place value. Returning to the set of 35 objects, the expression

$$3(10) + 5$$

shows up very clearly the digits 3 and 5 that are used in the compact notation for 35. A child will then write

$$35 = 3(10) + 5.$$

The expanded notation to the right of the = symbol becomes very useful when the addition and subtraction algorithms are taught.

Another very useful application of this process occurs when the need arises to change from one unit of measure to another. Suppose, for example, a child has an object 35 inches in length and is instructed to change to a description involving feet and inches. If his teacher had been using the grouping-process ideas to good advantage, the child would say to himself, "Group by 12." Grouping either by a manipulation of physical objects or by a counting process of some sort, he would write:

$$35 = 2(12) + 11.$$

The numbers 2 and **11 are the** ones of interest, and he can read **directly** that "35 inches is the same as 2 feet 11 inches."

The unit of measure is ounces and you want to convert to pounds. How? Group by 16.

$$35 = 2(16) + 3.$$

35 ounces is the same as 2 pounds 3 ounces.

The unit of measure is quarts and you wish to convert to gallons. How? Group by 4.

$$35 = 8(4) + 3.$$

35 quarts is the same as 8 gallons 3 quarts.

To exchange pennies for nickels, you group by 5.

$$35 = 7(5) + 0.$$

35 pennies can be exchanged for 7 nickels.

How many weeks are there in 35 days? Group by 7.

$$35 = 5(7) + 0.$$

There are 5 weeks in 35 days.

Next, suppose you wish to introduce the concept of odd number versus even number. One method that children can use to test a given number is to subject that number to the grouping process, using 2

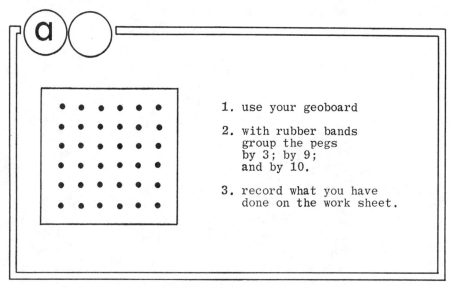

1. use your geoboard

2. with rubber bands group the pegs by 3; by 9; and by 10.

3. record what you have done on the work sheet.

Fig. 4

as the number to group by. If the remainder is 1, the number is odd; if the remainder is 0, the number is even. For example,

$$35 = 17(2) + 1$$

tells us that 35 is an odd number.

Focusing on the remainders after grouping by a particular number has been performed on a set of whole numbers is at the heart of modular arithmetic and numerical congruences, a topic that has been introduced with some popularity in the intermediate grades.

The use of the word *remainder* conjures up thoughts of division. And well it should! The equations we have been writing, such as

$$35 = 8(4) + 3,$$

are examples of the division algorithm. Basically, a statement of a theorem about the division algorithm says that there will exist numbers called the quotient (8) and the remainder (3) when a given number (35) is divided by a second number (4), which is called the divisor. Moreover, the remainder will always be less than the divisor. Thus, the grouping process has as its foundation the division process.

The claim to be made is that if a child is exposed early—in the first or second grade—to the grouping process and is continually given the opportunity to use it, the concept of division will not be difficult for him to learn. It is not suggested here that the child will more easily learn the long-division algorithm, but rather that the idea of division as a model of the grouping process will be more reasonable to him. For example, if a question such as "How many 5s are there in 35?" is raised, the child may be likely to respond, "That means to me, How many groups of 5 are there? I will group by 5 to get my answer."

Mathematically, it is also proper to think of division as the inverse process of multiplication. Happily, this aspect is also an outcome of a use of the grouping process in a mathematics program. Rather than give a child a set of objects and ask how many groups of a certain size there are, it is very natural to say to him: "You have seven groups of 3. How many objects are there in all?" Last year some second graders who had been brought up on a diet of grouping as part of their involvement in our Developing Mathematical Processes field test were observed answering such a question. They simply made marks on a paper to represent seven groups of 3 and then decided they had to regroup the objects into groups of 10. They did so very quickly, finding that there were two groups of 10 and 1 left over; from this, they immediately decided that seven groups of 3 were equal to 21, writing:

$$7(3) = 21.$$

The thing that was most striking was not so much that they got the answer but rather that they were able to decide on a method of figuring it out by themselves. According to their teacher, these children were very ready to think about multiplication ideas.

And this is one of the very distinct advantages of using the grouping process together with the notation that has been used throughout this article. There is such a great deal of *pre*multiplication and *pre*-division ideas that multiplication and division become much easier to learn when they are formally studied. And when these two processes *are* studied, it is possible to use the language and ideas of grouping to explain what is happening.

The notation employed is easy for children to use and is consistent with accepted usage. Parentheses are the symbols used to represent a *group* or a set of symbols that are to be considered as a single entity. The × is not used to denote multiplication; and so the symbolism can be used with advantage long before multiplication is introduced, although the pupil will eventually learn that the symbolism does represent multiplication.

some elementary concepts of logic

James J. Roberge

IN RECENT YEARS, various mathematics curriculum study groups and conferences on school mathematics have recommended that instruction in logic be incorporated in the elementary school mathematics curriculum. For example, the Report of the Cambridge Conference on School Mathematics stated that "while extensive formal study of logic in the elementary grades is not favored by most mathematicians, it is hardly possible to do anything in the direction of mathematical proofs without the vocabulary of logic and explicit recognition of the inference schemes."[1]

The purpose of this article is to discuss some basic concepts of logic which should prove enjoyable for elementary school pupils, and will provide them with some insights into deductive reasoning. In the first section, these concepts will be treated informally and the discussion will focus on teaching strategies which could be used to give pupils in the elementary grades an opportunity to learn the form and meaning of some common types of logical statements. In the second section, these concepts will be explored in greater depth, with increased emphasis on vocabulary and the examination of the validity of logical arguments. The latter part of each section describes some classroom activities which elementary school pupils should find interesting and challenging.

─────────── an informal introduction to logic ───────────

We can begin our informal study of logic with a discussion of categorical statements. Briefly, *categorical statements* are logical statements that make assertions which affirm or deny that one set is included in another set, either as a whole or in part. The four standard forms of categorical statements are illustrated by the following declarative sentences:

> *All* peaches are round.
> *No* squares are round.
> *Some* balloons are round.
> *Some* stones are *not* round.

Figure 1

We can discuss the meaning of the logical concepts *all*, *no*, and *some* in terms of the set theory concepts *subset, disjoint sets, empty set,*

[1] Cambridge Conference on School Mathematics. Report of the Conference. *Goals for School Mathematics* (Boston: Houghton Mifflin, 1963), p. 39.

Reprinted from *Timely Topics* (January/February 1971) by permission of Houghton Mifflin Company.

and *intersection*. In addition, Venn diagrams can be used to illustrate the correspondence between these concepts. For example, consider the categorical statement "All peaches are round." We can illustrate the meaning of this statement by using a Venn diagram:

The diagram indicates that *all* of the elements of {peaches} are also elements of {round objects}. Thus, {peaches} is a *subset* of {round objects}.

Similarly, we can illustrate the meaning of the categorical statement "No squares are round" by the following Venn diagram:

The diagram shows that there are *no* elements of {squares} that are also elements of {round objects}. That is, {squares} and {round objects} are *disjoint sets*. Hence their *intersection* is the *empty set*.

We can represent the meaning of the categorical statement "Some balloons are round" by the following Venn diagram:

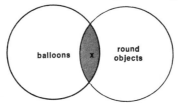

The diagram shows that *some* elements of {balloons} are also elements of {round objects}. More precisely, there is "at least one" element of {balloons}, call it x, that is also an element of {round objects}. Hence the *intersection* of the two sets is nonempty, and is represented by the shaded region in the diagram.

Finally, the meaning of the categorical statement "Some stones are not round" can be represented by the following Venn diagram:

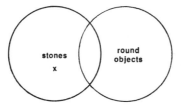

The diagram indicates that *some* elements of {stones} are *not* elements of {round objects}. Specifically, there is "at least one" element of {stones}, call it x, that is *not* an element of {round objects}. We should note, however, that a categorical statement of this form does not make any assertion about whether or not there are elements in the *intersection* of {stones} and {round objects}.

When the pupils understand these concepts we can begin our discussion of some simple logical arguments called syllogisms. Briefly, a *syllogism* is a logical argument consisting of two categorical statements called premises, and a conclusion related to these two premises. Syllogisms are basic arguments in deductive reasoning, that is, reasoning which proceeds from given information, called the hypothesis, to a result, called the conclusion. Moreover, if one accepts as true the information given in the premises of a syllogism, and the conclusion is the only one which can be arrived at by reasoning logically from the premises (that is, which is *logically acceptable*), then the syllogism is said to be valid.

We can use Venn diagrams to help us to decide whether or not the conclusion of a syllogism is logically acceptable and the syllogism is valid. For example, consider the following syllogism and the accompanying Venn diagram:

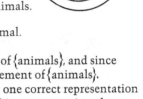

Hypothesis: All dogs are animals.
            Rover is a dog.
Conclusion: Rover is an animal.

Reasoning logically, we state that {dogs} is a subset of {animals}, and since Rover is an element of {dogs}, he must also be an element of {animals}. Referring to the diagram, we note that there is only one correct representation of the conditions of the hypothesis. Furthermore, this representation also satisfies the conditions of the conclusion. Therefore, the conclusion is logically acceptable and the syllogism is *valid*.

Now consider the following syllogism:

Hypothesis: All cats are animals.
            Snowball is an animal.
Conclusion: Snowball is a cat.

For this syllogism, there are two correct representations of the conditions of the hypothesis:

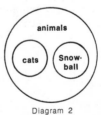

Diagram 1   Diagram 2

However, only Diagram 1 also satisfies the conditions of the conclusion, "Snowball is a cat." Diagram 2 shows that a possible conclusion is "Snowball is not a cat." Therefore, since more than one logically acceptable conclusion is possible, we say that the syllogism is *invalid*.

Using the preceding discussion as a springboard, the elementary school teacher could develop a variety of classroom activities which would be of interest to pupils and would reinforce their understanding of the relationships between the concepts of logic and those of set theory. For example, pupils could be given sets of objects (blocks, cut-outs, etc.) which differ in terms of various attributes (color, size, shape, etc.), and could be asked to use the terms "all," "no," or "some" to make statements, such as "Some circles are blue" or "All red squares are large," which describe subsets of these objects. The fringe benefits of such an activity are obvious.

As a second classroom activity, the teacher could draw a diagram such as the following on the chalkboard:

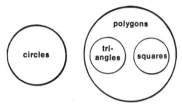

Then, the teacher could make statements, such as "All squares are polygons" or "No circles are triangles," about the sets shown in the diagram, and pupils could be asked whether each statement is true or false. Moreover, pupils could be asked to write categorical statements, such as "No triangles are squares," about the sets shown in the diagram.

We should note that the statements which are used to supplement classroom discussions should be drawn from the everyday experiences of the pupils, and should be stated so that they can be readily shown to be true or false.

an extension of the basic ideas

We can extend our earlier description of categorical statements by including the fact that every standard form categorical statement has two basic characteristics, that is, quantity and quality. The *quantity* of a categorical statement is universal or particular depending upon whether the statement refers to all of the elements or only to some of the elements of the set named in the subject of the statement. For example, statements such as "All peaches are round" and "No squares are round" are universal statements, whereas statements such as "Some balloons are round" and "Some stones are not round" are particular statements. The *quality* of a categorical statement is affirmative or negative according to whether set inclusion is affirmed or denied by the statement. For example, statements such as "All peaches are round" and "Some balloons are round" are affirmative statements, whereas statements such as "No squares are round" and "Some stones are not round" are negative statements. Thus, we can uniquely describe each of the categorical statements shown in Figure 1 by naming its quantity first and its quality second. Moreover, we can use a coding scheme, developed for this purpose by logicians, to simplify our task. The coding scheme uses the first four vowels of the alphabet. These ideas about categorical statements are summarized in Figure 2.

Categorical Statement	*Name*	*Symbol*
All peaches are round.	Universal affirmative	A
No squares are round.	Universal negative	E
Some balloons are round.	Particular affirmative	I
Some stones are *not* round.	Particular negative	O

Figure 2

We may also extend our earlier description of simple syllogisms to a more general type called categorical syllogisms. *Categorical syllogisms* consist of two premises and a conclusion, each of which is a categorical statement. In addition, they contain three set names, each of which occurs in exactly two of the constituent categorical statements.

Again, we can use Venn diagrams to help us to decide whether or not categorical syllogisms are valid. As an example, consider the following syllogism and the accompanying Venn diagram:

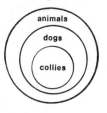

Hypothesis: All dogs are animals.
　　　　　　All collies are dogs.
Conclusion: All collies are animals.

Since this is the only way that we can draw a Venn diagram to represent the conditions of the hypothesis, and the closed curve which represents {collies} lies entirely within the closed curve representing {animals}, the diagram also satisfies the conditions of the conclusion, "All collies are animals." Therefore, the conclusion is logically acceptable and the syllogism is *valid*.

 Now consider the following syllogism:

Hypothesis: All canaries are animals.
　　　　　　All birds are animals.
Conclusion: All birds are canaries.

For this syllogism, each of the following diagrams correctly represents the conditions of the hypothesis:

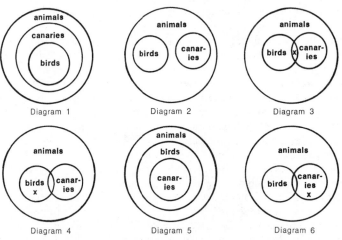

However, only Diagram 1 also satisfies the conditions of the conclusion, "All birds are canaries." Diagram 2 shows that a possible conclusion is "No birds are canaries" or "No canaries are birds," Diagram 3 shows that a possible conclusion is "Some birds are canaries" or "Some canaries are birds," Diagram 4 shows that a possible conclusion is "Some birds are not canaries," Diagram 5 shows that a possible conclusion is "All canaries are birds," and Diagram 6 shows that a possible conclusion is "Some canaries are not birds." Therefore, since more than one logically acceptable conclusion is possible, we say that the syllogism is *invalid*, even though the categorical statements for the conclusions shown in Diagrams 3, 4, and 5 are, in fact, true statements.

As a classroom activity, pupils could be presented with categorical syllogisms, and could be asked to draw Venn diagrams to determine whether or not the syllogisms are valid.

For a second classroom activity, pupils could be asked to state three-letter arrangements, such as EAO, which would identify the form of each of the premises, and that of the conclusion, for given categorical syllogisms. Moreover, since the form of each of the premises, and that of the conclusion, must be one of the four standard forms of categorical statements (see Figure 2), pupils could be shown how to use simple counting principles to determine the number of possible three-letter arrangements. Finally, pupils could be shown how to count and list the possible outcomes by using a "tree" diagram.

These activities are but a few of the wide variety of interesting activities which imaginative and creative elementary school teachers could use to introduce some basic concepts of logic.

References

Cambridge Conference on School Mathematics. *Goals for School Mathematics*. Boston: Houghton Mifflin, 1963.

Copi, I. M. *Introduction to Logic*. New York: Macmillan, 1961.

Dienes, Z. P. and Golding, E. W. *Learning Logic, Logical Games*. New York: Herder and Herder, 1966.

National Council of Teachers of Mathematics. *More Topics in Mathematics for Elementary School Teachers*. Thirtieth Yearbook of the National Council of Teachers of Mathematics. Washington, D. C.: The National Council of Teachers of Mathematics, 1968. Pp. 225-230.

Smith, E. P. and Henderson, K. B. "Proof." *The Growth of Mathematical Ideas Grades K-12*. Twenty-Fourth Yearbook of the National Council of Teachers of Mathematics. Washington, D. C.: The National Council of Teachers of Mathematics, 1959. Pp. 111-181.

Swain, R. L. "Logic: For Teacher, for Pupil" *Enrichment Mathematics for the Grades*. Twenty-Seventh Yearbook of the National Council of Teachers of Mathematics. Washington, D. C.: The National Council of Teachers of Mathematics, 1963. Pp. 282-301.

activities for building concepts of elementary logic

The use of *all*, *some*, or *none* in mathematics is common and important. For example, we may say that *all* prime numbers have only two factors. If we could find a prime number which has more than two factors, we would have a contradiction, and our first statement would be incorrect. Similarly, we may say that *all* prime numbers are odd. But once we recall that 2 is a prime number and that 2 is even, we have a contradiction.

Syllogistic reasoning is useful in mathematics and in everyday life. Understanding of syllogisms is best developed by working with sets and subsets. Some of the following activities, taken from the Teacher's Annotated Editions of the Houghton Mifflin *Modern School Mathematics* program, may help you.

● On three different parts of the chalkboard, draw a number of circles, each group forming a set. Ask one student to color all the circles in the left-hand set, and one student to color some of the circles in the center set. Leave the right-hand set uncolored. Ask the students to use the words *all*, *some*, or *none* in describing the part of each set that was left uncolored.

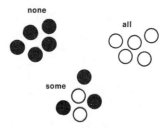

● On the chalkboard, draw circles to illustrate one set as a subset of another, one set as the intersection of two sets, and two disjoint sets. Point out that the first diagram shows that all members of *A* are in *B*, since *A* is a *subset* of *B*; in the second diagram, some members of *B* are in *A*, since the sets *intersect*; in the third diagram, no members of *B* are in *A*, since the sets are *disjoint*.

● Have the pupils disagree with a statement such as "All children like candy." From the variety of responses, emphasize the ones similar to "Some children do not like candy" and "There is at least one child who does not like candy."

● On the chalkboard, draw the set diagram shown at the right, and ask the students to write some *if* and *if-then* statements from the diagram. Lead them to discover that the syllogism, *If all bears are animals, and Smokey is a bear, then Smokey is an animal*, is valid.

● Have children draw Venn diagrams to illustrate their own syllogisms.

Numeration—a fresh look

WAYNE PETERSON

Numeration is the topic most closely associated with the "new mathematics" in the minds of many. This is evidenced by the rash of articles which has appeared in The Arithmetic Teacher and elsewhere.

Few people need to be convinced that numeration belongs in the mathematics curriculum. Increased understanding of base ten is reason enough for including the topic in a modern program. There are, however, two questions that need to be resolved: (1) At what grade level may optimum results be expected? (2) How extensive should the treatment be?

My own experience as a teacher indicates that treatment of the topic at the seventh-grade level yields best results with the least expenditure of time. While this position is subject to modification in light of local conditions, ability of the students, and other special considerations, this is where treatment of numeration should be placed in the general design of a course of study. Although children at an earlier age can be taught to count and compute in various bases, this does not mean that anything of lasting significance has been accomplished. Indeed, unless they possess enough sophistication to draw sound generalizations from specific cases, all the teacher has done is spoil the subject for a meaningful development at a later date. Most seventh graders have the requisite level of sophistication, and there seems little reason to postpone treatment of the topic to the eighth grade or beyond.

It is necessary to agree, first of all, on how much understanding we should expect of the students. This in turn depends upon whether we are talking in terms of enrichment or in terms of the basic program. My feeling is that much of what is done in the classroom should be considered "for enrichment purposes only." This would include the study of ancient systems of numeration, algorisms for translating from one transdecimal base to another (even from base ten to a given transdecimal base), the relationship of base two to base eight, negative bases, and extensive development of computation techniques within the various bases.

In contrast I believe that the *basic program* should include only those activities which emphasize the importance of base and place value in a numeration system, and thus deepen the students' appreciation of the decimal system. Keep the rest as enrichment material, but don't let the associated activities preempt an inordinate amount of time, as is presently the case in too many classrooms. If this advice is followed, only four or five lessons should be required to consider the topic of numeration adequately.

The remainder of this article will be devoted to a discussion of an approach that minimizes the number of lessons required while it provides for the necessary understandings.

Place-value pocket chart

The activities outlined below depend upon the use of classroom sets of two devices, both of which are inexpensive and easy to prepare. The first of these is simply

Reprinted from *The Arithmetic Teacher* (May 1965), 335–338, by permission of the author and the publisher.

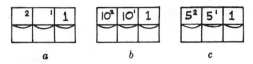

Figure 1

a place-value pocket chart of the type used in primary classrooms (Fig. 1a). Note that the place-value designations on the chart are incomplete. Also note that the use of this device and the activities that follow presuppose a familiarity with exponential notation.

Each student will need a chart and a generous supply of paper tabs. The sequence of steps for using this chart is as follows:

1 Display a number of objects on a table at the front of the classroom. Tell the students that they are to determine the number of objects without counting in the usual manner (this would defeat the purpose). Instead, as you move objects one by one from one side of the table to the other, the students are to match their tabs one-to-one with the objects.
2 Tell the students that they are to put the tabs in the pockets of the chart in such a way that the numeral for the number of objects will be indicated. Discuss with them the base ten place-value designations of the pockets to the left of the ones' pocket, and have them mark the pockets as indicated in Figure 1b. (So that the pocket chart may be reused for other bases, have the students do this on pocket-size card inserts instead of directly on the chart itself.)
3 Remind the students that ten is the grouping number, and discuss with them the proper packaging of the tabs. Have them put tabs one by one in the ones' pocket until the grouping number is reached and then package and "carry" until the process is completed.
4 Make sure that each student can "read" the numeral thus indicated. Emphasize that they have thus been able to determine the number of objects without

resorting to the usual counting process, the only counting having been that necessary in reaching the base number. Repetition of this activity should bring a realization that the counting activity is a result of the construction of the system of numeration, not the other way around.

With or without prompting, some students should then be ready to suggest that some other number might serve equally well as a grouping number. For convenience we will use *five*, and proceed as follows:

1 Discuss with the students the necessary place-value designations, and have them mark the chart as indicated in Figure 1c.
2 Then ask, "If you have a single tab, into which pocket will it go? Where will you put a pair of tabs? A trio of tabs? A quartet of tabs?"
3 Discuss with them what they should do if another tab is added to the quartet in the ones' pocket. They should see that having reached the grouping number, they must package the tabs and "carry" so that subsequently there will be a package of five tabs in the middle pocket. Then ask, "What numeral is now represented by your chart arrangement?" (Since "ten" is a base-ten word, avoid using it when referring to the numeral 10. Instead, adopt the accepted practice of saying "one 0.")
4 Make sure that the students recognize that the *numeral* 10 names the base or grouping *number*, which in this case is the number five. Then continue this activity, adding tabs one at a time to the ones' pocket, grouping, and "carrying" as necessary. As this is being done, the students should be required to write the numeral indicated for each change in the display. Encourage them to look for a meaningful pattern so that sooner or later they will be able to continue writing numerals for successive numbers without using the chart.
5 Now, follow the sequence of activities

60

that you employed with base ten. After each set of objects has been considered and tabs are properly packaged and located, ask, "What base-five numeral names the number of objects displayed on the table?" They may wish to translate to base ten, but the only reason for so doing would be to satisfy their curiosity—they shouldn't be expected to "know" the number thus named unless the translation is made to the familiar base-ten language.

6 Let the students choose other numbers to use as grouping numbers. They should then follow the sequence of activities suggested above for base five.

7 Help the students understand that, although they may not immediately know what number a specific transdecimal numeral names, a "Dozenlander" would automatically make the proper association between a base-twelve numeral and the number named.

8 The important goal of all these activities should remain the appreciation of the structural identity of these various numeration systems, regardless of base. This can be represented nicely by the use of a variable for the base number (see addendum).

The one fringe activity I would include as a part of the regular program is the invention of a phonetic scheme for some selected transdecimal base. Construction of such a scheme could follow a discussion of the imperfections in our own decimal scheme: "eleven" might become "onety-one," and so on.*

It should be clear that, aside from the construction of a phonetic scheme, the activities described above would require

only one class period for their completion. It is the use of the pocket chart that makes this much progress possible in such a short time.

Multibase abacus

The second device that I recommend for work in numeration is a multibase abacus. In its simplest form, this may be a block of wood with a number of washers held on each of several nails driven into the block (Fig. 2). The number of washers on each nail is determined by your choice of bases—two less than twice the largest base number will do.

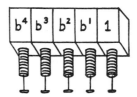

Figure 2

Here are several uses to which the abacus can be put:

1 Have each student use his abacus to determine the numeral for a number of objects in several selected bases. Relate this to the work the students did with the pocket charts. They should see in particular that, instead of packaging tabs and "carrying" by moving the package to the next pocket to the left, they now trade the base number of counters on one nail for one counter on the next nail to the left.

2 Next, have the students use the abacus to work several simple addition exercises in base ten. Encourage them to rely on the abacus instead of their prior knowledge of the basic addition facts. Provide them with a form on which they can record the numerals for the addends and for the sum.

3 Let the students choose some other base; then provide them with the numerals for addends in the selected base. Instruct them to write the numerals on

* One phonetic scheme for base five is given below:

0	nuttin	20	doofy	40	furfy
1	oom	21	doofy-oom		.
2	doo		.		.
3	fee		.		.
4	fur		.	100	oom hut
10	fit	30	feefy	101	oom hut oom
11	oomfy-oom	31	feefy-oom		.
12	oomfy-doo	32	feefy-doo		.
13	oomfy-fee		.		.
14	oomfy-fur		.	1000	oom tut

the form provided and then record the first numeral on the abacus by pulling down the required number of counters on each nail. Now, have them pull down enough counters so that the second numeral is recorded along with the first. Most students should be able to proceed without further instruction.

For any students having difficulty, review the "carrying" procedure in a base-ten context, and remind them that the steps are the same in each base—only the grouping number is different. They should then write the numeral for the sum on the form provided. As before, the students may wish to translate to base ten so that they may "know" the number named.

4 Sufficient repetition of this procedure in various bases will make clear that the addition process is not unique to any one base, but dependent only upon the structure of the generalized base-place system of which base ten is just one example.

Subtraction can be treated in similar fashion. Many students will soon be able to complete computations, both addition and subtraction, without the abacus. Be sure, however, that they are not computing mentally in base ten and then translating to the base being used. Use the abacus only as long as it seems profitable. It is easy to overdo with such devices.

A basic fact chart for addition should subsequently be made in each of the several bases the students have considered. Then, treating multiplication as repeated addition, students might also make multiplication fact charts. Allot some time to computation exercises with the charts, but

leave the development of the division algorism for enrichment.

Implicit in all the foregoing suggestions is the importance of relating the activities to the generalized base-place concept. A deeper understanding of base ten, the end product toward which this work is directed, should be automatic. Even though I have touched only upon a basic sequence of activities, I hope it is clear that no essential principles have been omitted and that these activities need consume no more than four or five class periods.

Addendum

Some writers question the purity of an approach that uses base-ten numerals to name place values in other bases. Some have adopted the convention of using the base-ten word instead of the corresponding symbol for the base number. However, these same writers see no objection to using base-ten numerals to indicate exponents. This seems to me to indicate a singular lack of consistency.

My position is that, since we have borrowed the familiar Hindu-Arabic digits as a convenience, it is just as sensible to borrow base-ten numerals to indicate place values, whatever the base. It is a matter of using the familiar to help explore the unfamiliar—not much different from using a familiar language to help in the process of learning a foreign tongue.

While a case could be made for the use of the numerals 1, 10, 100, 1000, and so on, to designate place values in any base, this is an understanding which should evolve in any case. I think it is a mistake from a pedagogical point of view to begin the work in numeration with an insistence upon the understanding of this idea. Instead, if the suggestions in this article are followed, students should arrive at the generalization indicated below:

$$\cdots b^5 \ b^4 \ b^3 \ b^2 \ b^1 \ 1 \cdots$$

From this they may then obtain

$$\cdots 100{,}000 \ 10{,}000 \ 1000 \ 100 \ 10 \ 1 \cdots$$

From this point forward they would be in a position to use either method for indicating place value.

Understanding different number bases

HAROLD F. RAHMLOW

Introduction

It is now common practice in many of the elementary schools to introduce the students to numeration in bases other than ten so they may appreciate and understand base ten more fully. This paper presents a methodology, or more correctly a teaching aid, for presenting numeration systems using different bases.

A teaching aid

For centuries the Orientals have used the abacus as an aid to calculation. The abacus has also been used of late in some schools as a device for helping children understand the nature of our numeration system. Children are motivated to higher levels of learning by the novelty and manipulative aspects of this mechanical contrivance. A variable abacus with which students can discover the principles of numeration in any base can easily be constructed. Once this device has been constructed, its use is limited only by the imagination of the pupils and the teacher. Figure 1 is the schematic of this device.

A wooden frame holds the thin metal rods used to mount the wooden counters. The device can be constructed in any size

desired. The hinges and clasps at both top and bottom of the abacus permit the addition or subtraction of counters so that the device can be used for any desired base.

Working with even bases

The same principle can be followed in computation on the abacus for any even-numbered base. Figures 2, 3, and 4 show some possible arrangements of the abacus. In each example the number 3172 (base ten) is shown as it appears on an abacus of the base under consideration. With

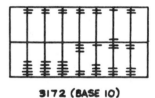

3172 (BASE 10)

Figure 2. Base ten abacus.

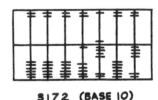

3172 (BASE 10)

Figure 3. Base twelve abacus.

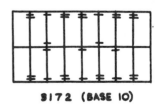

3172 (BASE 10)

Figure 4. Base four abacus.

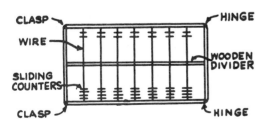

Figure 1. Variable abacus.

Reprinted from *The Arithmetic Teacher* (May 1965), 339–340, by permission of the author and the publisher.

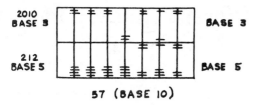

Figure 5. A comparison of numerations.

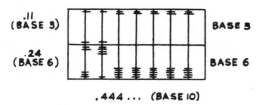

Figure 6. A comparison of numerations.

these abacuses and other similar ones the elementary school child can discover principles of calculation and numeration by his own manipulation.

Comparing notations

The variable abacus can also be used as a dual counting frame to compare the representation of numbers in different bases. When used in this manner, the abacus is considered as two separate counting frames, each of a different base. The comparison of notation for the two bases is immediately evident. Two examples of this use of the variable abacus are shown in Figures 5 and 6.

Let the top half of the variable abacus represent base three and the lower half represent base five. Compare the form of the number 57 (base ten) in base three and base five. The written form of the comparison of these two numerations is shown to the left of the abacus in Figures 5 and 6.

Decimals can also be compared on the variable abacus by choosing the left edge of the abacus as the decimal point. Thus, the problem—compare the form of the rational number .444 . . . (base ten) in base three and base six notation—would appear as in Figure 6. This particular problem raises the further question of infinite sequences of numbers. At this point the limitations of a finite counting frame could be discussed.

Summary

The variable abacus is an easily constructed and highly versatile teaching aid which provides a graphic means for illustrating arithmetic in different bases. Arithmetic calculation in even bases and comparisons of the representation of quantities in different bases are only two of the possible uses of the device. As pupils and teachers work with this device, they constantly discover new uses for it, but, more important, the pupils discover basic principles of numeration.

Let's use our checkers and checkerboards to teach number bases

LUCILE LaGANKE

In this day of teaching modern mathematics, one notices the tremendous numbers of manipulative materials being manufactured for the sole purpose of teaching number concepts. However, the teacher might find interesting math materials at home.

Why shouldn't the teacher and his pupils bring their checkers and checkerboards to school, to use in teaching the idea of various number bases to fifth and sixth graders.

A checkerboard has sixty-four squares and twenty-four checkers. The pupil, by drawing four rows of eight squares, the size of those on his checkerboard, on an appropriate size of paper, can change the checkerboard into an arithmetic tool.

The squared piece of paper can be placed at the top of the checkerboard. The first row should be marked from right to left with the number base being used, each number having the proper exponent above the base number, to indicate its value. The second row should have the numerical equivalent of each base number, written in base ten. Then, two rows could be used for manipulating the checkers.

A pupil could place checkers on each space of the fourth row, and then push the checkers into the proper spaces in the third row, to build numbers. This is really using the checkerboard as an abacus.

Figure 1 is a sample of what a paper guide for base two or the binary system would look like.

To express the base-ten number 21, in base two a pupil would push the checkers into the proper squares in the third row.

If checkerboards would take up too much space in the classroom, the teacher could make hectograph checkerboard sheets, and as each number base was studied, the pupil could work out his own number bases and write the correct numbers in the proper squares. (See Fig. 2.) Cardboard or paper disks could be used in lieu of checkers.

These number boards could be used to build numbers, to add numbers, to subtract numbers, and to compare how a certain number appears in different num-

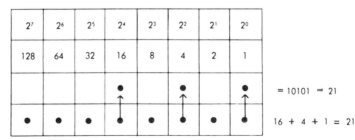

2^7	2^6	2^5	2^4	2^3	2^2	2^1	2^0
128	64	32	16	8	4	2	1
			●		●		●	= 10101 = 21
●	●	●	●	●	●	●	●	16 + 4 + 1 = 21

Figure 1

Reprinted from *The Arithmetic Teacher* (November 1967), 573–575, by permission of the author and the publisher.

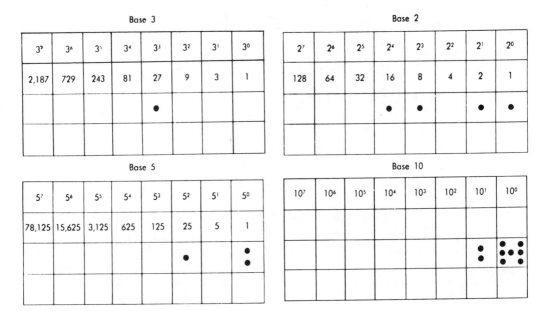

Figure 2

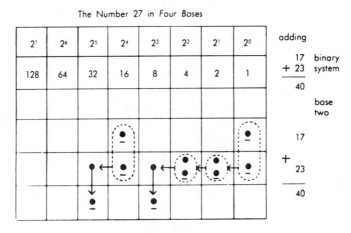

The Number 27 in Four Bases

Figure 3

ber bases. Drills might be given to see how quickly the children have learned to perceive the component parts of a number.

Adding several numbers in a different number base should be fun, for as soon as a pupil obtained the maximum number of checkers permissible in each square in the number base being studied, he would then discard the excess number of checkers, and move one checker into the next column to the left. He would really be "carrying." (See Fig. 3.)

It would probably be wise for the child to start with base ten, and then learn to work in base two, base three, and base five. The number of symbols or checkers that could be used in each number system could be found out by the "discovery" method. Other bases might be presented in the junior high grades.

There are many possibilities in teaching various aspects of number bases, by using a checkerboard and checkers. Probably all the members of the family will join in with the children in playing checkers, a math game for the twentieth century.

SELECTED BIBLIOGRAPHY FOR CHAPTER 3
Sets, Logic and Numeration Systems

Aichele, Douglas B. " 'Pica-Centro'—A Game of Logic." *The Arithmetic Teacher* (May 1972), 359–61.

Barney, LeRoy. "It's O.K. to Count on Your Fingers." *Parents* (September 1968), 70.

Barrett, M. J. "A Method for Changing Numerals in Certain Nondecimal Bases to Numerals in Other Certain Nondecimal Bases, Directly." *The Arithmetic Teacher* (May 1968), 453.

Breithaupt, Keith. "The Key to Roman Numerals." *The Arithmetic Teacher* (April 1968), 374.

Byrkit, Donald R. "Changing Bases—A Method for Squares." *School Science and Mathematics* (June 1971), 513–17.

Dubisch, Roy. "Set Equality." *The Arithmetic Teacher* (May 1966), 388.

Geddes, Dorothy and Lipsey, Sally. "Sets—Natural, Necessary, (K)nowable?" *The Arithmetic Teacher* (April 1968), 337.

Goutard, Madeleine. "A Study of Classification." *Mathematics Teaching* (Spring 1970), 46–50.

Griffin, Harriet. "Discovering Properties of the Natural Numbers." *The Arithmetic Teacher* (December 1965), 627.

Henry, Boyd. "Zero, the Troublemaker." *The Arithmetic Teacher* (May 1969), 365.

Husk, James. "Changing Bases Without Using Base Ten." *The Arithmetic Teacher* (May 1968), 461.

Jansson, Lars C. "Judging Mathematical Statements in the Classroom." *The Arithmetic Teacher* (November 1971), 463.

Lenchner, Mitchell. "A New Base Notation." *New York State Mathematics Teachers' Journal* (April 1971), 77–79.

Morgenstern, Frances B. "Using Classroom Lights to Teach the Binary System." *The Arithmetic Teacher* (March 1973), 184–85.

Muente, Grace. "Where Do I Start Teaching Numerals?" *The Arithmetic Teacher* (November 1967), 575.

Muller, Adelyn. "Fraught With Naught." *The Arithmetic Teacher* (January 1966), 51.

Niman, John. "A Game Introduction to the Binary Numeration System." *The Arithmetic Teacher* (December 1971), 600–01.

Perkins, James R. "A Short Note on Number Bases." *New York State Mathematics Teachers' Journal* (June 1972), 114.

Ranucci, Ernest. "Tantalizing Ternary." *The Arithmetic Teacher* (December 1968), 718.

Rinker, Ethel. "Eight-Ring Circus: A Variation in the Teaching of Counting and Place Value." *The Arithmetic Teacher* (March 1972), 209–16.

Rudd, Lonie E. "Non-Decimal Numeration Systems." *Enrichment Mathematics for the Grades.* National Council of Teachers of Mathematics, *27th Yearbook* (1963), 41–63.

Rudnick, Jesse A. "Numeration Systems and Their Classroom Roles." *The Arithmetic Teacher* (February 1968), 138.

Sanders, Walter J. "Equivalence and Equality." *The Arithmetic Teacher* (April 1969), 317.

——. "Cardinal Numbers and Sets." *The Arithmetic Teacher* (January 1966), 26.

Schlinsog, George. "The Effect of Supplementing 6th Grade Instruction with a Study of Nondecimal Numbers." *The Arithmetic Teacher* (March 1968), 254.

Smith, Frank. "Odd, Isn't It?" *The Arithmetic Teacher* (May 1968), 462.

Smith, Joe. "A Method for Converting from One Nondecimal Base to Another." *The Arithmetic Teacher* (April 1968), 344.

Smith, Lewis. "Venn Diagrams Strengthen Children's Mathematical Understanding." *The Arithmetic Teacher* (February 1966), 92.

Steinberg, Zina. "Will the Set of Children . . .?" *The Arithmetic Teacher* (February 1971), 105–08.

Thompson, Gerald A. "Computers and the Use of Base Two in the Memory Unit." *The Arithmetic Teacher* (March 1969), 179.

Vaughan, Herbert E. "What Sets Are Not." *The Arithmetic Teacher* (January 1970), 55–60.

CHAPTER 4

ADDITION AND SUBTRACTION OF WHOLE NUMBERS

Even in this age of the computer, the student who has mastered the basic facts of addition and subtraction has an adequate background and advantage as he prepares for further work with these fundamental operations. Children spend considerable time finding solutions to addition and subtraction problems; the selections in this chapter will add variety and extensions to many of their classroom activities.

Margaret Hervey and Bonnie Litwiller, in "The Addition Table: Experiences in Practice-Discovery," refer to mathematics as a study of relations and patterns. The addition table is used as an effective vehicle for the discovery of some number patterns. While searching for patterns, students are engaged in purposeful practice, which results in a "practice-discovery" activity.

Donald Meyers, in "A Geometric Interpretation of Certain Sums," illustrates Karl Friedrich Gauss's solution to the problem of finding the sum of the counting numbers from 1 to 100. This article, which relates arithmetic and geometry, suggests motivating children to think about alternate solutions to problems.

Irv King, in "Giving Meaning to the Addition Algorithm," suggests that according to Piaget and Bruner we should be seeking to give students more concrete explanations of the mathematical concepts they are studying. He demonstrates an effectiv model using the arithmetic blocks from the English Nuffield Project.

In "Subtraction Steps" by C. Winston Smith, Jr., a mathematical game is used which not only provides practice but also strengthens the child's understanding of the inverse operations of addition and subtraction.

E. W. Hamilton introduces a novel idea in "Subtraction by the 'Dribble Method'," which the author claims is not only a labor-saving method of subtraction but also one which avoids the so-called "hard combinations" in subtraction.

In the final article in this chapter, "Fourth Graders Develop Their Own Subtraction Algorithm," Ida Mae Silvey shows how careful planning, good questioning, and evaluation and re-examination of ideas lead children into doing some very good mathematical thinking at the fourth-grade level.

There is a multitude of good ideas and activities in these articles and those suggested in the bibliography, so READ ON!

The addition table: Experiences in practice-discovery

MARGARET A. HERVEY and
BONNIE H. LITWILLER

Mathematics is often referred to as a study of relations and patterns. Many elementary teachers would agree with this idea, but they may have difficulty in finding materials that will provide opportunities for their students to discover patterns. The addition table may be used as an effective vehicle for the discovery of some number patterns. While searching for patterns, the students are engaging in purposeful practice, which results in a "practice-discovery" activity.

Given an addition table like the one shown in figure 1, students in our elementary methods classes observed such patterns as the following:

1. Each row and column is a sequence of alternating even-odd or odd-even numbers.

2. The entries in any diagonal are either all odd or all even.

3. The entries in any diagonal from upper right to lower left are constant, whereas the entries in the diagonals from upper left to lower right are either consecutive even numbers or consecutive odd numbers.

4. The consecutive sums of the entries in the rows or columns have a constant difference of 10.

5. The sums of the diagonals parallel to the main diagonal (the diagonal containing the entries 0, 2, 4, 6, . . . , 18) are consecutive multiples of 9.

6. In any square drawn on the table, the sums of the two diagonals are equal.

+	0	1	2	3	4	5	6	7	8	9
0	0	1	2	3	4	5	6	7	8	9
1	1	2	3	4	5	6	7	8	9	10
2	2	3	4	5	6	7	8	9	10	11
3	3	4	5	6	7	8	9	10	11	12
4	4	5	6	7	8	9	10	11	12	13
5	5	6	7	8	9	10	11	12	13	14
6	6	7	8	9	10	11	12	13	14	15
7	7	8	9	10	11	12	13	14	15	16
8	8	9	10	11	12	13	14	15	16	17
9	9	10	11	12	13	14	15	16	17	18

Fig. 1. Addition table

In addition to these patterns, others may be found by drawing polygons on the addition table. In the table shown in figure 2, isosceles right triangles have been drawn so that the legs of the triangles lie in the rows and columns.

Our students worked through the following set of instructions for each triangle A, B, and C in figure 2. Each student had copies of the addition table so that triangles could be sketched on the table.

1. Find the sum of the numbers that represent the vertices and call this sum SV.

In $\triangle A$, the sum is $1 + 4 + 7 = 12$.
In $\triangle B$, the sum is $5 + 9 + 13 = 27$.
In $\triangle C$, the sum is $8 + 13 + 18 = 39$.

Reprinted from *The Arithmetic Teacher* (March 1972), 179–181, by permission of the authors and the publisher.

2. Find the sum of the numbers that are in the interior of the triangle and call this sum *SI*.

In ΔA, the sum is 4.
In ΔB, the sum is $8 + 9 + 10 = 27$.
In ΔC, the sum is $11 + 12 + 13 + 13 + 14 + 15 = 78$.

3. Find the ratios of the sums found in steps 1 and 2, that is, *SV/SI*.

From ΔA, the ratio is 12/4, or 3/1.
From ΔB, the ratio is 27/27, or 3/3.
From ΔC, the ratio is 39/78, or 3/6.

+	0	1	2	3	4	5	6	7	8	9
0	0	1	2	3	4	5	6	7	8	9
1	1	2	3	4	5	6	7	8	9	10
2	2	3A	4	5	6	7	8	9	10	11
3	3	4	5	6	7	8	9B	10	11	12
4	4	5	6	7	8	9	10	11	12	13
5	5	6	7	8	9	10	11	12	13	14
6	6	7	8	9	10	11	12	13	14	15
7	7	8	9	10	11	12	13	14	15	16
8	8	9	10	11	12	13C	14	15	16	17
9	9	10	11	12	13	14	15	16	17	18

Fig. 2. Addition table with triangles superimposed

The students observed that they had all used the same set of isosceles right triangles. They then drew triangles at random on the addition table so that the lengths of the legs measured 3 units, 4 units, and 5 units, as was the case in ΔA, ΔB, and ΔC. It was found that the ratios 3/1, 3/3, and 3/6 held regardless of the placement of the triangles on the addition table. Note, however, that the legs of the triangles lie in the rows and columns of the table.

Students then discovered that if right triangles are drawn so that the successive lengths of the legs are increased by 1, the sequence of ratios *SV/SI* that results is 3/1, 3/3, 3/6, 3/10, 3/15, 3/21, 3/28. What could this sequence represent? The constant 3 may represent the three vertices of a triangle. The sequence 1, 3, 6, 10, 15, 21, 28 may represent the number of interior numbers in the triangles and may be

recognized as the first eight triangular numbers. The arrangement of the set of interior numbers of the right triangles is similar to a geometric representation of the triangular numbers (see fig. 3).

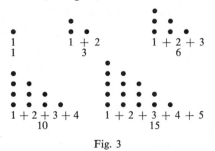

Fig. 3

Students were then given the following instructions to follow for each triangle in figure 2:

1. Find the sum of the numbers that lie on the sides of the triangle and call this sum *SP*.

In ΔA, the sum is $1 + 2 + 3 + 4 + 5 + 6 + 7 + 5 + 3 = 36.$
In ΔB, the sum is 108.
In ΔC, the sum is 195.

2. Find the ratios of *SP/SI*, where *SI* is the sum of the interior points.

The resulting sequence is 9/1, 12/3, 15/6, 18/10, 21/15, 24/21, 27/28. How could these ratios be interpreted? The sequence 9, 12, 15, 18, 21, 24, 27 may represent the number of numbers that lie on the triangles. It may also be noted that there is a constant increase of 3 in the sequence. Each time a new triangle is drawn with an additional number in each leg, the hypotenuse also contains an additional number, giving a total of three additional numbers. Again the triangular numbers may be seen.

3. Find the ratios *SV/SP* from *SV/SI* ÷ *SP/SI* = *SV/SP*.

The resulting sequence is 1/3, 1/4, 1/5, 1/6, 1/7, 1/8, 1/9, where 3, 4, 5, 6, 7, 8, and 9 represent the number of units in the successive legs of the triangles.

In $\triangle A$ the length of each leg is 3 units, in $\triangle B$ the length is 4 units, and in $\triangle C$ the length is 5 units.

Next the students were asked to draw squares on the addition table so that the sides of the squares lay on the rows and columns of the table as shown in figure 4.

+	0	1	2	3	4	5	6	7	8	9
0	0	1	2	3	4	5	6	7	8	9
1	1	2	3 D	4	5	6	7 E	8	9	10
2	2	3	4	5	6	7	8	9	10	11
3	3	4	5	6	7	8	9	10	11	12
4	4	5	6	7	8	9	10	11	12	13
5	5	6	7	8	9	10	11	12	13	14
6	6	7	8	9 F	10	11	12	13	14	15
7	7	8	9	10	11	12	13	14	15	16
8	8	9	10	11	12	13	14	15	16	17
9	9	10	11	12	13	14	15	16	17	18

Fig. 4. Addition table with squares superimposed

They were given the following instructions to follow for each square:

1. Find the sum of the numbers that represent the vertices and call this sum SV.

In square D, the sum is 12.
In square E, the sum is 28.
In square F, the sum is 36.

2. Find the sum of the numbers that are in the interior of the square and call this sum SI.

In square D, the sum is 3.
In square E, the sum is 28.
In square F, the sum is 81.

3. Find the ratios of the sums found in steps 1 and 2, that is, SV/SI.

From square D, the ratio is 12/3, or 4/1.
From square E, the ratio is 28/28, or 4/4.
From square F, the ratio is 36/81, or 4/9.

If squares are drawn on the addition table so that the successive lengths are increased by 1, the sequence of ratios $SV/SI = 4/1$, 4/4, 4/9, 4/16, 4/25, 4/36, 4/49, 4/64, 4/81. The constant 4 may represent the four vertices of a square, and the sequence 1, 4, 9,

16, 25, 36, $\cdots$, 81 is easily recognized as the first nine square numbers. The arrangement of the set of interior numbers of the square is similar to the geometric representation of the square numbers (see fig. 5).

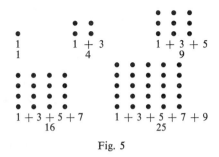

Fig. 5

The ratio SP/SI, where SP is the sum of the numbers that lie on the sides of the square and SI is the sum of the numbers that are in the interior of the square, results in the sequence 8/1, 12/4, 16/9, 20/16, 24/25, 28/36, 32/49, 36/64, 40/81 where 8, 12, 16, 20, $\cdots$, 40 may represent the number of numbers that lie on the sides of the square. A constant increase of 4 may be noted. Also, $SV/SI \div SP/SI = SV/SP$, or 1/2, 1/3, 1/4, 1/5, 1/6, 1/7, 1/8, 1/9, 1/10, where 2, 3, 4, 5, $\cdots$, 10 represent the number of units in each side of the squares.

An additional practice-discovery activity is to find the sequences when hexagons and octagons are drawn on the addition table. The ratios exhibit patterns that are similar to the relations previously shown for the triangles and squares.

It may be noted that one characteristic of the addition table is that the consecutive entries in the rows have a constant increase and the consecutive entries in the columns also have a constant increase. Since it is this idea of constant increases on which the table is built, the patterns for the polygons should hold on any table with a similar design, such as the 100 table, a page of a calendar, a table of consecutive odd numbers or even numbers, a table of multiples of n where n is a whole number, and a table formed by "skip counting" (beginning with 1).

A geometric interpretation of certain sums

DONALD E. MYERS

Elementary school teachers frequently complain that mathematics is abstract and makes too extensive use of symbols. Geometry can help reduce this dependence on symbols. We shall use a story about a famous mathematician to illustrate how some abstractions can be formulated in terms of geometric figures.

Karl Friedrich Gauss (1777–1855) is credited with using an idea, at an early age, that is now used to find an answer to such sums as $1 + 2 + 3 + 4 + \ldots + 100$. Such a sequence is called an arithmetic progression.

As is frequently the case with bright, inquisitive children, Gauss was troublesome for his teachers. On one occasion, his teacher assigned the following addition problem, believing that it would keep the young Gauss occupied for a considerable period of time:

Find the sum of the counting numbers from one to one hundred inclusive.

We could approach this problem in an obvious way, as shown in figure 1. This is a tedious method, and an elementary school child is likely to make an error. Purportedly, Gauss did not use this method. He gave the correct answer after just a moment's thought. His method may be visualized as shown in figure 2.

To use Gauss's method, we write the numbers to be added twice, once in increasing order and once in decreasing order. Reading across, we see that each pair has a sum of 101 and that there are 100 such pairs. The sum of the right-hand column, then, is found by the product 100 times 101, or 10,100. This sum, however, is twice the desired sum, since the right-hand column represents adding each number twice.

$$
\begin{aligned}
1 &= 1 \\
1+2 &= 3 \\
(1+2)+3 = 3+3 &= 6 \\
(1+2+3)+4 = 6+4 &= 10 \\
(1+2+3+4)+5 = 10+5 &= 15 \\
(1+2+3+4+5)+6 = 15+6 &= 21 \\
21+7 &= 28 \\
&\;\;\vdots \\
4950+100 &= 5050
\end{aligned}
$$

Fig. 1

$$
\begin{aligned}
1 + 100 &= 101 \\
2 + 99 &= 101 \\
3 + 98 &= 101 \\
4 + 97 &= \cdot \\
5 + 96 &= \cdot \\
6 + 95 &= \cdot \\
\vdots \;\;\;\;\; \vdots \;\;\; &\;\;\; \vdots \\
100 + 1 &= 101
\end{aligned}
$$

Fig. 2

The answer to the original problem, then, is

$$10{,}100 \div 2 = 5{,}050.$$

The most important idea here is that it isn't really necessary to write all the sums down; it's only necessary to think about them, since we know that the number of pairs is the same as the original number of

Reprinted from *The Arithmetic Teacher* (November 1971), 475–478, by permission of the author and the publisher.

addends and that for each pair the sum is 101.

To see that the idea is a workable one, consider all the counting numbers from 1 to 10,000 and ask what is their sum. Again, pairing each number with its opposite (in the sense of reversed counting), we would have 10,000 pairs; for each pair the sum is 10,001, so to get the answer to our problem we multiply 10,000 by 10,001 and divide this product by 2. Thus the sum of the first 10,000 counting numbers is 50,005,-000. We see that we have replaced 9,999 additions by an addition followed by a multiplication and then division by two.

One way to make Gauss's arithmetic problem less abstract is to replace it by a geometry problem. Let us make associations as shown in figure 3.

Such representations of the counting numbers can be placed adjacent to each other so that they look like the bar graph formed by the solid lines in figure 4. A second "bar graph" of the same size is then made by inverting the first and placing it as shown by the dotted-line portion of figure 4.

The resulting rectangle now represents twice the sum of $1 + 2 + 3 + 4 + 5 + 6$. The area of this rectangle is 42, and this is twice the sum $1 + 2 + 3 + 4 + 5 + 6$. Notice that

$$1 + 2 + 3 + 4 + 5 + 6 = \frac{6 \times 7}{2} = 21.$$

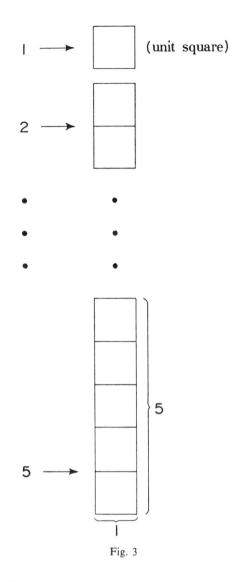

Fig. 3

In figure 4 we have shown how pairing the numbers corresponds to constructing a rectangular region, the area of which is twice the desired sum. Again we note that it's necessary only to visualize the rectangular region, not to construct it, since the necessary information—the length and width—is readily available. The reader might want to try the "think" method to find the sum of the first one thousand counting numbers.

It is possible, and profitable, to relate this problem to the ordinary addition table. Such a table is frequently used to note that addition is commutative by observing that

the table is symmetric about a diagonal drawn from the upper left corner to the lower right corner.

Diagonally across the array of squares there are chains of squares, each with the same entry. In figure 5, we have shaded such a chain. Each block in this chain corresponds to those pairs of counting numbers whose sum is three.

If a child is to construct such a table, it will be necessary to specify the largest number that is to appear in the table. In figure 5, this specified largest number is 5.

The natural question to ask when laying out such a table is how many squares

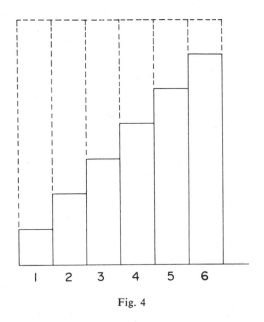

Fig. 4

+	0	1	2	3	4	5
0	0	1	2	3	4	5
1	1	2	3	4	5	
2	2	3	4	5		
3	3	4	5			
4	4	5				
5	5					

Fig. 5

are required. In figure 5, as we count the filled-in squares beginning with the bottom row, we see that the total can be represented by $1 + 2 + 3 + 4 + 5 + 6$. This is exactly the problem that Gauss solved so readily. If we make two geometric forms like the one shown by the heavy lines in figure 5 and fit them together, we get the same 6×7 rectangle as in our previous solution. The sum of the first five counting numbers is 21.

An alert teacher would now ask about such sums as follows:

a) $1 + 2 + 3 + 4 + 5 + 6 + 7$

$$\left(\frac{7 \times 8}{2}\right)$$

b) $1 + 2 + 3 + 4 + 5 + 6 + 7 + 8 + 9$

$$\left(\frac{9 \times 10}{2}\right)$$

c) The sum of the first one hundred counting numbers

d) The sum of the first N counting numbers $\left(\dfrac{N(N + 1)}{2}\right)$

The last exercise is rather difficult but worth thinking about for the students who find symbolization rather easy and challenging.

For the resourceful teacher, it is often easy to relate numbers to various geometric figures. This technique has the distinct advantages of (1) showing that mathematical problems can be solved in different ways, (2) relating arithmetic and geometry, and (3) motivating children to think about alternate ways to solve problems. Since geometry is receiving more emphasis in the elementary school, this approach to problem solving should be given serious consideration by classroom teachers.

Giving meaning to the addition algorithm

IRV KING

In the February 1971 issue of the ARITH- METIC TEACHER, Henry Van Engen stated his belief that the formalism of the new mathematics programs is inappropriate for elementary school children. My experiences over the past few years lend strong support to Van Engen's statement, and in this article I examine one specific use of such formalism—the attempt to give meaning to the algorithms of arithmetic. After the attempt is analyzed, an alternative approach will be presented. Although my comments apply to all four basic operations, the discussion will be restricted to the addition algorithm.

Prior to the advent of the modern mathematics movement, it was common practice to teach the algorithms of arithmetic by rote. To the reformers this was an intolerable situation, and an attempt was made to "explain" the algorithms. Their efforts resulted in the following kind of "explanation":

$$
\begin{aligned}
27 &= 20 + 7 \\
+36 &= 30 + 6 \\
\hline
&= 50 + 13 \\
&= 50 + (10 + 3) \\
&= (50 + 10) + 3 \\
&= 60 + 3 \\
&= 63
\end{aligned}
$$

My experiences with both children and teachers indicate that instead of clarifying matters, such procedures usually create a great deal more confusion and frustration than was formerly the case. But why should this be the case? Why should an explanation confuse rather than clarify?

Perhaps a clue to the answer can be found in the writings of psychologists such as Piaget and Bruner. They have amassed a good deal of empirical evidence that indicates that children progress through various stages of mental development. From the ages of two to seven children learn best from concrete materials and the physical world; from seven to twelve years of age they are capable of reasoning with pictures and mental images of objects; and after the age of twelve they are able to think and learn in terms of abstract symbols. As I interpret the theory, the explanation of the addition algorithm cited above is confusing to many youngsters because it is too abstract for them, it does not match the cognitive structures of their young minds.

An analysis of the explanation reveals what the child must be able to do in order to understand it. In addition to having a good understanding of place value, the child must possess an understanding of the renaming principle (number-numeral dis-

Reprinted from *The Arithmetic Teacher* (May 1972), 345–348, by permission of the author and the publisher.

tinction), the meaning and purpose of expanded notation, repeated equalities, the parentheses, regrouping (the associative law for addition), and on top of all this, he must be able to follow the overall derivation while paying attention to the seven steps involved. This is asking quite a bit of the average child. And, if he does not have these basic understandings, exercises based on such derivations are a far more tortuous form of rote than the traditional algorithm ever hoped to be!

Does this mean that we are whipped, that we are forced to teach the algorithm by rote? Not at all, for there are other explanations available. According to the theories of Piaget and Bruner, we should seek more concrete explanations. Although there are a number of physical models available for teaching addition, I have found the most effective model to be the Arithmetic Blocks used by the Nuffield Project in England. The blocks have the unique advantage of having the idea of place value built into them (see fig. 1).

The set consists of unit-blocks, ten-blocks, hundred-blocks, and thousand-blocks. Ten unit-blocks contain as much wood as one ten-block, ten ten-blocks as much wood as one hundred-block, and ten hundred-blocks as much wood as one thousand-block. These relationships can be easily verified by placing the blocks beside one another. The exchange principle is fundamental: equal amounts of wood may be exchanged for one another. For example, ten ten-blocks may be exchanged for one hundred-block.

The child is first taught the meaning of place value in terms of the wooden blocks. When given a pile of blocks, such as those shown in figure 2, he should be able to write the numeral that represents the pile. In this particular case, there are three hundred-blocks, one ten-block, and two unit-blocks; so the child should write the numeral 312. Conversely, if given a number, the child learns to represent that number by selecting the corresponding amount of wood. The written work is viewed as a way

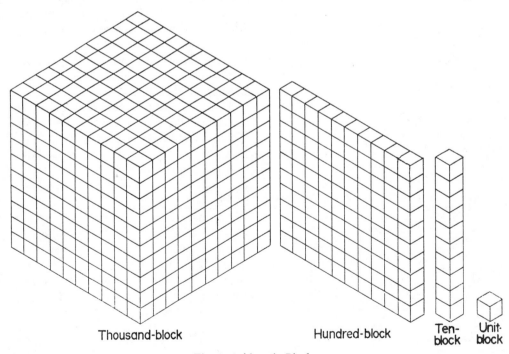

Thousand-block   Hundred-block   Ten-block   Unit-block

Fig. 1. Arithmetic Blocks

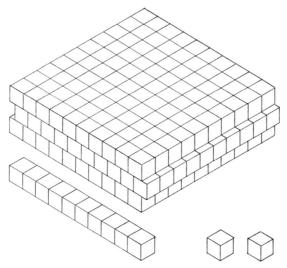

Fig. 2. Arithmetic Blocks representing the number 312

of recording and keeping track of what is done with the blocks. With these understandings, the child is ready to learn the addition algorithm.

The original problem was to find the following sum:

$$\begin{array}{r} 27 \\ +36 \\ \hline \end{array}$$

To solve the problem, the child represents each number in wood (fig. 3). Since adding means "putting together," the child physically combines the two sets of blocks. He then applies the exchange principle to trade ten unit-blocks for one ten-block. This leaves three unit-blocks, so the child records a 3 in the units column on his paper. He writes a 1 above the tens column to record the fact that ten unit-blocks were traded for one ten-block. There are now six ten-blocks, so the child records a 6 in the tens column. The sum is 63. Note that it makes perfectly good sense to say "carry the one," for in a very real sense the child carries one ten-block to the pile of ten-blocks.

I have found that young children take readily to the blocks, enjoy working with them, and are able to use them to add correctly. Even older students and teachers

have found the algorithm more meaningful after working with the blocks. To be

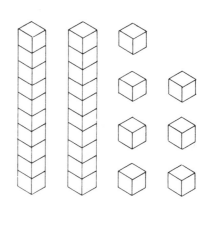

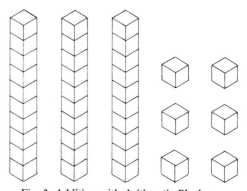

Fig. 3. Addition with Arithmetic Blocks

sure, an understanding of the algorithm based on the Arithmetic Blocks is not polished and mature. There is still a long way to go. But the algorithm does have meaning for the child. As he matures he can be exposed to more sophisticated explanations, such as the formal one that utilizes expanded notation.

If a school is unable to purchase a set of blocks, simple copies for the units, tens, and hundreds can be cut from tagboard. Although such models are not as appealing to children as the wooden blocks, a good deal of learning can be derived from them.

References

Bruner, Jerome S. *Toward a Theory of Instruction.* Cambridge: Harvard University Press, Belknap Press, 1966.

Flavell, J. H. *The Developmental Psychology of Jean Piaget.* Toronto: D. Van Nostrand Co., 1963.

Nuffield Mathematics Project. *Computation and Structure, 2.* New York: John Wiley & Sons, 1969.

Subtraction steps

C. WINSTON SMITH, JR.

While games that provide *practice* in subtraction are easily located, those that also build an understanding of the inverse relationship between addition and subtraction are not easily found. A few years ago we developed an activity the children called "Subtraction Steps" that seemed to do just that. It was fun. It provided practice and yet strengthened the children's understanding of the inverse operations of addition and subtraction. And, as an added dividend, children soon found they could easily make their own "steps" to contribute to the arithmetic table.

While the "steps" can be developed in various ways and many variations can be used, the following dialogue provides a good beginning procedure.

"Today we're going to walk down a set of stairs right in our own classroom! If we were to use the stairs in the hall, we would need to be very careful so that we would not stumble and fall, so we would take only one step at a time. Here, however, we can take giant steps of two, three, five, even ten steps at a time—without fear of falling!

"Look at our first step in Figure 1. From the top of the stairs to the first landing

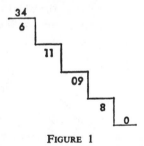

FIGURE 1

is 6 steps. How many steps to the bottom now? [28] Our next jump is 11 steps. Where would we land? [17] Then where would a jump of 9 steps put us? [8] Eight more will take us to the bottom. Could we walk back up the stairs? If we were at landing 17, would a giant step of 11 take us to 28?"

After several different staircases of this

Reprinted from *The Arithmetic Teacher* (May 1968), 458–460, by permission of the author and the publisher.

type the class is ready for variations. Giving the class a set of stairs with the landings marked instead of the steps presents a different situation. (See Fig. 2.) Where

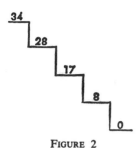

FIGURE 2

the child was first required to think in terms of the number sentence $34 - 6 = n$, he now finds himself thinking in terms of $34 - n = 28$.

He is asked, "If we start at the top, 34, and jumped down to landing 28, what would be the size of the jump?" This is a little more challenging situation. Some of the class will find the answer by subtracting 28 from 34, while others will look for a number they can add to 28 to make 34 (they "run back up the stairs").

A third variation gives the class the size of the steps but neither the starting point nor the landings. Here the student must work his way back up the stairs, undoing all of his jumps down the stairs (Fig. 3). The subtracting sentence that

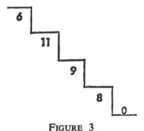

FIGURE 3

most nearly fits this situation might be $n - 9 = 8$. Most children, however, think in terms of addition, reasoning that if they

are now at 8 and they "jumped down" 9, they'll have to go back up again.

The variation most enjoyed by the students was class construction of the "steps." Here the teacher gives the top landing, such as 43, then calls upon a member of the class to give a size for the first step. This, the students find, is quite easy—it just must be less than the top landing! (See Fig. 4.) However, as more steps are

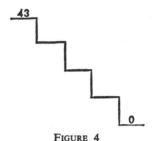

FIGURE 4

added, the task becomes much more difficult, for the last step must bring them to zero. Once the steps are completed, the class fills in the landings to see if they "ended on the ground floor," are "left up in the air," or are "digging a cellar." Some students quickly realize that the sum of the "steps" must equal the "top landing" and become quite proficient at tackling the tough job of naming the size of the last step.

Once the "steps" are constructed as a class activity, students find it quite easy to make up their own "Subtraction Steps" and challenge their classmates to "jump down MY steps."

EDITOR'S NOTE.—Devising good practice activities is always a challenge to teachers. The one described by Mr. Smith is designed to provide practice *to the point of need*. It not only develops skill in non-pencil-and-paper computation but emphasizes relational thinking. Could something similar be developed in multiplication and division?—CHARLOTTE W. JUNGE.

Subtraction by the "dribble method"

E. W. HAMILTON

Most of us have spent some time since our public school days developing short-cuts—labor-saving ways of looking at things—most of which find no place in formal instruction.

It is to be hoped that there is some trend now toward encouraging children to develop such schemes when they see sense and patterns to numbers and computation, but the occasions seem rare in which a so-called shortcut makes even better sense than the more respectable, if longer and harder, original method.

Such is the case with the "dribble method" of avoiding the so-called hard combination in subtraction. This is very attractive to children who have not reached full command of such combinations as 13 − 8 or 15 − 9. Also, those children who have been encouraged to avoid 8 + 6 by thinking of it as 10 + 4—and there are several programs that do this—have a natural and convenient way of doing subtraction without the trouble and delay of going back and learning the "hard" subtractions even though they avoided the "hard" additions.

Briefly then, instead of "borrowing" or "changing" 3 tens to 2 tens and conveying the 10 ones thus obtained over to the ones column to be added to the 4 ones—making 14 ones from which 8 ones must be subtracted—we merely "borrow" or "change" as before; but while trudging across from the tens column to the ones column we let the 8 dribble through our fingers and arrive with 2 to be added to 4, yielding 6 with no reference to 14 − 8, a "hard" combination.

This will appeal particularly to students of any age who are faced with subtracting in a different, unfamiliar base wherein the "facts" are not already at the tip of the tongue. For example:

Here one is tempted to fudge or to violate the base by thinking of 9 when there is no 9 in the system. Merely letting the 5 drib-

Reprinted from *The Arithmetic Teacher* (May 1971), 346–347, by permission of the author and the publisher.

ble out of the bundle as you move to the ones position avoids both the temptation and the unfamiliar fact $13 - 5 = 4_{(six)}$.

To be really prissy about it, of course, there is no 6 in the base-six system, but neither is there a 10 in the base-ten system. In either case, when we "change" the base number and speak of moving it, we merely think of it as a collection.

None of the computations were completed above. "Borrowing" in the second and third position to complete them will give the reader enough experience to go on to bigger and better examples at will.

I have heard a perhaps apocryphal story to the effect that during the 1800s this method was made mandatory by law in the state of New York as a concession to a certain governor's son who couldn't learn the then popular equal-additions method. Perhaps they had something after all.

EDITOR'S NOTE. I have heard of this approach as "long, long subtraction," and I have seen young children use it with success. This is the solution employed by many children when solving subtraction with objective materials. In the example: $\$ .42$ $-.17$,

the child uses 4 dimes and 2 pennies and subtracts 1 dime and 7 pennies by changing one dime into pennies. He then "takes away" 7 pennies *from the 10 pennies,* putting them aside. The remaining 3 pennies are gathered together with the 2 pennies and he finds that he has 5. The subtraction of one dime completes the operation. He thinks: "I can't take 7 from 2. I change a dime to 10 pennies; $(10 - 7) + 2 = 5$; 1 dime from 3 dimes leaves 2 dimes; so $\$.42 - \$.17 = \$.25$." Watch the child solve subtraction problems using objects, and you will observe that he almost invariably subtracts from the 10 (he "borrowed") and combines the remaining units with those of the original example.—CHARLOTTE W. JUNGE.

Fourth graders develop their own subtraction algorithm

IDA MAE SILVEY

Children in the schools of today are being encouraged to experiment, to estimate, to conjecture, and to verify ideas. The development of a spirit of adventure and excitement through such activities has been noted by many teachers. However, it is easier to talk about such a spirit than to create it. Certainly it is more easily attained if guidelines for solutions to problems are kept loose and many possible avenues to these solutions are explored.

I should like to share with you my experiences in helping fourth-grade pupils develop this "spirit" while working with subtraction. The lesson was started by putting this portion of the number line on the board:

25 26 27 28 29 30 31 32 33 34 35 36 37

A line segment beginning at 28 and ending at 32 as shown was drawn, and the following question was asked, "Can you make another line segment in a different position that is the same length as mine?" This provided a method of visual response that enabled me to evaluate the amount of comprehension taking place.

After some discussion as to how we could be sure that theirs was the same

Reprinted from *The Arithmetic Teacher* (March 1970), 233–236, by permission of the author and the publisher.

length as mine, we decided that if it began at a point on the line and ended at a point four units away, it was the same length as our original line segment. We then had a situation such as:

25 26 27 28 29 30 31 32 33 34 35 36 37

I chose the members at the ends of the line segments and wrote them on the board in the form of subtraction examples, hoping that this would suggest examples of the type they had seen before.

$$32 \qquad 35 \qquad 37$$
$$28 \qquad 31 \qquad 33$$

The children were then asked to name two numbers that would satisfy our rule, and our board looked like this:

$$32 \quad 35 \quad 37 \quad 34 \quad 38 \quad 28 \quad 30 \quad 29$$
$$28 \quad 31 \quad 33 \quad 30 \quad 34 \quad 24 \quad 26 \quad 25$$

This portion of the activity allowed many chances for testing and "proving" or "disproving" that a certain pair met the criteria. It was necessary for one boy to mark the line segment represented by his pair

$$32$$
$$29$$

before he would concede that he was wrong.

Subtraction signs were placed on each of the examples, and the children were asked which of the examples they thought would be easiest to solve. The choices ranged all of the way from René's, who chose

$$34$$
$$30$$

"because it is easy to subtract zero from anything," to Dan's

$$32$$
$$28$$

"because I already know the answer and see no point changing the problem." After reasons for many of the choices were given, we voted and

$$34$$
$$30$$

won the election.

After much attention was given to the fact that the ones place in the subtrahend contained a zero, the class was asked to write a pair of numbers such that the bottom one would end in zero that would have the same answer as

$$51$$
$$-46$$

One boy went all out on this:

$$55 \qquad 65 \qquad 75 \qquad 85 \qquad 15$$
$$-50 \quad -60 \quad -70 \quad -80 \quad -10$$

When asked how they were doing this, most of the class were able to tell that all they were doing was adding or subtracting the same number to both of those given. We looked at our examples to check the idea.

$$32 \qquad 34 \qquad 51 \qquad 55 \qquad 43 \qquad 46$$
$$-28 \quad -30 \quad -46 \quad -50 \quad -37 \quad -40$$

Added 2   Added 4   Added 3

When asked how they decided how much to add, they explained that they were subtracting the subtrahend from the next multiple of ten and adding the difference to the minuend to get the new minuend. The new subtrahend was the multiple of ten greater than the original subtrahend. After a few more examples, they decided that it was necessary to subtract the digit in the ones place from only ten in order to decide how much to add to the minuend. Mike said this was all right if they would remember to "bump-up" the next digit to the left by one.

The next task was to see if they could think the example through and write the answer without rewriting to a second example. The example

$$53$$
$$-28$$

was explained this way:

"Subtract 8 from 10 and add that answer to 3."

82

$$\begin{array}{r} 53 \\ -28 \\ \hline 5 \end{array}$$

"Raise the 2 to 3 and subtract from 5."

$$\begin{array}{r} 53 \\ -28 \\ \hline 25 \end{array}$$

They were using the algorithm that involves the complementary method of subtraction. This is not a new method, but was new to the fourth grade.

A second choice the class had made for the example

$$\begin{array}{r} 32 \\ -28 \\ \hline \end{array} \quad \text{was} \quad \begin{array}{r} 30 \\ -26 \\ \hline \end{array}$$

By this time they were freely using the idea that

$$a - b = (a + n) - (b + n)$$

to write a new pair. The new restriction was that our second pair must have a top number ending in zero. They decided that it was easier to use the idea that

$$a - b = (a - n) - (b - n)$$

here and write

$$\begin{array}{r} 23 \\ -17 \\ \hline \end{array} \quad \text{as} \quad \begin{array}{r} 20 \\ -14 \\ \hline \end{array}$$

instead of

$$\begin{array}{r} 30 \\ -24 \\ \hline \end{array}$$

Again we wrote examples with the same answer side by side.

32	30		51	50
−28	−26		−43	−42
	4			8

45	40		31	30
−37	−32		−26	−25
	8			5

The children were able to tell that they got the number in the ones digit of the second example by subtracting in the first example

the ones digit in the top number from that in the bottom. Subtracting upside down! They really were. Of course, we agreed that in order to get the answer to our example, we then needed to subtract from ten. Our next task was to try writing our method of thinking with one operation. These were some of their attempts:

10	10	70
4	4	64
74	74	74
68	−68	−68
6	6	6

Because all of the examples given so far were from successive decades, we had never been concerned with the tens digits. I should have been prepared, when

$$\begin{array}{r} 44 \\ -27 \\ \hline \end{array}$$

was solved

$$\begin{array}{r} 10 \\ 3 \\ 44 \\ -27 \\ \hline 7 \end{array}$$

By counting (in groups) from 27 to 44 we decided that 17 must be the correct answer. Then the question arose as to what we could do with

$$\begin{array}{r} 10 \\ 3 \\ 44 \\ -27 \\ \hline 7 \end{array}$$

to get

$$\begin{array}{r} 10 \\ 3 \\ 44 \\ -27 \\ \hline 17 \end{array}$$

The class exhibited true fourth-grade sympathy and helpfulness. They suggested either changing the 4 to a 3 such as

$$\begin{array}{r} 10 \\ 33 \\ 44 \\ -27 \\ \hline 17 \end{array}$$

or changing the 2 to a 3 as

$$\begin{array}{r} 10 \\ 34\cancel{4} \\ -\cancel{27} \\ \hline 17 \end{array}$$

I am not sure that their suggestions were based on anything more than their urgent desire to get a 1 in the tens place of the answer. One boy suggested that the ten from which we had subtracted our 3 could be thought of as having been subtracted from 40, thus making the 4 a 3. Another reminded us that when we were doing the complementary method we had "bumped" the digit to the left by one everytime we added a complement of ten to the subtrahend. After a few trials, we found that it works either way.

By this time the fourth graders were "drunk" with their newfound power, and they were ready to stand behind any choice they had made in the first exercise of choosing two numbers that had the same answer as

$$\begin{array}{r} 32 \\ -28 \end{array}$$

Mike suggested that we always add enough to the minuend and subtrahend to make the ones digit in the subtrahend 1, because he thought 1 was the easiest number to subtract. His suggestion went something like this: Write

$$\begin{array}{r} 32 \\ -28 \end{array} \quad \text{as} \quad \begin{array}{r} 35 \\ -31 \end{array}$$

by subtracting 8 from 11 and adding that answer to 2, or

$$\begin{array}{r} 45 \\ -37 \end{array} \quad \text{as} \quad \begin{array}{r} 49 \\ -41 \end{array}$$

by subtracting 7 from 11 and adding that answer to 5. Of course, we noted that in this case it was also necessary to "bump" the tens digit. He was using the complementary method by adding the complement of the ones digit in 11 instead of the complement in 10.

EDITORIAL NOTE. Miss Silvey shows how careful planning, good questioning, steady evaluation, and reexamination of ideas lead children into doing mathematical thinking at the fourth-grade level. This was not accomplished, as you know, in one day or one lesson. It *built* over a period of time. I think you will want to read and reread Miss Silvey's report to find all the "leads" which she gives for this kind of teaching.
CHARLOTTE W. JUNGE

SELECTED BIBLIOGRAPHY FOR CHAPTER 4
Addition and Subtraction of Whole Numbers

"Arithmetic With Frames." *The Arithmetic Teacher* (April 1957), 119–24.

Ashlock, Robert B. "Teaching the Basic Facts: Three Classes of Activities." *The Arithmetic Teacher* (October 1971), 359–64.

Broadbent, Frank W. " 'Contig': A Game to Practice and Sharpen Skills and Facts in the Four Fundamental Operations." *The Arithmetic Teacher* (May 1972), 388–90.

Brumfiel, Charles and Vance, Irvin. "On Whole Number Computation." *The Arithmetic Teacher* (April 1969), 253.

Caldwell, J. D. "Just for Fun." *The Arithmetic Teacher* (May 1968), 465.

Cleminson, Robert A. "Developing the Subtraction Algorithm." *The Arithmetic Teacher* (December 1973), 634–38.

D'Augustine, Charles H. "Multiple Methods of Teaching Operations." *The Arithmetic Teacher* (April 1969), 259.

Easterday, Ken and Easterday, Helen. "A Logical Method for Basic Subtraction." *The Arithmetic Teacher* (May 1966), 404.

Heckman, M. Jane. "They All Add Up." *The Arithmetic Teacher* (April 1974), 287–89.

Litwiller, Bonnie and Duncan, David. "Patterns on the Subtraction Table." *New York State Mathematics Teachers' Journal* (April 1974), 95–98.

Lowrey, Charlotte. "Making Sense of the Nines Check." *The Arithmetic Teacher* (March 1967), 222–24.

Marion, Charles F. "How to Get Subtraction Into the Game." *The Arithmetic Teacher* (February 1970), 169–70.

Sanders, Walter J. "Let's Go One Step Farther in Addition." *The Arithmetic Teacher* (October 1971), 413–15.

Weaver, J. Fred. "Some Factors Associated with Pupils' Performance Levels on Simple Open Addition and Subtraction Sentences." *The Arithmetic Teacher* (November 1971), 513.

CHAPTER 5

MULTIPLICATION AND DIVISION OF WHOLE NUMBERS

Children need many meaningful experiences and models to work with if mathematics is to be understood and relevant to them. Merely presenting rote mathematical combinations or exercises is not adequate for good learning, or, for that matter, good teaching. Mathematical experiences, in multiplication for example, may include activities with arrays, equidistant moves on a number line, repeated addition experiences, Cartesian products with sets, working with ordered pairs, and so on.

Many of the articles in Chapter 4 suggested ways in which children could have meaningful experiences with the addition and subtraction operations. This chapter extends those experiences to include the operations of multiplication and division with the set of whole numbers. For example, Alistair McIntosh, in "Learning Their Tables—A Suggested Reorientation," feels that "table facts" should not be taught as such (rote) because this obscures many fundamental concepts, or at least does not build the kind of understanding necessary for good learning. Children need visual or tactile pegs on which to hang abstractions. Numbers in particular, being abstract, need this visual or tactile representation to make them accessible to young children.

"Learning Multiplication Facts—More Than a Drill" by Masue Ando and Hitoshi Ikeda and "Presenting Multiplication of Counting Numbers on An Array Matrix" by Merry Schrage add excellent reinforcement to McIntosh's introductory article.

If you haven't thought of multiplying on your fingers (many school children have tried to over the years), read "Finger Reckoning" by Robert W. Prielipp. Those who are really interested in this technique should also read the article listed in the bibliography by Louisa Alger.

Frances B. Cacha, in "Understanding Multiplication and Division of Multidigit Numbers," extends the ideas of learning mathematics, in this case multiplication and division, to working with large numbers.

Lola May in "Making Division Easier to Grasp" and Joseph Di Spigno in "Division Isn't That Hard" follow with some interesting strategies and experiences to add greater meaning and insight into the division operation.

These articles will stimulate many useful and meaningful activities for use in your classroom. For further ideas, be sure to see the suggested bibliography at the end of the chapter.

Learning Their Tables— A Suggested Reorientation

ALISTAIR MCINTOSH

Introduction

In spite of the advent of modern mathematics in all its many-headed, misunderstood, mind-blocking glory, in the middle of any discussion about primary mathematics comes the *cri de coeur*, 'and what about learning their tables?' Hackles rise amongst the traditionalists; eyebrows rise amongst the moderns. The primary teacher cries from the bowels of his Nuffield-fed being: 'Understanding is all: rote-learning is an abomination.' 'Rubbish!' mutters the secondary teacher from the ordered chasm of his brain, 'they don't realise the pressures we face'—and another attempt at primary-secondary liaison falters and dies. 'Well, we tried,' says the primary teacher, 'but they don't understand.' 'Well, we understand', says the secondary teacher, 'but they are trying—very!'

Some assumptions

I want to make some suggestions which I hope may clear the air, suggestions based on a number of assumptions which I believe would be generally accepted. The assumptions are these:

[1] Both points of view contain a degree of truth, but they are in both cases presented in too extreme a form. Thus on the one hand we can be tolerably certain, both from the results of research and from our own individual observation of children in our classrooms, that children do better what they understand, enjoy it more, and can build more upon it. On the other hand the bonds between numbers, whether addition/subtraction bonds or multiplication/division bonds are essential bricks *on* which to build, and essential tools *with* which to build. For example there is no doubt that number patterns provide a rich, fertile field of exploration in the junior and lower secondary school; but if these patterns depend solely on addition and subtraction they are very restricted; if they are to include patterns involving multiplication and division, then some immediate recognition of certain 'table' facts is necessary.

For example, here is a series of numbers:

0, 2, 6, 12, 20, 30, 42 . . .

What is the pattern? Most people I have asked see this in terms of successive additions of 2, 4, 6, 8, 10 . . . What is the 8th number? Easy, add 14, it is 56. What is the 10th number? Easy, add 16, then 18 . . . 90. What is the 100th number? Wait a minute, I'm not going to add all those up.

Can you see the pattern in some other way? Think about 42, or 56 or 90. *If we know the multiplication facts*, we may re-see the pattern as:

$0 \times 1, 1 \times 2, 2 \times 3, 3 \times 4, 4 \times 5, 5 \times 6 \ldots$
Then the 100th number is easy.

[2] These table-facts are *only* bricks or tools—they are not an end in themselves. They are useless except in so far as they are used.

[3] It is the facts themselves ($3 \times 4 = 12$, $12 \div 4 = 3$, the factors of 12 are 1, 2, 3, 4, 6, 12) which are important; the table form ($3 \times 1 = 3$, $3 \times 2 = 6, 3 \times 3 = 9 \ldots$) is quite unimportant. We do want children to know these facts, and the interconnections between them; we will not mind if they are not learned in table form.

[4] The table form of learning these facts in many ways obscures the interconnections; at the least, it is not helpful in highlighting these connections. For example, consider the following statements:

$3 \times 4 = 12$
$4 \times 3 = 12$
$3 \times 4 = 4 \times 3$
3 is a factor of 12
4 is a factor of 12
12 is a multiple of 3
12 is a multiple of 4
$12 \div 4 = 3$
$12 \div 3 = 4$
$12_{base\ ten} = 110_{base\ 3} = 30_{base\ 4}$

All these statements are very closely connected, and all are important: but the table form highlights only the first two statements and ignores or obscures all the rest.

Other disadvantages of concentration on the table form are:

This article is reproduced from *Mathematics Teaching* (Autumn 1971), 2–5, the Journal of The Association of Teachers of Mathematics, by permission of the author and the editor.

(a) It obscures the relationship
      between multiples of 2, 4 and 8
      between multiples of 3, 6 and 9
and between multiples of 5 and 10.

(b) 63 and 64 are, after all, next to each other on the number line, yet they appear in quite different tables. The connection between 8×8 and 7×9, as one instance of the general case $x^2 - 1 = (x + 1)(x - 1)$ comes, I find, to most teachers as a complete surprise; and yet the tables conceal many instances of this general form: 3 and 4, 8 and 9, 15 and 16, 24 and 25, 35 and 36, 48 and 49, etc.

(c) Commutativity is not made explicit: for example 4×7 and 7×4 appear in quite different tables. Of course one points out the connection, but one shouldn't have to.

(d) 24 appears in many different tables. In general, the table form concentrates on multiplication and multiples, and is unhelpful about division and factors; and yet the set of factors of a number is at least as important as the set of multiples of a number. I have recently asked ten separate groups of primary teachers to write down (a) the multiples of 3 as far as 24 and (b) the factors of 24. In every case part (b) took twice as long, caused twice the number of worried looks, and was incorrectly done by most.

(e) The table form makes the 10 times table look easy: it is, of course, easy to memorise this particular set of facts; but it completely obscures the point that these facts are one of the corner stones of *place value* rather than of simple multiplication. Keeping only to base ten, it does not seem to me that 'three tens make thirty' and 'ten threes make thirty' are equally obvious. The first *is* simple: the second is not at all so obvious.

(f) The multiples of 11 and 12 are simply not worth the trouble involved. The 11 times table is useless, and the 12 times table is better seen as a question of long multiplication, particularly as metrication and decimalisation will make these facts less commonly used.

[5] Children learn better by starting with a small collection of things which they understand and then, when these are familiar, by building on these facts and extending them, using similar methods. For example larger numbers should be seen as extensions of smaller numbers which behave in a similar way, just as negative numbers and rationals are an extension of the natural numbers. The table form, however, keeps leap-frogging up the number line, and then leaping back to square one in a manner calculated to bemuse the clearest young mind.

[6] Children of primary age need visual or tactual pegs on which to hang abstractions. Thus numbers in particular, being totally abstract, need some visual or tactual representation to make them accessible to young children.

[7] The reasons why a teacher may wish children to know the multiplication/division bonds between numbers may well be different from the reasons which may impel *children* to want to know them. The teacher's reasons may be more remote and long-term; children will need more immediate satisfactions, and these satisfactions will be more successful if they are intrinsic to the field of study than rewards/marks/punishments, which are extrinsic, which say in effect, 'I can't think of a reason why you should want to know those facts, so I'll make you learn them regardless'. Such an attitude is at best an admission of failure; at worst it is an insult to the child's ability and desire to learn.

[8] A personal belief, but one I know is shared by others: The numbers from 1 to 100 are a much under-used source of excitement for children. The introduction of place-value opens up the possibility (or the threat) of using much larger numbers; it often tends to obscure the over-riding importance of familiarity with the smaller numbers. It is, for example, on an investigation of patterns and connections between these numbers that the generalisations of the algebra of numbers will be properly based. Too often the generalisations precede the investigations; too often the investigations are never undertaken.

For example, the series 0, 2, 6, 12, 20, 30, 42 . . . may be seen by some as

0×1 and by others as	$0^2 + 0$
1×2	$1^2 + 1$
2×3	$2^2 + 2$
3×4	$3^2 + 3$
4×5	$4^2 + 4$
etc.	etc.

Clearly these sets of expressions are equivalent, and we can *go on* to write this equivalence more neatly as $n \times (n+1) = n^2 + n$.

That is the right order of events
— play with numbers
— find some patterns
— generalize by algebra.

Some proposals

Now to some concrete proposals, first about restructuring the order in which these facts should be met by children, then about methods.

My proposal about the order in which children should meet these facts is very simple: instead of separating the facts into ten subsets—the ten tables—I suggest that they should be built up in three stages:

Stage 1—Acquaintance with numbers up to 20 (*roughly* the infant stage).

Stage 2—Acquaintance with numbers up to 50 (*roughly* the lower junior stage).

Stage 3—Acquaintance with numbers up to 100 (*roughly* the upper junior stage).

Thus all the following are examples of Stage 1 facts, and all are equally important:
    $3 \times 5 = 15$

$12 \div 4 = 3$
$2 \times 3 = 3 \times 2$
multiples of 4 are 4, 8, 12, 16, 20
factors of 18 are 1, 2, 3, 6, 9, 18
13 is a prime number.

The general intention would be that children should initially become friendly with the numbers up to 20; this familiarity would then be extended to the numbers up to 50, and finally to 100.

The two main advantages which I see in such a restructuring would be first, that each stage would build on the previous stage, extending and enlarging while including it, and second that at any stage teachers and children would have a coherent set of numbers within which most aspects of number work could be explored. At present children are more than likely to meet numbers like 27, 40, even 100 as numbers appearing in tables, *before* they meet them as numbers which they can use for addition and subtraction.

Methodology — the basic concept of the rectangle

How then could children explore these numbers, bearing in mind that they need concrete representations of them? My suggestions here are not radically new, but I do believe that they bring together a number of strands present in number work but not hitherto unified from the start.

An important starting point is that we are here primarily concerned with multiplication and division. So long as addition and subtraction are the main concern, the number line is a useful metaphor. Addition and subtraction can usefully be seen as moves to the right and left respectively along the number line. Moreover the number line (or more accurately the free vectors associated with addition and subtraction) can be made concrete in the form of wooden or plastic strips—as with Unifix, Cuisenaire, Colour-Factor or Stern materials.

However, when we come to multiplication and division, the use of the number line tells only half the story. The two operations *can* be seen as repeated addition and subtraction ('so many lots of 3' or 'how many threes in . . . '). But in fact we can see in any classroom that the usual representation of 4×3 is not

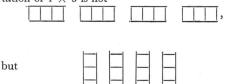

but

While up to this point the rods are seen as strips, only incidentally having width as well as length, children are suddenly assumed to accept that the width, being one unit, is also important, and the rectangle is introduced as the model.

I suggest that we make quite explicit the implication of this changeover, and see multiplication as a two-dimensional question—the making of rectangular arrays—as opposed to addition and subtraction whose analogy is the one-dimensional idea of length.

Whereas the addition and subtraction bonds will be seen in terms of *strips*, the multiplication and division bonds will be seen primarily in terms of *rectangles*. This idea is incorporated in the Dienes AEM material, but where this is used it is not introduced until the age of 9 or 10, and is then introduced in a rather perfunctory manner at a late stage in the child's development. It seems so fundamental and fruitful an idea that it should be introduced at a much earlier stage.

This, incidentally, would make much more sense of the Cartesian product, introduced to many teachers by the Nuffield 5-13 Mathematics Project in a way calculated to deter them from understanding it. For, in practical terms, the idea of the Cartesian product is the notion of the rectangle in which both measurements are equals, as opposed to the traditional view of multiplication as repeated addition on the number line where, for example, in 'three lots of four' the 'three' and the 'four' are two different sorts of numbers. Secondary teachers will no doubt be aware of the difficulties caused by the idea of multiplication as repeated addition when fractions are introduced.

Finally, here are a series of jottings about the rectangular array, which may suggest activities at various levels.

1. Man seems to like the rectangular array: milk bottles in a crate, windows in office blocks, floor tiles, banks of switches, pegboard, graph paper, calendars (I can see at a glance from my calendar for June that 30 is not divisible by 7).

2. Logic blocks:

	Triangle	Square	Circle	Oblong
Red	△	□	○	▭
Yellow	◸	▨	◍	▨
Blue	◢	■	●	▬

Four shapes; *three* colours; *twelve* pieces.

3. Factors of 6:

```
O O O O O O    O O O    O O    O
                O O O    O O    O
                         O O    O
                                O
                                O
                                O
```

'Obviously' multiplication is commutative.

4. Prime numbers:

O O and O O O and O O O O O

89

5. Square numbers:

O and O O and O O O
       O O   O O O
             O O O

'Obviously' squares are rectangles—special rectangles.

6. Multiples of 3:

O and O O and O O O
O   O O   O O O
O   O O   O O O

7. Consolidation

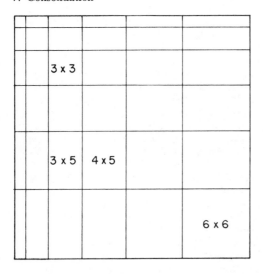

All the possible rectangles up to 6 by 6. Where are the multiples of 4? Where are the squares?

8. Consolidation again.

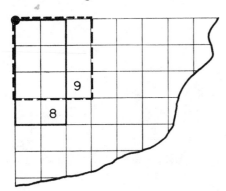

Take a piece of squared paper 10 by 10. Place a dot in the top left hand corner as shown. This will form the top left hand corner of a lot of rectangles which you can see on the squared paper. Two of these have been outlined on the diagram. Can you see why 8 and 9 have been written in two of the small squares? Write a number for the same reason in each of the small squares on your squared paper.

9. An extension to long multiplication: 27×13

$$27$$
$$13$$
$$\overline{}$$

$81 \rightarrow$ rectangle 3×27
$270 \rightarrow$ rectangle 10×27

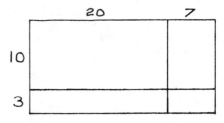

10. An extension to fractions.

● ● O represents $\frac{2}{3}$ (the ratio of shaded to the whole)

● ● ● O represents $\frac{3}{4}$ (the ratio of shaded to the whole)

● ● O represents $\frac{2}{3} \times \frac{3}{4}$, the result of which
● ● O is clearly $\frac{6}{12}$ (the ratio of shaded
● ● O to the whole).
O O O

Clearly $\frac{2}{3} \times \frac{3}{4}$ is the same as $\dfrac{2 \times 3}{3 \times 4}$

(the comparison of two rectangles)

Conclusion

I hope these suggestions are seen to be both intelligible and practical, not as a radical change suggesting that the facts are unimportant or as a denial of what is useful in present practice, but as a sensible shift of emphasis which incorporates this important aspect of primary work in a wider and more mathematically coherent context.

Now, primary teachers, please comment—you are in the classrooms, not me.

Learning multiplication facts— more than a drill

MASUE ANDO and
HITOSHI IKEDA

Three stages are commonly observable in working with multiplication facts:

Stage 1: Learning the concept of multiplication by making equal groups with manipulative materials while determining products

Stage 2: Organizing the determined facts in a table

Stage 3: Memorizing the facts through random drill

While random drill for mastery of the multiplication facts is appropriate, other factors may contribute toward this objective and should be considered. Underlying these factors is the notion of order in the way digits appear in the multiplication facts. To this end, the following sequence of activities coincident with stages 1 and 2 (stated above) deserves consideration.

Stage 1

In stage 1, counting discs and a mat with 100 squares (10 squares by 10 squares) can be used to show a number of equal groups (see fig. 1). An advantage of using this type of mat is that when the total number is asked for, the student can move counters from lower rows to fill in the rows above, starting at the top. In doing this he may become aware of the number of groups of a particular size that can "fit"

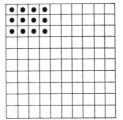

Fig. 1

into a ten and can begin to place his fact in an appropriate decade. For instance, he can see that it takes five twos to fill a ten-row, or that only three threes can fit into a ten-row, or that only one nine can fit into a ten-row.

In working with groups of nines, moving one counter from the second group to fill the first row, then moving two from the third group to fill the second row, then moving three from the fourth group to fill the third row, and so on (fig. 2), shows the student that the units digit is decreasing one at a time as increasing multiples of nine are considered. The student can also see that two nines are 20 less 2, three nines are 30 less 3, four nines are 40 less 4, and so on.

Working with groups of fives, the student can see that it takes two fives to fill each row of ten; so the number of tens is half the number of fives being considered.

Reprinted from *The Arithmetic Teacher* (October 1971), 366–369, by permission of the authors and the publisher.

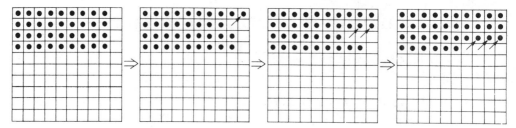

Fig. 2

Stage 2

After all the facts are determined on the mat, the facts can be arranged in a multiplication table. One of the first considerations for making the job of mastering facts simpler is the reduction of the number of facts to be memorized. Here the mat can be useful again. For example, by giving the mat a quarter of a turn one can see that four groups of six is the same number as six groups of four. This commutative property of multiplication can be observed by students as they work with the facts involving one, two, and five.

The idea of skip counting, using only even numbers (or only odd numbers), can be brought in while working with the facts involving two. By using the mat as well as the multiplication table for fives, students can record their findings in a table, considering the even multiples and the odd multiples of five separately. (See fig. 3.) Odd multiples of five can then be shown as one five greater or one five less than a given number of tens. For example, 5 × 7 can be thought of as one five more than six fives or as one five less than eight fives.

Number of fives	Number of tens
2	1
4	2
6	
8	
10	

Fig. 3

The facts involving nine are also easily approached by using the mat and by looking at the way the digits appear in the products. Again, a table such as that in figure 4 can be used by the students to see the relationships involved. Three ideas can be discovered from such a table: (1) the number of tens in the product is one less than the multiplier; (2) the units digit decreases by one as each additional nine is considered; and (3) the sum of the digits in the product is always nine. Using ideas 1 and 3, the student can easily determine the facts involving nine.

n	$9 \times n$
1	9
2	18
3	
4	

Fig. 4

The facts involving three, four, six, seven, and eight still remain to be considered. While no direct rule can be given for determining these facts, a study of the facts presents many interesting ideas. Consider the even numbers two, four, six, and eight. As even numbers, five times any of these numbers is a given number of tens. Hence, the units digit for each of these facts will be even and the sequence will repeat after reaching zero. The mat can also be used to show the distributive nature of multiplication over addition. For example, students can first show four groups of seven (fig. 5). They can then be directed to move two from each row over to the right to show each seven as five and two (fig. 6). Not having increased or decreased the number of counters on the mat, they can see that four groups of seven is the same number as four groups of five plus four groups of two. Thus 4 × 7 has the same units digit as 4 × 2 and is 4 × 5, or 20, greater than 4 × 2.

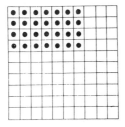

Fig. 5

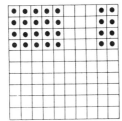

Fig. 6

This idea can also be shown by a pentagonal arrangement of numbers as in figure 7. One way to remember this arrangement of numbers is to draw a "star" (without lifting the pencil point off the paper) by beginning on a point labeled 0 and directing your first stroke down and toward the left. The points of the star can thus be labeled 2, 4, 6, and 8 in the order created (see fig. 8).

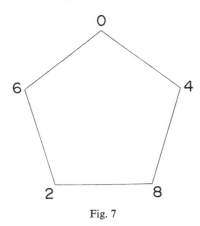

Fig. 7

By considering the units digits in the facts involving four, students can see that the facts are even and that they are located

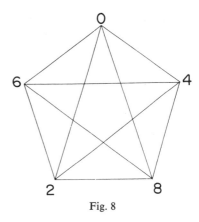

Fig. 8

in clockwise order around the pentagon, repeating this order after reaching 0.

By considering the units digits in the facts involving six, students can see that the facts are even and that they are located in counterclockwise order (6-2-8-4-0), repeating this order after reaching 0. Notice that 6 and 4 are numbers that add up to ten and that on this pentagon they are located across from each other horizontally, as are 8 and 2. Recalling that the multiples of two were used to outline the star within the pentagon, the units digits in the facts involving eight can now be considered. The order of these digits also outlines the star, starting at 8 and repeating after reaching 0 again.

Now consider the facts involving three and seven. Since they are odd numbers, some of their multiples are odd. Write in the odd numbers on the pentagon, as shown in figure 9. Notice that 3 and 7, which when

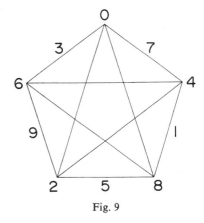

Fig. 9

93

added give ten, are placed across from each other horizontally again, as are 9 and 1. By considering the units digits in the facts involving three, the student can see the consecutive counterclockwise order around the pentagon. Considering the units digits in the facts involving seven, he can also notice the consecutive clockwise order around the pentagon.

A chart summarizing the directional order of these sequences may be constructed for the classroom as in figure 10. The single arrow ($\rightarrow$) would denote following the digits as they appear in the direction indicated; the double arrow ($\Rightarrow$) would indicate following every other of the digits as they appear in the direction indicated; and the triple arrow ($\Rrightarrow$) would indicate following every third of the digits as they appear in the direction indicated. The triple arrow may not be necessary if students have developed other cues for determining facts involving

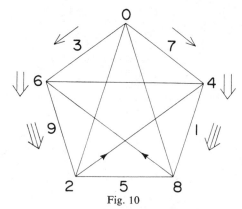

Fig. 10

nine. Other pentagonal arrangements of numbers may also be used.[1]

The activities mentioned in this article can help children develop insights into the order existing in number work and assist them in understanding and recalling the multiplication facts.

1. Joan Herold, "Patterns for Multiplication," ARITHMETIC TEACHER 16 (October 1969):498–99.

Presenting multiplication of counting numbers on an array matrix

MERRY SCHRAGE

Multiplication tends to imply an avenue of excitement and adventure for children. During my ten years as a grade three teacher, I have yet to find a year when some eager child hasn't asked the question, "When are we going to learn multiplication?" Prior to third grade the children are exposed to readiness activities in preparation for multiplication. However, the real drive to master facts and algorisms is in the third-grade mathematics sequence in Dade County, Florida.

A visual technique was developed to be used primarily during the introductory stages of multiplication. With the use of a matrix and individual cards showing the various arrays, it is possible to present and strengthen a number of basic understandings peculiar to multiplication.

The basic construction of the matrix can vary in size. I use a large one (36" × 36") because it is utilized and displayed for a period of time. Figure 1 shows the blank matrix chart in preparation for the multiplication facts through seven times seven.[1] A set of separate cards, which fit in the blanks of the matrix, can be easily made by using a magic marker, gummed circles, stars, or any other convenient item. The cards can be attached to the matrix with tape or other types of

X	1	2	3	4	5	6	7
1							
2							
3							
4							
5							
6							
7							

FIG. 1.—Blank matrix

affixing material. Making the arrays on separate cards enables development of the matrix chart in a sequential manner according to readiness, properties, and facts. Different colors may be used for the arrays, making the commutative facts the same in color. Figure 2 shows how the array cards look.

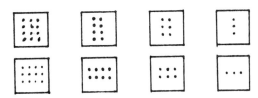

FIG. 2.—Sample array cards

---

[1] The matrix has not included the zero facts since the array would show only blank cards.

Reprinted from *The Arithmetic Teacher* (December 1969), 615–616, by permission of the author and the publisher.

The primary concept that needs to be developed prior to using the matrix is the meaning and use of the array. Visual activities should be provided that enable the children to experience the manipulation and development of arrays. After activities of this type have been accomplished, the matrix chart can be introduced as a means of recording our findings in an orderly fashion.

During the initial introduction of the matrix, the children will readily recall using a matrix for the purpose of recording addition and subtraction facts. It is usually enough to indicate that we can use the matrix to find out some important ideas that the arrays show us in multiplication. Point out that the numerals across the top and along the left side of the matrix relate directly to the definition of an array. Pass out a few selected array cards to the class. Ask someone to develop a specific array on the chalkboard. "Does anyone have an array which shows the same array John has made for us on the board?" There will be two children with array cards that match the one made on the chalkboard unless the teacher has selected an array showing the product of two identical factors. Also, if some type of color scheme has been utilized in making the array cards, the two cards for the specific array will be the same color.

The next question that needs to be discussed is, "Where should these array cards be placed on the matrix?" Lead the children to a consideration of the definition of an array and how it relates to the numerals on the matrix. The children should also be encouraged to see that a quarter turn of one of the array cards produces the commutative fact.

The arrays to be developed should be approached in the manner just briefly out-lined. Of course, there are many more questions that can be asked to stimulate discovery on the part of the children. The ability of the class will also be indicative of the duration that this activity will take. As the pattern of the matrix emerges through the development of the facts in terms of arrays, the matrix will lend itself to a visual understanding of the identity element of one, the commutative property and the multiplicative property of zero. Figure 3 shows an array-matrix that has been completed through the facts of five.

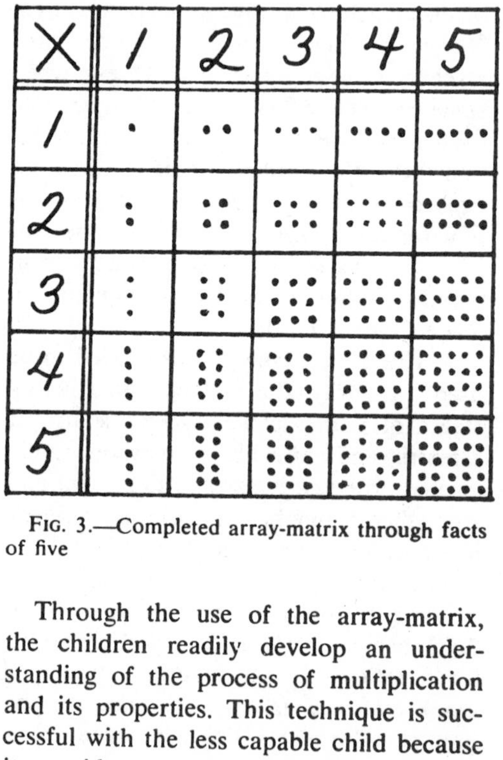

Fig. 3.—Completed array-matrix through facts of five

Through the use of the array-matrix, the children readily develop an understanding of the process of multiplication and its properties. This technique is successful with the less capable child because it provides an opportunity for manipulation and visualization of the concepts. The more capable child is greatly stimulated toward independent discovery of additional patterns and concepts.

FINGER RECKONING

ROBERT W. PRIELIPP

OVER the years many young school children (and some not so young) have discovered how helpful their fingers can be in performing simple addition problems. One can't help but wonder how many people have ever given serious thought to *multiplying* numbers on their fingers. Can such a thing be done? Well, let's see.

Suppose we begin by assuming that we know how to count and add and that we know the multiplication facts from 0×0 through 5×5.

For the moment, let's restrict our attention to multiplying two natural numbers both of which are greater than 5 and less than or equal to 10. For example, let's try 6×8. Begin by opening wide both of your hands, palms upward. On your left hand bend one finger in toward the palm of your hand, one being the excess of 6 over 5. (Your little finger will do nicely.) On your right hand, for the excess of 8 over 5, bend three fingers, say your little finger and the two fingers adjacent to it, in toward the palm. Now count the total number of fingers you have folded in. You should count four. They represent 4×10 or 40. Next multiply 4×2, where four is the number of fingers which remain outspread on your left hand and two is the number of fingers that remain outspread on your right hand. This gives you 8. $40 + 8 = 48$, which should be your answer.

Now, try to multiply 8×9 on your fingers, using the same method.

Why does this work? Perhaps you can gain some insight by noticing the identity

$$a \cdot b = [(a - 5) + (b - 5)]10 + (10 - a)(10 - b).$$

(To verify this, multiply out the right-hand side, and simplify.)

But what if the numbers involved are not both greater than 5 and less than or equal to 10? Suppose, for example, we want to multiply 14×12.

How can we do this on our fingers? We begin by bending four fingers on our left hand in toward the palm, where four, of course, is the excess of 14 over 10. Next on our right hand we bend two fingers (the excess of 12 over 10) in toward the palm. We now find the total number of fingers which are folded inward, in this case six. This represents 6×10, or 60. Next we multiply 4×2, where four is the number of fingers folded inward on our left hand and two is the number of fingers folded inward on our right hand. $4 \times 2 = 8$, $60 + 8 = 68$ and $100 + 68 = 168$. This is our answer. In fact, this method will work whenever both numbers are greater than 10 and less than or equal to 15.

Just like black magic, isn't it? Maybe the identity

$$a \cdot b = [(a - 10) + (b - 10)]10 \\ + (a - 10)(b - 10) + 100$$

will be helpful in explaining why this procedure works.

Suppose we consider 19×17. It will turn out that our procedure will work whenever both numbers are greater than 15 and less than or equal to 20. We begin by bending four (the excess of 19 over 15) fingers in toward the palm of our left hand.

Reprinted from *The Mathematics Teacher* (January 1968), 42–43. ©1968 by the National Council of Teachers of Mathematics. Used by permission.

On our right hand we bend two (the excess of 17 over 15) fingers inward. These fingers represent 6 × 20 or 120. Next we multiply 1 × 3 where one is the number of fingers extended on the left hand and three is the number of fingers extended on the right hand. 1 × 3 = 3 and 120 + 3 + 200 = 323, which is our answer.

The identity

$$a \cdot b = [(a - 15) + (b - 15)]20$$
$$+ (20 - a)(20 - b) + 200$$

may be of assistance in investigating why our method gave us the correct answer.

At this point you are probably wondering whether this approach works in cases such as 8 × 17. If we express 17 as 9 + 8, for example, then 8 × 17 = 8 × (9 + 8) = (8 × 9) + (8 × 8); using the left distributive property for multiplication over addition, we can apply our procedure to each of the addends, 8 × 9 and 8 × 8.

And now a challenge! Can you find a way to multiply 24 × 23 on your fingers? Can you show how to multiply 17 × 12 in this manner?

For those who would like to investigate this topic more fully and who, in particular, might like to become acquainted with its historical background, we present the references below.

REFERENCES

DANTZIG, TOBIAS. *Number, the Language of Science* (3rd ed.). New York: Macmillan Co., 1945, p. 11.

RICHARDSON, LEON J. "Digital Reckoning Among The Ancients," *American Mathematical Monthly*, XXIII (January 1916), 7–13.

Understanding multiplication and division of multidigit numbers

FRANCES B. CACHA

Anyone who has taught multiplication and division of large numbers knows that they are difficult processes for children to learn. Even though learning mathematics through understanding, rather than by rote, has been emphasized for decades, methods of teaching multiplication and division of multidigit numbers have been abstract. Because of the lack of understanding, these concepts are not retained.

Children learn the beginnings of multiplication and division by using concrete objects (colored rods and disks) and semiconcrete materials (the number line and arrays). It seems to be assumed that when children have learned the basic concepts, including the distributive property, they are ready to learn the next stages without using concrete or semiconcrete materials. Rather than engaging in purposeful discovery activities, much time is spent on meaningless drill. As a result, children often do not understand what they are doing but eventually learn the algorithms by rote.

The most common error children make with multiplication of larger numbers is

Reprinted from *The Arithmetic Teacher* (May 1972), 349–354, by permission of the author and the publisher.

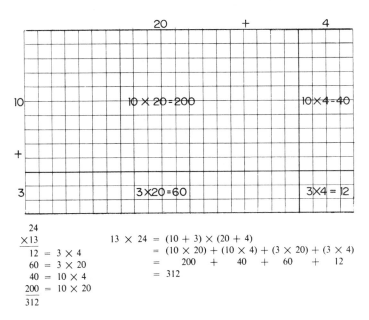

$$
\begin{array}{r}
24 \\
\times 13 \\
\hline
12 \\
60 \\
40 \\
200 \\
\hline
312
\end{array}
\quad
\begin{array}{l}
= 3 \times 4 \\
= 3 \times 20 \\
= 10 \times 4 \\
= 10 \times 20
\end{array}
$$

$$
\begin{aligned}
13 \times 24 &= (10 + 3) \times (20 + 4) \\
&= (10 \times 20) + (10 \times 4) + (3 \times 20) + (3 \times 4) \\
&= \quad 200 \quad + \quad 40 \quad + \quad 60 \quad + \quad 12 \\
&= 312
\end{aligned}
$$

Fig. 1

leaving out one or more partial products. By using guided-discovery questioning techniques, the distributive property, and arrays made of graph paper, children can understand what they are doing when they multiply multidigit numbers.

You can start by giving each child a piece of graph paper cut into a 13 × 24 array. With questions, develop the solution with your students by having them express each factor in expanded notation and partition the array accordingly; then relate each section of the partitioned array to the partial products in the vertical algorithm and in the equation. See figure 1.

After additional similar experiences, ask your pupils if they can suggest a quicker way to solve the example by combining steps. Since they are already familiar with the longer and shorter methods of solving examples with one-digit factors, they should be able to answer your question. Give them a 13 × 24 array and ask them to partition it by combining steps similar to those they have used previously with one-digit multipliers. See figure 2. Compare the two examples and equations with the partitioned arrays, and let your students generalize about the number of partial products in each form.

Division is usually introduced as undoing what multiplication does; yet with harder problems the two operations have not been taught as division being the inverse of multiplication. Children who study and verbalize the relationship of the two operations retain what they learn. Therefore, soon after your students understand the standard short form of the multiplication algorithm, introduce the division algorithm that is directly related to it. If you have been working with multiplication of a two-digit number by a two-digit number, proceed with division by a two-digit divisor that gives a two-digit quotient—for example, 224 ÷ 16 = □. Since the quotient is 14, distribute a 14 × 16 array to each student. See figure 3. Assuming that your students are familiar with finding partial quotients in multiples of 10 with a one-digit divisor, proceed to develop the answer with them. In the first step, can we subtract at least 10 subsets with 16 in each subset? Yes. Ask students to partition the array accordingly. How many are left? 64. How many subsets of 16 can we subtract

99

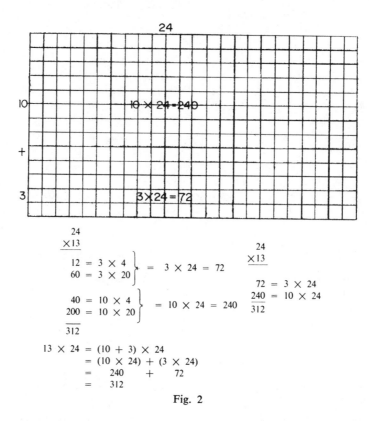

$$24$$

$$\begin{array}{r} 24 \\ \times 13 \\ \hline \end{array}$$

$$\left. \begin{array}{l} 12 = 3 \times 4 \\ 60 = 3 \times 20 \end{array} \right\} = 3 \times 24 = 72$$

$$\left. \begin{array}{l} 40 = 10 \times 4 \\ 200 = 10 \times 20 \end{array} \right\} = 10 \times 24 = 240$$

$$\overline{312}$$

$$\begin{array}{r} 24 \\ \times 13 \\ \hline 72 = 3 \times 24 \\ 240 = 10 \times 24 \\ \hline 312 \end{array}$$

$$\begin{aligned} 13 \times 24 &= (10 + 3) \times 24 \\ &= (10 \times 24) + (3 \times 24) \\ &= \quad 240 \quad + \quad 72 \\ &= \quad 312 \end{aligned}$$

Fig. 2

from 64? 4. Have the children simultaneously verify each step with both the array and the algorithm and check by adding the partial products.

After a few similar problems using the arrays made of graph paper, encourage your students to explore the concept independently. Children working together in small groups or individually can use pencils and rulers to outline the array. Ask them to hypothesize about the length—that is, the quotient—by estimating a reasonable answer. In this activity the outline of the array and the algorithm are developed at the same time to an accurate completion of the problem.

When children deal with larger products and dividends, it is still important for them to have meaningful experiences with semiconcrete materials, to make generalizations about the number of partial products needed to complete the multiplication problem, and to continue to relate the two operations of multiplication and division.

Even with large numbers it is possible to provide additional practice with partitioning arrays by either (1) obtaining graph

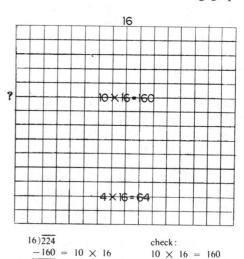

$$\begin{array}{r} 16 \overline{\smash{)}224} \\ -160 = 10 \times 16 \\ \hline 64 \\ -64 = 4 \times 16 \\ \hline 0 \quad 14 \end{array}$$

check:

$$\begin{aligned} 10 \times 16 &= 160 \\ 4 \times 16 &= 64 \\ \hline 14 \times 16 &= 224 \end{aligned}$$

Fig. 3

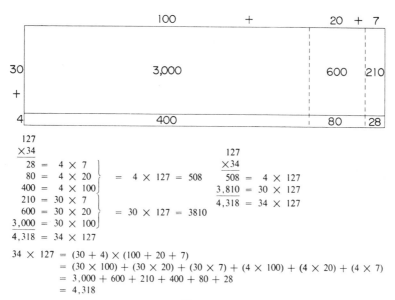

127
×34
28 = 4 × 7 ⎫
80 = 4 × 20 ⎬ = 4 × 127 = 508
400 = 4 × 100 ⎭
210 = 30 × 7 ⎫
600 = 30 × 20 ⎬ = 30 × 127 = 3810
3,000 = 30 × 100 ⎭
4,318 = 34 × 127

127
×34
508 = 4 × 127
3,810 = 30 × 127
4,318 = 34 × 127

34 × 127 = (30 + 4) × (100 + 20 + 7)
= (30 × 100) + (30 × 20) + (30 × 7) + (4 × 100) + (4 × 20) + (4 × 7)
= 3,000 + 600 + 210 + 400 + 80 + 28
= 4,318

Fig. 4

paper with smaller squares, (2) taping sheets of graph paper together, or (3) drawing an array to a scale that would represent larger numbers—for example, one in which half an inch represents 10 units.

In solving the equation $34 \times 127 = \square$, follow the above procedures, except instead of using graph paper draw an array to scale that will represent 34×127. See figure 4.

When the short, standard form of multiplying a three-digit number by a two-digit number is understood, comparable division problems can be introduced; for example $6,235 \div 43 = \square$. See figure 5.

An example of multiplication of a three-digit number by another three-digit number is shown in figure 6, and an example of division by a three-digit number is shown in figure 7.

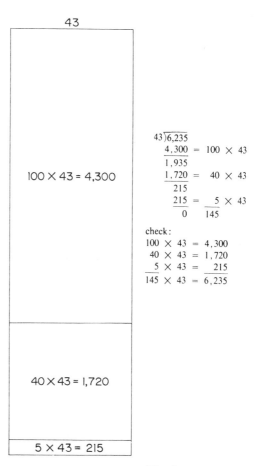

```
  43)6,235
     4,300  = 100 × 43
     1,935
     1,720  =  40 × 43
       215
       215  =   5 × 43
         0    145
```

check:
100 × 43 = 4,300
40 × 43 = 1,720
5 × 43 = 215
145 × 43 = 6,235

Fig. 5

101

$112 \times 127 = (100 + 10 + 2) \times (100 + 20 + 7)$
$= (100 \times 100) + (100 \times 20) + (100 \times 7) + (10 \times 100) + (10 \times 20) +$
$(10 \times 7) + (2 \times 100) + (2 \times 20) + (2 \times 7)$
$= 10{,}000 + 2{,}000 + 700 + 1{,}000 + 200 + 70 + 200 + 40 + 14$
$= 14{,}224$

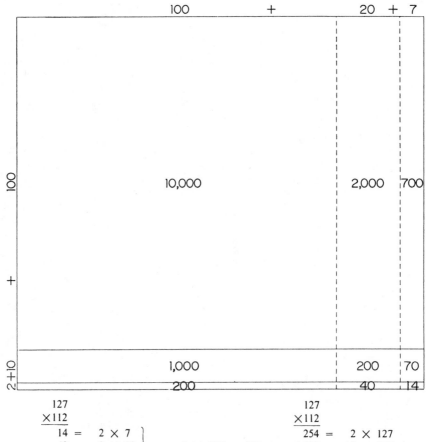

$$
\begin{array}{l}
\quad\;\; 127 \\
\underline{\times 112} \\
\quad\;\; 14 = \;\; 2 \times 7 \\
\quad\;\; 40 = \;\; 2 \times 20 \\
\;\; 200 = \;\; 2 \times 100 \\
\quad\;\; 70 = 10 \times 7 \\
\;\; 200 = 10 \times 20 \\
1{,}000 = 10 \times 100 \\
\;\; 700 = 100 \times 7 \\
2{,}000 = 100 \times 20 \\
\underline{10{,}000 = 100 \times 100} \\
\overline{14{,}224 = 112 \times 127}
\end{array}
$$

$= 2 \times 127 = 254$

$= 10 \times 127 = 1{,}270$

$= 100 \times 127 = 12{,}700$

$$
\begin{array}{l}
\quad\;\; 127 \\
\underline{\times 112} \\
\quad\;\; 254 = \;\; 2 \times 127 \\
1{,}270 = \;\; 10 \times 127 \\
\underline{12{,}700 = 100 \times 127} \\
14{,}224 = 112 \times 127
\end{array}
$$

Fig. 6

$20,864 \div 128 = ?$

$$
\begin{array}{r}
128\,)\overline{20,864} \\
\underline{12,800} = 100 \times 128 \\
8,064 \\
\underline{7,680} = 60 \times 128 \\
384 \\
\underline{384} = \underline{3} \times 128 \\
0 \quad 163
\end{array}
$$

check:

$100 \times 128 = 12,800$
$60 \times 128 = 7,680$
$3 \times 128 = \underline{384}$
$20,864$

128

```
┌──────────────────────────────────┐
│                                  │
│                                  │
│         100 × 128 = 12,800       │
│                                  │
│                                  │
├──────────────────────────────────┤
│                                  │
│          60 × 128 = 7,680        │
│                                  │
├──────────────────────────────────┤
│          3 × 128 = 384           │
└──────────────────────────────────┘
```

Fig. 7

103

Making Division Easier to Grasp

Lola J. May

MOST teachers feel that division and subtraction are harder for children to understand than the operations of addition and multiplication. This may be true because they are the inverse operations, and undoing is more difficult than putting together. This means that division cannot be understood until multiplication is understood.

Multiplication can be introduced by the use of arrays. An array is an arrangement of objects in rows with equal numbers of objects in each row. This is a 3 x 8 array:

The first factor in the multiplication fact is the number of rows, and the second factor is the number of objects in each row. The answer—24—is the product. The problem 24÷3 means you know the product and one factor and are seeking the other factor. The problem can be restated as: "What number times 3 equals 24?" If you know the multiplication fact, then you also know the division fact.

By using the array to introduce the multiplication facts, children learn to draw a picture of the fact. This gives meaning to the concept before memorization and drill are started.

The number line can also be used to help give understanding to division and multiplication. Playing the cricket game on a number line can be fun and will provide some groundwork needed before the drill stage. Crickets are named by the number of spaces they jump. For example, a "2" cricket always jumps two spaces on the number line:

If a "2" cricket starts on zero and jumps five times to the right, where does he land? The child can draw the arcs from 0 to 2, 2 to 4, 4 to 6, 6 to 8, and 8 to 10. The cricket jumps five times. What multiplication fact does this show? The child can see that 2 times 5 equals 10. If the "2" cricket starts on 10 and jumps to the left until he lands on zero, how many jumps will he make?

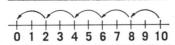

The children start on 10 and draw the arcs showing the path of the cricket. They draw from 10 to 8, 8 to 6, 6 to 4, 4 to 2, and 2 to 0. This is a picture of the problem 10÷2 equals what number? Children can see that division is the operation of subtracting the same number over and over just as multiplication is the operation of adding the same number over and over.

After the multiplication facts and division facts have been introduced and are under control, then the next step is uneven division. Uneven division can be introduced after even division by using the number line. The children can be told that the crickets are having a convention at the point called 20 on the number line. All the crickets are bragging they can jump to zero. Some of them are telling the truth and some are not. Can the children name a cricket that is telling the truth? One answer may be that the "10" cricket is telling the truth. The child can show he is right by going to the number line and drawing arcs from 20 to 10 and from 10 to 0, as shown at the bottom of the page.

Another answer may be that the "6" cricket is not telling the truth. The child can show this by drawing arcs from 20 to 14, 14 to 8, and 8 to 2. What would happen if the cricket jumped again? The child can see that if the cricket jumped again he would go beyond zero to the left. A few bright children may say he would land on a negative four. Then the problem can be written as 20÷6 and the children can see that the answer is 3 (jumps) and a remainder of 2.

In teaching any new concept, you should work on the prerequisites first. In the problem 27÷5, the question that needs to be answered first

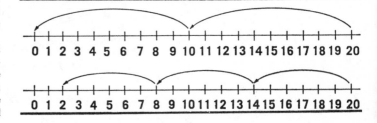

is: "What is the largest digit times 5 that is less than 27?" The child has to know his multiplication facts of 5 before he can do division with problems having 5 as a factor and another number as the product.

The first step is for children to know the multiplication facts. The next step is to practice problems like those below that ask the question: "What is the largest digit that will make each sentence true?"

$\square$ x 6 $<$ 44	$\square$ x 4 $<$ 35
$\square$ x 8 $<$ 50	$\square$ x 9 $<$ 59

When children are faced with problems such as $44 \div 6$, $35 \div 4$, $50 \div 8$, and $59 \div 9$, they have a method to help them solve the problems. Now they can ask the leading question: "What is the largest digit times the given factor that is less than the given product?"

After the multiplication facts are under control, the next step is to practice multiplying by multiples of 100. Draw an array of 6×10. What is the product of 6×10? What is the sum when you use 10 as an addend 6 times? This approach is better than having the teacher tell the children that when you multiply by 10 just annex a zero.

The zero in the numeral 20 is a giveaway that 10 is one factor. In multiplying 4×20, the 20 can be rewritten as $4 \times 2 \times 10$, which is 8×10 or 80. Much practice is needed with the multiples of 10 before children try division problems with more than two-digit products. Problems like these are needed:

4 x 2 = _____	5 x 7 = _____
4 x 20 = _____	5 x 70 = _____
4 x 200 = _____	5 x 700 = _____

After practicing like this, a problem such as $800 \div 4$ can be solved by asking: "What number times 4 equals the product 800?" If a teacher skips this part of the preparation, many children will never understand what they are doing in a division problem. There is a method of division called scaffolding that many teachers use. Students learning this

method have a better understanding of the operation of division than adults who know only the short algorithm. In the example at left below, the student looks at the product

243 R 3		243 R 3	
4) 975		4) 975	
800	200	400	100
175		575	
160	40	400	100
15		175	
12	3	80	20
3		95	
		80	20
		15	
		12	3
		3	

975 and the factor 4 and asks himself: "What is the largest multiple of hundred times 4 that is less than 975?" The numeral 200 is written to the right, and the product of 200 times 4 is written below 975. Eight hundred is subtracted from 975.

The next question is: "What is the largest multiple of 10 times 4 that is less than 175?" The numeral 40 is written to the right. The product of 40 times 4—160—is subtracted from 175.

The last question is: "What is the largest digit times 4 that is less than 15?" The numeral 3 is written to the right, and the product of 3×4 is subtracted from 15. The 3 is a remainder because it is less than 4. The sum of 200, 40 and 3 is now written on top with the remainder 3.

This method of working the problem is longer, but it gives meaning to the problem and stresses place value. A child who does not see that 200 times 4 is 800 at first could write 100 and multiply this by 4 and continue the problem as in the example at right. This is a still longer method, but to the child who feels secure in multiplying by 100 and by 20 instead of 200 and 40, it is a solution and he understands what he is doing.

The short method says: "How many times does 4 go into 9? Write the 2 above the 9. How many times does 4 go into 17? Write the 4 above the 7. How many times does 4 go

into 15? Write the 3 above the 5." But what does the phrase "go into" mean in mathematics? Does a child who learns the "go into" method first really know what he is doing?

Those students who look at a problem and, instead of saying "go into," say: "How many hundreds will *come out of*" will be much closer to understanding the operation. The shorter method of division can be taught later after the children understand what the process means.

The transfer from a one-digit factor to a two-or-more digit factor in division can be made with little difficulty if the student has been taught the scaffolding method. Take this example:

	34 R 5	
23) 787		
	690	30
	97	
	92	4
	5	

The student knows he cannot use a multiple of a hundred because 23 times a hundred is 2,300, which is much larger than 787. The first question is: "What is the largest multiple of 10 times 23 that is less than 787?" The child tries 30 and writes the numeral 30 to the right and multiplies 30 times 23. The product is subtracted from 787. The next question is: "What is the largest digit times 23 that is less than 97?" The numeral 4 is written to the right, and the product of 4 times 23 is subtracted from 97.

Division will always be difficult for pupils who lack the necessary background. They must know their facts, how to subtract, and how to multiply by the multiples of 10. Teachers who force all children to go through the process of division when they are not ready will reinforce failure that is most difficult to change in later years. Why subject children to failure when with a little more time and effort success can be the end result? There is no law that says all children must work division by the short method on their first try.

Division isn't that hard

JOSEPH DI SPIGNO

Learning to work division problems is probably the most difficult arithmetic task for elementary school children to handle. When they master the mechanics of division, they may have little or no understanding of the division process. Try this method, beginning at perhaps the fourth-grade level, to help students visualize what is involved:

Direct the children to set up division problems (6$\overline{)745}$, for example) in the following way:

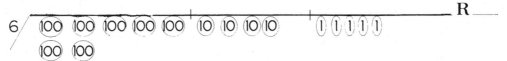

Ask the children to find out how many groups with six members can be extracted from 745, starting in the hundreds column.

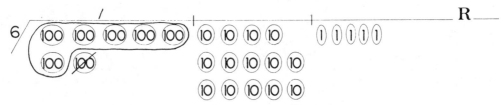

There is 1 group of six hundreds, and 1 hundred left over, which must be converted to 10 tens.

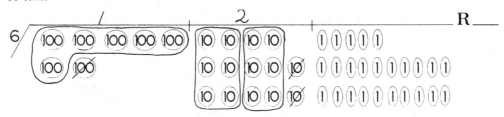

There are 2 groups of six tens and 2 tens that must be converted to ones.

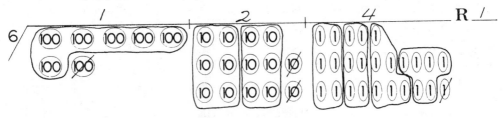

There are 4 groups of six ones with 1 unit remaining.

In 745, then, there are: 1 group of six hundreds, 2 groups of six tens, and 4 groups of six ones, with a remainder of 1.

Reprinted from *The Arithmetic Teacher* (October 1971), 373–377, by permission of the author and the publisher.

After the children can handle division in this fashion, duplicate the following form:

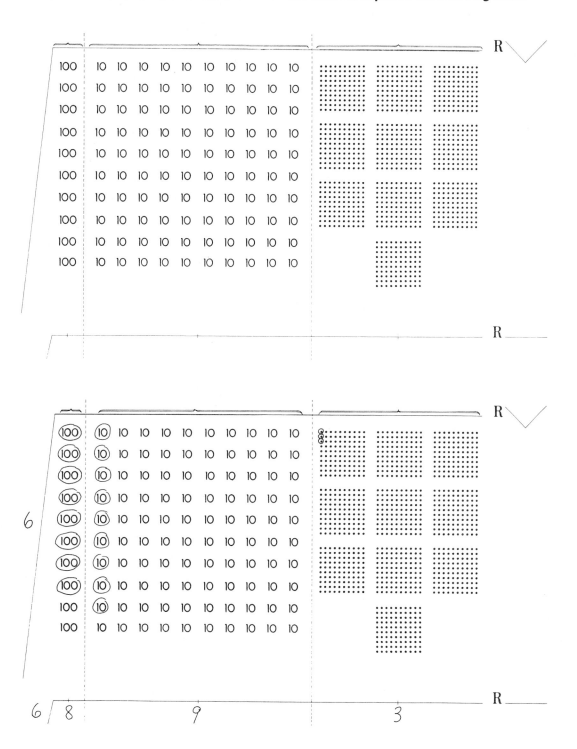

When this form is used, most of the work is done for the children. Two examples follow: $6 \overline{)893}$ and $16 \overline{)515}$.

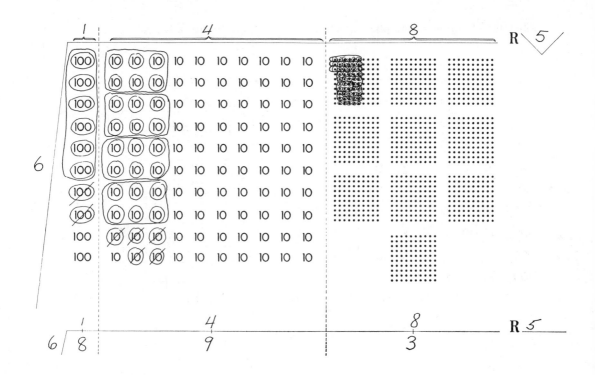

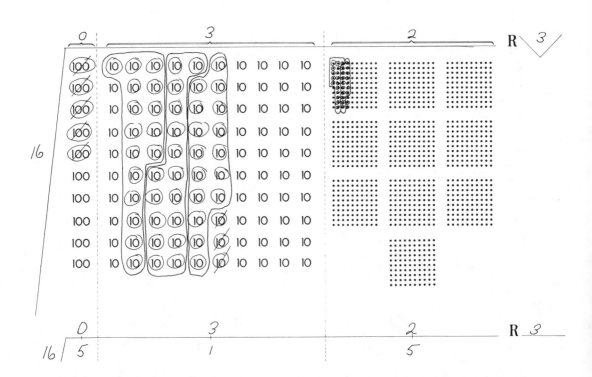

SELECTED BIBLIOGRAPHY FOR CHAPTER 5
Multiplication and Division of Whole Numbers

Alger, Louisa. "Finger Multiplication" *The Arithmetic Teacher* (April 1968), 341.

Bernard, Sister Mary. "An Open Letter: 6x9 and the 'Critical Triangle'." *The Arithmetic Teacher* (May 1968), 431.

Boomer, Lyman W. "An Intuitive Approach to Square Root." *The Arithmetic Teacher* (October 1969), 463.

Callahan, LeRoy. "A Romantic Excursion into the Multiplication Table." *The Arithmetic Teacher* (December 1969), 609.

Connelly, Ralph and Heddens, James. " 'Remainders That Shouldn't Remain." *The Arithmetic Teacher* (October 1971), 379.

Deans, Edwina. "Early Development of Concepts of Multiplication and Division." *The Arithmetic Teacher* (February 1965), 143-50.

——. "Grouping—An Aid in Learning Multiplication and Division Facts." *The Arithmetic Teacher* (January 1961), 27-31.

Dilley, Clyde and Rucker, Walter. "Teaching Division by Two-Digit Numbers." *The Arithmetic Teacher* (April 1969), 306.

Duncan, Hilda F. "Division by Zero." *The Arithmetic Teacher* (October 1971), 381.

Ginther, John L. "Some Activities With Operation Tables." *The Arithmetic Teacher* (December 1968), 715-17.

Herold, Joan. "Patterns for Multiplication." *The Arithmetic Teacher* (October 1969), 498.

Hullihan, William. "Multiplication Unlimited." *The Arithmetic Teacher* (May 1968), 460.

Jarosh, Sharon C. "617—The Number Line and Division." *The Arithmetic Teacher* (November 1970).

Kurtz, Ray. "Fourth Grade Division: How Much Is Retained in Grade Five?" *The Arithmetic Teacher* (January 1973), 65-71.

McDougall, Ronald V. "Don't Sell Short the Distributive Property." *The Arithmetic Teacher* (November 1967), 570.

McLean, Robert C. "Estimating Quotients for the New Long Division Algorithm." *The Arithmetic Teacher* (May 1969), 398.

Mehl, William G. "Where, On the Number Line, Is the Square Root of Two?" *The Arithmetic Teacher* (November 1970), 613.

Rappaport, David. "Multi-Logical or Pedagogical?" *The Arithmetic Teacher* (February 1968), 158.

Reeve, Olive R. "The Missing Factor in Division." *The Arithmetic Teacher* (March 1968), 275.

Rogers, Frank. "Divisibility Rule for Seven." *The Arithmetic Teacher* (January 1969), 63.

Schloff, Charles E. "A Pictured Approach to an Idea for Division." *The Arithmetic Teacher* (May 1969), 403.

——. "Double and Double Again." *The Arithmetic Teacher* (November 1970), 613.

——. "8 = Turkey." *The Arithmetic Teacher* (April 1971), 268-70.

Shafer, Dale M. "Multiplication Mastery Via the Tape Recorder." *The Arithmetic Teacher* (November 1970), 581.

Smith, Frank. "Divisibility Rules for the First Fifteen." *The Arithmetic Teacher* (February 1971), 85-87.

Stenger, Donald. "Prime Numbers From the Multiplication Table." *The Arithmetic Teacher* (December 1969), 617-20.

Stephens, Lois. "An Adventure in Division." *The Arithmetic Teacher* (May 1968), 427.

Stern, Jane L. "Counting: A New Road to Multiplication." *The Arithmetic Teacher* (April 1969), 311.

Swart, William L. "Teaching the Division-by-Subtraction Process." *The Arithmetic Teacher* (January 1972), 71-75.

Traub, Raymond G. "Napier's Rods: Practice With Multiplication." *The Arithmetic Teacher* (May 1969), 363-64.

Tucker, Benny F. "The Division Algorithm." *The Arithmetic Teacher* (December 1973), 639-46.

CHAPTER 6

NUMBER OPERATIONS AND THE SET OF INTEGERS

Understanding the fundamental operations with negative numbers is usually difficult for children. Rather than exploring models and providing experiences which should build understanding, unfortunately teachers have often given a computation algorithm which the children must learn by rote. The selections in this chapter have been chosen to provide models and activities that will help lead to a meaningful understanding of operations with signed numbers.

Laurence Sherzer's article, "Adding Integers Using Only the Concepts of One-to-One Correspondence and Counting," describes an excellent activity which starts building foundations for later work in higher mathematics.

"A Model for Arithmetic of Signed Numbers" by Lois M. Luth is an interesting and creative approach through which children can visualize operations with signed numbers.

Stanley Cotter's article, "Charged Particles: A Model for Teaching Operations With Directed Numbers," not only gives a suitable model for the concrete stage of learning but also supplies a foundation for understanding the concept of ion exchange in the solution of chemical equations. Those who are interested in the correlation of mathematics-science activities will want to read this selection.

In "A Rationale in Working With Signed Numbers," Louis S. Cohen uses the "Postman Stories" model devised by the Madison Project. This activity allows children to become personally involved in the learning situation.

Understanding the operations with negative numbers may be difficult for children, but it is a necessity if more complex mathematics are to be done and understood. The ideas and exercises presented in these articles not only aid the children's understanding but make teaching this difficult concept much easier.

Adding integers using only the concepts of one-to-one correspondence and counting

LAURENCE SHERZER

The following is a method for teaching the addition of integers which requires only that the students have developed the concepts of one-to-one correspondence and counting. This means that the student need not be familiar with the operations of whole numbers.

The students are shown a statement such as

$$^-4 + {}^+3$$

and told that since $^-4$ is equal to four negative ones, they can write the negative four as

$$^-1$$
$$^-1$$
$$^-1$$
$$^-1.$$

Similarly, since the $^+3$ is equal to three positive ones, they can write $^+3$ as

$$^+1$$
$$^+1$$
$$^+1.$$

But instead of writing these ones separately, they should match them, i.e., put them in one-to-one correspondence like this:

$$^-1 \; ^+1$$
$$^-1 \; ^+1$$
$$^-1 \; ^+1$$
$$^-1 \quad .$$

It is then explained that $^-1 + {}^+1 = 0$ and, therefore, all we need do to find the sum of $^-4 + {}^+3$ is to count those numbers that are not matched. In this case there is only one $^-1$ unmatched, so the sum is $^-1$, i.e., $^-4 + {}^+3 = {}^-1$.

As another example, consider $^+4 + {}^-7$. The student writes

$$^+1 \; ^-1$$
$$^+1 \; ^-1$$
$$^+1 \; ^-1$$
$$^+1 \; ^-1$$
$$\quad ^-1$$
$$\quad ^-1$$
$$\quad 1.$$

Counting the unmatched numbers, he finds that $^+4 + {}^-7 = {}^-3$.

Of course, it is not absolutely necessary that the students write $^+1$ and $^-1$. For simplicity in practice they may just write down the positive signs and the negative signs, as long as they understand that these stand for the numbers $^+1$ and $^-1$. Then for $^+4 + {}^-7$ they would write

$$+ \; -$$
$$+ \; -$$
$$+ \; -$$
$$+ \; -$$
$$\quad -$$
$$\quad -$$
$$\quad -$$

and conclude, by counting, that $^+4 + {}^-7 = {}^-3$. As stated, the only concepts required are those of one-to-one correspondence and counting.

We almost might say that all the students need to know is how to count, for the idea of the correspondence is implicit

Reprinted from *The Arithmetic Teacher* (May 1969), 360–362, by permission of the author and the publisher.

in the demonstration. They count the four positive signs as they write them out, then they count the seven negative signs as they set them into one-to-one correspondence with the positive signs, and finally they count the signs that are left over.

Note that the procedure is consistent when involving integers that do not differ in sign. For example, in $^-2 + {}^-1$, if we place all the negative signs in a row we have

$$-$$
$$-$$
$$-,$$

therefore $^-2 + {}^-1 = {}^-3$. Similarly, $^+3 + {}^+2$ becomes for them:

$$+$$
$$+$$
$$+$$
$$+$$
$$+$$

and $^+3 + {}^+2 = {}^+5$. The students need to know only how to count!

From these experiences, the student comes logically to an intuitive understanding of the usual definition of adding integers. Namely, if the signs are the same, add the numbers and prefix the common sign; if the signs are different, subtract the smaller number from the larger and prefix the sign of the larger. These words usually do not have much meaning for the student until he has had some specific experiences. Therefore, introducing the concept through this language is inefficient anyway, and almost useless in the case of very young students.

By introducing the student to the addition of integers by this method of counting and one-to-one correspondence, he will come to learn the procedures indicated in the definition through the process of discovery. Once he has himself taken the step from actually writing the positive and negative signs to mental addition and subtraction, a verbal definition can be introduced and will be readily accepted and understood.

The origin of this idea has its roots in the set-theory foundations of mathematics. In modern set theory and/or foundations of mathematics we construct the integers from the whole numbers, 0, 1, 2, 3, etc. The integers are the set of all ordered pairs of numbers that can be made from the whole numbers. Namely, (0,0), (0,1), (0,2), (0,3), etc.; (1,0), (1,1), (1,2), etc.; (2,0), (2,1), (2,2), etc., and so on. The number (0,0) turns out to be 0, (0,1) is $^-1$, (0,2) is $^-2$, (0,3) is $^-3$, (0,4) is $^-4$, (1,0) is 1, (2,0) is 2, (3,0) is 3, etc. and each ordered pair in which both members are non zero is equivalent to exactly one of these pairs.[1]

Two definitions are significant: one is that of equivalence and the other of addition. Two integers (a,b) and (c,d) are said to be equivalent if $a + d = b + c$; therefore (3,2) is equivalent to (1,0) since $3 + 0 = 2 + 1$. The sum of the integers (a,b) and (c,d) is $(a + c, b + d)$; therefore the sum of (1,0) and (5,1) is $(1 + 5, 0 + 1)$ or (6,1).

Now the significant thing is that nowhere in this latter definition do we find any mention of subtraction. Yet our traditional definition of the addition of integers requires subtraction, "subtract the smaller number from the larger and prefix the sign of the larger." But if our theoretic foundations do not require it, it must be possible to add integers without the use of subtraction. Let us examine the addition of two integers, say $^-2$ and 3. Writing these as ordered pairs of whole numbers, they are (0,2) and (3,0) and their sum is by definition $(0 + 3, 2 + 0)$ which is (3,2).[2] This number is equivalent to (1,0), since $3 + 0 = 2 + 1$, and (1,0) is the integer 1. We note again that at no time was subtraction required. But what did we do? First we added the cardinal numbers of which the integers are constructed: (0,2) + (3,0) = $(0 + 3, 2 + 0)$ = (3,2). But this was trivial, for there was practically

[1] For a fuller discussion of this material, see Howard Levi, *Elements of Algebra* (New York: Chelsea Publishing Co., 1954).

[2] In this scheme of writing integers as ordered pairs of whole numbers, the ordered pair $(a,0)$ represents the integer a and $(0,b)$ represents the integer ^-b.

nothing to do since the addition in both cases involved zero. What else was involved? The number (3,2) had to be reduced to an equivalent form involving zero, namely (1,0). How is this done? Do we ask what number (x,y) exists such that $3 + y = 2 + x$? Certainly not. What we do is simply reduce both numbers until one is zero. Of course we might subtract (although really we cannot, since subtraction has not yet been defined at this stage of the mathematical construction), but the point is we need not!

We can put sets that each number represents into one-to-one correspondence and count the balance. Let us look at this. There are two places. Let us represent the set in the first by x's and the set in the second by y's. Therefore (3,2) would be represented by three x's and two y's. Putting these into one-to-one correspondence, we have

$$x \; y$$
$$x \; y$$
$$x \; \; .$$

We are left with one x and no y's. Translating this back into integer notation we have (1,0).

How does this translate into our conventional notation? We have (3,0) + (0,2) = (1,0), which is conventionally written $^+3 + {}^-2 = {}^+1$. (The upper case positive and negative signs are, of course, not really "conventional," but they are finding their way more and more into the literature and are used here for emphasis and clarity. Our negative and positive signs are now a logical tool to use in the one-to-one correspondence or, if you will, matching process, so that we write three positive signs for $^+3$, two negative signs for $^-2$, and since we end up with one positive unmatched, we have $^+3 + {}^-2 = {}^+1$.

We see then that this is not just a simplified method for learning, but also a procedure that is consistent with the logical structure of modern mathematics. Therefore, it will have a heuristic value as well when the student encounters higher mathematics.

A model for arithmetic of signed numbers

LOIS M. LUTH

To enable elementary school children to perform the arithmetic of signed numbers easily and correctly, I use the story of "Hy in the Sky." Hy is a man living in a house in the sky to which are attached an undetermined number of balloons and sandbags. A helicopter, from which Hy can buy more balloons and sandbags, comes each morning.

The balloons come in different sizes and are marked to show the number of feet a particular balloon can raise the house. For instance, a five-foot balloon will raise the house five feet. It would be marked $^+5$, since all balloons are marked with a positive sign.

The sandbags are also sold in different sizes, and they are marked to show the number of feet a particular sandbag will lower the house. For instance, a five-foot sandbag will lower the house five feet. It would be marked $^-5$, since all sandbags are marked with a negative sign.

As Hy buys balloons or sandbags, he is adding to his house, so buying illustrates addition. If he buys a five-foot balloon and two-foot sandbag, it is shown as

$$+ \, {}^+5 + {}^-2.$$

The first "+" can be omitted, since it can be understood that he is *buying* first. A

Reprinted from *The Arithmetic Teacher* (March 1967), 220–222, by permission of the author and the publisher.

question could then be asked, "Is Hy higher or lower than he was the day before?" Of course, he is higher. "How much higher?" is then asked. The answer is three feet. Since he is *higher*, a "+" sign is written in front of the numeral. Higher by three feet is written as $^+3$. The sentence for this story would be

$$^+5 + {}^-2 = {}^+3.$$

If Hy buys a two-foot sandbag and a three-foot sandbag, he will be five feet lower. To illustrate his being *lower*, a "—" sign would be written in front of the numeral. Five feet lower is written as $^-5$. The sentence for this would be

$$^-2 + {}^-3 = {}^-5.$$

To illustrate subtraction, the story is continued. Later in the day Hy is able to free both balloons and sandbags from his house. Since this is subtracting from his house, freeing illustrates subtraction. If he frees a two-foot balloon, it is shown as $-({}^+2)$. As a result, the house is lowered, so

$$- ({}^+2) = {}^-2.$$

If Hy buys a two-foot sandbag and later frees a three-foot balloon, he is lowered by five feet. The sentence showing this is

$$^-2 - {}^+3 = {}^-5.$$

This question could also be asked: "What did Hy do here—

$$^-3 - {}^-2?"$$

(*Answer:* He bought a three-foot sandbag and later freed a two-foot sandbag, so he is now one foot lower. The sentence for this is

$$^-3 - {}^-2 = {}^-1.)$$

A vertical, unmarked number line could be used to show the results of the stories. If Hy buys a three-foot balloon and a two-foot sandbag, he is one foot higher.

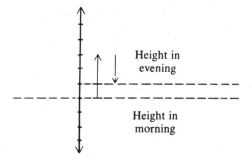

Height in evening

Height in morning

The sentence for this story would be

$$^+3 + {}^-2 = {}^+1.$$

After the students are thoroughly familiar with the stories of addition and subtraction, multiplication can be introduced. In this story, Hy is able to buy more than one balloon or sandbag with the same marking. For instance, he could buy two of the five-foot balloons. As a result, he would be ten feet higher, and the sentence would be

$$^+2 \times {}^+5 = {}^+10.$$

If he frees two of the four-foot balloons, he is then eight feet lower, and the sentence is

$$^-2 \times {}^+4 = {}^-8.$$

Other stories and their sentences are as follows:

If he buys three of the two-foot sandbags, Hy is six feet lower.

$$^+3 \times {}^-2 = {}^-6.$$

If he frees three of the two-foot sandbags, Hy is six feet higher.

$$^-3 \times {}^-2 = {}^+6.$$

These are also handy balloons and sandbags, in that Hy can buy or free a part of one of them and then re-mark them. This is done by letting out gas or sand before or after purchase. For instance, he can buy one-half of a six-foot balloon by letting out one-half of the gas. He is then three feet higher, and the sentence is

$$\frac{+1}{2} \times {}^+6 = {}^+3.$$

The other stories are the same with fractions as they were with integers.

Another question to ask would be, "Is Hy higher or lower if he buys a two-foot balloon and a two-foot sandbag?" The answer, of course, is neither. Since balloons and sandbags are opposites, so are ${}^+2$ and ${}^-2$, and their sum is zero. The sentence is

$${}^+2 + {}^-2 = 0.$$

It is interesting to let one student tell a story about Hy and let another write the sentence, or vice versa. Such sentences as

$${}^-2 \times \square = {}^+4$$

or

$${}^+2 - \square = {}^-4$$

could be written on the board to let the students fill in the place holders, using the story. Another use of Hy would be to ask how many different ways he could lower his house six feet, buying or freeing any number of balloons or sandbags or both, and to make sentences to illustrate this.

At a later time, division can be introduced if desired. "To go up six feet using three-foot balloons, will Hy have to buy or free them? How many?" are the questions that could be asked. He will have to buy two, shown as ${}^+2$. If the division sign ($\div$) is used in place of the word "using," the sentence is

$${}^+6 \div {}^+3 = {}^+2.$$

Other stories and sentences of division are as follows:

To go down six feet using three-foot balloons, Hy frees two.

$${}^-6 \div {}^+3 = {}^+2.$$

To go up six feet using three-foot sandbags, Hy frees two.

$${}^+6 \div {}^-3 = {}^-2.$$

To go down six feet using three-foot sandbags, Hy buys two.

$${}^-6 \div {}^-3 = {}^+2.$$

With "Hy in the Sky," addition, subtraction, multiplication, and division of positive and negative numbers can be illustrated. I find it very meaningful to the students, because they can easily visualize Hy's going up or down by buying or freeing balloons and sandbags.

In brief, I like this story because it gives meaning to signed numbers and because, once the story is learned, it helps students remember the results of operations with them.

Charged particles: a model for teaching operations with directed numbers

STANLEY COTTER

In preparing to teach mathematical concepts, a good deal of a teacher's time is spent in finding suitable models for the concrete stage of learning. In no other area of elementary mathematics is this more critical than in the presentation of operations with directed numbers. Whether the topic is presented for the first time in the sixth grade or occurs in a remedial course in algebra at the college level, there is always a need for some students to relate directed number operations with a physical interpretation. Many such models have been devised, and the literature abounds with helpful suggestions in this area.[1] The usefulness of any particular model is directly related to the success of the student in conjuring a vision of the model when it is needed. It is also helpful to the teacher when he can evoke the image of the model with a simple verbal reminder.

Another criterion for determining the value of a model is the length of time it serves as a crutch to the student. The rapidity of the transition involved in moving from dependence upon a physical interpretation to understanding and using the related mathematical abstraction is a true measure of a model's worth. In this respect, a mathematical model should be designed to be discarded by the student as soon as possible.

For several years I have been using a model for introducing the operations with directed numbers which successfully meets the criteria mentioned above. Borrowing from my own limited experience in the physical sciences, I have constructed a model that gives solid foundation to the mathematical concepts involved and does not violate the scientific evidence. The classroom approach to the presentation of this model might occur as in the monologue that follows.

Addition

"Consider a field that is empty at the moment." (Draw the circular region labeled Figure 1).

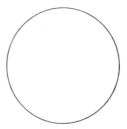

FIGURE 1

"I have a special bucket containing positive particles. From this bucket I take four

[1] Robert B. Ashlock and Tommie A. West, "Physical Representations for Signed-Number Operations," THE ARITHMETIC TEACHER, XIV (November 1967), 549–54.

Reprinted from *The Arithmetic Teacher* (May 1969), 349–353, by permission of the author and the publisher.

116

of them and place them in the field (Fig. 2). The field now contains a charge of positive four.

"If I add three more positive particles, the charge on the field will increase to positive seven." (Fig. 3.)

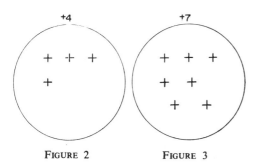

FIGURE 2   FIGURE 3

"Now, I have another bucket containing negative particles. What do you think will happen if I place five negative particles in the field?" (Here is the opportunity to present the concept of neutralization, drawing upon appropriate examples from physics and chemistry, such as the acid-base reaction or electromagnetic theory. It may be suitable to have an actual demonstration with magnets or the litmus-paper test. It is probable that at the level when directed numbers are introduced, almost all the students will have had some experience with positives attracting negatives and neutralizing each other.)

"We see that each negative particle will attract a positive particle, and they will

"The five negative particles have neutralized five of our seven positive particles after being added to the field. Will someone come to the board and write a symbol to represent the resultant charge on the field?" (The reader should append a "+2" over the picture of the field in Figure 4.)

"Now it is time to summarize what has occurred, using mathematical symbols. Starting with an empty field that had a zero charge, we added both positive and negative particles, changing the charge on the field as follows." (Board display)

$$0 + {}^+4 = +4$$
$${}^+4 + {}^+3 = +7$$
$${}^+7 + {}^-5 = +2$$

(The last step may be elaborated upon by showing that "neutralization" occurs mathematically as ${}^+5 + {}^-5 = 0$.) Thus, in detail,

$${}^+7 + {}^-5 = ({}^+2 + {}^+5) + {}^-5 \quad \text{Renaming}$$
$$= {}^+2 + ({}^+5 + {}^-5) \quad \text{Associativity}$$
$$= {}^+2 + 0 \quad \text{"Neutralization" now; "Additive Inverse Property" later}$$
$$= {}^+2 \quad \text{Addition Property of Zero}$$

"Referring again to our field (Fig. 4), what kind of particles and how many of them must be added to the field to bring it back to a charge of zero?" The answer (hopefully offered by one of your students) is "two negative particles" (Fig. 5). "We have a field in which all of the negative particles are joined to and neutralized by positive particles."

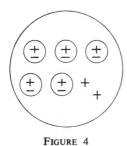

FIGURE 4

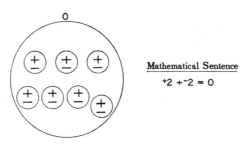

Mathematical Sentence
${}^+2 + {}^-2 = 0$

FIGURE 5

neutralize each other's charge. Let's show this neutralization by drawing a zero around each pair." (Fig. 4.)

"But now my buckets are empty. The only way I can fill them is to *take away* positive and negative particles from the

117

field. To do this, I will have to split up the neutralized particles." (Remove the circles around four of the neutralized particles and detach four of the negatives. Also erase the charge of zero above the field.) "The field now contains four unneutralized positive particles, which gives the field a charge of positive four (Fig. 6). Since the operation of subtraction is related to the process of 'taking away,' we can symbolize what has been done by the following mathematical sentence." (Fig. 6.)

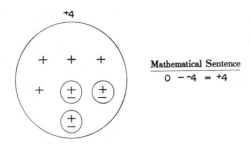

Mathematical Sentence

$0 - {}^-4 = {}^+4$

FIGURE 6

"Notice that I started with a charge of zero and subtracted four negative particles. This resulted in a field charge of positive four. Is there any other way that you could begin with a zero charge and end up with a charge of positive four?" The last question begins the development of the concept of addition and subtraction as inverse operations. That is, for the particular question under discussion, that

$$0 - {}^-4 = 0 + {}^+4 = +4.$$

"You remember that when we began to add particles to our field that four positive particles added to an empty field resulted in a charge of positive four." (See Fig. 2.) "Now, what do we have to take away from the field in order to increase the charge to positive seven?" (Figure 7 indicates the demonstration required to answer this question.)

"We have seen that *subtracting* three negative particles from a field with a charge of positive four results in a field with a charge of positive seven. The same change

in field charge can be brought about by *adding* three positive particles to a field with a charge of positive four. Therefore, since the results are the same, we say that ${}^+4 - {}^-3$ and ${}^+4 + {}^+3$ are *equivalent*."

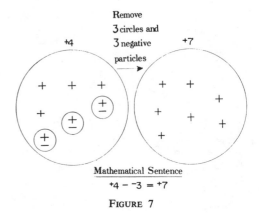

Mathematical Sentence

${}^+4 - {}^-3 = {}^+7$

FIGURE 7

"What should we subtract from the field to reduce the charge from ${}^+7$ to ${}^+2$?" (Fig. 8.)

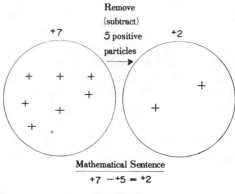

Mathematical Sentence

${}^+7 - {}^+5 = {}^+2$

FIGURE 8

"What addition fact that we have seen before is equivalent to the subtraction fact that is written on the board?" (Answer: ${}^+7 + {}^-5$ is equivalent to ${}^+7 - {}^+5$.)

"Finally, what are the two ways we can use to bring our field back to charge of zero?" (Fig. 9.)

The structured lesson above attempts to show in cyclical form the connection between addition and subtraction of directed numbers. Building on the foundation of

118

addition as the joining of sets of objects, and subtraction as the removal of sets of objects, the model provides a physical representation for the following generalizations:

1. The additive inverse property: $a + {}^-a = 0$.
2. The definition of subtraction in terms of addition: $a - b = a + {}^-b$.

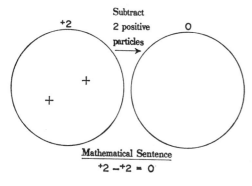

Subtract
2 positive
particles

Mathematical Sentence
${}^+2 - {}^+2 = 0$

or

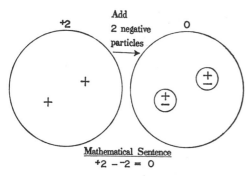

Add
2 negative
particles

Mathematical Sentence
${}^+2 - {}^-2 = 0$

FIGURE 9

The model is easily visualized and can be suitably presented on the chalkboard or with cut-outs for a flannel board. The number of illustrative examples can be varied depending upon the grade level and abilities of the students. An extension to the rules for multiplying directed numbers can be made with the model.

Multiplication

In using this model for teaching multiplication of directed numbers, the instructor can rely upon the student's former ex-

perience with this operation as it was applied to the set of natural numbers. The student has already learned that the phrase "three groups of four objects" can be translated into the mathematical expression "3 × 4." We can combine this concept with that of producing a change in the resultant charge of our field.

To begin, the field model must have a charge of zero whenever it is used for multiplication only. This is best suggested to the student by a field containing a large complement of neutralized particles as in Figure 10.

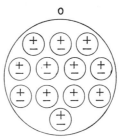

FIGURE 10

To add three groups of two positive particles each, the instructor should go through the actual process of successive addition. After this task is accomplished, the field should appear as in Figure 11.

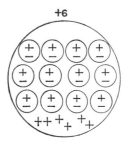

FIGURE 11

The phrase, "add three groups of two particles each" should be translated: $+3 \times {}^+2$. This will underscore the different meanings assigned to the plus sign.[2] In

---

[2] The changeover to the more convenient notation; i.e., "$(+4)(-2) = -8$," should be left for later when the process of multiplication is formalized into rules.

119

this way, when a student is confronted with the sentence: $+2 \times {}^-3 = -6$, he can translate it into the meaningful statement, "If we add two groups of three negative particles each to a field with zero charge, the charge on the field will become negative six." (Fig. 12.)

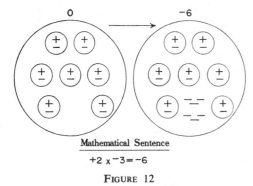

Mathematical Sentence

$+2 \times {}^-3 = -6$

FIGURE 12

If the multiplier is negative, the interpretation can be made in terms of successive subtraction. Thus, "$-2 \times {}^+3$" is translated as "subtract two groups of three positive particles each from the neutral field." In carrying out this process by demonstration, the teacher will have liberated six negative particles from their neutralized bonds, thereby yielding a field charge of -6 (Fig. 13.)

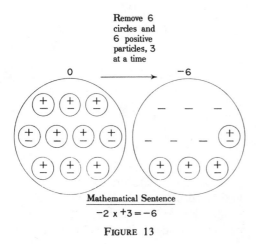

Mathematical Sentence

$-2 \times {}^+3 = -6$

FIGURE 13

Finally, the case of multiplying two negative numbers is illustrated in the ana-

logous manner of successive subtraction of negative particles from the field. Here, the subtraction of two groups of three negative particles each from the field results in liberating six positive particles from their neutral bonds, thereby changing the field charge from zero to positive six (Fig. 14).

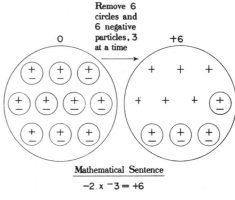

Mathematical Sentence

$-2 \times {}^-3 = +6$

FIGURE 14

To the student who delights in intellectual consistency, this revelation by the field model of the mysterious mathematical "trick" of multiplying two negative numbers to yield a positive product is an ultimate satisfaction. For this reason, it provides a fitting conclusion to the exploitation of the model. The rules for division of directed numbers are more appropriately taken up after formalization of multiplication. This occurs in the customary manner of relating the open sentence, $a/b = q$ with $a = bq$. Thus, $-12/-4 = +3$ because it has been established that $-12 = (-4)(+3)$.

One of the fringe benefits of using this model is the foundation it provides for understanding the concept of ion exchange in the solution of chemical equations. As a practical suggestion, you might bring this model to the attention of your colleagues in the Chemistry Department of your school. They may return the compliment by providing you with the materials needed to put on a colorful and dramatic demonstration.

A rationale in working with signed numbers

LOUIS S. COHEN

In the teaching of signed numbers to children in Grades 4–8, I have used the "Postman Stories" model devised by the Madison Project.

I have used this lesson many times with children in both the St. Louis and Chicago Public Schools, and each time I have sincerely felt that the "message" has been put across not by heavy-handed exposition nor by adherence to "rules," but rather by personal involvement in the situation. The time allowance for such a lesson should, of course, be determined by the grade and ability of the class. I have done this complete lesson in a forty-five minute mathematics period for fifth graders, though at other times two or three class meetings have been needed.

Postman stories
for addition and subtraction

Let us pretend that we live in a city inhabited by people who do things very strangely. For instance, this city has a postman who drops off letters at the different houses quite arbitrarily. He quite often leaves mail at the wrong address and then comes back on a later date to take them back. But what is even more strange, the people of this city really don't seem to mind. They take the letters, read them, and then set them aside without noticing to whom they were really meant to be delivered.

Besides receiving letters from friends and relatives, one also receives two other types of mail—checks and bills. The company which employs your father may send his earnings through the mail in the form of a check. Your mother, in her shopping, might not pay cash for the articles she bought. In that case the store will send her a bill for the amount she owes.

Let's put the last two types of letters in a mathematical notation. Then we can use symbols instead of words to describe what has happened. Let us write a check for, say $6.00, like this: (+6). We will call this "positive 6." A bill for the same amount will be written as (−6) and called "negative 6."

Let us further write the following:

"DELIVER A CHECK FOR $6.00"
AS + 6

and say "plus a positive 6."
"Deliver a bill for $6.00" as +−6.
"Take back a check for $6.00" as −+6.
"Take back a bill for $6.00" as −−6.

How would you feel, happy or sad, if the postman delivered you a check? (Happy) How would you feel, happy or sad, if the postman delivered you a bill? (Sad) How would you feel, happy or sad, if the postman took back a check? (Sad, because now you cannot cash the check) How would you feel, happy or sad, if the postman took back a bill? (Happy, because now you do not have to pay the bill)

Now, the housewife has a given amount of money in the bank for her weekly needs,

Reprinted from *The Arithmetic Teacher* (November 1965), 563–567, by permission of the author and the publisher.

and, even though she doesn't cash any of the checks or pay any of the bills she receives, she does keep track of her financial position by adding the amounts of the checks to and subtracting the amounts of the bills from her balance.

Let us assume that our housewife has $20 as her balance early Monday morning.

Monday: $20

$$+2 + {}^+6$$

What did the postman do?

(He delivered a check for $2 and a check for $6.)

Is she happy or sad?

(Happy)

Is our housewife richer or poorer after this transaction?

(Richer)

How much richer?

($8 richer)

Let me write "$8 richer" as $^+8$, therefore,

$$+2 + {}^+6 = {}^+8.$$

Then on Tuesday, how much money will she have to work with?

($28)

Tuesday: $28

The postman came and what did he do?

$$-3 + {}^-4$$

(He delivered a bill for $3 and a bill for $4.)

Happy? Sad?

(Sad)

Is our housewife richer or poorer after this transaction?

(Poorer)

How much poorer?

($7 poorer)

Let me write "$7 poorer" as $^-7$, therefore,

$$-3 + {}^-4 = {}^-7.$$

Then on Wednesday, how much money will she have to work with?

($21)

Wednesday: $21

The postman came and what did he do?

$$+4 + {}^-6$$

(He delivered a check for $4 and a bill for $6.)

Happy? Sad?

(Sad)

Richer? Poorer?

(Poorer)

How much poorer?

($2 poorer)

How should I write that she is $2 poorer?

($^-2$)

$$+4 + {}^-6 = {}^-2.$$

Then on Thursday, how much money will she have to work with?

($19)

Thursday: $19

The postman came and what did he do?

$$+7 + {}^-3$$

(He delivered a check for $7 and a bill for $3.)

Happy? Sad?

(Happy)

Is our housewife richer or poorer after this transaction?

(Richer)

How much richer?

($4 richer)

How should I write "$4 richer"?

($^+4$)

$$+7 + {}^-3 = {}^+4.$$

Then on Friday, how much money will she have to work with?

($23)

Friday: $23

The postman came and what did he do?

$$+3 - {}^+6$$

(He *delivered* a check for $3 and *took back* a check for $6.)

Happy? Sad?

(Sad)

Which part made her happy?

(Delivering the check for $3)

Which part made her sad?

(Taking back the check for $6)

Is our housewife richer or poorer after this transaction?

(Poorer)

How much poorer?
(\$3 poorer)
How should I write "\$3 poorer"?
($^-3$)

$$^+3-{}^+6={}^-3.$$

Then on Saturday, how much money will she have to work with?
(\$20)
Her balance is the same now as it was when she started. Why?
(All bills and checks cancel; discuss.)

Saturday: \$20
The postman came and what did he do?

$$^-4-{}^-6$$

(He *delivered* a bill for \$4 and *took back* a bill for \$6.)
Happy? Sad?
(Happy)
Which part made her happy?
(Taking back the bill for \$6)
Which part made her sad?
(Delivering the bill for \$4)
Is our housewife richer or poorer after this transaction?
, (Richer)
How much richer?
(\$2 richer)
How should I write "\$2 richer"?
($^+2$)

$$^-4-{}^-6={}^+2.$$

Then on the next delivery day, Monday, how much money will she have to work with?
(\$22)

Monday: \$22
The postman came and what did he do?

$$^+7-{}^-4$$

(He *delivered* a check for \$7 and *took back* a bill for \$4.)
Happy? Sad?
(Happy)
Which part made her happy?
(Both transactions)
Which part made her sad?
(Neither)

Is our housewife richer or poorer after this transaction?
(Richer)
How much richer?
(\$11 richer)
How should I write "\$11 richer"?
($^+11$)

$$^+7-{}^-4={}^+11.$$

Then on Tuesday, how much money will she have to work with?
(\$33)

Tuesday: \$33
The postman came and what did he do?

$$^-8-{}^+5$$

(He *delivered* a bill for \$8 and *took back* a check for \$5.)
Happy? Sad?
(Sad)
Which part made her happy?
(Neither)
Which part made her sad?
(Both transactions)
Is our housewife richer or poorer after this transaction?
(Poorer)
How much poorer?
(\$13 poorer)
How should I write "\$13 poorer"?
($^-13$)
Then on Wednesday, how much money will she have to work with?
(\$20)
Her balance is the same as it was when she started. Why?
(Again, checks and bills cancel; discuss.)

Wednesday: \$20
The postman came and what did he do?

$$^+3-{}^-3$$

(He *delivered* a check for \$3 and *took back* a bill for \$3.)
Happy? Sad?
(Happy)
Is our housewife richer or poorer after this transaction?
(Richer)

How much richer?

(\$6 richer)

How should I write "\$6 richer"?

(+6)

$$+3 - {}^-3 = {}^+6.$$

Then on Thursday, how much money will she have to work with?

(\$26)

Thursday: \$26

The postman came on this day, like always, but this time when I asked the housewife what happened, this is what she told me:

$${}^+6 + \square = {}^-8.$$

What had to happen here so that she would end up \$8 poorer?

(He delivered a check for \$6 and *a bill for \$14.*)

Then on Friday, how much money will she have to work with?

(\$18)

Friday: \$18

This is what she told me happened on Friday.

$${}^+3 - \square = {}^+5.$$

What happened to make her \$5 richer?

(He delivered a check for \$3 and *took back a bill for \$2.*)

Then on Saturday, how much money will she have to work with?

(\$23)

Saturday: \$23

What would have had to happen here?

$$\square - {}^-4 = {}^-10.$$

(*He delivered a bill for \$14 and took back a bill for \$4.*)

How much money will be available next Monday?

(\$13)

Monday: \$13

What would have had to happen here?

$$\square - {}^-4 = {}^+10.$$

(*He delivered a check for \$6 and took back a bill for \$4.*)

What if the postman came, delivered a check for a certain amount, and delivered another check for another amount? Let me symbolize the last sentence in this manner.

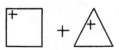

We know the answer will be a certain amount—but I'm not really interested in exactly how much it is. What I am interested in is whether our housewife will be richer or poorer. I will use a rectangle

as the answer to our problem—and your task will be to tell me what symbol to place in the circle in the rectangle to show whether she is richer or poorer.

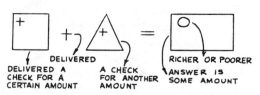

Is she richer or poorer?

(Richer)

What symbol should I place in the circle?

(+)

What happened here?

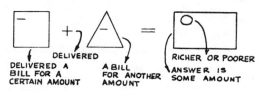

Is she richer or poorer?

(Poorer)

What symbol should I place in the circle?

(−)

What happened here?

(A check for a certain amount and a bill for another amount were delivered.)

Is our housewife richer or poorer?

(You cannot tell.)

Why? What would have to happen to make her richer? To make her poorer?

[Richer—the amount of the check (`☐`) would have to be **larger** than the amount of the bill (`△`). Poorer—the amount of the bill (`△`) would have to be larger than the amount of the check (`☐`).]

What happened here?

(A check for a certain amount was delivered and a check for an amount was taken back.)

Is our housewife richer or poorer?

(You cannot tell.)

Why? What would have to happen to make her richer? Poorer?

(If the `☐` check is more than the `△` check, she will be richer. If the `☐` check is less than the `△` check, she will be poorer. If the `☐` check and the `△` check are the same amounts, she will be neither richer nor poorer. She will have the same as she started out with that morning.)

Can you tell me what happened here?

(A bill for a certain amount was delivered and a check for an amount was taken back.)

Can you tell me if our housewife is richer or poorer from this transaction?

(Yes, she is poorer.)

What symbol shall I place in the circle?

(⁻)

What symbol shall I place in this circle?

(You cannot tell.)

If we want to make our housewife richer, what would have to happen?

(The `△` bill would have to be more than the `☐` bill.)

If she ends up poorer, what would have to happen?

(The `☐` bill would have to be more than the `△` bill.)

Is it possible for our housewife to break even—be neither richer nor poorer?

(Yes, the `☐` bill and the `△` bill would have to be for the same amounts.)

What sign goes in the circle then?

(No sign is necessary here.)

What symbol shall I place in this circle?

(⁺)

Does this mean that she will be richer or poorer?

(Richer)

The story, though odd in plot, gives the children a rationale in working with signed numbers. The solution of $^-3+{}^-6={}^+9$ will just not hold water because the children can really feel that if two bills are delivered, a person must end up poorer—not richer.

Try this—I think you'll like it!

SELECTED BIBLIOGRAPHY FOR CHAPTER 6
Number Operations and the Set of Integers

Ashlock, Robert and West, Tommie. "Physical Representations for Signed-Number Operations." *The Arithmetic Teacher* (November 1967), 549.

Becker, Stanley. "Elevator Numbers." *The Arithmetic Teacher* (October 1971), 422.

Cantlon, Merle Mae, Homan, Doris and Stone, Barbara. "A Student-Constructed Game for Drill With Integers." *The Arithmetic Teacher* (November 1972), 587–89.

Cochran, Beryl. "Children Use Signed Numbers." *The Arithmetic Teacher* (November 1966), 587.

Coon, Lewis. "Number Line Multiplication for Negative Numbers." *The Arithmetic Teacher* (March 1966), 213.

Demchik, Virginia C. "Integer Football." *The Arithmetic Teacher* (October 1973), 487–88.

Frank, Charlotte. "Play Shuffleboard With Negative Numbers." *The Arithmetic Teacher* (May 1969), 395.

Havenhill, Wallace P. "Though This Be Madness. . ." *The Arithmetic Teacher* (December 1969), 606–08.

Hill, Warren H., Jr. "A Physical Model for Teaching Multiplication of Integers." *The Arithmetic Teacher* (October 1968), 525.

Hollis, Loye Y. "Multiplication of Integers." *The Arithmetic Teacher* (November 1967), 555.

Mauthe, Albert. "Climb the Ladder." *The Arithmetic Teacher* (May 1969), 354.

Mehl, William and Mehl, David. "Grisly Grids." *The Arithmetic Teacher* (May 1969), 357.

Milne, Esther. "Disguised Practice for Multiplication and Addition of Directed Numbers." *The Arithmetic Teacher* (May 1969), 397.

——. "Subtraction of Integers–Discovered Through a Game." *The Arithmetic Teacher* (February 1969), 148.

Peterson, John C. "Fourteen Different Strategies for Multiplication of Integers Or Why $(-1)(-1) = +1$." *The Arithmetic Teacher* (May 1972), 396–403.

Schultz, James E. "Why I Don't Have Any Examples of Negative Numbers." *The Arithmetic Teacher* (May 1973), 365.

CHAPTER 7

NUMBER OPERATIONS AND THE SET OF RATIONAL NUMBERS

Like negative numbers, fractions and the fundamental operations with rational numbers are also difficult for children to understand. Many educators have seriously questioned whether much work should be done with fundamental operations on the set of rational numbers in the elementary school. Some of this questioning may be well substantiated, but in reality, many pages and chapters on fractions appear in most current mathematics programs. This is probably because the expansion from the set of nonnegative integers to the nonnegative rationals requires no new or extended concepts (with the exception of place-value columns for decimal fractions). In addition to the social utility of fractions, the inclusion of the rationals adds to the learner's expanding number concepts and mathematical perspective.

The selections in this chapter have been chosen to give additional meaning, depth, and perspective to the learner's mathematical domain. Eve Omejc's article, "A Different Approach to the Sieve of Eratosthenes," is both interesting and helpful. Although prime numbers are usually discussed along with number theory, they are being presented here because primes are most often utilized in elementary school programs with the computation of rational numbers.

Ernest R. Duncan's "Using Sets to Teach Fractional Numbers, Ratios, and Rates" is exceptionally well done and offers many excellent ideas for classroom use. Briefly, Dr. Duncan indicates how a limited number of set concepts can be used in the classroom to develop basic ideas in three areas: renaming fractional numbers, probability, and rate. Be sure to take time to read and internalize this article.

Have you been searching for a game or interesting activity with fractions? If so, you will want to read both "Make A Whole—A Game Using Simple Fractions" by Joann Rode and Harry Bohan's "Paper Folding and Equivalent Fractions—Bridging A Gap." In fact, you may wish to extend the ideas presented in these articles and develop some independent learning activities of your own.

In the last article in this chapter, "Division With Fractions—Levels of Meaning," Harry C. Johnson summarizes the commonly used methods of dealing with division by a fraction. This article should indeed be a useful reference for this troublesome issue.

A different approach to the sieve of Eratosthenes

EVE OMEJC

The sieve of Eratosthenes, as a method of determining prime numbers by elimination of all multiples of primes, has been known and used by teachers for many years. The process usually presented becomes tedious as the numbers become larger. For instance, in crossing out all multiples of 29, one must count twenty-nine numbers each time or use multiplication or repeated addition to locate each succeeding multiple.

There is a different form of the sieve that not only is easier to use but, in addition to locating primes, has other possible uses. In this form, the counting numbers are listed in order across six columns (fig. 1). To locate the primes, proceed in the following manner:

1	2	3	4	5	6
7	8	9	10	11	12
13	14	15	16	17	18
19	20	21	22	23	24

Fig. 1[1]

By definition, the number 1 is not a prime; so it is crossed out.

The number 2 is a prime; so it is circled. All multiples of 2 are composite and should be crossed out. A study of the chart in figure 2 shows that the even numbers, which are multiples of 2, are in the second, fourth,

and sixth columns. With a straightedge, draw lines through these columns (excluding the circled 2 in the second).

The next prime is 3; so it is circled. All the other numbers in the third column are multiples of 3; so the third column can be ruled out.

This leaves only the first and fifth columns in which to find the rest of the primes.

The next prime is 5; so it is circled. Locate the multiples of 5. The multiples 10, 15, 20, and 25 are so placed that a slant line through 5 and 25 will cross out 10, 15, and 20. (See figure 3.) Similarly, a slant line through 30 and 55 will eliminate the multiples 35, 40, 45, and 50. Continue this process to the bottom of the page. Observe that the slant lines that have been drawn through the successive multiples of 5 are parallel to one another.

Continue in the same manner for the other primes in order. Circle 7 and locate its multiples. They are on a slant to the right as far as 42. The next line of multiples of 7 begins with 49 and goes through 84. Again, continue the process of crossing out the multiples of 7 to the end of figure 3. And again, observe that the lines are parallel. However, the slant of the lines for the multiples of 7 is greater than the slant of the lines through the multiples of 5.

The next number not crossed out is 11; so it must be prime. Circle 11 and eliminate its multiples, which are on a slant line to the left through 55. The next line of mul-

---

1. This chart appears in the teacher's edition of *Seeing through Arithmetic, 6,* by Maurice L. Hartung, Henry Van Engen, and Lois Knowles.

Reprinted from *The Arithmetic Teacher* (March 1972), 192–196, by permission of the author and the publisher.

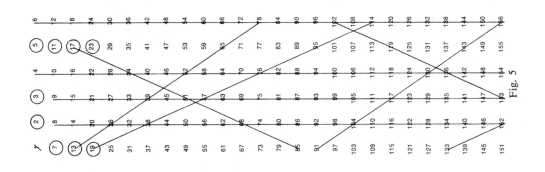

Fig. 5

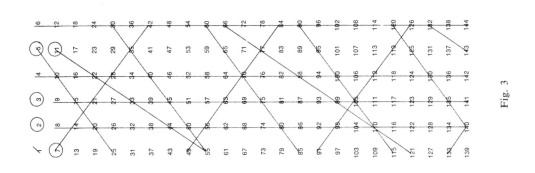

Fig. 4

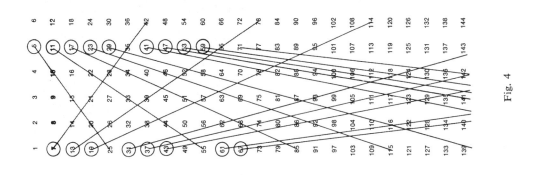

Fig. 3

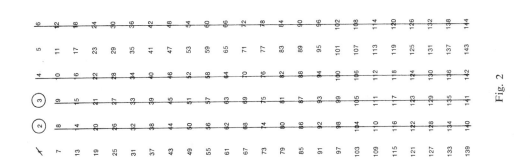

Fig. 2

129

tiples of 11 begins with 66 and goes through 121. Again, these lines are parallel but have a greater slant than those through the multiples of 7.

(As the process is continued through the larger primes, the slant of the lines approaches the perpendicular. (See figure 4.) This suggests a good question for class discussion: If we go far enough, will there be a prime large enough that the line will be perpendicular? And there is a related question: If the line of multiples were perpendicular, what would this imply? Some prime number would have to have all its multiples in the same column—that is, either all in the first column or all in the fifth column. But both these columns contain only odd numbers. However, since any number can be multiplied by an even number, some of the multiples of any prime number would have to be even numbers and therefore in even columns. In other words, the multiples of a prime could not all be in the first or fifth column.)

To find both ends of each succeeding line of multiples for a given prime after the first line has been established, count down in the first column and in the sixth column the number of rows that corresponds to that particular prime number. For example, the first line of multiples of 13 begins at 13 and ends at 78. The next line will begin 13 rows below the 13, namely, at 91, and will end 13 rows below 78, at 156. (See figure 5.) This method becomes rather time-consuming as the primes increase. There is a faster method using formulas that can be determined by writing tables like those shown in figure 6 on the chalkboard.

From the tables, students can discover that to locate each succeeding line of multiples for a given prime you add $6p$ to the first and last numbers of the line just completed. Classes that have worked with number patterns should find it challenging to derive the following formulas for locating any given line of prime multiples:

If the prime is in the fifth column,

Let p = the prime number
n = the number of the line of multiples
k_1 = the first number on the line
k_2 = the last number on the line

Example: $p = 5$

n	k_1	k_2
1	$5 = 5 \times 1$	$25 = 5 \times 5$
2	$30 = 5 \times 6$	$55 = 5 \times 11$
3	$60 = 5 \times 12$	$85 = 5 \times 17$
4	$90 = 5 \times 18$	$115 = 5 \times 23$

Example: $p = 7$

1	$7 = 7 \times 1$	$42 = 7 \times 6$
2	$49 = 7 \times 7$	$84 = 7 \times 12$
3	$91 = 7 \times 13$	$126 = 7 \times 18$
4	$133 = 7 \times 19$	$168 = 7 \times 24$

Example: $p = 11$

1	$11 = 11 \times 1$	$55 = 11 \times 5$
2	$66 = 11 \times 6$	$121 = 11 \times 11$
3	$132 = 11 \times 12$	$187 = 11 \times 17$
4	$198 = 11 \times 18$	$253 = 11 \times 23$

Fig. 6

$$k_1 = p[6(n-1)]$$
$$k_2 = p(6n-1)$$

If the prime is in the first column,

$$k_1 = p[6(n-1)+1]$$
$$k_2 = p(6n)$$

If students make their own charts for this exercise, it is advisable that they use graph paper, for the numerals in the chart must be evenly spaced both vertically and horizontally. To make it easier to see the different patterns, the lines can be drawn with felt pens or colored pencils.

In the same six-column array of counting numbers, patterns in the lines through the multiples of composite numbers can also be explored. For example, the lines of the multiples of 4 go from right to left. Those for the multiples of 8 go from left to right but have the same angle of slant as the lines for the multiples of 4. The lines through the multiples of 9 zigzag back and forth.

What patterns can your class find?

130

Using Sets to Teach Fractional Numbers, Ratios, and Rates

ERNEST R. DUNCAN

USING sets has great advantages from a teaching as well as a mathematical point of view. It enables the teacher to give pupils appropriate experiences with objects and then translate those experiences into general terms which can be used in many different situations. It also provides a method of explaining and illustrating concepts in their simplest form.

Sets are basic to the development of many mathematical skills and concepts. They should be used, for example, to teach fractional numbers, ratios, and rates. In preparation for these ideas, pupils need to be familiar with the following:

1. A *set* is a *collection of objects*. We may identify a set by *listing* or *describing* its elements. We usually enclose the list or description in braces, as shown here. Set B = {Sunday, Monday, Tuesday, Wednesday, Thursday, Friday, Saturday} = {the names of the days of the week}.

2. A *subset* is *part of a set*. Set M is a subset of set B if it contains only elements which are in B. For example, M = {Monday, Friday} is a subset of B above. A subset may contain no elements, in which case it is the empty set; it may contain all the elements of the set, in which case it is equal to the set.

3. The *cardinal number of a set* tells how many elements there are in the set. For example, the cardinal number of B above is 7. We may write this as an equation: $n(B) = 7$. Two sets with the same cardinal number are *equivalent*. For example, B above is equivalent to Q = {a, b, c, d, e, f, g}.

4. Two sets may be joined to form a new set. This set is called the *union* of the sets and contains all the elements in the sets. If a set of 3 is joined with a separate set of 2, the cardinal number of the union is 5. We see that addition is the number operation corresponding to the joining of sets and we may use the cardinal numbers of the sets to write the equation: $3 + 2 = 5$.

5. The elements of two separate sets may be matched. The matching may be of two kinds. Firstly, it may be one-to-one; that is, the elements may be matched in such a way that for every element in one set there is just one element in the other. Consider these two examples:

Figure 1   *Figure 2*

Reprinted from *Timely Topics* (January/February 1967) by permission of Houghton Mifflin Company.

131

Figure 1 shows that for every circle in set A there is a triangle in set B. Figure 2 illustrates an everyday example; for every child there is a chair. Sets whose elements can be matched one-to-one are equivalent.

Secondly, matching may be many-to-one. In this instance, the elements in the larger set are grouped to form equivalent subsets and each of these subsets is matched with an element in the other set. Consider these two examples:

Figure 3   *Figure 4*

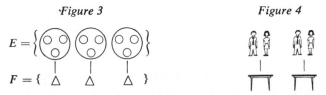

Figure 3 shows that for every subset of 3 circles in E there is a triangle in F. The circles and triangles are matched three-to-one. Figure 4 illustrates an everyday example; for every two children there is a table. This is a two-to-one matching.

Of course, there are many other set concepts the pupils should learn. But the five listed here are all that are needed to develop the basic ideas of fractional numbers, ratios, and rates.

FRACTIONAL NUMBERS

A fractional number may be used to compare the cardinal numbers of two sets, one of which is a subset of the other. Consider these two examples:

Figure 5   *Figure 6*

In Figure 5, 3 of the 5 shapes are triangles. We may use the fractional number $\frac{3}{5}$ and say that $\frac{3}{5}$ of the shapes are triangles. If the set of triangles is T and the set of shapes is S, T is a subset of S. The cardinal number of S is 5, of T is 3. We may write the equation:

$$\frac{n\,(T)}{n\,(S)} = \frac{3}{5}$$

In Figure 6, the subsets are equivalent and the objects in 3 of the 5 subsets are triangles. Again we may use a fractional number and say that $\frac{3}{5}$ of the objects are triangles.

The two numbers that are used to form a fractional number are called the numerator and denominator. The denominator tells how many elements (or equivalent subsets) there are in the set with which we are making our comparison, and names them fourths of the set, fifths of the set, sixths of the set, and so on. The numerator tells how many elements (or equivalent subsets) are being compared with the total set. In our examples above, there are 5 elements (or equivalent subsets) and each of them is one-fifth of the set; the denominator is 5. We are comparing 3 elements (or equivalent subsets) with the set; the numerator is 3.

Every number has many names, and fractional numbers are no exception to this principle. The name (symbol) for a fractional number is called a fraction, and every fractional number has many fractions. For example, $\frac{1}{2}$, $\frac{2}{4}$, $\frac{5}{10}$, .5, 50% are all names (fractions) for the same fractional number.

If two fractions name the same fractional number, they are equivalent. We may teach the concept of equivalent fractions by means of sets, as shown here.

Figure 7   *Figure 8*

Four of the six shapes shown in Figure 7 are squares. That is, $\frac{4}{6}$ of the shapes are squares. In Figure 8, we see the same set with the shapes grouped to form 3 equivalent subsets, two of which contain squares. Figure 8 shows that $\frac{2}{3}$ of the subsets contain squares. But Figures 7 and 8 refer to the same set and so $\frac{4}{6}$ and $\frac{2}{3}$ state the same comparison. They are equivalent fractions, $\frac{4}{6} = \frac{2}{3}$.

If we look again at Figures 7 and 8, we see how grouping objects in sets may be used to teach students how to rename fractional numbers. In Figure 7, we compare the number of squares with the cardinal number of the total set and write $\frac{4}{6}$. In Figure 8, we group the shapes into sets of two. We compare the subsets containing squares with all the subsets and write $\frac{2}{3}$. In this way, we see that $\frac{4}{6}$ and $\frac{2}{3}$ are names for the same fractional number, but $\frac{2}{3}$ is a simpler name because its denominator is less.

If the pupil works with other examples like this, he should develop the generalization that we simplify a fraction by dividing the numerator and the denominator by the same number. Later, he should discover that dividing the numerator and the denominator by the same number is the same as dividing the fractional number by 1. For example,

$$\frac{4 \div 2}{6 \div 2} = \frac{4}{6} \div \frac{2}{2} = \frac{4}{6} \div 1.$$

To simplify $\frac{4}{6}$ we divide by 1 in the form of $\frac{2}{2}$.

$$\frac{4}{6} \div \frac{2}{2} = \frac{4 \div 2}{6 \div 2} = \frac{2}{3}.$$

We use fractional numbers to compare sets in many everyday examples, for instance, in problems like the following: If we have 12 flowers and give one-third of them away, how many do we give away? In this instance, we know that the initial set contains 12 objects and is separated into 3 equivalent subsets. We also know that the number operation corresponding to the set operation of separation is division. So the number of flowers we give away is $12 \div 3 = 4$.

In this simple problem, there would appear to be little need to introduce the words *set* and *subset* but the terminology is valuable even here. It enables the pupil to use a generalized vocabulary so that he need not think in terms of special referents, flowers in our example, but only in terms of objects. Consequently, he can use the same thinking when he meets the problem: If you have 12 cents and want to spent a third of your money, how much would you spend? Again, he has a set of 12 objects which is separated to form 3 equivalent subsets. His answer again is $12 \div 3 = 4$, but this time as cents.

If we use set concepts with fractional numbers, we may introduce topics which have not ordinarily been introduced in elementary classrooms. For example, we may readily teach some basic concepts in probability.

Consider the following example: Suppose you had 8 marbles in a bag — 3 black, 3 red, and 2 white. If a marble drops out of the bag, what is the chance (probability) that it is black? In this instance we need to compare the number of black marbles with the cardinal number of the total set. The probability that the marble that falls out is black is 3 out of 8, that is, $\frac{3}{8}$. We may write this as an equation in which P represents the probability of an event:

$$P \text{ (black marble falling out)} = \frac{\text{n (subset of black marbles)}}{\text{n (set of marbles)}} = \frac{3}{8}.$$

Consider the marbles situation again. What would be the probability that the marble that falls out of the bag is black *or* red? In this case the subset is increased. It is, in fact, the union of 2 subsets — the black marbles and the red marbles. The probability of having a black marble fall out is $\frac{3}{8}$, the probability of its being a red marble is also $\frac{3}{8}$, and the probability of its being black or red is $\frac{3}{8} + \frac{3}{8} = \frac{6}{8}$.

The second marble problem indicates how we may use the joining and separating of subsets to teach adding and subtracting fractional numbers. Consider these two examples:

Figure 9   *Figure 10*

$C = \{ \triangle \quad \triangle \quad \square \quad \square \quad \bigcirc \}$ $D = \{ \blacksquare \quad \blacksquare \quad \square \quad \bigcirc \quad \bigcirc \}$

In Figure 9, $\frac{2}{5}$ of the shapes in set C are triangles, $\frac{2}{5}$ are squares, and $\frac{2}{5} + \frac{2}{5} = \frac{4}{5}$ of the shapes are triangles *or* squares. In Figure 10, $\frac{3}{5}$ of the shapes in set D are squares and $\frac{2}{5}$ of the shapes are shaded squares, so $\frac{3}{5} - \frac{2}{5} = \frac{1}{5}$ of the shapes are unshaded squares.

RATIOS

The idea of using a fractional number to compare a subset and a set may be extended to the comparison of two separate sets of similar objects. A fractional number used in this way is called a *ratio*. Consider these two examples:

Figure 11   *Figure 12*

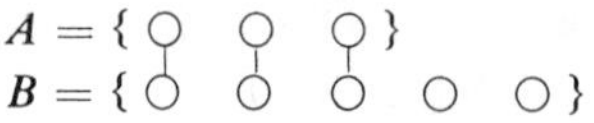

$A = \{ \bigcirc \quad \bigcirc \quad \bigcirc \}$ $P = \{ \bullet \bullet \bullet \}$
$B = \{ \bigcirc \quad \bigcirc \quad \bigcirc \quad \bigcirc \quad \bigcirc \}$ $Q = \{ \bullet \bullet \bullet \bullet \bullet \}$

The circles in A may be matched with the circles in a subset of B. If we then think of A as replacing the matched subset of B, we can now compare A with B in the same way that we previously compared a subset with a set, by means of a fractional number. We say that the number of circles in A is $\frac{3}{5}$ of the number of circles in B or that the circles in A and B are in the *ratio* of 3 to 5.

134

Figure 12 illustrates an everyday example of a ratio. If set P represents the records Betty has and set Q represents the records Sue has, then we may say that Betty has $\frac{3}{5}$ as many records as Sue or that their records are in the ratio of 3 to 5.

The interpretation of any ratio depends upon a set comparison and if the sets are changed in any way, this will be reflected in the ratio. For example, if both Betty and Sue bought two more records, the ratio of Betty's records to Sue's records would be named by reference to the new sets, a set of 5 and a set of 7, and the new ratio would be $\frac{5}{7}$.

RATES

To develop a ratio, the elements of two separate sets are matched one-to-one. The elements of two separate sets may also be matched many-to-one. Consider these two examples:

Figure 13 *Figure 14*

In Figure 13, the circles in set G are grouped to form 4 equivalent subsets, each containing 3 circles, and each of these subsets is matched with one of the stars in set H. We see that for every 3 circles in G there is 1 star in H. The elements in the sets have been matched three-to-one, that is at the *rate* of 3 circles *per* star.

Figure 14 illustrates an everyday example of a rate. The pennies have been matched with the candy bars at the rate of 5 pennies per candy bar.

This kind of matching, called a rate, is very important mathematically and in everyday situations. With very few exceptions, practical problems which require multiplication or division for their solution involve the concept of a rate, as in these examples: What is the cost of 7 ties if one tie costs $1.25? What is the cost of 1 tie if 7 ties cost $8.75? How far do I travel in 5 hours at 45 miles an hour?

Very briefly we have seen how a limited number of set concepts can be used to develop basic ideas in three limited areas: renaming fractional numbers, probability, and rate. The set concepts which apply in one area also apply in each of the others.

Make a whole—a game using simple fractions

JOANN RODE

The game called "Make a Whole" helps develop the concept of fractional numbers by using concrete examples; it gives reinforcing experience with the use of fractions and their equivalents. It is suggested for use at third-grade level and above, depending on the achievement of the students.

The material required for the game is pictured in figure 1 and described more completely in a later section. The sections are made in four different colors, equally distributed; that is, of 8 sections representing $\frac{1}{2}$, 2 could be red, 2 blue, 2 green, and 2 yellow. The area for the "Whole" is also prepared in four different colors. The faces of the die are marked with a star and the following fractions: $\frac{1}{2}$, $\frac{1}{3}$, $\frac{1}{4}$, $\frac{1}{6}$, and $\frac{1}{8}$.

Rules of play

The object of the game is to use the fractional sections to construct a multicolored disk, the "Whole," in the black mat frame. The Whole may be constructed with any of the differently colored sections so long as they fit together to be equivalent to a whole.

Two to eight children may play the game in individual competition (in which case the color of the area for the Whole makes no difference), or they may play on color teams as determined by the color of that area.

The fractional sections are placed in a box, from which they are to be selected by the children as they roll the die, in clockwise order.

To determine who goes first, each child

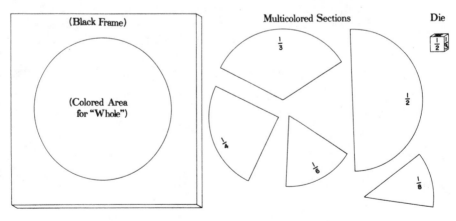

Fig. 1

Reprinted from *The Arithmetic Teacher* (February 1971), 116–118, by permission of the author and the publisher.

rolls the die seen in figure 1. The child rolling the star or the fraction with the greatest value goes first.

At each turn the child must decide whether he can use one of the sections represented by the fraction showing on top of the die.

If the player cannot use the section, he passes.

If he decides to take the section and can use it, he places the section in his mat. (If the section taken completes the Whole for that player, then the game is over and that player has won. If it does not, the roll of the die passes to the next player.)

If he decides to take the section and cannot use it, he loses his next turn.

If he passes up a section he could have used, he loses his next turn.

If the star is rolled, the player may choose any section he can use.

The winner is the first to construct a multicolored Whole.

The game may be played in various other ways, according to rules designed by the teacher. For example, a child might be asked to construct five Wholes with no more than one of a kind of fractional section ($\frac{1}{2}$, $\frac{1}{3}$, etc.) to go into each. For advanced students, the numeral identification may be omitted from the sections.

The materials may also be used by the teacher to show equivalent fractions in classroom demonstrations and by the individual student in working with equivalences at his desk.

Construction of game materials

A complete list of pieces follows, the color of the mat disks and sections being equally divided among whatever four colors are used.

Mats, 8
$\frac{1}{2}$ sections, 8
$\frac{1}{3}$ sections, 12
$\frac{1}{4}$ sections, 16
$\frac{1}{6}$ sections, 24
$\frac{1}{8}$ sections, 32
Die, 1

The material used for the black mat frame and the fractional sections is inexpensive tagboard (poster board) in different colors. Because the Wholes constructed by the students are to be multicolored and because the fractional sections come in equal numbers of the different colors, the colors give no clues to value.

The mats were made first by cutting 8 eight-inch squares from black tagboard. Disks six inches in diameter were cut from the center of these squares, and the frames were then glued to other eight-inch squares, 2 of each of the four colors chosen.

Five disks in each of the four colors were then constructed, with a diameter of six inches. A compass was used for accuracy. The disks were then divided, with the aid of a protractor, into sections corresponding to $\frac{1}{2}$, $\frac{1}{3}$, $\frac{1}{4}$, $\frac{1}{6}$, and $\frac{1}{8}$. The central-angle measurements for the five sections are $180°$, $120°$, $90°$, $60°$, and $45°$, respectively. Depending on the level of the children, the sections may or may not be marked with the fractions to which they correspond, as seen in figure 1.

The die was made from a cube cut at a lumber yard so that each edge had a measurement of three-fourths of an inch. The five fractions were painted, one to a face on each of five faces of the cube. A star was placed on the sixth face. The cube was sprayed with a clear varnish to prevent smudging.

EDITOR'S NOTE. Here is a device that can be used for independent learning activity. I dare say the game could be varied by the construction of rectangular sections and mats, so that children would view fractional parts in a broader spectrum. CHARLOTTE W. JUNGE.

Paper folding and equivalent fractions—bridging a gap

HARRY BOHAN

Much emphasis has long been placed on the value of using concrete models and manipulative devices in the early stages of the development of mathematical ideas. As Johnson suggests, however, it is entirely possible for a teacher to use instructional aids and get little benefit from them. "Some use them but fail to take students from the concrete representation to the concept behind it. Others use an inadequate or inappropriate aid" (Johnson 1967, p. 20). Brownell emphasizes that the processes children use at the concrete and semiconcrete levels must be closely related to the processes they will use later in dealing with abstract symbols (Brownell 1928, p. 24).

In selecting a model, one must be sure that the choice will provide transfer of training from the concrete model to the abstract concept it was chosen to support. This transfer can be facilitated by having students manipulate the concrete model in a manner as much like the later manipulation of concepts and symbols as possible. Van Engen and Gibb suggest that "good instruction produces good fits between the physical structured elements and the abstract principles which the child is to draw from the experience" (Van Engen and Gibb 1960, p. 59).

An investigation of the approach used by most modern textbooks in leading students to the generalization

$$\frac{a}{b} = \frac{a \times n}{b \times n}$$

will reveal a situation where there does not seem to be such a fit. The usual approach begins with the use of diagrams. Several appropriately shaded congruent rectangular regions are pictured. (For an example, see fig. 1.)

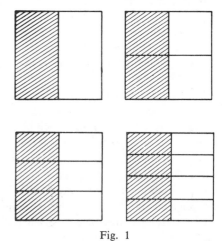

Fig. 1

Then the child is told, or is led to discover—or, it is hoped, discovers for himself—that although the fractions being pictured are different, each tells him that the same portion of the region is shaded. Fractions that do such things as this are identified as equivalent fractions.

The generalization

$$\frac{a}{b} = \frac{a \times n}{b \times n}$$

is then developed. In this procedure some lowest-terms fraction has its numerator and denominator successively doubled, tripled,

Reprinted from *The Arithmetic Teacher* (April 1971), 245–249, by permission of the author and the publisher.

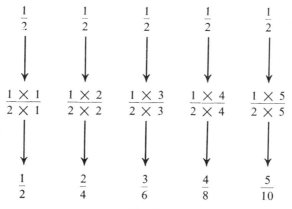

$$\frac{1}{2} \qquad \frac{1}{2} \qquad \frac{1}{2} \qquad \frac{1}{2} \qquad \frac{1}{2}$$

$$\frac{1 \times 1}{2 \times 1} \qquad \frac{1 \times 2}{2 \times 2} \qquad \frac{1 \times 3}{2 \times 3} \qquad \frac{1 \times 4}{2 \times 4} \qquad \frac{1 \times 5}{2 \times 5}$$

$$\frac{1}{2} \qquad \frac{2}{4} \qquad \frac{3}{6} \qquad \frac{4}{8} \qquad \frac{5}{10}$$

Fig. 2

and quadrupled, as in figure 2. Textbooks vary in the amount of this computation that is to be done by students. Then the student is told, asked, or led to compare—or, it is hoped, on his own will compare—the fractions in this set with those previously shown, by using diagrams, to be equivalent. Thus he learns, or is led to learn, or discovers, that one can find a fraction equivalent to a given fraction by multiplying its numerator and its denominator by any natural number.

If you have ever tried such an approach to the generalization

$$\frac{a}{b} = \frac{a \times n}{b \times n},$$

you have probably been struck by a feeling of shallowness when the time came to begin multiplying the numerator and the denominator by the same number. This act is usually accomplished without explanation at all, or as "something interesting at which to look," or as an activity that "produces an interesting pattern."

It was the conjecture of the author that in such an approach there is a huge gap between the conceptual work previously done with equivalent fractions and the generalization

$$\frac{a}{b} = \frac{a \times n}{b \times n}.$$

There is not a good "fit" between the concrete models and the generalization they are designed to support.

A recent research study by the author lends support to the position that paper-folding activities can be used to bridge this gap and help build a more solid learning sequence for equivalent fractions and for the generalization

$$\frac{a}{b} = \frac{a \times n}{b \times n}.$$

The idea was to introduce the concept of equivalent fractions with paper-folding activities that would lead directly to the generalization

$$\frac{a}{b} = \frac{a \times n}{b \times n}.$$

For example, a rectangular region, half of which was shaded, was provided for each child. The following questions were asked:

"What fraction is pictured by this diagram?" ($\frac{1}{2}$)

"How many parts in all?" (2)

"How many shaded parts?" (1)

The students were then directed to fold their paper into two parts as indicated by the broken line through the region shown in figure 3. The child was then asked to respond to the following questions before unfolding his paper.

Fig. 3

139

"What fraction will be pictured when the paper is unfolded?" ($\frac{2}{4}$)

"How many parts in all?" (4)

"How many shaded parts?" (2)

"Will more of the paper be shaded than was shaded before the folding?" (No)

"What could be said about the fractions $\frac{1}{2}$ and $\frac{2}{4}$?"

The students were then directed to unfold their papers and verify their decisions.

"How many parts do you see in all?" (4)

"How many parts did we have in all before the folding?" (2)

"How could you explain what happened to the total number of parts?" (Doubled)

"How many shaded parts now?" (2)

"How many before the folding?" (1)

"What happened to the number of shaded parts?" (Doubled)

Another paper was then folded into three parts and the same series of questions asked. Folding the paper into three parts (fig. 4) had the effect of tripling the total number of parts and tripling the number of shaded parts.

Fig. 4

Still another paper was folded into four parts and the same series of questions asked. Folding the paper into four parts had the effect of what one child, and soon everyone in one class, called "fourpling" both the total number of parts and the number of shaded parts. The coinage of the word "fourple" led to many new words like "fiveple," "sixple," and even "nineteenple." These words not only were fun for the students but also provided a simplified vocabulary for the author. The same procedure would then be followed with several other fractions.

It is hoped that one can see in these activities manipulation of a model (papers) very similar to the later manipulation of numbers; that is, doubling, tripling, and "sevenpling" both the numerator (shaded parts) and the denominator (total parts) of a fraction.

When the time came to make the move to the generalization

$$\frac{a}{b} = \frac{a \times n}{b \times n},$$

it was a relatively simple matter to get the students to generalize from the earlier activities, possibly by asking the following questions.

"Suppose we were asked to find fifty fractions equivalent to two-thirds. Could we find them by folding papers?" (Yes)

"What would you have to do?"

"Wouldn't this take a great deal of time?"

"Can anyone think of a faster way of finding these fifty fractions?"

Since knowledge of this generalization is in the "need to know" rather than a "nice to know" category of objectives, it is essential that individual children get an opportunity to work the generalization out for themselves. An excellent teacher provides for this very nicely by allowing children, individually, to whisper to her their response to a question. Thus she reacts to the correct responses of individuals without cheating other class members of making the discovery for themselves.

There are those who might argue for an alternative approach to the generalization

$$\frac{a}{b} = \frac{a \times n}{b \times n}$$

based on the "property of one"; in other words,

$$\frac{a}{b} = \frac{a}{b} \times 1 = \frac{a}{b} \times \frac{n}{n} = \frac{a \times n}{b \times n}.$$

Here is a specific example:

$$\frac{2}{3} = \frac{2}{3} \times 1 = \frac{2}{3} \times \frac{2}{2} = \frac{4}{6}.$$

Hence, $\frac{2}{3} = \frac{4}{6}$ because $\frac{2}{3}$ was multiplied by the number 1. If you use a textbook that follows this approach, it might prove interesting to ascertain the point in the book where multiplication of fractional numbers is introduced. Don't be too surprised to find it is *after* the introduction of equivalent fractions. Thus the "property of one" approach to the generalization

$$\frac{a}{b} = \frac{a \times n}{b \times n}$$

often uses **multiplication** of fractional numbers *before* **it has been** taught.

The research study by the author mentioned earlier compared these three approaches to the generalization

$$\frac{a}{b} = \frac{a \times n}{b \times n}.$$

For the "property of one" approach, the order of topics was altered to begin with multiplication so that it could be used to get at equivalent fractions.

In an attempt to make all treatments alike except for the variable being tested, all units but the equivalent-fractions unit were identical for all three approaches. Thus the "usual" and "paper folding" treatments were identical except for six lessons related to equivalent fractions. The "property of one" treatment differed from the other two treatments only in the six lessons related to equivalent fractions and the *order* in which the units were arranged. All six classes participating in the study were taught by the author. Because of initial differences in the treatment groups, analysis of covariance was used with arithmetic achievement as covariate. Figure 5 is a schematic diagram of the treatments.

Findings

No significant differences were found between adjusted means on either immediate posttests or retention tests on addition, subtraction, or multiplication of frac-

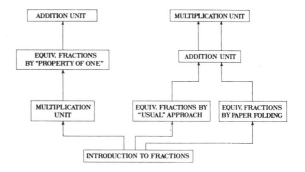

Fig. 5

tional numbers. On an immediate posttest of concepts and skills related to equivalent fractions, both the "usual" and "paper folding" approaches showed a significant superiority over the "property of one" approach. However, on a retention test given three weeks after completion of the study the results were somewhat different. The "paper folding" approach was then superior to *both* the "usual" and "property of one" approaches, and the "usual" approach was no longer superior to the "property of one" approach.

In summary, the research evidence supports strongly the idea that paper folding as described in this article helps promote better attainment of objectives related to equivalent fractions.

References

Bohan, Harry. "A Study of the Effectiveness of Three Learning Sequences for Equivalent Fractions." Ph.D. dissertation, University of Michigan, 1970.

Brownell, William A. *The Development of Children's Number Ideas in the Primary Grades.* Supplementary Educational Monograph, no. 35. Chicago: The University of Chicago Press, 1928.

Johnson, Donovan A. "Instructional Materials in the Mathematics Classroom." *NEA Journal* 56 (May 1967): 39–40.

Van Engen, Henry, and E. Glenadine Gibb. "Structuring Arithmetic." *In Instruction in Arithmetic.* Twenty-fifth Yearbook of the National Council of Teachers of Mathematics. Washington, D.C.: The Council, 1960.

Division with fractions—
levels of meaning

HARRY C. JOHNSON

Over and over during the years writers
in the field of mathematics teaching have
expressed concern about the difficulty of
evolving a meaningful method of teaching
the division of a whole number by a fraction or a fraction by a fraction. There are
those who point out the low utilitarian
value of the operation and argue for its
elimination from the curriculum. Others
have attempted to analyze the inherent nature of the operation, relating it
structurally to its inverse operation, multiplication, and suggesting what they believe to be a meaningful approach to the
traditional algorism. Some authorities continue to present approaches which reduce
the fractions to those with like denominators, compare numerators or express the
two fractions as one complex fraction,
and then multiply each fraction by the
reciprocal of the denominator. In some
treatments the attempt is made to develop the answer intuitively by a subtractive process, usually through the use of
some type of geometric approach of a
linear or circular nature, or the subtraction of fractional parts from larger areas.
After the reasonableness of the answer is
recognized, the traditional algorism is developed somewhat deductively, or presented as a rule. One may even find those
who see the "inversion" process as a trick
or a "gimmick" devoid of mathematical
logic, and would rely only on meaningful
presentations of the algorism.

In his review of the literature relating
to the operation, Krich[1] points out that

"there does not appear to be general agreement as to what method of teaching is
truly meaningful." He goes on to say that
the

consensus of the articles written on teaching
division of fractions was that since the division
of a fraction by a fraction process is rarely used
in real life situations, the time needed for teaching the reasons why the act of "inverting the
divisor and multiplying" works is a waste.

The general recommendation from the
review of the literature was that one might
just as well state the rule and ignore the
explanation, then provide practice and
drill to insure at least temporary mastery.

Schools all over the country are in the
process of examining their present curricula in elementary mathematics, preparing new courses of study, and in many
cases analyzing up-to-date programs or
textbook series with the thought of making systemwide adoptions. That the newly
adopted programs will largely determine
teaching procedures as well as content is
almost a certainty. It is quite unlikely
that the typical teacher will deviate from
a type of development found in the book,
even though he himself may prefer a different approach.

The author of this article suggests very
strongly that committees of teachers del-
egated to make recommendations to their
superintendent or faculty look carefully
and critically not only at the format of
the materials and the underlying philosophy of the authors, stated and implied,
but also at the introduction and development of major concepts or understandings. The final choice of a new textbook
series or a modern program with a new

[1] Percy Krich, *An Experiment to Compare Two Methods of
Teaching Division of Fractions* (unpublished Ed.D. dissertation, University of California at Los Angeles, 1962), p. 15.

Reprinted from *The Arithmetic Teacher* (May 1965), 362–368, by permission of the author and the publisher.

format should certainly depend on the approval of teachers who must use the materials.

This article will deal particularly with the division of a whole number by a fraction or a fraction by another fraction. The article has two purposes: first, to point up in some detail the typical approaches to the operation, and second, to provide an examining ground for textbook committees as they consider this particular operation.

These various approaches will be presented in three groups, according to the kind and degree of understanding that they seem to generate: the "drill psychology" approach, the intuitive approach, and the mathematically meaningful or structurally logical approach.

Authoritative identification— memorization of the rule

This is the traditional "drill psychology" approach, instructing the pupil to "invert the divisor" and proceed as in multiplication. The procedure is mechanical, simple in nature, and can be learned and reinforced in a matter of moments. There is no reason given for the method, nor is any attempt made to "rationalize" the answer. Perhaps this is why children sometimes "invert" the wrong fraction before they multiply.

This method has no defense or justification today. Arithmetic is not just finding answers to exercises or problems; it is a language of quantity and comparison which provides a meeting ground for the stimulation and development of critical thinking.

But even when this method is used, there is an erroneous suggestion about what really occurs when one "inverts" the divisor. The reader himself needs no illustration to see what happens when a fraction is turned "upside-down." What one should really say is that you interchange the positions of the numerator and denominator of the second fraction, and then multiply the result by the first frac-

tion. Since the "inversion" language is shorter, less involved, and more readily interpreted into practice, those using this method are willing to sacrifice accuracy of statement for brevity.

A modification of this method involves the previous development of the term "reciprocal." The pupil is then told to multiply the first fraction by the reciprocal of the second. This slight deviation from tradition, if taught completely in the vacuum of connectionism, has no more justification than the one it is designed to replace.

Intuitive discovery of the answer, followed by application of the rule

This procedure is based upon one version of the assumption that "the end justifies the means." If the result of an operation can be "discovered" intuitively or by some nonalgorithmic method, and if the result makes sense, then a rule which yields the same result or solution would be comfortable and acceptable to the learner. He need not fully understand the rule or why it works. All he knows is that the rule does work, and yields a sensible answer. To be sure, this is mechanical learning, but at a somewhat higher level than the purely rote. The learner may "raise his intellectual eyebrows," but it is hoped he will "buy the rule" and be content to apply it.

Around this point learning theorists and curriculum designers might well draw up battle lines. On the one hand it can be argued that certain concepts are too difficult to understand, or that the steps toward understanding are so intricate or frustrating in their self-evidence it is better to learn a rule without bothering to understand it. Others would say that nothing can truly function effectively in life unless it is understood, and unless what is taught can be taught with meaning and understanding, it should not be taught at all. Then, of course, some people would say that a concept, if not within the comprehension of a child at a given level

of maturation, could certainly not be useful to him at that level, and should be
postponed to a later year when his mental
powers and breadth of experience have
matured.

The method of teaching division with
fractions described in this section seems
to strike a middle ground between these
battle lines. It seems to rest on the
premise that, if an operation fits logically
and sequentially into the total curriculum
at a certain point but is too difficult to understand, we should do the best we can
under the circumstances, and try to make
the operation sensible to the learner, if
not mathematically meaningful.

Basically, this approach looks at division as a process of repeated subtraction,
and builds up to the idea by relating it to
operations with whole numbers, such as
"how many twos in eight?"

Here are some examples of intuitively
developed experiences which can lead inductively to the generalization embodied
in the conventional rule.

Using comparisons involving money
How many quarters in one dollar?

$$1 \div \tfrac{1}{4} = 4$$

How many halves in 3 dollars?

$$3 \div \tfrac{1}{2} = 6$$

How many halves in $4\tfrac{1}{2}$ dollars?

$$4\tfrac{1}{2} \div \tfrac{1}{2} = 9$$

Examples like the above can fairly
readily be reasoned out informally by
noting, for example, in the last illustration
that, since each dollar is equivalent to two
half dollars, four dollars would be equivalent to eight half dollars. In four and one-
half dollars, then, there would be nine half
dollars.

Using linear measures
Mark off a definite, workable length on
the chalk tray or floor, or on a number
line, and then "count out" the number of

times a fractional distance is contained in
this length.[2]

How many "thirds of a yard" are contained in four yards?

$$4 \div \tfrac{1}{3} = 12$$

How many "two-thirds of a yard" are
contained in six yards?

$$6 \div \tfrac{2}{3} = 9$$

Using circular measures for divisors
How many times will a wheel with a
circumference of one-fourth of a yard turn
rolling a distance of two yards?

$$2 \div \tfrac{1}{4} = 8$$

How many turns will a wheel three-
quarters of a foot in circumference make
rolling a distance of one foot?

$$1 \div \tfrac{3}{4} = 1\tfrac{1}{3}$$

In solving the latter example experimentally the pupil would note that the
wheel would make one complete revolution with a part left over equal to one-
third of the distance covered by one turn
of the wheel. Therefore, a wheel three-
quarters of a foot in circumference would
make one and one-third complete turns
over a distance of one foot.

It is argued by the proponents of circular measure[3] that the pupil is less
likely to answer $1\tfrac{1}{4}$, as he might do in
comparing linear measures.

*Interpreting the fractions
as geometric surfaces*
For sake of brevity other illustrations
will not be shown,[4] but it is understood

[2] See, for example, Edwina Deans, *et al.*, *Unifying Mathematics, Book Six* (New York: American Book Company,
1963), pp. 213–214; E. T. McSwain, *et al.*, *Arithmetic 6* (River
Forest, Ill.: Laidlaw Brothers, 1963), p. 93; School Mathematics Study Group, *A Brief Course in Mathematics for Elementary School Teachers*, Studies in Mathematics, Vol. IX
(Rev. ed.; New Haven, Conn.: Yale University Press, 1963),
pp. 276–278.
[3] See, for example, George H. McMeen, "Division by a
fraction—A new method," THE ARITHMETIC TEACHER, IX
(March, 1962), 122–126.
[4] See, for example, Theodore Kolesnik, "Illustrating the
multiplication and division of common fractions," THE ARITH-
METIC TEACHER, X (May, 1963), 268–271, and Robert Lee
Morton, *et al.*, *Modern Arithmetic Through Discovery, Grade
Six* (Morristown, N.J.: Silver Burdett Company, 1963), p. 86.

that good teachers would lead to the more complicated examples by introducing illustrations of gradually increasing complexity.

Consider, then, the example:

$$4\tfrac{1}{5} \div \tfrac{4}{5}.$$

How many sections with a surface of four-fifths of a given rectangle would be equivalent to four and one-fifth rectangles having the same dimensions?

Example:

Solution:

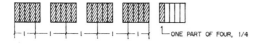

ONE PART OF FOUR, 1/4

The answer is, therefore, $5\tfrac{1}{4}$.

It is argued that the subtractive interpretation of division has meaning to the child, and can readily be applied to situations involving fractions which are within his experience and comprehension level. It is admitted that these assumptions have not yet been thoroughly tested and subjected to experimental investigation.

However, it could be argued that working out many intuitive solutions to examples involving fractions might prove to be much more challenging, interesting, and exciting than computing a larger number of exercises by the rule only. In any event it is argued that the rule, if used at all, can be stated, if not "discovered," by the pupils after a few experiments have been observed.

Those who see little merit in the above approaches to division with fractions object on the grounds that the ideas lack general applicability, particularly in those instances where the first fraction (dividend) is smaller than the divisor and has a different denominator.

Mathematically meaningful or structurally logical approaches

The above heading is used advisedly. First, some will question whether each of the following approaches is truly mathematically meaningful. Second, the author cannot supply evidence that any one method is more meaningful than another.

Perhaps we have been using the term "meaningful teaching" too freely. What really makes something meaningful—the fact that it can be logically and rigorously deduced, the fact that previously learned and comprehended operations on numbers when applied to a new operation yield a result which makes sense, the fact that one intuitively sees a relationship which leads to an intellectually satisfying conclusion? Just what takes place when something is learned meaningfully? Does this activity occur only in the mind, guided by naked symbols, either spoken or written, or do experiences with things—manipulation of physical objects of various shapes, sizes, or colors—contribute to mathematical meaning?

Authorities may differ on whether or not any experience of a physical, nonintellectual kind can be looked upon as a contribution to the mathematical meaning of an operation. One might also argue that no experience could be purely physical or social in nature and not relate in some degree with one's mental processes.

Having presented this background, and raising questions which the author feels are not yet fully answered, he will now enumerate a few approaches, which in and of themselves are divorced, at least to a degree, from more physically developed approaches.

Application of two inverse operations

Chabe has proposed a rationale for the "inverting and multiplying" process which is not readily comprehended.[5] He reminds the reader that two approaches are ap-

[5] Alexander M. Chabe, "Rationalizing inverting and multiplying," THE ARITHMETIC TEACHER, X (May, 1963), 272–273.

plied to an example like

$$\tfrac{3}{4} \div \tfrac{2}{3}.$$

Inverse operation is involved in the substitution of the multiplication process for division. The other approach is the substitution of the reciprocal of the divisor for the original fraction. These approaches cause a retention of mathematical balance in any given example calling for a division by a fraction." Thus, "the arithmetic value of the given example is unchanged when both processes are sequentially applied."[6]

Make fractions similar,
then compare numerators

When this method is used, it is pointed out that the dènominator of a fraction indicates the denomination, or names *what* we are talking about. This part of the fraction represents the part of the whole that is being considered.

The numerator of the fraction also indicates exactly what the term implies— it numerates or identifies how many parts of the whole we are talking about.

This means an example such as

$$\tfrac{3}{4} \div \tfrac{2}{3}$$

is changed to read

$$\tfrac{9}{12} \div \tfrac{8}{12}.$$

Now asking the question, "How many eight-twelfths are contained in nine-twelfths?" and relating the operation to situations, such as "How many groups of four pupils could be formed from a group of twelve children?" the teacher leads the pupils to the generalization that one divides the first numerator by the second, yielding the answer $1\tfrac{1}{8}$.

Make fractions similar, then divide
numerators and denominators separately

In this type of treatment,[7] the following rationale is presented to the pupil:

---

[6] *Ibid.*, p. 273.
[7] See, for example, E. T. McSwain, *et al., loc. cit.*

We want the division of fractional numbers to agree as much as possible with that of whole numbers. You know that $15 \div 3 = 5$ and that $1 \div 1 = 1$. Then, $\dfrac{15}{1} \div \dfrac{3}{1} = \dfrac{15 \div 3}{1 \div 1} = \dfrac{5}{1} = 5$. A procedure that is correct for the denominators of 1 should also work for any fractions with like denominators. Study the solution below.

$$\frac{15}{16} \div \frac{3}{16} = \frac{15 \div 3}{16 \div 16} = \frac{5}{1} = 5$$

Reduce the divisor to one

When this method is employed, it is suggested that one should first express the division example as a complex fraction. Thus,

$$\tfrac{3}{4} \div \tfrac{2}{3} \text{ becomes } \frac{\tfrac{3}{4}}{\tfrac{2}{3}} .$$

Then, according to some presentations,[8] the multiplication identity 1 is introduced and related to the solution. Recognizing that the fraction

$$\frac{\tfrac{3}{2}}{\tfrac{3}{2}}$$

names the same number as 1, it is possible to multiply the given fraction by it, with the result that one obtains

$$\frac{\tfrac{3}{4}}{\tfrac{2}{3}} \cdot \frac{\tfrac{3}{2}}{\tfrac{3}{2}} = \frac{\tfrac{9}{8}}{1} = \tfrac{9}{8} = 1\tfrac{1}{8}.$$

In other approaches, the child's attention is directed more particularly to the denominator of the complex fraction (in the above case, $\tfrac{2}{3}$) and is led to see, or is told, that he needs to get 1 for the denominator to facilitate the solution.[9]

In some textbook approaches the child is not encouraged to discover what he must do to accomplish this objective. Instead, he is told rather authoritatively that he needs to multiply the $\tfrac{2}{3}$ by $\tfrac{3}{2}$. (It is pre-

---

[8] See, for example, Leo J. Brueckner, *et al., Moving Ahead in Arithmetic, Book 6* (New York: Holt, Rinehart & Winston, Inc., 1963), p. 196, and Edwina Deans, *et al., Unifying Mathematics, Book Six* (New York: American Book Company, 1963), p. 216.
[9] See, for example, Maurice Hartung, *et al., Seeing Through Arithmetic, Grade 6* (Chicago, Ill.: Scott, Foresman & Company, 1963), p. 198.

sumed that he knows the meaning of a reciprocal, and the effect of multiplying one number by its reciprocal.) He then is led to see that if he multiplies the divisor by $\frac{3}{2}$, he must also multiply the dividend by $\frac{3}{2}$. The example then takes on this form:

$$\frac{\frac{3}{4}\times\frac{3}{2}}{\frac{2}{3}\times\frac{3}{2}} \text{ or } \frac{\frac{3}{4}\times\frac{3}{2}}{1} \text{ or } \frac{3}{4}\times\frac{3}{2} \text{ or } \frac{9}{8} \text{ or } 1\frac{1}{8}.$$

It seems that this approach provides a real opportunity for the teacher to challenge the pupils to verify or discover errors in hunches, and test hypotheses about what kinds of operations are permissible and what kinds are not.[10]

For example, having suggested that the exercise above be expressed in the form,

$$\frac{\frac{3}{4}}{\frac{2}{3}},$$

the teacher would then point out that the simpler the divisor, the easier the operation might become. The class would then be led by skillful questioning to see that what one really would like to have in the divisor is not $\frac{2}{3}$, but 1. Now, what can be done with $\frac{2}{3}$ to change it to 1?

The teacher should hope that someone will suggest that $\frac{1}{3}$ be added. Now, is this permissible—may one add $\frac{1}{3}$ to $\frac{2}{3}$ to get the value of 1? Of course, it is permissible. One may ordinarily add any two numbers he wishes to get a new number. But may it be done here?

Now, someone will suggest that to be permitted to do it here one must also compensate in a way by doing the same thing to the dividend, with the result that

$$\frac{\frac{3}{4}+\frac{1}{3}}{\frac{2}{3}+\frac{1}{3}} = \frac{\frac{3}{4}+\frac{1}{3}}{1} = \frac{3}{4}+\frac{1}{3}, \text{ and so forth.}$$

This hypothesis, that one may add the same quantity to both terms of a fraction, must be tested. The usual way is to try it out in a simpler situation—in this case with whole numbers, where the quotient is

already apparent. Two or three well-chosen examples should convince the class that, except for one unique case, the operation suggested is not permitted.

So, now, the class is encouraged to try something else. Someone might suggest that one multiply the divisor $\frac{2}{3}$ by $\frac{3}{2}$. Through questioning, setting up an hypothesis, and testing the hypothesis the teacher can show that it is permissible to multiply both dividend and divisor of a given fraction by the same number (zero excepted) without changing the value of the fraction.

Someone else might suggest during this exploratory-discovery process that by multiplying the $\frac{2}{3}$ by 3, the divisor is changed to 2, which is much easier to operate with than the original number $\frac{2}{3}$. It is then seen that the example reduces to

$$\frac{\frac{3}{4}\times3}{\frac{2}{3}\times3}, \text{ which becomes } \frac{\frac{9}{4}}{2}$$

Recognizing that dividing by two yields the same result as multiplying by one-half, the final result becomes:

$$\frac{9}{4}\times\frac{1}{2}=\frac{9}{8} \text{ or } 1\frac{1}{8}.$$

Other approaches might also be explored. Children should be given freedom of direction in their learning. Exploration of possibilities should be encouraged, and hypotheses should be tested and verified. Only then will confidence with numbers be generated, and only then can the pupils become masters of the subject, arithmetic —a useful tool in their hands. Less and less will they utter in surprise, "Oh can you do *that* here—I didn't know that before!"

Reasoning by analogy

This method relates division by a fraction to division by whole numbers, making as meaningful as possible the notion of a reciprocal.

10 A modest attempt to do this appears in Claud J. Bray, "To invert or not to invert," THE ARITHMETIC TEACHER, X (May, 1963), 274–276.

For example, $4\frac{1}{2}\div7$ is the same in meaning as $\frac{1}{7}\times4\frac{1}{2}$. Thus, since $4\frac{1}{2}\div7$ is the same as $4\frac{1}{2}\div\frac{7}{1}$, which is the same in value as $\frac{1}{7}\times4\frac{1}{2}$, it follows that $4\frac{1}{2}\div\frac{2}{3}$ would have the same value as $\frac{3}{2}\times4\frac{1}{2}$.

Then by applying the commutative principle, $4\frac{1}{2}\div\frac{2}{3}$ becomes $4\frac{1}{2}\times\frac{3}{2}$, and hence the rule is established.

Division as the inverse of multiplication

This method is relatively new and has found its way in one form or another into some of the newer programs of arithmetic.[11] It states that division can be thought of as a process of finding an unknown factor of a product when one factor is known. Also, if it can be shown by analogy that the number sentence $12\div3=n$ can be restated by the number sentence $3\times n=12$, it follows that the number sentence $\frac{2}{3}\times n=\frac{3}{4}$ states the same relationship as the number sentence $\frac{3}{4}\div\frac{2}{3}=n$.

Thus, if $\frac{3}{4}\div\frac{2}{3}=n$, it follows that $\frac{2}{3}\times n=\frac{3}{4}$. Then it follows that $\frac{3}{2}\times\frac{2}{3}\times n=\frac{3}{2}\times\frac{3}{4}$, from which $n=\frac{9}{8}$ or $1\frac{1}{8}$.

In essence, then, this method suggests that the division exercise be treated as a number sentence in division, which is changed into a sentence on multiplication with an unknown factor.

Method of "multividing"

For want of a more concise and descriptive term the author has coined the expression, "multivide," to describe a process which has been reported in the literature.[12] The dividend of the example is first multiplied by the denominator of the second fraction, then divided by the numerator. The rationale is essentially as follows:

Consider the example $1\div\frac{2}{3}$. We know that, since there are three-thirds in one, $1\div\frac{1}{3}=3\times1$, and since $\frac{2}{3}$ is 2 times as

much as $\frac{1}{3}$, $1\div\frac{2}{3}$ must be only $\frac{1}{2}$ of 3, or $1\frac{1}{2}$. Thus, $1\div\frac{2}{3}=\frac{1}{2}\times3\times1$ or $1\frac{1}{2}$.

Similarly, in $2\frac{1}{4}\div\frac{3}{4}$ there would be nine one-fourths in $2\frac{1}{4}$ (4 times $2\frac{1}{4}$), but only one-third as many three-fourths. Therefore, $2\frac{1}{4}\div\frac{3}{4}=\frac{1}{3}\times(4\times2\frac{1}{4})=3$.

It would follow then that this has the same effect as multiplying by the reciprocal of the divisor.

Summary

The purpose of this article has been twofold.

1 It has attempted to point out that many ways are being suggested by textbook writers, developers of modern programs in mathematics, and research workers for teaching the operation of division by fractions. These methods vary in their degrees of meaningful presentation, reliance on intuition, and concrete approaches. The reader is encouraged to judge these approaches by standards in keeping with good teaching procedures. There has been no attempt in this article to judge the relative merits of those newer procedures which obviously are not identified as drill-oriented approaches.

2 It provides an examining ground on which committees of teachers and others in the process of selecting textbooks or deciding upon a choice of a newer program might analyze, at least somewhat critically, one narrow phase of the whole mathematics curriculum.

Even though diversity of presentation is not to be discouraged, one of the above methods, or some appropriate combination of two or more, should eventually become the pattern followed somewhat universally in the future. Perhaps, this area of research, which has been relatively barren of activity over the years because of its apparent futility or hopelessness, might yet prove to be a field for profitable and engaging investigations.

---

[11] See, for example, School Mathematics Study Group, *Student's Text, Grade 6, Part II* (New Haven and London: Yale University Press, 1963), Unit 34, p. 344, and *Studies in Mathematics*, 279–280.

[12] See, for example, Percy Krich, *loc. cit.*, and George H. McMeen, *loc. cit.*

SELECTED BIBLIOGRAPHY FOR CHAPTER 7
Number Operations and the Set of Rational Numbers

Adachi, Mitsuo. "Addition of Unlike Fractions." *The Arithmetic Teacher* (March 1968), 221–23.

Allen, Ernest E. "Bang, Buzz, Buzz-Bang, and Prime." *The Arithmetic Teacher* (October 1969), 494.

Anderson, Rosemary C. "Suggestions from Research—Fractions." *The Arithmetic Teacher* (February 1969), 131.

Bates, Thomas. "The Road to Inverse and Multiply." *The Arithmetic Teacher* (April 1968), 347.

Bell, Kenneth M. and Rucker, Donald D. "An Algorithm for Reducing Fractions." *The Arithmetic Teacher* (April 1974), 299–300.

Brousseau, Roland L., Brown, Thomas A., and Johnson, Peter J. "Introduction to Ratio and Proportion." *The Arithmetic Teacher* (February 1969), 89–90.

Christensen, Anina and Silvey, Ida Mae. "Using Prime Numbers to Teach Mathematics in the Elementary School." *School Science and Mathematics* (March 1971), 247–56.

Cohen, Louis S. "The Board Stretcher: A Model to Introduce Factors, Primes, Composites, and Multiplication by a Fraction." *The Arithmetic Teacher* (December 1973), 649–56.

Cole, Blaine L. and Weissenfluh, Henry S. "An Analysis of Teaching Percentages." *The Arithmetic Teacher* (March 1974), 226–28.

Constantine, Deane G. "An Approach to Division with Common Fractions." *The Arithmetic Teacher* (February 1968), 176.

Cunningham, G. S. and Raskin, D. "The Pegboard As a Fraction Maker." *The Arithmetic Teacher* (March 1968), 224–27.

Dubisch, Roy. "The Sieve of Eratosthenes." *The Arithmetic Teacher* (April 1971), 236–37.

Fehr, Howard. "Fractions As Operators." *The Arithmetic Teacher* (March 1968), 228.

Hales, Barbara Budzynski and Nelson, Marvin N. "Dividing Fractions With Fraction Wheels." *The Arithmetic Teacher* (November 1970), 619.

Hannon, Herbert. "All About Division With Rational Numbers—Variation on a Theme." *School Science and Mathematics* (June 1971), 501–07.

Hewitt, Frances. "Pattern for Discovery: Prime and Composite Numbers." *The Arithmetic Teacher* (February 1966), 136

Hildebrand, Francis H. "An Ordered Pair Approach to Addition of Rational Numbers in Second Grade." *The Arithmetic Teacher* (February 1965), 106–08.

Holdan, Gregory. "Prime: A Drill in the Recognition of Prime and Composite Numbers." *The Arithmetic Teacher* (February 1969), 149.

Hyde, David and Nelson, Marvin N. "Save Those Egg Cartons!" *The Arithmetic Teacher* (November 1967), 578.

Immerzeel, George and Wiederanders, Don. "Ideas." *The Arithmetic Teacher* (October 1972), 457–65.

Junge, Charlotte W. "A Game of Fractions." *The Arithmetic Teacher* (October 1966), 494.

——. "Now Try This—Division of Fractions." *The Arithmetic Teacher* (February 1968), 177.

Matthews, Warren. "Teaching Comparison of Common Fractions." *The Arithmetic Teacher* (March 1968), 271.

May, Lola J. "How to Give Meaning to Rational Numbers." *Grade Teacher* (January 1968), 62–64.

——. "Adding and Subtracting Rational Numbers." *Grade Teacher* (February 1968), 74.

Nelsen, Jeanne. "Percent: A Rational Number or a Ratio." *The Arithmetic Teacher* (February 1969), 105.

Nelson, Diane and Mariur, A. "Pegboard Multiplication of Fraction by a Fraction." *The Arithmetic Teacher* (February 1969), 142–44.

Pincus, Morris. "Addition and Subtraction Fraction Algorisms." *The Arithmetic Teacher* (February 1969), 141.

Prielipp, Robert W. "Teaching One of the Differences Between Rational Numbers and Whole Numbers." *The Arithmetic Teacher* (May 1971), 317.

Rasof, Elvin. "Prime (Candy Bar) Numbers." *The Arithmetic Teacher* (January 1968), 67.

——. "The Fundamental Principle of Counting, Tree Diagrams, and the Number of Divisors of a Number (The Nu-Function)." *The Arithmetic Teacher* (April 1969), 308.

Romberg, Thomas A. "A Note on Multiplying Fractions." *The Arithmetic Teacher* (March 1968), 263.

Rowland, Rowena. " 'Fraction Rummy'—A Game." *The Arithmetic Teacher* (May 1972), 387–88.

Stenger, Donald. "Prime Numbers From the Multiplication Table." *The Arithmetic Teacher* (December 1969), 617.

Wassmansdorf, M. "Reducing Fractions Can Be Easy, Maybe Even Fun." *The Arithmetic Teacher* (February 1974), 99–102.

Wilson, Patricia, Mundt, Delbert and Porter, Fred. "A Different Look at Decimal Fractions." *The Arithmetic Teacher* (February 1969), 95.

Winzenread, Marvin R. "Repeating Decimals." *The Arithmetic Teacher* (December 1973), 678–82.

Zink, Mary. "Greatest Common Divisor and Least Common Multiple." *The Arithmetic Teacher* (February 1966), 138.

Zytkowski, Richard T. "A Game With Fraction Numbers." *The Arithmetic Teacher* (January 1970), 82–83.

PART III

DATA, SIZE AND SHAPES

CHAPTER 8

ELEMENTARY STATISTICS AND PROBABILITY

Many of the problems of everyday life can be better understood with the aid of probability and statistics. For this reason you may want to consider teaching these ideas to pre-high school students.

The articles and activities in this chapter will add meaning and understanding to some of the problems, experiences, and information encountered in daily living. For instance, Ida Mae Heard's article, "Making and Using Graphs in the Kindergarten Mathematics Program," and Robert C. Pierson's "Elementary Graphing Experiences" provide many ideas for use in the classroom and should stimulate the teacher to develop further his own ideas.

"Checking the Calculated Average Through Subtraction" by Irwin Albert and "Finding Averages With Bar Graphs" by Joyce Ball provide some excellent ideas for working with averages. Bar graphs are a major means of scientific communication, and Joyce Ball's article not only helps children understand but also lends reality and significance to mathematical content.

Linda Ball's article, "Pascal's Triangle," offers a wealth of interesting number patterns. You and your students should be able to find many more patterns after reading this article.

Bruce C. Burt gives an interesting approach to the teaching of simple sampling ideas in "Drawing Conclusions from Samples (An Activity for Low Achievers)." The experiments discussed in this article are excellent material for the mathematics laboratory. Classroom teachers will surely want to try some of these experiments to satisfy their own curiosity.

Finally, Ernest R. Ranucci's article, "Teaching Permutations," may sound like a weighty subject for young children, but it's easy—and fun. How many ways can we arrange the letters of the alphabet? Read this excellent article and find out.

Making and using graphs in the kindergarten mathematics program

IDA MAE HEARD

During the 1967/68 school term, the writer used a faculty research grant to initiate a modern mathematics program in the morning kindergarten class at the Laboratory School of North Texas State University at Denton. This was the first time a structured mathematics program at the kindergarten level had been tried at the Laboratory School. The kindergarten teacher, Mrs. Margaret Eden, worked closely with the author in planning learning activities and in selecting materials of instruction. The college students in the writer's courses entered wholeheartedly into the spirit of our adventure—"Operation Kindergarten." They made charts and manipulative materials to be used by the children and the investigator.

There are seventeen boys and eight girls in the class. These twenty-five children are from advantaged homes, and their parents are business and professional people.

One of our most rewarding experiences was the use of graphs to picture number relations.

The block graph

The children encountered the ideas of (1) one-to-one correspondence, (2) many-to-one correspondence, and (3) none-to-one correspondence when they made a block graph.

After labels for each of the twelve months were placed on a separate tile on the floor of the classroom, each child stacked his block by the month of his birthday.

After the blocks were placed, the children made these observations: (1) There were more birthdays in February than in any other month. (2) There were no birthdays in May. (3) There were more birthdays in the first half of the year than in the last half. (4) There were the same number of birthdays in some of the months —one birthday in January, July, and August; two birthdays in April, July, September, and December; and three birthdays in March, October, and November.

The children also had the experience of coloring one-inch square blocks on graph paper to show this relationship. We were pleased that every child was able to successfully follow these directions: "Choose any color you like, but do not use the same color for any two blocks that are side by side or above each other." They got the idea that a careful placement of colors was necessary for each child's birthday

Reprinted from *The Arithmetic Teacher* (October 1968), 504–506, by permission of the author and the publisher.

block to really stand out. No two of the graphs were colored alike. The children would color a block and say, "That's my birthday," or "That's your birthday."

The bar graph

After the heights and weights of the children were determined, the bar graph and the picture graph (shown in photos) were designed by Mrs. Charlotte Peters in the writer's methods class.

The children learned about scale drawings from the bar graph. They knew that their actual height could not be shown on the poster. We let one inch on the poster represent two inches of their height. Using a tape measure, each child was allowed to measure the length of the bar on the chart and then to double this length for his real height. The range of heights was not great.

The picture graph

The weight of the children was pictured to the nearest five pounds. Since the children could already count by 5's to 100, they enjoyed counting by 5's to determine the approximate weight of their friends. They also counted on from 5 to find out how many children had about the same weight.

This is their summary:

Number of children		Approximate weight
8	. . . .	40 lb.
7	. . . .	45 lb.
8	. . . .	50 lb.
1	. . . .	55 lb.
1	. . . .	65 lb.

The children discovered that the taller children were, in most cases, the children who had the longer feet and were the heavier. They decided that this made good sense.

The activities described above illustrate only a few ways in which graphs might be used with young children. Teacher-pupil planning at the college level made these experiences possible for an eager group of learners in the morning kindergarten class.

Elementary graphing experiences

ROBERT C. PIERSON

In this modern day and age people are confronted with various pictorial representations of data and, in the future, children will be faced with an ever-increasing mass of pictorial information. It is urgent that students be able to understand these pictorial representations and from this understanding make intelligent interpretations.

The following series of exercises is intended to be a guide to the teacher. It is hoped that this sequence of graphs could be used as a starting point and that many subsequent exercises on graphing would follow.

The beginnings of graphing are very basic and unsophisticated. The student who has seemingly no understanding of pictorial representations might be guided to do something like these exercises. The ideas presented are only samples of what might be done. The students are encouraged to think of things they might graph themselves.

1. Make a graph of the people in the room who wear glasses and those who don't. To emphasize the one-to-one correspondence between the student and the square which represents that student, names are written in each square.

2. The next type of graph would have a greater variation, yet the one-to-one correspondence can clearly be seen.

Make a pictorial representation of the months of the birthdays of the people in the room.

Birthdays

3. A slight sophistication could be made by labeling both vertically and horizontally on the graph. The following type of representation might result from a survey of shoe sizes of the boys in the room.

Boys' Shoe Sizes

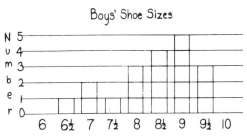

4. The next type of representation would require more time-consuming research.

Take one column of newsprint and make a graph of the occurrences of *a*'s, *e*'s, *i*'s,

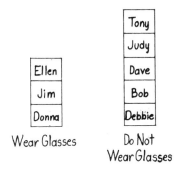

Reprinted from *The Arithmetic Teacher* (March 1969), 199–201, by permission of the author and the publisher.

o's, and *u*'s. An even more ambitious task might be to graph the frequency of all the letters of the alphabet which occurred. Or, as a short exercise, the students might graph the letters in the sentence, "The brown fox jumps quickly over the lazy dog."

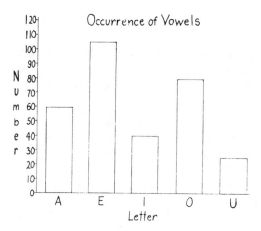

5. The next step in graphing would be to represent random occurrences. For example, the students could record the frequency of various vehicles that travel past the school in a given half hour.

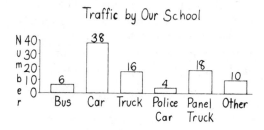

6. A graph of the temperature at a given time each day for one week might

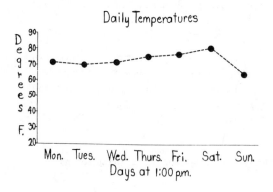

be made. An interpretation might be derived from this graph. (e.g., "The temperature got hotter and hotter during the week, but the storm Saturday evening cooled things off on Sunday.")

7. The students might throw a pair of dice 200 times and record the results as follows.

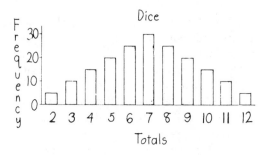

What number would you pick to predict the result of a throw of a pair of dice? Which total occurred the greatest number of times?

8. After much activity like that which has been described, the students can enter into a more abstract type of graphing. Give the students a "machine" that has a rule. This machine has an input and an output.

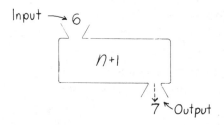

Graphs such as the following can be made.

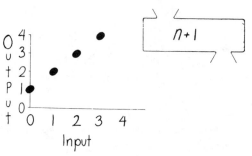

156

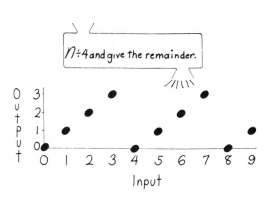

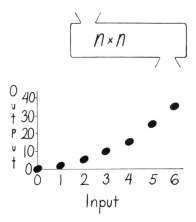

Other possible graphs are these:

1. Number of girls and boys in a class
2. Colors of shoes
3. Color of hair of people in class
4. Color of eyes of people in class
5. Color of family cars
6. Coins held by various students
7. Distance students live from school
8. Machine rule: $(n \times 2) + 1$
9. Machine rule: If even, add 1; if odd, subtract 1.

157

Checking the calculated average through subtraction

IRWIN ALBERT

How often have you tried to involve your students in some meaningful practice in adding and subtracting negative numbers? One such application is checking the average through subtraction.

Suppose 75 has been calculated in the traditional way as the average for the set {60, 100, 65, 75}. It can be checked in the following way:

1. *Separate* the set {60, 100, 65, 75} into any two subsets—for example,

    A {60, 100}   B {65, 75}

2. *Test* your solution of 75 by subtracting it from each member of the two subsets:

$$A \quad \begin{array}{cc} 60 & 100 \\ -75 & -75 \\ \hline -15 & +25 \end{array} \qquad B \quad \begin{array}{cc} 65 & 75 \\ -75 & -75 \\ \hline -10 & 0 \end{array}$$

3. *Combine* the differences within each subset. Thus in subset A $(-15, +25) = +10$, while in subset B $(-10, 0) = -10$.

4. *Combine* the new values of the subsets. Thus $(-10, +10) = 0$, or the final value.

By achieving a final value of 0 we indicate that 75 is the zero point between two subsets of numbers.

Teachers will be pleasantly surprised when students use varying subsets to check their calculation of the average. One student may rearrange the set {60, 100, 65, 75} into two other subsets: {60, 65} and {75, 100}. Thus:

$$\begin{array}{cccc} 60 & 65 & 75 & 100 \\ -75 & -75 & -75 & -75 \\ \hline -15 & -10 & 0 & +25 \\ \underbrace{}_{-25} & & \underbrace{}_{+25} \end{array}$$

By combining the remaining values $(+25, -25)$ the student can reach a final value of zero.

Another student may rearrange the set {60, 100, 65, 75} into subsets {60} and {100, 65, 75}. Thus:

$$\begin{array}{cccc} 60 & 100 & 65 & \mathbf{75} \\ -75 & -75 & -75 & -75 \\ \hline -15 & +25 & -10 & 0 \\ & \underbrace{}_{+15} & & \end{array}$$

By combining the remaining values $(-15, +15)$ this student also can reach a final value of zero.

Of course, this system also works for numbers that are not multiples of five. For example, the average of {60, 71, 85} is 72. The check would be:

$$\begin{array}{ccc} 60 & 71 & 85 \\ -72 & -72 & -72 \\ \hline -12 & -1 & +13 \\ \underbrace{}_{-13} & & \end{array}$$

$$0 = \text{the final value}$$

This system of checking will provide a new and enriched dimension to your teaching of averages, plus some needed practice in operations with negative numbers.

Reprinted from *The Arithmetic Teacher* (November 1971), 499–500, by permission of the author and the publisher.

Finding averages with bar graphs

JOYCE BALL

Our school still uses the numerical marking system for reporting to students and parents. I have found over a period of several years that pupils in the fourth grade are apt to look at the single mark on papers that are returned daily or weekly, and have little sense of any cumulative effect over the ten weeks' marking period. This is true both of students doing superior work and those doing below-average work. While the process of averaging marks is not a complicated one, it presents two kinds of problems for the typical fourth-grader: (1) It may involve more complicated arithmetic than he is able to handle at the early stages of the fourth grade; and (2) The possibility of error both in addition and division is so large that the end result may be far from accurate.

At the time we were studying bar graphs, I hit upon the idea of using them as a device whereby the pupils would do some manipulating in order to find the average of a series of marks. The class members had learned at the beginning of the year to make graphs of their scores in the SRA Reading Laboratory, and we spent a day or two reviewing the making and interpretation of graphs, using the material in our mathematics book. In explaining the meaning of the term "average" I used the members of the class as examples. I suggested that if we wanted to make everyone in the class the same size, thus finding the "average" size of a fourth-grader in our class, we would have to cut off part of the head of the tallest person and add it on to the smallest person!

Since our school ranks four subjects— reading, English, mathematics, and social studies—as the "major" subjects to determine a pupil's progress, we used the marks in those subjects for the seven weeks after the first reporting period to make our graphic averages. I had done some preliminary work figuring out weekly averages in order that the pupils would not be faced with the problem of making too large a number of bars for each subject.

Each pupil was given sheets of graph paper, with instructions to number the lines by 2's. When this was done, he drew a red line across the paper between 74 and 76, making a line to stand for 75, the "passing mark" in our school. Each student was given a sheet of paper with his weekly averages in the four subjects and made his own bar graph. Figure 1 shows a section of the graph of one child in one of the subject areas. I "read" each graph with each pupil to make sure there were no mistakes at this first stage of the process.

With a grid chalkboard, I then made a bar graph of the marks of one of the pupils.

In order to "even them up" (cutting off the heads of the tall people, etc.) we took some squares from his highest mark and added them on to his lowest. We continued this process until they were as nearly even as it was possible to get them, and determined that this was his "average" for the seven weeks' period. We did this with two or three sets of marks, including some very low ones, as well as some very high ones, to show the "leveling-off" process that took place.

Each pupil then proceeded to do the same thing with his own bars. In order to make the changes more apparent, they were made with colored pencils. As one square (2 points) was taken from the highest bar, an X was put across it, and another square was added to the lowest bar with the colored pencil. In this way the preliminary

Reprinted from *The Arithmetic Teacher* (October 1969), 487–489, by permission of the author and the publisher.

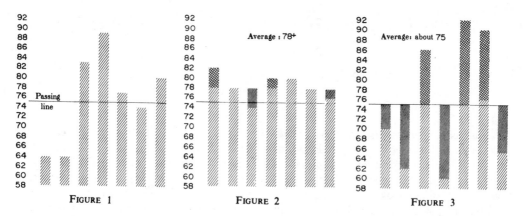

FIGURE 1   FIGURE 2   FIGURE 3

figures showed clearly in relation to the final average.

In some cases there was little variation among the marks, leading to finding the average relatively quickly (fig. 2).

In other cases, a wide range from the highest to the lowest required a great deal of manipulation to even up the bars (fig. 3).

When they were finished with the graphs, the pupils compared their averages at that point with the marks they had had on their report cards for the first ten weeks. I felt that it gave them a somewhat clearer idea of what went into the determination of their marks, and that a long period of failing work could not be quickly overcome by a few 100's. At the same time I took the occasion to discuss with them the other factors, aside from the strictly numerical, which went into the determination of their report card marks.

Many of the comments made by the pupils as they worked indicated how surprising the results were to them. "That shot my 91!" remarked one pupil as he had to keep crossing out squares in that bar to "even up" a 48. One girl was observed trying vainly to manipulate her bars so that she could make bars of 73, 68, 77, and 72 average to a 75. Several pupils remarked that it showed them what subjects they needed to improve in, and how much they would need to improve in. One boy was interested in seeing that a completely different kind of information can be obtained from a graph. They obviously enjoyed the manipulation involved.

Later in the year, when their arithmetic ability had enlarged, they learned the traditional process of finding the mean, but they still felt that it had been a great deal more fun and much more meaningful to cross out and add on squares in a graph in order to carry out this leveling-off process. Although we did not try it, it would also be possible to have the pupils cut out the squares and actually manipulate them physically.

EDITOR'S NOTE.—A project of this type not only helps children understand the "grade earned" as reported on his marking card but also lends reality and significance to mathematical content. I wonder if "averaging" could be introduced with movable objects, such as books, blocks, felt pieces, etc. It might help children to visualize the "evening-up" process!—CHARLOTTE W. JUNGE.

PASCAL'S TRIANGLE

Linda V. Ball

What could be useful to a gambler figuring odds, a high school freshman struggling with the coefficients for the expansion of $(a + b)^5$, and a math teacher seeking an interesting classroom diversion? The answer is Pascal's triangle — an arrangement of numbers developed by Blaise Pascal.

Many teachers of mathematics are familiar with Pascal's triangle and marvel at the patterns one can find in its structure. Unfortunately, a classroom attempt to actually construct the triangle is seldom made until the students are doing advanced work at the secondary level. Pascal's triangle *can* be developed by several methods appropriate for the elementary school. The purpose of this article is to suggest one method, the one that involves a listing of the number of possible arrangements of objects.

For our discussion here, we will use objects which are either red or black[*] and will specify the number of each. A simple classroom model for this exercise can be made using colored cubes. If you do not have cubes or rods in your classroom, you can construct your own by making a pattern like that pictured below on construction paper. To get a convenient one-inch cube, make the sides of each square in the pattern one inch long. Cut along the solid lines and fold along the dashed lines, securing the edges with tape.

Consider the following question. How many arrangements of three cubes can we make by using red and/or black cubes? We could begin by simply placing cubes in a row and keeping a record of the different arrangements as we shift the cubes around. However, as we increase the number of cubes to four, five, or more, we may find it difficult to decide when we have exhausted all the possibilities. To avoid this confusion, we should develop some sort of system. One solution is to list all the possible sets of a given number of cubes and consider each set as a separate case. Since we agree that four or five would be confusing, let's begin with groups of three. The three cubes can be any of the following sets: *(a)* three red cubes, *(b)* two red and one black, *(c)* two black and one red, or *(d)* three black cubes. So, to determine the number of arrangements of three cubes, we must decide

- *(a)* in how many ways we can arrange three red cubes;
- *(b)* in how many ways we can arrange two red cubes and one black cube;
- *(c)* in how many ways we can arrange one red cube and two black cubes;
- *(d)* in how many ways we can arrange three black cubes.

Our models might look like this:

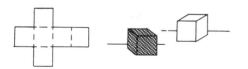

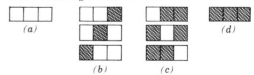

*In this book, the colors used in the original article could not be reproduced. The "red" cubes are represented here as white squares, and the "black" cubes are shaded.

Reproduced with permission from "Timely Topics in Modern Mathematics," © 1970 by Houghton Mifflin Company.

The resulting tally of cases is

$$(a) \quad (b) \quad (c) \quad (d)$$
$$1 \quad\quad 3 \quad\quad 3 \quad\quad 1.$$

This tally for arrangements of three cubes furnishes us with one row of numerals in Pascal's triangle.

The number of possible arrangements of two cubes can be determined by finding the arrangements of the following:

(a) two red cubes;
(b) one red cube and one black cube;
(c) two black cubes.

The models for these arrangements would be:

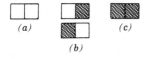

The tally is

$$(a) \quad (b) \quad (c)$$
$$1 \quad\quad 2 \quad\quad 1.$$

In exploring arrangements of one cube, it becomes obvious that we can have only one red or one black, and our tally is

$$red \quad\quad black$$
$$1 \quad\quad\quad 1.$$

We can now begin to construct Pascal's triangle.

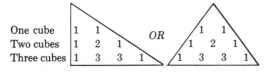

If all number patterns are not yet obvious, our next step is to find the possible arrangements of four cubes. To do this, we consider the following cases:

(a) four red cubes;
(b) three red cubes and one black cube;
(c) two red cubes and two black cubes;
(d) one red cube and three black cubes;
(e) four black cubes.

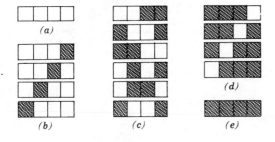

The resulting tally of arrangements is

$$(a) \quad (b) \quad (c) \quad (d) \quad (e)$$
$$1 \quad\quad 4 \quad\quad 6 \quad\quad 4 \quad\quad 1.$$

When we add this row, Pascal's triangle takes on the form of *Figure 1*.

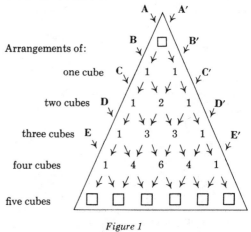

Arrangements of:
one cube
two cubes
three cubes
four cubes
five cubes

Figure 1

As we continue to add rows, the patterns become more obvious. In fact, we are now able to predict the top number and all the members of the five-cubes row without further help from our cubes. Since the "sides" of the triangle are "1's", it seems that our top number should be one, and the end numbers in the five rows should also be ones.

To aid our discussion of the remaining patterns, we can name the diagonals as in the figure. When talking with students, you may wish to refer to the diagonals as streets. Looking down *B Street* or *B' Street*, we see a pattern of counting numbers or natural numbers in ascending order. This enables us to predict two more entries in the five-cubes row.

C and *C' Streets* give us a different pattern. The numbers in this sequence are called triangular numbers because each can be represented by a triangular array. *(See illustration at top of next page.)*

162

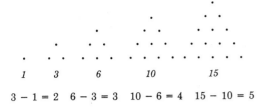

$$1 \quad 3 \quad 6 \quad 10 \quad 15$$

$$3 - 1 = 2 \quad 6 - 3 = 3 \quad 10 - 6 = 4 \quad 15 - 10 = 5$$

We can also note that the differences between successive numbers in the sequence increases by one each time. For example, $10 - 6$, or 4, is one greater than $6 - 3$, or 3.

This pattern of differences gives us the last two numbers in this row and a new version of Pascal's triangle *(Figure 2)*.

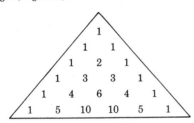

Figure 2

So far, we have ignored the pattern which is perhaps the most useful one. It will enable us to find any entry in the triangle and thus develop Pascal's triangle to unlimited dimensions. To find it, choose any two adjacent numerals from one horizontal row of the triangle. Let's refer to them as **a** and **b**. Below and equidistant from **a** and **b** is a third number, **c**. We notice that $c = a + b$ for all possible selections of **a**, **b**, and **c**.

$$a + b = c$$

We can use this pattern to obtain entries in the next row of the triangle.

$$1 \quad 5 \quad 10 \quad 10 \quad 5 \quad 1$$
$$1 \quad 6 \quad 15 \quad 20 \quad 15 \quad 6 \quad 1$$

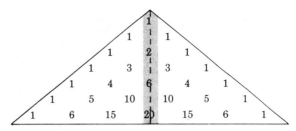

You may have noticed the *axis of symmetry* which is marked on the triangle above. A fold along this axis would match identical entries.

Let's take one more look at our triangle. This time add up all the numbers of any one row and note the developing pattern.

1	$=$	1	$= 2^0$
$1 + 1$	$=$	2	$= 2^1$
$1 + 2 + 1$	$=$	4	$= 2^2$
$1 + 3 + 3 + 1$	$=$	8	$= 2^3$
$1 + 4 + 6 + 4 + 1$	$=$	16	$= 2^4$
$1 + 5 + 10 + 10 + 5 + 1$	$=$	32	$= 2^5$
$1 + 6 + 15 + 20 + 15 + 6 + 1$	$=$	64	$= 2^6$

If your students have not yet studied exponents, they can still notice that the sum is doubled each time.

Pascal's triangle offers a wealth of interesting number patterns. We have looked at only a few of them. You will be pleasantly surprised at how many of these patterns your students can discover. Work with Pascal's triangle can introduce the concept of symmetry or merely provide a break from the routine. The only prerequisites are the ability to add and the curiosity which most elementary school children possess. They will gain practice in looking for patterns, developing their own theories, checking to see if their theories are valid, and perhaps revising those which are not. The students who are unable to see the more subtle patterns can contribute to the group by actually making models from the colored cubes and verifying the predictions of other students. Why not give it a try? Challenge your class to find a pattern that is new to you.

163

Drawing conclusions from samples (An activity for the low achiever)

BRUCE C. BURT

The 1963 Cambridge Report[1] suggests that more applications of mathematics will be possible in the upper elementary grades due to the amount of science that has been and probably will continue to be introduced into primary school. Some of the most important applications involve probability and statistics, which should be considered purely empirical subjects at this level.

Sampling from a finite population is one concept of probability and statistics that can be introduced in the intermediate grades. A population[2] consists of any set or collection of objects relevant to a particular question. For example, if we are interested in determining the average number of transistor radios per household in the United States, the totality of these figures, one for each household, constitutes the population for this study. A sample is any part or subset of a population that is used to infer results pertaining to the entire population. A common use for samples is to obtain an estimate for a population characteristic—the percentage of color-blind men, the number of trees in a particular area, or the height of ten-year-old boys in the United States. For example, to estimate the mean amount of money car-

ried by students in a school, we can take a sample of students and use the mean amount carried by these as the estimate. Similarly, to estimate the mean growth of bean plants under particular conditions, we take a sample and use the mean growth of the sample. Clearly, these estimates depend upon the samples chosen. When the sample is large, the estimate is likely to be close to the true population value. Thus a large sample usually produces quite accurate information about the population characteristic. These estimates also depend upon the method of drawing the sample. Simple random sampling, where every possible sample has an equal opportunity of occurring, is the safest way to select a sample.

Student activities

One application of probability with which students may be acquainted is opinion polls. The following activities give the student an opportunity to draw conclusions based on sampling.

WHICH CONTAINER DO YOU HAVE?

Begin with five containers containing beads:

A	25 white	5 black
B	20 white	10 black
C	15 white	15 black
D	10 white	20 black
E	5 white	25 black

Explain the following problem to the class. There are five containers each with

---

[1] *Goals for School Mathematics,* The Report of the Cambridge Conference on School Mathematics (Boston: Houghton Mifflin Co., 1963).

[2] For further information on populations and samples see Fehr, Bunt and Grossman, *An Introduction to Sets, Probability and Hypothesis Testing* (Boston: D. C. Heath & Co., 1964).

Reprinted from *The Arithmetic Teacher* (November 1969), 539–541, by permission of the author and the publisher.

30 beads—some black, some white. (Do not indicate which containers contain which combinations.) The class is to sample the containers by drawing beads, recording whether the bead is black or white and then replacing it. After recording 20 beads they are to guess which container they have.

Assign the students to five groups and have each perform the experiment with one of the containers. Then rotate the containers and have each group perform the experiment with a second container. Continue (if time permits) until each group has sampled all five containers and recorded their results on the data sheet.

Following the experiment, have each group record the results of their drawings and their guess for each container on the board. If there is disagreement, the drawings could be totaled for a given container and an attempt made to agree on a conclusion. If the groups agree, ask pupils if they think they could predict the contents of the containers with fewer samplings. If so, they can repeat the experiment with a different number of samplings. Finally, students should count the number of beads in each container.

Sample Data Sheet: Which container do you have?

Directions: Shake the container and draw a bead. Record whether the bead is black (*B*) or white (*W*). Replace the bead. Do this 20 times. Then guess the number of black and white beads in the container. Remember that there are 30 beads in the container.

Container _____

Shake	1	2	3	4	5	6	7	8	9	10
Color										

Shake	11	12	13	14	15	16	17	18	19	20
Color										

Your Guess: The container has

_____ black beads

_____ white beads

How Many of Which?

For this activity you need sets of 50 beads of three different colors, or sets of 50 bottle caps of three different products (students might be able to supply these) in boxes or bags.

Ask students to make an estimate by sampling, of the distribution of the colors of beads or kinds of bottlecaps. Begin by telling them how many objects there are in the container, then ask them how many "drawings" (replacing the object drawn) they would need to make to get a reasonably good indication of the distribution. Try to have them agree on a minimum number of drawings that would be needed.

Assign students to groups and have them perform the experiment, recording their results on the data sheet.

Following the experiment, have each group place the results of its drawings and estimates of the contents of the containers on the board. Find out how some of the groups arrived at their estimates, then average the totals of the drawings to get class averages and use these to make refined estimates of the contents of the containers. Direct some students to count the contents of the containers and compare with the estimates. Finally ask the class if, based on the experiment, they should change their estimate of the minimum number of drawings needed to arrive at a reasonably good estimate of the contents of the container. If so, the containers could be changed and they could try the experiment again with the new minimum.

As a follow-up activity, have students make a survey of opinion on a student council election or similar situation and predict the outcome. Compare with actual results.

Sample Data Sheet: How many of which?

Directions: Draw an object from the container, record which kind or color it is and replace it. Do this the number of times agreed upon. Then estimate the contents of the container without looking in it. Try it again and see if you get the same results.

Drawing	1	2	3	4	5	6	7	8	9	10
Kind drawn										

Drawing	11	12	13	14	15	16	17	18	19	20
Kind drawn										

Estimate: In the container there are

_____

_____

_____

Additional problems and questions for the students might include:

1. A town has a population of 2,400. In an election for mayor, a survey of 400 people showed that out of every 16 people, 10 favored Mr. Smith, 4 favored Mr. Brown, and 2 favored Mr. Jenkins. If the election were held and all 2,400 people voted,

   a) about how many votes would Mr. Smith get?

   b) about how many votes would Mr. Brown get?

   c) about how many votes would Mr. Jenkins get?

2. When the election in the town in Problem 1 was held, the results were

Candidate	Votes
Mr. Smith	1,350
Mr. Brown	775
Mr. Jenkins	275

   a) How do the results compare with the survey?

   b) Out of every 16 votes how many did Mr. Smith get, Mr. Brown get, and Mr. Jenkins get?

3. Imagine that we ran an experiment and, with a sampling of 200, found that of every 10 beads we drew out there was an average of 5 red, 3 yellow, and 2 white.

   a) If there are 1,000 beads in the container, how many would you predict are red, how many yellow, and how many white?

   b) After making the estimates in (a), someone counts the beads in the container and finds that there are 486 red beads, 309 yellow beads, and 205 white beads. Was the prediction accurate? Explain.

All of these activities acquaint the students with probability through the use of already familiar opinion polls.

Teaching permutations

How many different ways can we arrange the letters of the
alphabet? It sounds like a weighty subject for
young children, but it's easy—and fun

ERNEST R. RANUCCI

MANY IMPORTANT mathematical ideas of the 20th Century stem from the fields of permutations, combinations and probability. Although the study of combinations and probability is much too advanced for young children, the study of permutations—call them "arrangements" if you wish—can be introduced in the third or fourth grade.

There's a very good reason why you should consider including a unit on permutations in your curriculum. The basic concepts that youngsters learn when they're eight or nine years old will pay off later on when and if they study probability. And the chances are very good that they *will* study probability in one form or another—it's one of the vital unifying structures in all of the sciences. (For more on probability see the enrichment project for the upper grades, "Probability—Chance for a Change" by Dr. Lola J. May, *GT, Jan. '69*, p. 31.)

Permutations sounds like rather a terrifying subject for both teacher and child, but actually just the opposite is true. First of all, the children do not need to know a lot of arithmetic facts in order to understand the basic concepts of permutations—that's a plus right there. Then there's a certain fascination that even young children feel about the many ways we can arrange a series of numbers, letters or objects. Finally, you can inject a lot of physical movement into a unit on permutations—children going to the

chalkboard to rearrange a series of letters, children changing places with each other and so on. It all adds up to fun for everybody in the classroom.

The following unit consists of 10 related activities. Since the permutations involved get progressively more complex, it might be best not to schedule too many activities for one day after you get past the first two or three.

1. Print the letters A and B on two pieces of cardboard and prop them up in front of the chalkboard. (Make sure the letters are large enough so that all the children can see them easily.) Now have a child go to the chalkboard and arrange the letters in as many different ways as he can. (No fair turning the letters upside down or putting them on their sides—everyone must be able to read them.) There are, of course, only two possibilities: AB or BA.

2. Say to the children, "Suppose we do the same experiment with three letters—A, B and C (see Figure 1). In how many different ways can these three letters be placed next to each other? The answer is that there are six possibilities: ABC, ACB, BAC, BCA, CAB and CBA.

3. Pose this problem: "Johnny's house has three doors (see Figure 2). If he is outside the house, in how many different ways can he enter through one of the doors, leave through a second and reenter through a third?" Again, there are six possibilities and if you have your

youngsters list all the entrances and exits that Johnny makes (using a letter for each door), they'll see for themselves that the six possibilities are identical with the ones in the previous problem.

You can get a little more mileage out of this activity by asking the children whether Johnny would end up inside or outside the house if he began his journey inside. Would that make the permutations any different? (Obviously, the permutations are the same no matter where Johnny starts out, but by beginning inside, he ends up outside looking in.)

4. Suppose we take the letters A, B and C and place just two of them in different arrangements. How many arrangements are possible? The magic number is still six: AB, BA, AC, CA, BC and CB. Ask the children if they find anything strange about this result. Someone will surely point out that the same result is obtained regardless of whether it's two or three letters that are being arranged. This is absolutely right, but do the children know *why* it's right? (See box on p. 168 for a simplified answer.)

Figure 1

Figure 2

Figure 3

5. Make three large cardboard signs, mark them 1, 2 and 3, and attach a loop to each sign (see Figure 3). Give a sign to each of three students and tell them to leave the room and file back in with the signs hung around their necks. Have them continue to leave and file back into the room, lining up in a different order each time they return until all possible permutations have been accounted for. Have a child tally the different permutations on the chalkboard. Ask the class which of the permutations (123, 132, etc.) represents the largest number? Which represents the smallest number?

6. The problem posed in Exercise 5 can be solved in a different way—through a device known as a *tree* (see Figure 4). The tree, customarily read from left to right, is extremely valuable for conveying mathematical ideas. Have the children list vertically the three numbers which are to be arranged—in this case, 1, 2 and 3. Then have the children draw two "branches" from each of these three numbers to show that each has two possibilities for the second choice of numbers. The final branches are then drawn to show that there is but one possible choice for the third number in any given three-number series.

7. Your children are now ready to work with four units. Prepare four large cards—A, B, C and D—and display them prominently. Let the children take turns arranging the cards in new and different combinations. Have one child act as secretary with the responsibility of keeping a record of the different arrangements and vetoing any that have been recorded.

Things will get hectic before too long. After all, there are 24 different permutations of the four letters. Per-

haps some child will suggest that the class draw a tree (see Figure 5) to solve the problem. If no one suggests it, *you* do so before things get completely out of hand.

If a hectic classroom doesn't faze you—and you have at least 24 children in your class—it might be fun to have the youngsters line up in front of the room in the 24 different permutations of the four letters. It will cause some confusion, but I doubt that the children will forget the lesson in a hurry.

8. There is still another way to obtain the solution to the previous problem. It goes like this:

Any of the four letters could have occupied the first of the four positions: A, B, C, D.

Any of three letters could have occupied the second of the four positions: AB, AC, AD, BA, BC, BD, CA, CB, CD, DA, DB, DC.

Any of two letters could have occupied the third position: ABC, ACB, ADB, ABD, ACD, ADC, BAC, BCA, BDA, BAD, BCD, BDC, CAB, CBA, CDA, CAD, CBD, CDB, DAC, DCA, DBA, DAB, DBC, DCB.

This leaves only one letter, in each case, which can occupy the fourth position. The 24 permutations must be: ABCD, ACBD, ADBC, ABDC, ACDB, ADCB,

BACD, BCAD, BDAC, BADC, BCDA, BDCA, CABD, CBAD, CDAB, CADB, CBDA, CDBA, DACB, DBAC, DCAB, DABC, DBCA, DCBA.

9. Suppose we have three cards marked 2, 4 and 6. How many three-digit numbers would we have if we listed all the permutations of 2, 4 and 6? Which is the largest number and which is the smallest? How many of the numbers are odd and how many are even?

10. Suppose we have four cards marked 1, 2, 3 and 7, and wished to list only the two-digit permutations. How many such permutations are there? (12) How many of these would be odd and how many would be even? What are the largest and the smallest odd numbers? The largest and the smallest even numbers?

Permutations principles

There's at least one mathematical principle behind every operation in mathematics—and permutations are no exception. To understand how and why we arrive at the number of permutations which are possible for any particular problem, take a good look at the "tree" (see Figure 5 at left and read Exercise 8 again.)

You'll see that the number of possible choices diminishes by one for each succeeding unit of an arrangement (whether the unit happens to be numbers, letters or ways of entering and leaving a house). If we wish to arrange three letters, we have three possible choices for the first letter in an arrangement, two possible choices for the second letter and one possible choice for the third letter. This can be expressed in the following principles:

Principle 1. If the first of a series of maneuvers can be done in *m* ways and the second can be done in *n* ways, then together they can be done in *m* × *n* ways.

Thus, if a house has three doors, you can come in one door and go out another in six ways (3 × 2).

Principle 2. If *more* than two possibilities exist, simply multiply the number of possibilities for each choice.

Thus, if there are five different slips of paper in five different colors, the slips can be arranged in 120 ways (5 × 4 × 3 × 2 × 1). If three colors are selected from the five, they can be arranged in 60 ways (5 × 4 × 3). All of the letters of the alphabet can be arranged in 26 × 25 × 24 × 23 . . . × 3 × 2 × 1 ways. This value is written *25!* and is read as *25 factorial*.

In essence, these principles are shortcuts. They are given here primarily for the teacher's benefit. Under no circumstances should the shortcuts be taught to elementary school children unless the children have figured out the principles for themselves.

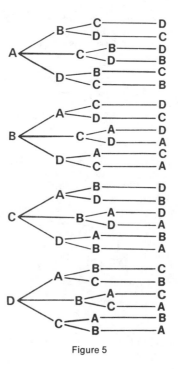

Figure 5

Figure 4

SELECTED BIBLIOGRAPHY FOR CHAPTER 8
Elementary Statistics and Probability

Baker, Dianne M. and Overholser, Jean S. "IMF for Grades 3 to 6." *The Arithmetic Teacher* (May 1969), 400.

Buxton, R. "Probability and Its Measurement." *Mathematics Teaching* (Spring 1970), 56–61.

Clarkson, David M. "A Number Pencil." *The Arithmetic Teacher* (November 1967), 557.

Coppola, Jean N. "Graphs Tell a Story." *The Arithmetic Teacher* (April 1969), 305.

Driscoll, Lucy E. "Ordered Pairs, Patterns, and Graphs in Fourth Grade." *The Arithmetic Teacher* (March 1961), 127–30.

Gold, Sheldon. "Graphing Linear Equations—A Discovery Lesson." *The Arithmetic Teacher* (May 1966), 406.

Grass, Benjamin A. "Statistics Made Simple." *The Arithmetic Teacher* (March 1965), 196–98.

Hall, Wendell C. "Pascal's Triangle in the Classroom." *New York State Mathematics Teachers' Journal* (April 1967), 56–63.

Helmes, Allen and Simon, Julian L. "A New Way to Teach Probability Statistics." *The Mathematics Teacher* (April 1969), 283–88.

Higgins, James E. "Probability With Marbles and a Juice Container." *The Arithmetic Teacher* (March 1973), 165–66.

Junge, Charlotte W. "Dots, Plots, and Profiles." *The Arithmetic Teacher* (May 1969), 371.

Kleber, Richard S. "A Classroom Illustration of a Nonintuitive Probability." *The Mathematics Teacher* (May 1969), 361–62.

Niman, John and Postman, Robert D. "Probability on the Geoboard." *The Arithmetic Teacher* (March 1973), 167–70.

Overholser, Jean S. "Hide-a-Region—$N \geqslant 2$ Can Play." *The Arithmetic Teacher* (October 1969), 496.

Rosser, Barbara. "Take a Chance With the Wheel of Fortune." *The Arithmetic Teacher* (November 1970), 614.

Schell, Leo. "Horizontal Enrichment With Graphs." *The Arithmetic Teacher* (December 1967), 654.

Sullivan, John J. "Statistics in Baseball." *New York State Mathematics Teachers' Journal* (April 1973), 75–77.

Wilkinson, Jack D. and Nelson, Owen. "Probability and Statistics—Trial Teaching in Sixth Grade." *The Arithmetic Teacher* (February 1966), 100.

CHAPTER 9

MEASUREMENT AND GEOMETRY

Measurement and geometry are major topics in elementary school mathematics programs with some concepts from each area appearing at the different grade levels. There is considerable and growing evidence* that measurement and geometry should be experienced early in one's school career.

Measurement is important not only because of its wide application (social utility), but also because the basic mathematical ideas involved in measurement are studied throughout school at all grade levels. A well-balanced program in elementary mathematics must include many opportunities for children to acquire concepts of measurement, and a variety of suggestions and ways to provide first-hand experiences are offered in this chapter.

Geometry, the study of spatial relationships, is all around us. When we observe different shapes in our environment (stop signs, leaves, etc.) or describe a path to the playground, we are doing geometry. Children are exposed to and aware of spatial relationships from their earliest days. Many of the activities in this chapter will help children realize that geometry is something that plays an important role in the world in which we live, work, and play.

In "Geometry in the Elementary School," Joseph N. Payne claims that the geometry taught must be based on (1) the maturity of the learner, (2) the interest of the learner in a given topic, (3) the need for geometry in everyday life situations, and (4) the use geometry has in teaching other topics. Dr. Payne's article reflects these goals. Be sure to read it for some teaching suggestions that should be fun and fascinating for teachers and pupils alike.

T. E. Liedtke and W. Kieren's article, "Geoboard Geometry for Preschool Children," deals with the ways preschool children react to a geoboard and the sort of learning you can expect from these youngsters when they are given geoboards. The observations in this article should stimulate those working with young children as well as provide some extra insights for those working with older children.

John J. Sullivan provides some interesting exercises in his article, "Intuitive Notions of Diameter," which should help children develop a meaningful notion of pi, the ratio of the circumference to the diameter. Try some of these worthwhile exercises with your learners.

William L. Swart claims that a measurement laboratory can be developed inexpensively (under $10) and still include all the material needed to provide mean-

---

*For additional information, see reports from studies associated with the Developing Mathematical Processes Program, Wisconsin Research and Development Center for Cognitive Learning, University of Wisconsin, Madison.

ingful experiences in measurement for all the children in a classroom. His article, "A Laboratory Plan for Teaching Measurement in Grades 1–8," should help you get started on laboratory activities, especially in measurement.

The next five articles all provide excellent backup and supporting activities for Swart's article. V. K. Brown presents ideas for teaching some abstract measurement aspects in "A System of Measurement Evolved in the Classroom." "The Platform Balance" by Albert Eiss explores several ways to encourage students to use the balance to weigh objects and apply mathematics. Ernest R. Ranucci illustrates several meaningful activities with perimeter, area, and volume in his article, "P . . . A . . . V." Perhaps you might develop some of these for laboratory activities in your own classroom. Kathryn Strangman's article, "Grids, Tiles, and Area," offers a method for introducing area to children which is mathematically sound and readily teachable. This article should help your pupils develop the understanding that area means more than "length times width."

If you are interested in correlating science and mathematics, then read "Using Stream Flow to Develop Measuring Skills" by Carlton W. Knight II and James P. Schweitzer. This outdoor activity will provide a change of pace in the daily routine while demonstrating the value of measuring skills in a practical, real-life situation.

Finally, be sure to read the article by D. Richard Bowles, "Get Ready for the Metric System." We all should become acquainted with the metric system and this article can be a good beginning.

Geometry in the Elementary School

Joseph N. Payne

EOMETRY in the elementary school is the study of positions and locations in space. A position or location is viewed as a point. Space is the set of all points. For example, a line segment, a triangle, a closed path, and a prism are each viewed as a set of points. Relations between geometric figures — *parallel, perpendicular,* and *intersection* — are also studied. For example, the intersection of two lines, as when perpendicular lines intersect, contains one point; the intersection of two parallel lines contains no points.

In the elementary school, geometry is not a simplified version of the theorems from high school geometry. Neither is it oriented to proving theorems and to understanding a postulational system, as it is in the secondary school. While the vocabulary, symbolism, and informal concepts of geometry will provide useful background for the study of geometry in the junior and senior high schools, geometry in the elementary school is more intuitive and more informal than secondary school geometry.

Goals for geometry in the elementary school must be based upon (1) the maturity of the learner; (2) the interest for the learner of a given topic; (3) the need for geometry in everyday life situations; and (4) the use geometry has in teaching other topics. The following suggestions reflect these goals.

Reprinted with permission from "Teacher's Notebook in Elementary Mathematics,"
©1967 by Harcourt Brace Jovanovich, Inc.

Use the language of geometry to describe shapes and forms of physical objects in the environment of the learner. The edge of a book, the edge of a box, a rope suspended from the ceiling, a flagpole, and a baton have two properties in common: (1) straightness and (2) each has a beginning and an end. These common properties are abstracted and named *line segment*. The line segment is viewed as a set of points, and the set of points is the geometric model for the physical objects.

A sheet of paper lying flat on a desk, a window pane, and the ceiling of a room are physical objects that share a common property. All the objects have a property of flatness. Furthermore, their edges are rectangular. Each of the objects is a physical model of the mathematical concept of rectangular region. None of the objects is perfectly flat. None is bounded by a perfect rectangle. But the mathematical idea that is taught is one of "perfectly flat" and bounded by a "perfect rectangle." The rectangular region is the mental idealization of these properties.

A tennis ball, a soap bubble, and the rind of an orange are physical models for the geometric idea of sphere, although a sphere has none of the slight irregularities that occur on the surface of the physical objects. By describing each of the objects as spherical, one is choosing the mathematical idea from geometry that comes closest to fitting the physical object.

Use geometric ideas of line segment, plane region, and solid region to develop ideas of measurement. Geometry is a necessary background for a precise and correct development of concepts of area, and linear and volume measure. Linear measure is based on the concept of line segment. For example, the height of a table is determined by thinking of a line segment such as $\overline{AB}$* that joins a point at the edge of the top and a point on the floor such that this line segment is perpendicular to the floor *(Figure 1)*. The length

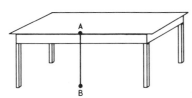

Figure 1. When we measure the height of a table, we mentally construct a line segment from a point on the edge of a table to a point on the floor.

of this segment is determined by comparing it with a given unit segment. Likewise, the area of a floor is viewed as the area of a rectangular region for which the floor is a model; the area is found by comparing a unit region with the rectangular region. Volume measures are established in the same way by comparing a solid region with a given unit region.

Use concepts of geometry to develop ideas of number and operations on number. The "number line" is well accepted as a useful tool in elementary school mathematics, but the line is a geometric idea quite distinct from number. The connection between the line and number is established by choosing a point

which can be thought of as a reference point or beginning point. Then, choosing another point to the right, a segment is determined and considered as a unit, or one *(Figure 2-a)*. By similarly marking other points to obtain other congruent segments, one has a model that is useful for ascertaining the number property of a discrete set where the objects are distinguishable, or a continuous set where some partitioning must be done before a number can be assigned. The final model, which is the number line with the numbers 0, 1, 2, 3, . . ., shows the number of segments or the number of steps from the starting place matched with 0 *(Figure 2-b)*.

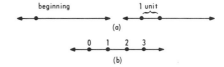

Figure 2. The number line, which is a model of a line segment, is useful in developing concepts of number.

The number line can also be used as a model when teaching concepts of fractional numbers, rational numbers, and integers. For example, three and two thirds is matched with point A because three and two thirds shows the number of units from zero, measuring to the right. The number ⁻2 is matched with point B because point B is two units to the left of the starting place, 0 *(Figure 3)*.

Figure 3. The number line is useful in teaching concepts of fractional and negative numbers.

Unit line segments and unit regions are particularly helpful in developing the concept of fractions and equivalent fractions. The shaded region in (a) shows that one out of two parts of equal size is being considered *(Figure 4)*. The shaded part of the same region, shown in (b), shows that two out of four parts of equal size are being considered. The concept of

Figure 4. The concept of region is useful in developing concepts of fraction and equivalent fraction.

equivalent fractions is best introduced as fractions that show *the same amount* of the region. Hence one half and two fourths name the same amount or the same part of the unit. Similarly, one half shows one of two equal sized parts of segment AB *(Figure 5)*.

*You read this as "line segment AB." When you use the words "line segment" in a written statement, you would not use the line over the letters.

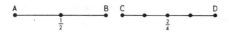

Figure 5. Line segments are also useful in developing concepts of equivalent fraction.

Two fourths shows two of the four equal sized parts of segment CD. Both fractions match the corresponding points because each names the same amount or same part of the line segment. The use of such geometric ideas to develop the concept of equivalent fractions supplements and provides a transition from work with physical objects.

Rectangular regions provide one model for finding the product of fractional numbers. For example, the shaded region at the right measures $\frac{2}{3}$ by $\frac{4}{5}$ *(Figure 6)*. Having established that the area of a rectangular region with measures that are whole numbers can be found by multiplication, it is sensible that multiplication will also produce the area of this region. The area is observed easily as being $\frac{8}{15}$. The model of a rectangular region also provides a visual model that shows why numerators are multiplied to get the number of parts and why denominators are multiplied to get the number of parts in all.

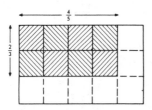

Figure 6. The area of the shaded region may be found by multiplying $\frac{2}{3}$ and $\frac{4}{5}$ to produce $\frac{8}{15}$.

A plane as a set of points is studied prior to the introduction of graphing in the plane where points of a plane are associated with pairs of numbers. For example, as pairs of numbers that make the equation $\square + \triangle = 6$ true are studied, a point can be associated with each pair as shown in the diagram *(Figure 7)*. The geometric representation helps in the perception of the number-pair idea and in determining the number of pairs that make the equation true.

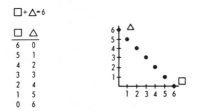

$\square + \triangle = 6$

$\square$	$\triangle$
6	0
5	1
4	2
3	3
2	4
1	5
0	6

Figure 7.

Use geometry as a vehicle for developing analytical and critical thinking. Because of the continuing use of visual models to teach geometric concepts, conjectures and discoveries are made which pupils

can test through their own drawings and observations. Making and testing conjectures is an essential ingredient of analytical or critical thinking. For example, the question of how many triangles are in figure 8 has a geometric goal of verifying the idea of tri-

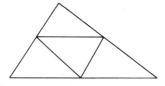

Figure 8. Close observation shows five triangles in this figure.

angle itself. However, a more important goal is to provide the opportunity for studying the figure and seeing the fifth triangle, which may not have been at all obvious at first glance.

After studying the area and perimeter of a rectangular region, the teacher might ask the following questions: "If I have 40′ of string and make a square that measures 10′ by 10′, would I enclose the same area if I took the same string and made a rectangle that measures 5′ by 15′? Tell me what you think before you do any computation." The surprise often registered by a pupil when he recognizes that the areas are not the same, results from his implicit conjecture that rectangular regions with the same perimeter have the same areas *(Figure 9)*.

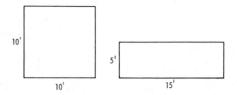

Figure 9. The area of the square is 100′ and the area of the rectangle is 75′. The perimeter of each figure is 40′.

Having disproved that conjecture with a single example, the question naturally arises, "What is the maximum area that I could enclose by using a rectangular region?" The stage is then set for further drawings and experimentation, culminating with a generalization that for a given perimeter the maximum area is enclosed by a square. A question such as "Can I make the area as small as I wish?" leads to the generalization that there is no minimum area for a given perimeter. Similarly, we may ask questions about the volume of a rectangular prism if a given surface is known. Intuition tends to support the notion that the surface area determines the volume. Advertisers often take advantage of this by packaging many products — grocery products, for example — so that the contents of a package are spread over a large area, giving the appearance of a greater volume.

Disproving conjectures such as these not only improves pupils' analytical or critical thinking, but also leads to results that have practical and utilitarian value.

174

Begin with physical models and make extensive use of physical models throughout the study of geometry. When teaching geometry, the teacher should keep in mind the goal of abstracting common properties of physical objects as the primary way geometric ideas are formulated in the minds of students. Physical models should be used to introduce an idea. As the content is developed, physical models and drawings can be used extensively to assist with spatial visualization so that students make conjectures and discoveries. Note, for example, how the idea of line segment may be introduced and developed.

Teacher: I am thinking of something, and I want you to guess what I am thinking. I'll show you some examples. (The teacher then runs her finger along the edge of a book, the edge of a window, the string that is hanging from a window shade, the edge of a board, and concludes by moving the tip of her finger from one point in the air to another point in the air in a straight path.) *Student:* You are thinking of something straight.

Teacher: That is part of the idea that I was thinking. What I am thinking about is straight, but there is another part of the idea. *Student:* Something straight and hard.

Teacher: When I moved my finger in the air, did I show something hard? *Student:* Yes. Your finger.

Teacher: No, I was not thinking of my finger as the example. I was thinking of the way the tip of my finger moved.

It is common for students to describe concrete properties of objects such as hardness, color, and texture. It is the responsibility of the teacher to direct the student's attention to the more abstract geometric idea.

The second idea that the teacher is trying to draw from the students is the notion of the beginning and the end of this "straightness." The following questions might help:

Teacher: Did I begin and end at a certain place? Did I begin at one corner of the book and move to another corner? Did I begin at one edge of the window pane and move to another edge? Did I begin at one place in space and move to another place? Did I begin at one point on the string and move to another point on the string? The two parts of the thing I was thinking about are (1) straightness and (2) it begins and ends at two different points.

Now that the essential characteristics have been abstracted from the physical objects, the teacher may proceed to introduce the name *line segment,* emphasizing that segment means part. Hence *line segment* means part of a line. The introduction of the symbol $\overline{AB}$ to designate the line segment that begins at A and ends at B will depend on the maturity of the learners and their facility with symbolization.

Now the concept that a line segment is straight, has two endpoints, and is a set of points can be summarized. Examples such as those below reinforce and deepen the idea.

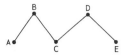

Example: Name the line segments in the figure above. How many line segments can be determined using only the letters shown? (Four are drawn: $\overline{AB}$, $\overline{BC}$, $\overline{CD}$, and $\overline{ED}$. Others could be named: $\overline{EC}$, $\overline{EB}$, $\overline{EA}$, $\overline{DB}$, $\overline{DA}$, $\overline{CA}$.)

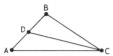

Example: How many line segments are shown in this figure? (6) Name each of them. ($\overline{AC}$, $\overline{AB}$, $\overline{BC}$, $\overline{DC}$, $\overline{AD}$, $\overline{DB}$.)

Example: How many line segments are there that contain E as one endpoint? (More than can be counted.)

Figure 10.

Following the same procedure, ideas such as circle, rectangle, rectangular region, prism, cylinder, and sphere can be developed. Physical models for most ideas are readily available in the classroom.

Introduce vocabulary and symbolism only after the geometric ideas are developed. While the proper use of the vocabulary and symbolism of geometry is a long-term goal of geometry in the elementary school, the more important goal is the understanding of the ideas which the words and symbols signify. The questions in the lesson suggested for the development of line segment clearly emphasized the idea of line segment before introducing the correct language and symbolism.

The precision of language which ultimately results from teaching geometry is a long-term goal reached after repeated and thorough development of the ideas. The emphasis on formal definition, on precise and short verbal descriptions of geometric ideas, has relatively little place in the elementary school, particularly in the middle and lower grades. The primary emphasis is on using the symbols and vocabulary to express already-developed ideas.

For example, upon introduction of sphere as the mathematical model that fits several physical objects, a teacher might ask how the idea could be described in words. At the middle-grade levels, the teacher might be satisfied with a description such as "perfectly round, no bulges." At the upper elementary and junior high school levels, the teacher might ask further questions leading to a more precise statement of the definition of a sphere. To the reply that a student might make that a sphere is perfectly round, the teacher could ask, "What do you mean — perfectly round? Is an egg perfectly round?" Through such questions, pupils can be led to see the need for greater precision of language and can be directed to the notion that a sphere is a set of points each of which is the same distance from a given point called the center. This formal definition is not likely to come from the students themselves, but they are more willing to accept it if the discussion has pointed to a need for greater precision.

Clearly the emphasis in teaching geometry is on the ideas, the mental manipulation of ideas, conjectures, and discoveries, with concomitant de-emphasis of formally precise statements to be stated verbatim by pupils.

Teach geometry using an intuitive and exploratory way to stimulate imagination and to improve spatial perception. An exploratory approach to the teaching of geometry involves the use of physical models and appropriate questions asked of the pupils in the class. For example, when teaching the notion of the intersection of two planes, a teacher might begin with this question, "Can two planes come together at just one point?" Pupils may reply "Yes" and use a sheet of paper and a desk as physical models to show a corner of the paper touching the desk in just one place. Then the teacher can ask, "Does the sheet of paper make you think of a plane?" To this, she might reply that both the sheet of paper and the plane are flat.

However, the plane goes on and on with no limit. The teacher may turn to examples within the classroom. The rear and side walls are parts of planes, for instance. She may ask, "If these are parts of planes, how do they intersect? Could they intersect at just one point?" This questioning leads to a full explanation of intersection of planes in a line and to the notion that either two planes intersect in a line or are parallel. In a similar way, a question such as the following can be explored: "How many planes contain two given points?" A teacher might choose a point at the top hinge of the classroom door and a point at the bottom hinge of the door. The door, in its given position, would represent a plane. "Can the door be in only one position? In how many positions could I put the door? Then how many planes contain two points?" This leads to the notion that two points, or a line segment determined by two points, contains more planes than can be counted, an infinite number of planes.

The concept of symmetry may be similarly explored. A plane figure is symmetric to a line if the line separates the figure so that one side is the mirror-image reflection of the other side. An ordinary hand mirror is a good device for testing symmetry to a line.

A closed surface is symmetric to a plane if the plane separates the figure so that one side is the mirror-image or reflection of the other. To understand symmetry with respect to a plane, the students can imagine a plane passing through the surface so that what is on one side is the mirror image of what is on the other. Finding objects in the physical environment whose geometric models have symmetry is interesting, imaginative, and improves the child's awareness of shape and form in his environment.

Developing the intersection of a sphere and plane is prerequisite to an adequate treatment of latitude and longitude, where circles of latitude and longitude are representations of the intersection of a plane through the center of a sphere. After this geometric idea is introduced, locating points on the earth becomes a simple adaptation of the work on graphing in a plane.

Expect different kinds of mastery for geometry than for number. Geometry teaching need not be as tightly sequenced as the teaching of number. It is, for example, difficult to develop the multiplication algorithms before pupils have mastered multiplication facts. However, pupils can study symmetry without having fully assimilated the notion that a geometric figure is a set of points. The geometry development is also more open for extension and refinement than is the idea of number.

It is reasonable to expect pupils to name geometric figures correctly, but it may be difficult for them to make formal definitions. Thus we should evaluate how well students have grasped ideas and how well they recognize and use terminology. Setting expectations for mastery that inhibit the learner or make him uncomfortable about the study of geometry will not achieve the goals.

Involve the students in physical activity, particularly when introducing an idea. To introduce the concept of area, you might ask children to cover regions with thin square blocks or paper squares. They may then easily imagine the notion of area within a plane.

To develop the idea that the sum of the measure of the angles of a triangle is 180°, you might ask children to fold a sheet of paper representing a triangular region as you ask the following questions:
(1) Find the midpoints of $\overline{AB}$ and $\overline{BC}$ by folding at the points marked (as shown in *Figure 11-a*).

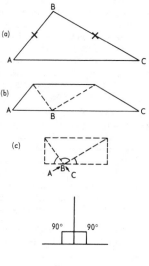

Figure 11.

176

(2) Then fold at the midpoints so that B touches $\overline{AC}$ (as in *Figure 11-b*).

(3) Now fold A over B and C over to B and crease (as in *Figure 11-c*).

The three original angles of the triangle form a figure whose base is a straight angle of 180°. Hence the sum of the measures of the angles of a triangle is 180°. The paper can be unfolded to verify that the three angles were the original angles of the triangle.

Constructions with a compass and straight-edge are fun for pupils, particularly at the middle- and upper-grade levels. They can construct angles, perpendiculars, circles, and intersecting geometric patterns. And they can use paper to make models of closed surfaces such as pyramids, cubes, and cones.

These are only a few teaching suggestions that are fun and fascinating for teachers and pupils. Many teachers will have their own to add to this list. Teaching methods should, however, reflect the unique goals for teaching elementary school geometry: that the language of geometry be used to describe shapes and forms of physical objects in the student's environment; that geometric concepts be used to help teach other mathematical topics; that we use geometry as a vehicle for developing analytical and critical thinking.

Geoboard geometry for preschool children

W. LIEDTKE and T. E. KIEREN

There is increasing emphasis today on preschool experiences in mathematics for children to capitalize on their eagerness to explore and interpret the world around them. This paper explores a wide variety of experiences based upon the use of a geoboard and rubber bands. The geoboard provides many opportunities to acquaint children with geometric concepts. Learning from their own play and from imitation of adults and other children, it is not long before they can recognize, correctly label, and form for themselves many common geometrical figures and instances of geometric properties, as well as such common shapes as letters of the alphabet. The illustrations that follow were drawn from the authors' observations of children aged 2–6 as they worked individually and in groups with the geoboard. Questions asked and possible suggestions given are classified under three main headings: *Familiar Shapes; Plane Figures;* and *Segments.*

Free activity

An excellent way to begin is by providing each youngster with a geoboard and a rubber band and letting him do whatever he wishes. While he works, he could be motivated to show and talk about what he has made. Recent experiences and objects from his immediate environment will be represented by various ingenious constructions and configurations. Some sample responses from a "free activity" period are presented on the following page (Our geoboards were 5 by 5 inches, and the pegs were about 1 inch apart.) The comments recorded represent the first responses of the children. Often, slight modification or even rotation of the geoboard led to the assigning of different names to very similar figures.

One sequence may begin by giving the children one rubber band, later increasing the number to two or even three, and challenging them to make something that was not possible before.

One rubber band:

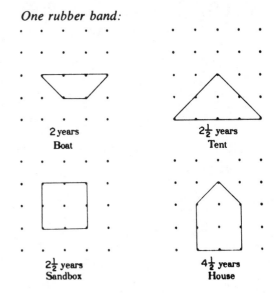

2 years
Boat

2½ years
Tent

2½ years
Sandbox

4½ years
House

Reprinted from *The Arithmetic Teacher* (February 1970), 123–126, by permission of the authors and the publisher.

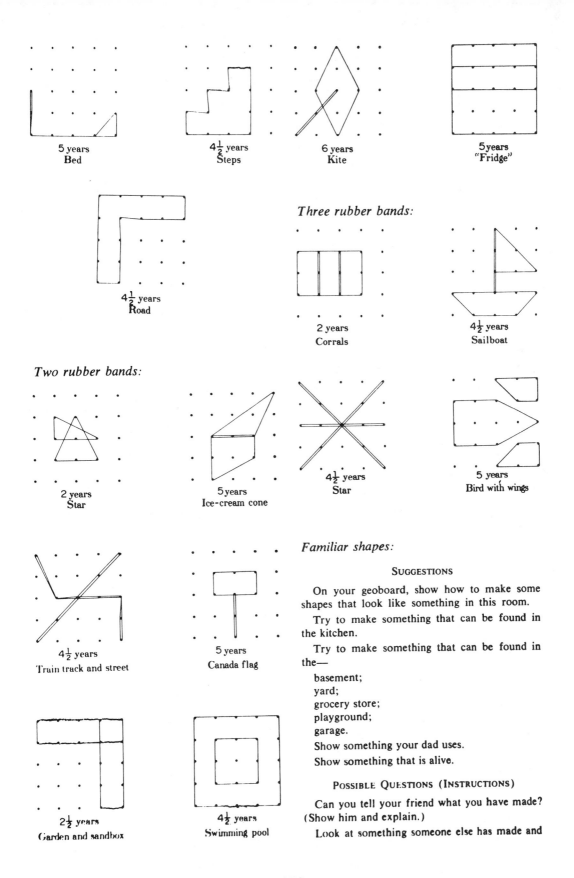

5 years
Bed

$4\frac{1}{2}$ years
Steps

6 years
Kite

5 years
"Fridge"

$4\frac{1}{2}$ years
Road

Three rubber bands:

2 years
Corrals

$4\frac{1}{2}$ years
Sailboat

Two rubber bands:

2 years
Star

5 years
Ice-cream cone

$4\frac{1}{2}$ years
Star

5 years
Bird with wings

$4\frac{1}{2}$ years
Train track and street

5 years
Canada flag

2$\frac{1}{2}$ years
Garden and sandbox

$4\frac{1}{2}$ years
Swimming pool

Familiar shapes:

SUGGESTIONS

On your geoboard, show how to make some shapes that look like something in this room.

Try to make something that can be found in the kitchen.

Try to make something that can be found in the—

basement;
yard;
grocery store;
playground;
garage.

Show something your dad uses.

Show something that is alive.

POSSIBLE QUESTIONS (INSTRUCTIONS)

Can you tell your friend what you have made? (Show him and explain.)

Look at something someone else has made and

179

try to guess what it is. (Ask for a hint where it can be found.)

Does your figure look the same if you turn the geoboard around?

How many sides does your figure have?

How many corners does your figure have? (Are there more corners or more sides?)

Sample responses:

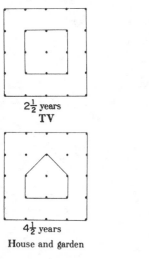

2½ years
TV

2½ years
Slide

4½ years
House and garden

4½ years
Hammer

Plane Figures:

SUGGESTIONS

Try to make figures with three sides that are—
small;
large;
"skinny";
"fat."

Try to make figures with four sides that are:
long;
short;
long and wide;
long and narrow;
short and wide;
short and narrow;
"like a square";
"not like a square."

Try to make figures with "many sides."

POSSIBLE QUESTIONS (INSTRUCTIONS)

What does the figure you have made remind you of? Does it look like anything that is familiar to you? (Where did you see something like it before?)

Does the figure change if you turn your geoboard?

Make two figures that: (1) do not touch; (2) touch; (3) cut into each other. Look at the figures you have made.

Can you make another one that looks just like it—but smaller, (or bigger)?

Make a triangle and a square. How are they alike? How are they different?

Sample responses:

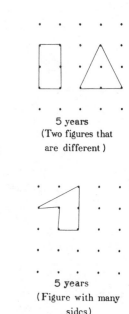

4½ years
(Make another
triangle like it,
but bigger)

5 years
(Two figures that
are different)

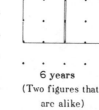

6 years
(Two figures that
are alike)

5 years
(Figure with many
sides)

Segments:

SUGGESTIONS

Try to make segments that are—
short;
long;
straight;
"crooked."

Try to make segments that—
do not touch;
touch;
cross each other (intersect);
will "never" touch (parallel);
are exactly on top of each other.

Try to make various segments—
leading to two (or more) points;
various numbers of segments, i.e., two that are equal;
two that are not equal;
many different segments.

POSSIBLE QUESTIONS (INSTRUCTIONS)

How would you make a road?

Can you make a very narrow road?

Can you make one that is long and narrow?

Make a railroad track. (If possible provide rubber bands of different colors.)

Can you make a road and a train track that cross? . . . do not cross? . . . will never cross?

Look at two pegs in different corners. How many different roads—crooked or straight; few or many corners—can you build between these two pegs?

Which road would you like to travel on? Why?

Sample responses:

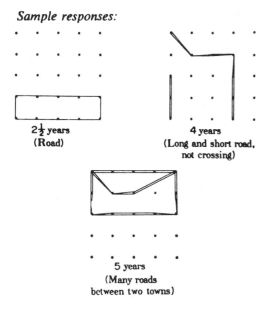

2½ years
(Road)

4 years
(Long and short road, not crossing)

5 years
(Many roads between two towns)

Summary and additional suggestions

The previously outlined activities present one possible way to begin a session with a geoboard. Since the children work with creations on their own that differ in many respects, the activities are open-ended. It will soon become evident that any session will be a combination of what has been suggested. Depending on the age and background of the children, they will interpret instructions and questions in various ways. They will give unique replies that often lead to some idea that was not intended at the outset. While working, some children will recognize configurations that suddenly remind them of something familiar. For example, while talking about roads and train tracks (segments), one girl looked at her constructed figure and remarked, "A 'T' on a line!" The question was raised, "Does it look like a 'T' if you turn your geoboard around?" "No, now it's an 'H'." The question "Can you make other letters on your

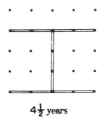

4½ years

geoboard?" resulted in having two children make the following:

I, D, L, M, N, 4.

The last response led to an attempt to build more numerals. Thus the topic of segments led to letters, numerals, and sets, and it could have also been used to discover something about angles (i.e., right angles).

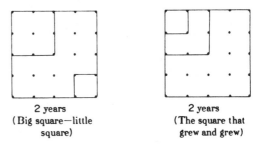

2 years
(Big square—little square)

2 years
(The square that grew and grew)

Similarly, the topic of "big and little" figures can lead to the discovery of some of the properties for similar figures (corresponding sides and vertices). Having children attempt to copy a figure can lead to discovery of some properties of congruence. Some children will exhibit an awareness of symmetry (e.g., "birds with wings"). Most of them will easily pick up such terms as triangle, square, and rectangle and use these terms correctly. Some will talk of polygons and angles, and a few might even be led to discover such polygons as parallelograms, quadrilaterals, or trapezoids.

In allowing for these developments it should be remembered that these children develop ideas through both imitation and free play. By imitating an adult or another child, the child may get an idea he never had before. But free play allows him to expand on the ideas and capabilities that he already possesses.

181

INTUITIVE NOTIONS OF DIAMETER

John J. Sullivan

On a visit to some mathematics classes participating in England's Nuffield Foundation Mathematics Teaching Project, this author noticed that investigation of circle properties using cylindrical food containers was a popular classroom activity in primary schools. Usually, each student was asked to bring a cylindrical can to class. Using strips of squared paper, each child measured various parts of a can — area of the circular base, lateral area, circumference, and diameter.

One of the principal aims of the exercise is to help children develop intuitively the notion of pi, the ratio of circumference to diameter. A table like the following would be constructed:

Circumference	4.3	5.2	6.4	...
Diameter	1.4	1.7	2.1	...

With entries from each of 30 or so class members, it is rather likely that the children will notice a definite relationship between measures of circumference and those of diameter.

The exercise is a good one for doing what elementary school geometry programs are supposed to do: introduce important mathematical concepts intuitively, preferably in interesting ways, making use of concrete materials.

The fact that measurement is always approximate receives needed emphasis in an exercise like this, since it is illustrated pointedly as children seek ways of finding an accurate measure of diameter. It is striking that the means children use to measure diameters of cans demonstrate intuitive understanding of the properties of diameters. The following are some of the methods they use; in brackets are statements of the geometric principles involved.

1. Use a protractor and straightedge as shown in Figure 1. [A line perpendicular to a chord at its midpoint contains the center of the circle.]

2. Use a strip of squared paper as one would use a tape measure (Figure 2). Move the strip across the upper base of the can trying to find a measure of the *longest* chord. A method very similar to this is to draw the circle on a piece of graph paper (Figure 3). This provides a permanent picture of some chords of the circle. "Counting squares" gives a measure of area. [A diameter is the longest chord of a circle.]

3. Use a card with a square corner, such as a 3 × 5 file card (Figure 4). If the vertex of a square corner is put on the circumference, the edges of the card adjacent to the vertex meet the circle in two endpoints of a diameter. [A right angle inscribed in a circle intercepts a semicircle.]

Reprinted with permission from the *New York State Mathematics Teachers' Journal* (January 1970), 2–3.

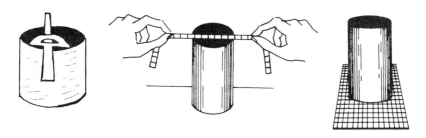

Figure 1.   Figure 2.   Figure 3.

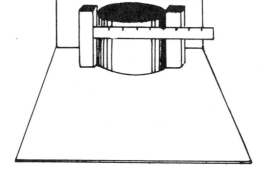

Figure 4.   Figure 5.

4. Use blocks of wood, as shown in Figure 5, and measure the distance between the faces of the blocks which are touching the can. [A tangent to a circle is perpendicular to the diameter drawn to the point of tangency.]

5. Draw the base circle of the can on a sheet of paper. Cut out the circular region and fold it so that the two parts match. [A diameter divides a circle into two congruent semicircles.]

This activity has the desirable feature of being open-ended. That is, it leads to other questions for interested students. Some of these are:

1. Is there a relationship between area of base and diameter?
2. Is there a relationship between lateral area and height?
3. Is there a relationship between volume and area of base?

183

A laboratory plan for teaching measurement in Grades 1-8

WILLIAM L. SWART

Give a layman the job of teaching an Australian aborigine how to weigh kangaroo steaks, and the layman's request for equipment will probably include a set of scales and various objects to weigh. When we, as professional teachers, are assigned a similar task with children, we request a book with pictures of scales and pictures of weights.

We claim to subscribe to these principles: (*a*) that the most fruitful instruction is based on meaningful *experience;* (*b*) that learning is best facilitated by proceeding from the *concrete* to the abstract; and (*c*) that the easiest and longest-lasting learning comes from *doing.* And yet topics in measurement, which lend themselves very well to these principles, are commonly taught abstractly and vicariously.

At worst, the child merely looks at pictures and diagrams in a text and does paper-and-pencil exercises. At best, the class observes the teacher (or another child) as measurement instruments are manipulated.

Developing a laboratory

A measurement laboratory can be developed inexpensively (for under $10) to include all the material needed to provide meaningful experiences in measurement for all the children in a classroom.

THE MATERIAL (A PARTIAL LIST)

Item	Source	Cost*
Thermometer	Supermarket or hardware	Under $1
Kitchen or baby scales	Used furniture store, Salvation Army, Goodwill	Under $1
Cubic-inch blocks	School Supply Co.	Under $3
Cubic foot	R. H. Stone Products, Box 414, Detroit	Under $5
Square foot	Cut cardboard	$0
Square inches	Tile store	Under $1
Milk containers (pt., qt., gal., etc.)	Home	$0
Gallon jug	Child donation	$0
Weights (of sand, etc.)	Various	$0
Baby bottle, with ounce graduations	Child donation	$0
Kitchen measuring cup	Supermarket	$ .25
Coffee cup	School kitchen	$0
Eye dropper	Medicine bottle	$0
Funnel	Supermarket	$0
Teaspoon	Kitchen	$0
Tablespoon	Kitchen	$0
Ruler	Teacher's desk	$0
Yardstick	Lumber company	$0
Tape measure	Hardware	$1

* The cost of the material is inversely proportioned to the teacher's scavenger instinct.

Reprinted from *The Arithmetic Teacher* (December 1967), 652–653, by permission of the author and the publisher.

THE PROCEDURE

Problem directions are typed on 3 × 5 cards, or printed on 5 × 8 cards, depending on grade level. Each card directs the child to go to the laboratory table (or shelves, or corner) and do a specific task. He then reports his findings to the teacher, either verbally or on paper. For instance, card 7 might say this:

Fill a glass with cold water at the sink. Put the thermometer in the water and leave it there until you think the mercury has gone up or down as far as it is going to go. Record the temperature on your paper. Now fill the glass with hot water and repeat the process.

The teacher might wish to provide another set of cards with answers. Of course, the answers to some, such as the above example, will vary. And we can even expect some degree of variation in answers to the problem of measuring the length of the teacher's desk.

Three aspects of this plan set it apart from the usual procedures for teaching measurement: (1) It is *not abstract* or pictorial. The child handles the measuring instruments and the objects to be measured. (2) The measurement activity is *not a unit* of work, but is accomplished gradually throughout the year. (3) It is *individual*. One child at a time removes a card from the box and does what he is directed to do.

SAMPLE EXERCISES

1. Place the square-inch tiles on top of a shoe box until it is covered. How many square inches does it take to cover the box?

2. Take the cubic foot to the filing cabinet. See if you can figure out how many cubic feet the cabinet would hold.

3. See how much your arithmetic book and reading book weigh together.

4. Find the milk carton that holds one pint. Fill it with water and pour the water into the gallon jug. How many pints does the jug hold?

5. Put cubic inches in the box marked "A" until it is full. How many does it hold? How many layers does it hold?

RECORD KEEPING

For those activities that the teacher feels should, if possible, be experienced by all the children, the following record might be maintained:

PUPIL	CARD NUMBER					
	1	2	3	4	5	...
Bobby	✓		✓	✓		
Sue		✓			✓	
Billy			✓	✗	✓	
James	✓	✗		✗		

A check mark indicates successful completion.

An "X" indicates the task is too difficult for that child and will not be expected of him.

SCOPE

The plan is extremely flexible. As the teacher prepares the cards, he will quite naturally use vocabulary appropriate to his pupils' grade level. For the primary grades, the instructions can be printed in large letters. When it becomes apparent that a certain task is inappropriate, it is a simple matter to throw the card away or revise it. For a task that should be done more than once to reinforce the concept, one can simply make more cards with some variation from one card to another.

It is not suggested that these activities should displace development of, and practice with, certain measurement formulas; but this kind of background will make such work more meaningful and better remembered.

Another virtue is that the laboratory needn't be completely developed to be useful. The file of exercises will grow as the teacher's ideas occur; and the collection of material will grow as the teacher comes across various items in the store, the attic, the junkyard, and sometimes the school supply catalog.

A SYSTEM OF MEASUREMENT EVOLVED IN THE CLASSROOM

V. K. Brown

This article presents ideas for teaching some abstract aspects of measurement at an elementary level. The aspects covered are:

A. Measurement is a descriptive process.

B. Objects are measured in terms of a fixed standard.

C. The standard used is an arbitrary one.

D. A standard can be interpolated if it is too large.

E. The logic used in the construction of a micrometer.

F. A standard of length gives us a unit of area and a unit of volume. These are called *derived* units.

Materials: C-clamps, rulers, paper, and micrometer (if available).

Procedure: Ask students to accurately describe the tops of their desks. To those who merely reply that it is a certain color and is wider than it is deep; the reply can be made, "So is a boxcar, or a matchbook." Probably one student will discover that he can measure his desk top and that he can include these dimensions in his description. Point out that the student made reference not only to numbers but to standards (feet and inches) in his description. Changing the standard changes the description; if a desk is 32 inches wide, it cannot be considered as 32 feet wide.

Have each student measure the width of his desk top in terms of a new standard, *i.e.,* the *hand.* The *hand* is just the width of the hand at the base of the fingers. When each has made his measurement, list the measurements obtained on the chalkboard. Point out that some measurements are different from others (probably the greatest difference will be between a large boy and a small girl). To be more accurate, therefore, we must decide upon just one person's hand as a standard. The teacher may select a student and mark the width of that student's hand on the chalkboard. Then have each student in the class bring up a piece of paper and pencil and mark that width on the edge of his paper. Using this standard, have each student remeasure his desk. Do the results from this measurement agree?

If the students will compare their pieces of paper, they will find that the marks they made do not agree precisely with those made by everyone else. This error can be accounted for in part by the width of the chalk lines—or the original standard—and the need for an even more accurate standard should be pointed out.

When measurements do not come out in an even number of hands, we can interpolate by using approximate fractions. (Perhaps some students did this during their first measurement of their desks, and gave the width as 8½ hands.) For more careful measurements, we may divide a hand into tenths quite precisely as follows: Make two dots, Z and Y, which are just one hand apart. Draw a line from Z to Y.

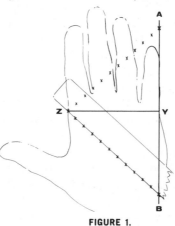

FIGURE 1.

Draw line AB through dot Y (this line may or may not be perpendicular to line ZY; the result is the same). Now place a centimeter ruler with the one centimeter mark at dot Z, and move the other end of the ruler downward on line AB until the eleven centimeter mark indicates exactly the intersection of the edge of the ruler with line AB. (See Figure 1.) Keeping the ruler in this position, mark the position of each centimeter mark on the paper (shown in x's in the diagram). Repeat the procedure, this time moving the ruler upward until the edge, the 11 cm line, and line AB all intersect. Connecting the corresponding x's will now produce nine parallel lines to AB, dividing the original distance of one hand into ten equal parts. (Figure 2.)

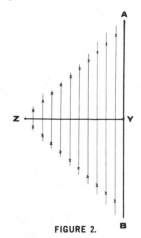

FIGURE 2.

Using a C-clamp, another method of interpolation can be introduced. Open the clamp until it is just one hand wide. Now close the jaws of the clamp, counting the number of turns (mark the threaded screw with paint or chalk to indicate when one turn of the bolt has been completed) needed to do so. If it takes 20 turns to close the clamp, then each turn must have brought the jaws 1/20th of a hand closer together. If we place the clamp snugly on a book, remove the book and then find that it takes 12 turns to close the jaws, then the book must have been 12/20ths of a hand thick. This can be further re-fined by dividing each turn into frac-tions. This is the basis for the opera-tion of the micrometer.

One further idea is to have stu-dents calculate the surface area of their desk tops in hands. (Multiply-ing the length by the width.) Point out that they have to multiply *hands* by *hands;* and, therefore, they now have *hands squared*. The unit of area must then be the *square hand*. When we do the same thing with inches, we get square inches. Using this process, what would the stu-dents have to do to calculate the volume of an object?

The Platform Balance

ALBERT EISS

THERE is ample justification for beginning to teach measurement to children before they understand all the underlying principles of the process. If we were to insist that children be capable of understanding all about the topic we are teaching before we begin, we would never be able to teach.

Concepts cannot be completely realized in a child's mind all at once. Instead, he must develop a perspective gradually through working and experimenting with many facets of major ideas. The yardstick, the meter stick, and the platform balance are elementary class-room instruments that can be used effectively to help children develop their understandings of concepts of measurement.

Linear Measurement

Children encounter fewer problems in measuring with a meter stick than with a yardstick. This is generally true because the meter stick is based on the metric system. As long as the teacher does not expect the children to convert their measurements from the metric system to the English system, they will have few difficul-ties. (When children are introduced to both systems at the same time, they will often wonder why anyone wishes to use the yardstick. Perhaps within the next generation we can dispense with it altogether.)

Linear measurement should parallel the student's study of mathematics. As soon as children have learned to multiply, they should learn to compute areas and volumes and begin, through the use of graph paper and blocks, to develop an understanding of such concepts as volume and area.

Platform Balance for Measurement

The platform balance can become an essential and useful tool in the elementary classroom for teaching concepts of measurement. Often, this instrument is not utilized for this purpose because the teacher does not know how to apply it to measurement lessons. There are, however, several ways in which students can be en-couraged to use the balance to weigh objects and apply mathematics.

Homemade balances and individual kits have their place in teaching children the principles of weighing. But, they are not substitutes for commercial balances to familiarize children with the metric system and the skill of precise measurement.

A double platform balance and set of weights can be purchased from a commercial supply house for approximately $25 - $30. The double and triple beam types eliminate the danger of losing weights essential to the weighing process. However, the danger of losing the small weights is not without its compensating features. Care and use of these small weights provide an excellent opportunity to emphasize to students that all work in science must be done carefully and that there is no place for careless, sloppy work.

Primary-Grade Activities

Most children in primary grades have had little experience in exploring the concept of weight and have not developed skills in weighing. However, primary

This balancing exercise enables students to vis-ualize a simple equation.

The weight of a small object can be calculated by dividing the total weight of a pile of objects by the number weighed.

students enjoy every type of activity, and will spend many interested sessions simply weighing various objects in the classroom. This activity will help them develop skill in the use of weighing equipment and will give them some basis for estimating the weight of various objects. They especially enjoy guessing contests, where they try to see who can make the closest guess of the weight of various objects.

The pupils can be encouraged to graph the rate of growth of a pet or to weigh the amount of food given to the pet each day. They may also try weighing a dish of ice and compare it with the weight after the ice has melted.

The following activity is very effective in teaching primary-grade children addition and subtraction as well as simple algebra. The materials needed are a dozen plastic bottles[1] (transparent), a pound of salt, and a platform balance.

Procedure:

1. Begin by finding two bottles that weigh the same. (Remember to include the cover when weighing every bottle.) Label both of these bottles No. 1.

2. Place the two No. 1 bottles on the left-hand pan of the balance. Place a bottle labeled No. 2 on the right-hand pan. Add just enough salt to this bottle to balance the two No. 1 bottles on the left. To insure accurate measurement, reverse the bottles on the two pans.

3. Place a No. 1 bottle and the No. 2 bottle on the left-hand pan of the balance and a bottle labeled No. 3 on the right and proceed as above.

4. Repeat this process until all the bottles are appropriately weighted and labeled. Be sure to switch the objects on the pans each time to insure balance.

You are now ready for mathematics class. Place bottle No. 3 and bottle No. 5 on one pan and ask the

[1] Plastic pill bottles can be obtained from the pharmacist at your local drug store.

children what they need to place on the other pan to balance them. They will be eager to try combinations, and eventually they will gain an understanding of equations; *i.e.,* weight of the objects in one pan equals the weight of the objects in the other. If you deliberately place unequal weights on the pans, you can have the children solve for the missing quantity ("X") necessary to achieve balance.

Intermediate-Grade Activities

For children who have learned to multiply and divide, an interesting activity can be introduced. Obtain several dozen of uniform size machine screws, wood screws, or washers. Ordinary nails are generally not uniform enough for this activity. Paper clips will do nicely. Weigh 20 of the objects to be used; then divide by 20 to find the weight of a single object. Now, put an unknown quantity of the objects on the left-hand pan and weigh them carefully. The children will be surprised to find that by dividing the total weight by the weight of one of the objects, they will be able to determine how many there are in the left-hand pan. (This technique is used commercially to count large quantities of small uniform objects.)

The children will learn how to use the small counterweight on the balance beam to weigh fractions of a gram. If they ask why the objects to be weighed are always placed on the left-hand pan, tell them to reverse the process and see if they can discover the reason.

They should discover that when the objects are placed on the right-hand pan, and the weights on the left, the correct weight will be the total of the weights on the pan *less* the weight indicated by the counterweight. As you move the counterweight, you are adding the relative amount to the right-hand pan. This is the same effect as subtracting weight from the left-hand pan.

There are an infinite number of activities that can result from your students' new facility with the use of the platform balance. Have them try as many of them as they can.

P...A...V...

Ernest R. Ranucci

Elementary school students, and some not so elementary, get confused by three ancient aspects of mathematical thought . . . P . . . A . . . V . . . Perimeter, Area and Volume

Perimeter is fundamentally an aspect of one dimensional thinking. A point generates a line; sometimes the line is straight:

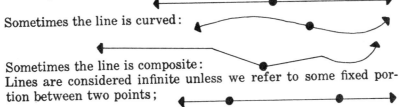

Sometimes the line is curved:

Sometimes the line is composite:

Lines are considered infinite unless we refer to some fixed portion between two points;

then we speak about line segments. The length of the segment is determined by reference to some fixed unit like the inch, centimeter, yard, foot, etc. When we say that a certain segment is four inches long, we mean that four of our standard inches will exactly encompass the length of the segment.

When we say that the *perimeter* of a circle is $3\frac{1}{2}$ inches, we mean that if the circle were broken, then straightened out, its perimeter would stretch along a measure of $3\frac{1}{2}$ inches.

Area deals fundamentally with material related to two dimensional thinking. When we say that the area of a figure is four square inches, we are referring to a particular unit called the square inch. This is a square one inch on a side.

Each of the following configurations has an area of 4 square inches.

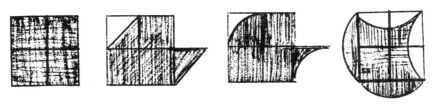

Reprinted with permission from the *New York State Mathematics Teachers' Journal* (January 1966), 33–37.

The *perimeters* of these geometric figures need not be the same. In fact, it would be quite unusual if the perimeters *were* the same. For example, the square one inch on a side (its diagonal has a length of $\sqrt{2}$ inches by use of the Pythagorean Theorem) may be deformed in many ways:

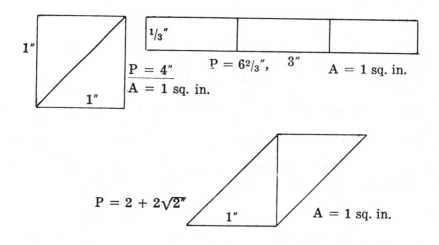

In some cases, configurations with the same *perimeter* may have completely different areas:

Perimeter and *area* are relatively unrelated. There is a legend related to this fact which illustrates the principle quite well. A certain man was to be given all the land he could cover with a cowhide. Nothing, apparently, was said as to whether or not he could cut up the cowhide, so he proceeded to cut up the cowhide into a very long strip of rawhide. He tied the ends, thus forming a long loop. The loop was then spread over as large an area as possible and the land thus covered was used as the site of a new city. The question is: what should be the shape of the area, thus encompassed, in order to cover the largest area? It can be shown through principles of plane geometry that the circle is the most economical in this respect. A loop with perimeter of one mile may be spread so as to cover the following areas:

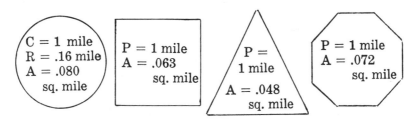

When the loop is drawn taut, the area covered becomes *zero*.

Volume is, fundamentally, an aspect of three dimensional thinking. When we say that a cube 3 inches on a side contains 27 cubic inches, we are referring to a particular standard of three dimensional measure called the cubic inch. This is a cube one inch on a side. Solids may, of course, contain a cubic inch of volume without themselves being cubical in shape.

Volume, area and perimeter are virtually independent of each other. A cube 12″ on a side has a volume of 1728 cubic inches and a total surface of 864 sq. in. The perimeter of all of its twelve edges is 144 inches. If the cube is cut apart and rearranged in various ways, the following table results. The *volume* has changed in no way; it has merely been rearranged into other shapes.

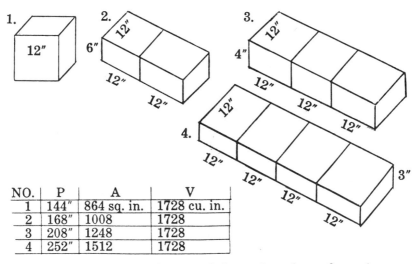

NO.	P	A	V
1	144″	864 sq. in.	1728 cu. in.
2	168″	1008	1728
3	208″	1248	1728
4	252″	1512	1728

It is possible for the perimeter of the twelve edges of a rectangular solid to remain constant while both total surface and volume change:

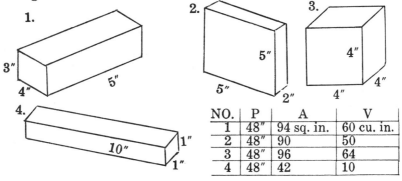

NO.	P	A	V
1	48″	94 sq. in.	60 cu. in.
2	48″	90	50
3	48″	96	64
4	48″	42	10

191

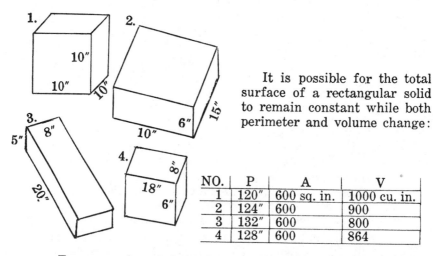

It is possible for the total surface of a rectangular solid to remain constant while both perimeter and volume change:

NO.	P	A	V
1	120″	600 sq. in.	1000 cu. in.
2	124″	600	900
3	132″	600	800
4	128″	600	864

To summarize, perimeter, area, and volume are attributes of one dimensional, two dimensional and three dimensional aspects of mathematical thought. Each is governed by its own mathematical principles and although each uses number as an indication of measure, virtual independence exists with reference to the others.

How Much Do You Know About Perimeter, Area and Volume? The answers are written below, but don't look too soon. . . .

1. May the perimeter of a square ever contain the same number of units, numerically, as its area?
2. May the volume of a cube ever contain the same number of units (numerically) as the total surface?
3. May the perimeter of all edges of a cube ever contain the same number of units (numerically) as the total surface?
4. May the perimeter of all edges of a cube ever contain the same number of units (numerically) as the volume?
5. Are there rectangles, with integral sides, whose areas have the same measures (numerically) as their perimeters?
6. Is there a square whose area is (numerically) twice its perimeter?
7. In a certain restaurant square tables, all the same size, are used. Four people may thus be accommodated: If other members of the same party arrive ,the waiter may place two or more tables in a line. What is the relation between the number of tables and the number of guests who can be accommodated?
8. Is the previous problem affected if tables are not necessarily placed in line? For convenience, let us assume that tables are always joined to others along a common edge.
9. Square channels are bored through the six faces of a cube as in the indicated sketch. What is the total surface of the solid thus formed? (Assume that the square channel is one inch on a side and that it is bored through the center of each face. Would it make a difference if it was *not* bored through the center?)

10. What is its volume? What is the perimeter of all of its edges?

3"
3" 3"

Key:

1. Yes, if the side is 4 units long.
   $P = 4s$, $A = s^2$, $s^2 = 4s$ and $s = 4$
2. Yes, if the side is 6 units long.
   $V = s^3$, $A = 6s^2$, $s^3 = 6s^2$ and $s = 6$
3. Yes, if the side is 2 units long.
   $P = 12s$, $A = 6s^2$, $6s^2 = 12s$ and $s = 2$
4. Yes, if the side is $2\sqrt{3}$ units long.
   $P = 12s$, $V = s^3$, $s^3 = 12s$, $s^2 = 12$, $s = 2\sqrt{3}$
5. There are three; 4 x 4, 6 x 3, and 0 x 0.
6. $s^2 = 2 \times 4s$, $s^2 = 8s$, $s = 8$

7. Let n = the number of tables.
   Then $G = 2n + 2$, $n > 0$

8. The problem *does* change in some cases. Here is just one of many possibilities:

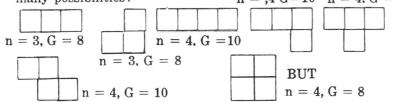

n = ,4 G=10   n = 4. G = 1

n = 3, G = 8

n = 4. G =10

n = 3, G = 8

n = 4, G = 10

BUT

n = 4, G = 8

9. S = 72 sq. in. (It makes no difference as long as the hole is bored vertical to the faces. Does the channel *have* to be vertical to the faces?)
10. V = 20 cu. in. and P = 96 inches.

193

Grids, tiles, and area

KATHRYN BESIC STRANGMAN

The purpose of this article is to report a method for introducing area to fifth-grade pupils which is mathematically sound and readily teachable.[1] Some programs begin their approach to area indirectly by making direct linear measurements and computing the area obtained from these. Some methods define the area of a rectangle as the product of the length and the width of the rectangle and then proceed to discover relationships between the areas of a rectangle and other polygons and, later on, other types of regions.

These indirect methods may mask the real meaning of area. Several questions arise here. What is area in general? Is area length times width? Are children learning key underlying concepts of area which will prove useful to them in their further study and encounters with the real world? What do we want to teach children about the concept of area? How can we go about doing this? We hope to answer these questions to some degree in the following discussion.

The approach reported here is direct, and it is analogous to the way in which linear measure is carried out. Thus it is a method that follows easily from the previous experience of the child. This, along with the fact that it is a direct method, provides an approach to area

measure which is easy to teach with meaning and understanding.

In this approach area is defined as a number; namely, as a number assigned to a region. We usually say that it is a measure of the amount of surface contained in some plane region. Length is defined as a number assigned to a line segment. It is a measure of a distance along some straight line. How does one proceed to determine the length of a line segment?

Suppose we wish to measure the line segment shown in Figure 1. The first step we take is to choose a unit of reference and agree to consider its measure as exactly 1 unit. Let us select the line segment in Figure 2(*a*) as the unit.

FIGURE 1

(*a*) ————

(*b*) ——

FIGURE 2

We next attempt to compare the unit segment with the line segment under consideration. This comparison can be accomplished by laying off copies of the unit segment, end-to-end, on the line containing the segment to be measured and counting the number of units needed to cover this segment. The units can be laid off on the line segment by using a com-

[1] This method was successfully used by approximately 2,500 pupils in the Madison, Wisconsin, area during the developmental year 1967/68 of the "Patterns in Arithmetic" program.

Reprinted from *The Arithmetic Teacher* (December 1968), 668–672, by permission of the author and the publisher.

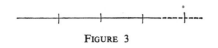

FIGURE 3

pass with the radius equal to the length of the unit. (See Fig. 3.)

It is impossible to determine the exact number of units needed to cover the segment, but we can easily determine two points between which the end of the segment lies, and thus two numbers between which is the exact length of the segment. It is customary to choose the number associated with the closer point as the length of the segment. In this instance the length is 3 units to the nearest unit. Obviously an approximation is involved. However, the approximation can be made more exact, if desired, by choosing a smaller unit of measure. For example, using the measuring unit in Figure 2(b), the length of the segment is 7 units, and the approximation is better. Figure 4 illustrates that the approximation was improved.

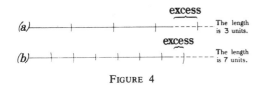

FIGURE 4

The procedure described above of laying off copies of the unit segment on a line amounts to the construction of an instrument for measuring length. This, of course, is a ruler, the instrument commonly used for measuring length.

It is possible to approach area measure in a similar, direct manner. The first step is to choose a measuring unit. For area the unit must be a certain amount of surface in a region. Although other units may be used, such as a triangle, rectangle, hexagon, or other region, it is convenient to choose some square region, which we will agree has an area of 1 unit. For illustration let us choose the unit shown in Figure 5 and determine the area of the closed plane region shown in Figure 6.

The next step is one of comparison. We must try to find how many square units

completely cover the plane region we wish to measure, and no more. Next, place copies of the measuring unit (Fig. 5) side

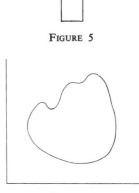

FIGURE 5

FIGURE 6

by side, row upon row, to obtain what we shall call a grid. (See Fig. 7.) Note that this works easily because of the shape of the measuring unit we chose. This grid is our measuring instrument. (Recall that in measuring length we placed copies of the unit line segment end-to-end to form a ruler.) As in working with length, the measurement problem now becomes that of determining how many of these measur-

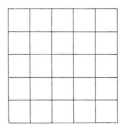

FIGURE 7

ing units are needed to completely cover the given region.

In the classroom one might proceed as follows. After discussing the term "amount of surface," choose a measuring unit of area (such as that in Fig. 5) and construct a grid. Duplicate the grid on translucent paper or plastic. Now place the grid over the region whose area is to be determined. (For illustration consider the area of the region in Fig. 6.) In order that each pupil obtain the same results, use guide marks on the lower left of the figure. (Note guide marks in Fig. 6.) We are interested, of course, in the number of measuring units that completely cover the region. We give the following instructions to the pupils.

1. Completely shade every unit of which at least part covers some of the surface of the region. (See Fig. 8.)

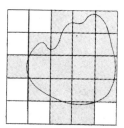

FIGURE 8

2. Count the number of units just shaded. (The answer is 18.) This number is called the *outer area* of the region. Note that the area of the region is less than this number, since more surface than the surface of the region itself was shaded.

3. Next, completely shade (using a different kind of shading or a new grid of the same unit) every unit that lies entirely inside the region. (See Fig. 9.)

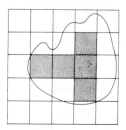

FIGURE 9

4. Count the number of units shaded this time. (The answer is 5.) This number is called the *inner area* of the region. Note that the area of the region is greater than this number, since 5 units are not enough to completely cover the region.

5. Now we put together the information

we obtained in questions 2 and 4 and conclude the following:

I. 5 < area of the region < 18;
that is, the area of the region is a number between 5 and 18.

We have gained a lot of information about the area of the region. (Note that if the grid had been placed in a different position the limits on the area might have been different.) However, we can do better. Next we use a different grid. For classroom use it is convenient to choose a grid in which four of the square units completely cover one larger unit in the first grid. (See Fig. 10.) Use the same region

FIGURE 10

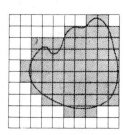

FIGURE 11

and repeat the entire process—steps 1–5 above. (See Figs. 11 and 12.) This time we conclude:

II. 31 < area of the region < 59.

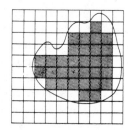

FIGURE 12

In order to compare the results from conclusions I and II we do the following. Consider covering the region with tiles that are the size of the large reference unit (Fig. 5). If each tile costs 8¢, how much will it cost to tile the entire region? Conclusion I tells us we need between 5 and 18 large tiles. Therefore, the cost will be between 40¢ and 144¢ ($1.44).

Next, suppose we use tiles the size of the smaller unit. Since each large tile covers as much of the region as 4 small tiles, each small tile would cost 2¢. Conclusion II tells us we need between 31 and 59 small tiles. Therefore, the cost will be between 62¢ and 118¢ ($1.18). In the second instance, we obtain an estimate of the cost that is better than the first one, since the interval 62–118 is smaller than the interval 40–144.

Following this, more examples and other grid sizes should be used. After the pupils have had considerable practice determining inner and outer areas of different regions and the costs of tiling such, we begin to generalize that *the smaller the grid size, the better the approximation for the area of the region.* Thus with a smaller size grid we have a better measure of the area.

Let us summarize the important concepts discussed so far.

1. Area is a number assigned to a given region.

2. Just as in linear measure, we cannot determine this number, i.e., the measure of a region, exactly, but we can make approximations to it.

3. Areas of different regions can be compared.

4. We determine limits for the area of a region by comparing the area of the region in question to the area of a measuring unit.

5. By choosing a smaller measuring unit it is possible to obtain a better approximation for the area.

Pupils should thoroughly comprehend these basic concepts so that they gain a clear understanding of area. Once this has been accomplished, the area of other special plane figures can be easily taught.

Consider a rectangle whose length and width measure 4 units and 3 units respectively—theoretically, of course. (See Fig. 13.) When we place the grid of Figure 7 on the rectangular region of Figure 13,

FIGURE 13

lining up the grid with the two adjacent sides of the rectangle, we note a very interesting thing. The inner area is 12 and the outer area is 12 also! From what we have seen before, we must conclude that $12 \leq$ area of the rectangular region ≤ 12. But this implies that the area of the rectangular region equals 12! It is further pointed out that since there are 3 rows of measuring units with 4 in each row, we have 3 sets with 4 in each set, and by the definition of multiplication (number of sets times the number in each set equals the total number of objects) we can obtain the area in this special case by multiplying 3 by 4 to obtain 12. The 3 rows correspond to the fact that the rectangle is 3 measuring units in width, whereas 4 in each row corresponds to the fact that the rectangle is 4 measuring units in length. After considering similar examples, we generalize that the area of a rectangular region can be found by multiplying the length of the rectangle by its width. This can be extended further when the pupils learn to multiply fractions.

One can proceed from rectangles to other special regions such as triangular regions and, later, circular regions. Standard measuring units, such as the square inch, can be introduced when the need for them is recognized by the pupils. These and other particular aspects and special cases of area should be presented to the pupils after a good conceptual foundation has been laid.

In summary, it is felt that the above approach to area can contribute a great deal to the understanding of what area really is. Area is measured directly, as length is, and the general region is considered first in an attempt to point out some of the basic general properties of area. The special cases, such as the area of a rectangular region, then fall into place very easily and simply. It is hoped that this aproach will lead students to the fact that area means more than "length times width."

Using stream flow
to develop measuring skills

CARLTON W. KNIGHT II and
JAMES P. SCHWEITZER

A small stream, irrigation canal, or drainage ditch can easily become a focal point of exciting mathematical field experiences. Students can demonstrate the practical application of various measuring skills by computing the surface velocity and volume flow of a nearby stream. These measurements can be related to ecological phenomena by studying (1) changes in the volume of flow by season, (2) the effects of different amounts of rainfall on the volume of flow, and (3) the effects of precipitation in more distant parts of the watershed.

Measuring the flow of a stream or similar small body of water provides opportunities for students of widely varying abilities to constructively participate in a group activity. Less able pupils may simply measure the width of the stream, while more able students can determine the surface velocity of the current or compute the volume of flow.

The following method of estimating the volume of water in a stream's flow requires simple arithmetic, few measurements, and no complicated equipment. The children will need a stopwatch or a watch with a sweep second hand, a tape measure or yardstick, nine to twelve small corks or blocks of wood, and three wooden stakes long enough to drive into the stream bottom and extend above water.

1. Select an area of a stream, small canal, or drainage ditch without bends or meanders where the stream bottom is nearly symmetrical.

2. Measure the width of the stream. Divide the width into three equal segments (W_1, W_2, W_3) that may be identified by driving wooden stakes into the bottom of the stream (see fig. 1).

3. Measure the depth of the water in the middle of each of the three equal segments (D_1, D_2, D_3).

Reprinted from *The Arithmetic Teacher* (February 1972), 88–89, by permission of the authors and the publisher.

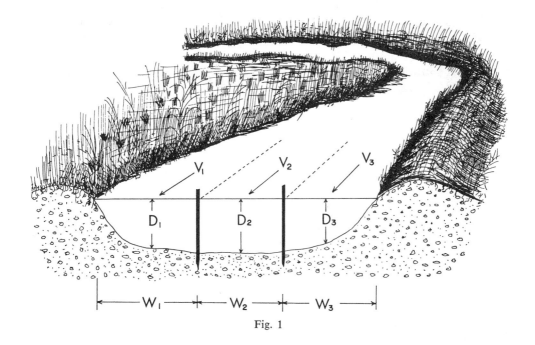

Fig. 1

4. Determine the surface velocity of each stream segment (V_1, V_2, V_3) by placing a small cork or block of wood in the middle of each segment and recording the time required for the cork to float ten feet down the stream. The pupils may wish to record the cork's velocity several times and then compute the average velocity of the current within each of the three segments or for the entire stream.

5. Compute the volume of flow in each of the three segments (F_1, F_2, F_3) by using the following formulas:

$$F_1 = W_1 \times D_1 \times V_1 \times A$$
$$F_2 = W_2 \times D_2 \times V_2 \times A$$
$$F_3 = W_3 \times D_3 \times V_3 \times A$$

Key:

F = Volume of flow in each segment
W = Width of each segment
D = Depth of stream at midpoint of each segment
V = Surface-current velocity of each segment
A = Bottom-factor constant; 0.8 if the stream bottom is rough (gravel or cobbles), 0.9 if the bottom is smooth (mud, sand, hardpan, or bedrock)

The total flow (F_t), expressed in cubic feet of water a second, is computed by adding the flows for all three segments of the stream.

$$F_t = F_1 + F_2 + F_3$$

The above method of measuring stream flow is appropriate for use in the elementary school. It is a modification of procedures employed by John A. Robins and Ronald W. Crawford while studying the ecology of numerous streams. A detailed explanation of the Robins and Crawford method can be found in the 1954 issue of the *Journal of Wildlife Management,* volume 18, pages 366–69.

The role of the teacher in providing such an activity is to lead a discussion of where the class might find a suitable stream, to organize group responsibilities, to gather equipment, and to obtain transportation and parental help, if needed. Then the teacher might engage the pupils in a discussion of the usefulness of measuring the volume of stream flow, letting them hypothesize and comment as they see fit. This out-of-door activity will provide a welcome change in the daily routine while demonstrating the value of measuring skills in a practical situation.

Get ready for the metric system

D. RICHARD BOWLES

I T's not whether, but when," states a newspaper headline for a story on the metric system. With an accelerated movement of the United States toward general use of the system, an intensive educational campaign will be demanded to acquaint pupils with it.

If we who work in the elementary schools are not to be faced with a crash program of in-service education on the metric system, we should begin to get ourselves acquainted with its background and characteristics.

How did the metric system originate?

With the growth of science, industry, and commerce in the eighteenth century, systems of measurement based on such variables as the distance from some king's nose to his fingertips were confusing and inadequate. French scientists were pursuing several possibilities for establishing a "natural and universal" unit when the French Revolution offered an opportunity to introduce new standards of measurement along with other changes.

The accounts of how two surveying teams spent six years in measuring the meridian through Paris from Dunkirk to Barcelona is a fascinating story of adventure and suspense. The survey was undertaken to establish accurately the length of one ten-millionth part of the distance from the equator to the north pole. When the survey was completed, France called an institute of international scientists to verify the measures. Their decision resulted in the preparation of standard meter bars, which were distributed among various nations and for many years served as models with which other measuring devices were compared.

Why did the United States not adopt the system at the beginning?

We very nearly did. George Washington in his first message to Congress called for the adoption of a uniform system of weights and measures, as directed by the Constitution. Thomas Jefferson, as Secretary of State, was appointed to make the study. He recommended a decimal system of measures, but one based on the seconds pendulum (the length of a pendulum needed to strike seconds at sea level at 45 degrees north latitude), the meter not having been defined by France until three years later. If Congress had acted on Jefferson's suggestion, we would have been the first nation to adopt a decimal system of measures. Ten years later, France dispatched to the United States a prominent scientist-diplomat with a copy of the model meter and an invitation to join in adopting the metric system. He was shipwrecked by a hurricane in the Caribbean, imprisoned by pirates, and died without reaching this country.

Later, John Quincy Adams, as head of another Congress-requested study, spoke highly of the metric system, but believed that it should be adopted by all nations outside France at the same time.

How widely used is the metric system outside the United States?

During the nineteenth century most of the nations of South America and the non-English-speaking countries of Europe adopted the meter. Most of Africa and Asia followed in the first half of this century. In 1965, Great Britain set into operation a plan designed to bring about a complete changeover in ten years. New

Zealand, Australia, and South Africa have been the last to move away from the English system.

Canada and the United States now stand as the only nations not having the metric system in general use.

Is the metric system in use at all in the United States?

Few Americans realize that in 1866 the United States, upon Abraham Lincoln's earlier recommendation, legalized the use of "the weights and measures of the metric system," and directed the Secretary of the Treasury to send to each state governor "one set of weights and measures of the metric system for the use of the states." Since 1893 the meter and kilogram have been the fundamental standards used by the U.S. Bureau of Standards; when the Bureau is called upon to verify length, weight, or volume, the measurements are certified with reference to metric standards, and then changed, if desired, to the common units.

Most scientific and engineering publications use the metric terms exclusively. Several branches of industry, pharmaceutical companies being among the first, have made the changeover. NASA and some branches of the armed services are using metric units. Photographic film sizes are in terms of metric units. Many sporting events have used metric standards for decades.

What are its advantages?

First, and most obviously, the decimal nature of the system makes for ease in learning the relationships of the units, and in performing operations on them. (The so-called "English system" probably should not be called a system at all, since there is very little about it that is systematic; there is no pattern among the various tables of length, weight, and volume.) Second, the great number of different units in each division of the metric system makes it very useful in describing small differences in quantity. Since all measurement is approximate, the degree of exactness desired depends upon the purpose of the measurement; with metric units, almost any desired degree of exactness can be expressed. Third, the nearly universal use of the system makes familiarity with it practically obligatory for one traveling in and communicating with other lands.

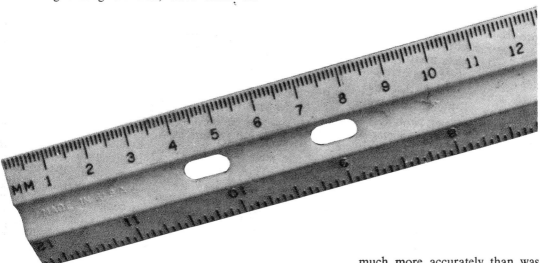

Exactly what is a meter?

Originally, a meter was defined as one ten-millionth of the distance from the equator to either pole. Later discoveries and technical developments make it possible to define measurements much more accurately than was done when the French surveyors established the meter. Now, by international agreement, a meter is defined in terms of the wave length of orange light from the gas krypton-86 at a specific temperature. In terms of our customary units, a meter is approximately 1.1 yards—or, 3.37 inches longer than a yard.

How is a meter related to other units in the system?

The base unit of volume is the liter. This is a cube whose sides are one-tenth of a meter (a decimeter). Its common-measure equivalent is another example of the confusion of the English system; the liter is smaller than a dry quart, larger than a liquid quart.

The basic unit of weight (or, more properly, mass) is the gram, originally defined as one one-thousandth of a liter of distilled water at a specified temperature. Because the gram represents such a small mass, the more commonly used unit of weight is the kilogram, or one thousand grams, which is about 2.2 pounds.

What is the meaning of the prefixes used with the basic units?

Greek prefixes *deca* (deka), *hecto* (hekto), and *kilo* are used to indicate units larger than the base unit. Thus, "decameter" means "ten meters," "hectogram" means "one hundred grams," and "kilometer" means "one thousand meters."

Latin prefixes *deci, centi,* and *milli* are used to signify units smaller than the base unit. One-tenth of a meter is a "decimeter," and so on.

In pronouncing the prefixes, the accent should be on the first syllable, regardless of the base term with which it is being used (except for *kilometer,* often accented on the second syllable).

How much of the system should be taught in the elementary school?

Probably the units just mentioned (from the one-thousandths to the thousands) are sufficient as content for the elementary grades. If pupils learn these well, they can easily add the other prefixes that might be needed in scientific work in later grades.

Should the metric system and the customary U.S. system be taught together?

For the immediate future, the English system of units should continue to be taught. At the same time, the metric system should be introduced, but not as a comparison with the common system. Other than a reference to the meter as being about 3⅓ inches longer than a yard, no attempt should be made to teach conversion factors. As in learning any second language, the objective should be to help the learner think in the new language, rather than to think in the first language and then have to translate to the second. This will best be accomplished if many activities are provided for both systems, so pupils will learn to think in whichever system they happen to be working.

How can I keep up with developments in the metric movement?

The Metric Association, Inc., 624 N. Drivey Lane, Arlington Heights, Illinois 60004, publishes a *Newsletter* three or four times a year, as well as bibliographies and a few teaching aids. (Dues are $2.00 a year.)

SELECTED BIBLIOGRAPHY FOR CHAPTER 9
Measurement and Geometry

Measurement

Alexander, F. D. "The Metric System—Let's Emphasize Its Use in Mathematics." *The Arithmetic Teacher* (May 1973), 395–96.

Bailey, Terry G. "Linear Measurement in the Elementary School." *The Arithmetic Teacher* (October 1974), 520–25.

Bourne, H. N. "The Concept of Area." *The Arithmetic Teacher* (March 1968), 233–43.

Bright, George W. "Bilingualism in Measurement: The Coming of the Metric System." *The Arithmetic Teacher* (May 1973), 397–99.

Brougher, Janet Jean. "Discovery Activities With Area and Perimeter." *The Arithmetic Teacher* (May 1973), 382–85.

Colter, Mary T. "Adapting the Area of a Circle to the Area of a Rectangle." *The Arithmetic Teacher* (May 1972), 404–06.

Feifer, Jeffrey P. "Using the Compass for Outdoor Mathematics." *The Arithmetic Teacher* (May 1973), 388–89.

Fisher, Ron. "Metric Is Here; So Let's Get On With It." *The Arithmetic Teacher* (May 1973), 400–02.

Hall, Harvey N. "Area Approximations by Use of Random Numbers." *The Arithmetic Teacher* (February 1974), 106–09.

Hawkins, Vincent J. "Teaching the Metric System As Part of Compulsory Conversion in the United States." *The Arithmetic Teacher* (May 1973), 390–94.

Immerzeel, George and Wiederanders, Don. "Ideas." *The Arithmetic Teacher* (May 1972), 363–73.

Krause, Eugene F. "Elementary School Metric Geometry." *The Arithmetic Teacher* (December 1968), 673–82.

Marks, John L. and Smart, James R. "Mathematics of Measurement." *The Arithmetic Teacher* (April 1966), 283–87.

McClintic, Joan. "The Kindergarten Child Measures Up." *The Arithmetic Teacher* (January 1968), 26–29.

——. "Capacity Comparisons by Children." *The Arithmetic Teacher* (January 1970), 19–25.

Peavler, Cathy Seeley. "Metricating—Painlessly, Cheaply, Cooperatively." *The Arithmetic Teacher* (October 1974), 533–36.

Reisman, Fredricka K. "Children's Errors in Telling Time and a Recommended Teaching Sequence." *The Arithmetic Teacher* (March 1971), 152–55.

Rosenberg, Howard. "What's the Area?" *The Arithmetic Teacher* (October 1971), 429.

Steffe, Leslie P. "Thinking About Measurement." *The Arithmetic Teacher* (May 1971), 332.

Strangman, Kathryn Besic. "The Sands of Time—A Sandglass Approach to Telling Time." *The Arithmetic Teacher* (February 1972), 123–25.

Swan, Malcolm. "How Long is 100 Feet?" *Science and Children* (December 1966), 17–18.

—— and Jones, Orville. "Preservice Teachers Clarify Mathematical Percepts Through Field Experiences." *The Arithmetic Teacher* (December 1969), 643.

Thompson, Mary Helen. "Smile When You Say Area!" *The Arithmetic Teacher* (October 1971), 430.

Tomjack, Kathleen and Reginald, Sr. Mary, O.P. "Tips on Teaching Time Telling." *Instructor* (January 1968), 33.

Geometry

Alspaugh, Carol Ann. "Kaleidoscopic Geometry." *The Arithmetic Teacher* (February 1970), 116–17.

Backman, Carl A. and Smith, Seaton E., Jr. "Activities with Easy-to-Make Triangle Models." *The Arithmetic Teacher* (February 1972), 156–57.

Bruni, James V. "A 'Limited' Approach to the Sum of the Angles of a Triangle." *The Arithmetic Teacher* (February 1972), 85–87.

Bush, Mary Thomas. "Seeking Little Eulers." *The Arithmetic Teacher* (February 1972), 105–07.

Clarkson, David N. "Taxicab Geometry, Rabbits, and Pascal's Triangle—Discoveries in a 6th Grade Classroom." *The Arithmetic Teacher* (October 1962), 309.

Coltharp, Forrest L. "Properties of Polygonal Regions." *The Arithmetic Teacher* (February 1972), 117–22.

Complo, Sr. Jannita Marie. "Teaching Geometry Through Creative Movement." *The Arithmetic Teacher* (November 1967), 576.

Dennis, J. Richard. "Informal Geometry Through Symmetry." *The Arithmetic Teacher* (October 1969), 433.

Dickoff, Steven S. "Paper Folding and Cutting a Set of Tangram Pieces." *The Arithmetic Teacher* (April 1971), 250–52.

Egsgard, John C. "Geometry All Around Us—K–12." *The Arithmetic Teacher* (October 1969), 437–45.

Farrell, Margaret A. "Patterns of Geometry." *The Arithmetic Teacher* (October 1969), 447.

Giddings, Marie. "Being Creative With Shapes." *The Arithmetic Teacher* (December 1965), 645.

Gogan, Daisy. "A Game With Shapes." *The Arithmetic Teacher* (April 1969), 283.

Grant, Nicholas and Tobin, Alexander. "Let Them Fold." *The Arithmetic Teacher* (October 1972), 420–25.

Forseth, Sonia and Adams, Patricia. "Symmetry." *The Arithmetic Teacher* (February 1970), 119–21.

Heard, Ida Mae. "Developing Geometric Concepts in the Kindergarten." *The Arithmetic Teacher* (March 1969), 229.

Immerzeel, George and Wiederanders, Don. "Ideas." *The Arithmetic Teacher* (May 1971), 310.

Inkeep, James, Jr. "Primary-Grade Instruction in Geometry." *The Arithmetic Teacher* (May 1968), 422.

Jacobs, Judith E. and Herbert, Elizabeth B. "Making $\sqrt{2}$ Seem 'Real'." *The Arithmetic Teacher* (February 1974), 133–36.

Krause, Eugene F. "Elementary School Metric Geometry." *The Arithmetic Teacher* (December 1968), 673.

Liedtke, Werner. "What Can You Do With a Geoboard?" *The Arithmetic Teacher* (October 1969), 491.

Lulli, Henry. "Polyhedra Construction." *The Arithmetic Teacher* (February 1972), 127–30.

May, Lola J. "Introducing the Compass As a Tool in Geometric Construction." *Grade Teacher* (March 1967), 98–100.

Moulton, J. Paul. "Some Geometry Experiences for Elementary School Children." *The Arithmetic Teacher* (February 1974), 114–16.

Neufeld, K. Allen. "Discovery in Number Operations Through Geometric Constructions." *The Arithmetic Teacher* (December 1968), 695.

Ogletree, Earl. "Geometry: An Artistic Approach." *The Arithmetic Teacher* (October 1969), 457.

Perry, E. L., Jr. "Integrating Geometry and Arithmetic." *The Arithmetic Teacher* (December 1973), 657–62.

Richards, Pauline L. "Tinkertoy Geometry." *The Arithmetic Teacher* (October 1967), 468–69.

Rosenberg, Howard. "What's the Area?" *The Arithmetic Teacher* (October 1971), 429–30.

Teegarden, Donald. "Geometry Via T-Board." *The Arithmetic Teacher* (October 1969), 485.

Viggiano, Joseph. "Constructing the Area of a Triangle from Its Medians." *New York State Mathematics Teachers' Journal* (June 1972), 115–19.

Vigilante, Nicholas J. "The Address of a Point." *The Arithmetic Teacher* (December 1968), 689.

Wahl, M. Stoessel. "Easy-to-Paste Solids." *The Arithmetic Teacher* (October 1965), 468–71.

Walter, Marion. "A Second Example of Informal Geometry: Milk Cartons." *The Arithmetic Teacher* (May 1969), 368.

Witt, Sarah M. "A Snip of the Scissors." *The Arithmetic Teacher* (November 1971), 496.

Wong, Ruth E. "Geometry Through Inductive Exercises for Elementary Teachers." *The Arithmetic Teacher* (February 1972), 91–96.

PART IV

LET'S SOLVE IT

CHAPTER 10

PROBLEM-SOLVING TECHNIQUES AND STRATEGIES

It has been said that verbal problem solving has attracted more attention from researchers than any other topic in the mathematics curriculum.* And it is certainly true that elementary school children find solving verbal problems in mathematics difficult.

Researchers have listed various factors that are associated with success in verbal problem solving, but there is often little agreement on which factors are of the greatest importance or what the teaching techniques should be. Among the factors which seem to characterize high achievers in problem solving are: ability to note likenesses, differences, and analogies; understanding of mathematical terms and concepts; ability to visualize and interpret quantitative facts and relationships; skill in computation; ability to select correct procedures and data; and comprehension in reading.†

Since problem solving is a multidimensional process, the articles in this chapter were chosen to give the reader a variety of ideas and techniques for developing problem-solving strategies. For instance, Mary M. Blatt, in "Problems of Problem Solving," approaches this topic as a science educator. She suggests that, since pages and pages about problem solving is not likely to get anyone started, perhaps some of the problem-solving situations she has listed will. Although these may be considered science activities, note the amount of *mathematics* that is involved.

C. Alan Riedesel's article, "Verbal Problem Solving: Suggestions for Improving Instruction," reports the results of a study that attempted to provide (1) verbal problems at two levels of difficulty for use in the same classroom and (2) specific experience with procedures that are highly recommended for improving learner achievement in the solution of verbal problems. You will certainly want to read this article for the activities and suggestions it provides as well as the results and conclusions from the study.

Anne Schaefer and Albert Mauthe, in "Problem Solving with Enthusiasm—The Mathematics Laboratory," give some very practical suggestions for those teachers interested in the laboratory approach. Ten good activity problems are reported in

---

*Marilyn B. Suydam and J. Fred Weaver, "Verbal Problem Solving," *Interpretive Study of Research and Development in Elementary School Mathematics* (Grant #OEG-0-9-480586-1352(010)), The Pennsylvania State University.

†*Ibid.*

detail in this article and it should be easy for interested teachers to reproduce these for their own classrooms.

The alert elementary teacher will find many worthwhile suggestions in "Improving Problem-Solving Skills" by Jacqueline Sims. She has included a series of experiences and exercises that can be used in the classroom with success. Activities are given to illustrate each technique that is suggested.

Cecil Trueblood presents an excellent way to develop meaningful problems in his article, "Promoting Problem-Solving Skills Through Nonverbal Problems." Teachers, especially those who have children with reading problems, will find this article stimulating and practical. Hopefully, it will encourage others to develop some of their own "nonverbal problems."

In "Using Flow Charts in Elementary School Mathematics Classes," James J. Roberge suggests that flow charts can provide a concrete means of illustrating a method of solution for a problem. Have you thought about using flow charts with your children? Read this fine article, then give it a try.

Students should be encouraged to dig deeply into the challenging and exciting area of problem solving. The articles presented here provide a glimpse at the many ways teachers can approach this topic in the classroom.

Problems of
PROBLEM-SOLVING

MARY M. BLATT

The available literature on teaching science from the problem-solving approach is prolific with reasons why it should be done. Descriptions of exactly "how" to go about teaching from this approach are meager.

Learning to solve problems in science, like learning to knit or to play golf, is done by trying to do it. (Learning by doing!) Perhaps this is why most descriptions of how-to-do-it really do not tell us how; therefore, we are not prompted to begin teaching science in this way.

How does the problem-solving method differ from the usual method of teaching? The chief difference may well be that you will surely "cover" less content than you would if you used a book. Science educators agree that students should "uncover" knowledge for themselves. The teacher's role is to provide an atmosphere conducive to thinking and to give guidance to that thinking. Teachers and children using this method will not be seeking verification of an answer which has been decided upon ahead of time. This may cause some difficulty if one is working with children who have not had previous problem-solving experience. There is evidence to suggest that children often change reports of observations if they do not agree with predetermined answers. Preconceived answers tend to curb any thinking on the part of the student.

What Is Your Task?

The hardest task to face is to resist giving away possible answers to the problem before students have used all the other resources at their command. To maintain your composure through the long silences which occur when children are desperately trying to think, is difficult. Practice is required to develop the habit of countering a student's question with another question. Another requirement of the problem-solving method is courage to embark upon an investigation of a problem to which neither you nor the students know the answer. If children are to become aware of the methods used by scientists, they must use these methods as the scientist does.

What Is a Problem?

Think of it as a question to which your students do not know the answer. A problem is not a topic heading in a textbook that happens to be stated as a question. These questions can be solved by reading the answer in a textbook. When the answer is known, the lesson is finished. Experience has shown that memorizing answers which have been read or heard by listening to a teacher's lecture are not long remembered, nor do the skills of problem-solving become integrated into the students personal work habits.

Another way to solve a problem is to allow students to become personally involved in trying to find possible answers by reading, experimenting, consulting with others, and by using their intuitive powers. The responsibility for learning is on the student, which is the proper place for it since teachers cannot do the learning anyway. This is the problem-solving, or the "discovery" method which we talk about, but seldom implement in practice.

How To Begin?

An elementary teacher is often enthusiastic and begins a new technique with the expectation of doing everything in a bigger and better fashion than before. Grand plans are made for complete units and whole semesters of work. After the initial novelty wears off and the enormity of the undertaking becomes apparent, the whole project becomes a staggering burden and the teacher abandons it. The memories of the attempt are unpleasant. This does not encourage the teacher to try another innovation.

Teachers who have been most successful with "problem-solving" have started in a small way. They usually decided to try one problem with the entire class. As teachers become more skilled in this method, other problems can be added to the science program.

Some time should be spent orienting the children before this approach is undertaken. Many of them will have little or no confidence that an idea they "think up" will have any value. If the students learn to exercise their own intuition, instead of depending on someone or something outside themselves, a great deal will be accomplished.

In the beginning, all the children may want to work on the same problem. When an investigative atmosphere has been developed, the student's interests will be diversified. There is no reason why a child or group of children cannot work on a topic of specific interest. Plans will have to be made for specific topics, so that they relate to the themes or courses of study set up for the grade. Permitting children to work in groups has the advantage of making it possible for teachers to allow for individual differences and to make better utilization of the time, equipment, and reference materials which they may have in the classroom.

Science problems, (which *are* problems, and not demonstrations or verifications of something already learned) usually require continued observation, measurement, and data collection. Important science skills and processes can be taught by using a few minutes each day, rather than planning for large blocks of time arbitrarily placed in the weekly schedule.

Since reading pages and pages about problem-solving is not likely to get anyone started, listed below are some problem-solving situations. Perhaps one of these activities will appeal to you or your students.

Activity I. A Mold.

Many children have had the experience of growing mold on bread. They may know that mold is a plant, and that it is spread by spores which are produced in spore cases. These spore cases can be observed by using a magnifying glass. After this background has been established ask:

1. What else will mold grow on? (cheese, jelly, fruit)

2. Are all molds the same? Is cheese mold the same as that which grows on jelly? Does bread mold grow on fruit?

   a. How can we find out? (Students collect moldy cheese, bread, jelly, or fruit; inoculate other specimens—cheese to jelly, bread to fruit, and so on. Several children can do combinations of their own. Duplicate experiments will add strength to evidence collected.)

   b. How do we know how our experiments are progressing? Help students to devise a chart to be used in recording daily observations, including the approximate diameter of any mold colony that appears. Here is an opportunity to introduce the metric system. This system is used by scientists and can be found in a high school physics book. Children should know about the metric system as well as the English system. Students can also record the dates, colors of mold, when spore cases appear, and evidence that the mold is dying, if this should happen.

3. What growing conditions provide the healthiest colonies of mold? (It is important to remember that the planning of experiments is as important as

getting answers. Let students plan to set up mold cultures with varying amounts of water, light, and heat present. If they are unable to figure out how to do this at first, let them mull over the possibilities for several days. If you plan the experiments for them, much of the instructional value is lost.)

Activity II. Simple Machines.

Traditional lessons can sometimes be converted into problem-solving by simply changing the sequence of the learning. Instead of reading the chapter on pulleys and then demonstrating how they work, try setting up a pulley arrangement on your science table. Let the children try out the pulleys and weights during spare moments over a period of several days. After each trial, the amount of weight lifted, the distance it moved, the force used (using a hand spring scale), and the distance the pull rope moved should be recorded in an orderly fashion. After many trials, have the children consider the evidence. Allow them to make a verbal conclusion about the advantages of using the arrangement of pulleys which was set up. Trials can then be made with different arrangements of pulleys. The children will begin to see for themselves that pulleys can change direction of force, increase the speed and distance, or increase the force, depending on how they are roped together.

In addition to learning about a simple machine, the children have had an opportunity to keep scientific records, measure, and draw conclusions based on their own findings. This unit offers the teacher a chance to practice skillful questioning, without lecturing and memorization.

If you are inexperienced in the use of pulleys, borrow a textbook from a junior high school teacher. Most secondary science teachers are happy to have the opportunity to give advice, books, and materials to elementary teachers. This gives them the opportunity to learn about the work done by the children in the early grades.

Activity III. Measurement.

Provide the children with a 12-inch ruler, yardstick, and a meter stick. Have them measure the

length of their room, length of desk, and other objects, using all three measures. The students should become familiar with the units involved and be able to recognize that there will be more yards in a given distance than meters, more centimeters than inches. A meter stick can be made from a slat which is 1½-inches wide and 40-inches long. Have it cut as close to 39⅜ inches as possible. Divide it into 100 centimeters. Note that centimeters can be written as decimals. This activity will allow children to become familiar with two systems of measurement, not to have them memorize equivalent values. When is the English system used? Why is the metric system used by scientists? How are measurement systems developed?

In this activity the teacher has an opportunity to teach the basic concepts that measurement systems are man made; and, therefore, subject to error.

Measurement is a basic skill needed in all science courses. Many of the newer elementary science curriculum projects contain activities that require students to be able to measure.

Summary

Why do science educators so strongly advocate teaching science by letting children learn for themselves? By now it is apparent that we can no longer teach children a body of knowledge that will serve them well throughout their lifetimes. No one can predict what science content will be valid or valuable ten years from now. Not even scientists can keep up with all the scientific developments in their specialty area. Our only alternative seems to be to familiarize students with the skills and processes which are used by scientists in solving problems as they appear. To do this, we use some of the basic concepts of science as vehicles for learning the skills of investigation.

There is general agreement that science is science, whether it be in kindergarten, high school, graduate school, or a research laboratory. There is no reason why a child cannot pursue a problem that is challenging (but not overwhelming) to him, at his level of understanding.

Verbal problem solving: suggestions for improving instruction*

C. ALAN RIEDESEL

The verbal or word problem has long been an area of arithmetic instruction of great concern to teachers and the cause of much pupil anxiety. This situation has led to the formulation of many proposals for improving the teaching of verbal problem-solving ability. Interest in most of these proposals has waned quickly after teacher tryout or after studies to test their validity failed to produce clear-cut evidence of the worth of the proposals. The overall result has been that during instruction on verbal problems the major portion of instructional time is devoted to the solution of the same verbal problems by all the students in a class and by means of general or poorly defined procedures. In view of the range of arithmetical ability found in most classrooms, it is necessary to have many pupils attempting to solve problems that are too difficult, while other more able pupils must be concerned with the solution of problems that are too easy. When the above situation is coupled with the fact that pupils are not guided into using specific suggestions which are believed to make for improved word problem-solving ability, there is little wonder that this area of arithmetic remains a matter of concern to teachers.

In an attempt to provide material better suited to pupil ability, some teachers have resorted to use of textbooks from different grade levels, and some textbook writers have included a few difficult problems (usually indicated by a star or other symbol) in each set of word problems. A little investigation of the use of these two procedures shows that neither procedure is regarded with much enthusiasm by leaders in the field of arithmetic instruction.

The suggestions for improving verbal problem solving used in the study reported here attempt to provide (1) verbal problems of two levels of difficulty for use in the same classroom and (2) specific experience with highly recommended procedures for improving pupil achievement in the solution of verbal problems.

The following five procedures were selected for use in this study: (1) writing the number question or mathematical sentence, (2) using drawings and diagrams, (3) having pupils formulate problems, (4) presenting problems orally, and (5) using problems that do not contain numerals.

The purpose of this study was to compare the effectiveness of the use of specific verbal problem-solving procedures in connection with the provision of two levels of problem difficulty to the problem-solving program followed in typical textbook instruction.

A total of thirty arithmetic problem-solving lessons were prepared. Each of the thirty lessons was written at two levels of difficulty. The problems for pupils of above-average problem-solving ability were reproduced on yellow paper while the problems for pupils of below-average

* This article is adapted from a section of the author's doctoral dissertation, "Procedures for Improving Verbal Problem Solving Ability in Arithmetic" (State University of Iowa, 1962).

Reprinted from *The Arithmetic Teacher* (May 1964), 312–316, by permission of the author and the publisher.

problem-solving ability were reproduced on white paper. The lessons made use of the five specific procedures referred to above. Following the problems for each lesson was a "How's Your P.Q.?" problem which was more difficult than the other problems in the lesson. These problems were for optional use. Copies of detailed answer sheets were provided so that each pupil was able to correct his own work. The lesson material for one day follows.

Problem-Solving Lesson 21—
Using Drawings and Diagrams
(white paper)

Read the problem carefully and then use a drawing or diagram to help you solve it. Try to check your work by using another method of solution.

1 When he was practicing basketball, Bill made 3 of every 5 free throws he attempted. If he continues at that rate, how many will he make out of 25 shots?
2 The Dover basketball team is playing at Mt. Royal, which is 135 miles away. The game begins at 7:30 P.M. The team starts at 4 P.M. and plans to arrive $\frac{1}{2}$ hour before game time. Will they arrive on time if they average 40 mi. per hr.?
3 Al bought 6 valentine cards marked "3 for 25¢." What was the cost of his purchase?
4 Lloyd needed to buy some fishhooks before going on a fishing trip with his father. If the fishhooks sell at 6 for 10¢, how much will 18 fishhooks cost?

How's your P.Q. ???? (*Lesson* 21)

Claudia was getting dressed for a party when a thunderstorm caused the lights to go out. She knew that she had only three colors of socks—white, yellow, and blue—but wanted to be sure she had a matching pair. Her brother said, "Just take several, and then when we get in the car you can pick out a matching pair." What would be the least number of socks that she could take to be sure to get a pair that matched?

Problem-Solving Lesson 21—
Using Drawings and Diagrams
(yellow paper)

Read the problem carefully and then use a drawing or diagram to help you solve it. Try to check your work by using another method of solution.

1 A baseball diamond is 90 ft. by 90 ft. (See diagram.) If a player runs a race of 100 yards in 12 seconds, how long will it take him to run around all of the bases? Assume that he ran on the base lines.

2 Billy wishes to cut a board that is 16 ft. 3 in. long into 5 equal parts. Ignoring the saw cuts, how long will each part be?
3 A nickel (5¢) is made of both nickel and copper. For every pound of nickel in the mixture there are 3 pounds of copper. How many pounds of copper will be needed to make 24 lbs. of coins?
4 The Cardinals won 5 of the first 8 games that they played. At that rate how many games would they win out of 56 games?

How's your P.Q. ???? (*Lesson* 21)

A train leaves Albertsville headed east for Bakersville every two hours. At exactly the same hour a train leaves Bakersville for Albertsville. The trip requires 12 hours. How many trains will be met by a train in going from Albertsville to Bakersville?

Correction sheet for Lesson 21—
Using Drawings and Diagrams
(white paper)

1 The following diagram can be used to find the number of free throws Bill will make out of 25.

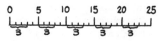

total of 15 free throws

2 If the team wants to arrive ½ hour before game time, it must arrive by 7:00 P.M. This allows from 4 P.M. to 7 P.M., or 3 hours for travel. If the team travels at 40 miles per hour, the diagram below shows that they will not arrive on time.

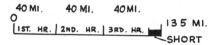

3 The drawing shows that if the cost of 3 cards is 25¢, then 6 cards will cost 50¢.

4 The problem can be diagramed in several ways; two of these ways are shown below:

(A)

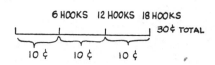

(B)

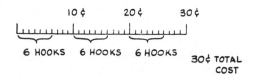

Lesson 21—"How's your P.Q.?"

You can draw a picture of a number of socks. Note that any time you pick out four socks you must have 2 of the same color. *Answer:* 4.

W = WHITE
Y = YELLOW
B = BLUE

Correction sheet for Lesson 21— Using Drawings and Diagrams (yellow paper)

1 First it is important to be sure we use the same units. Change the 90 ft. to 30 yards.

Checking the diagram we find that the total distance around the diamond is 120 yards. Thus, we know it will take longer than 12 seconds for the player to run around the diamond. We can find the length of time it takes by multiplying:

distance around the diamond—120
$$\frac{\quad}{100 \text{ yards}—100} \times 12$$

$$\frac{6}{5} \times 12 = \frac{72}{5} = 14\frac{2}{5} \text{ sec. Ans.}$$

2 Visualize cutting this into five equal parts:

4 FT.   8 FT.   12 FT.   16 FT.
                              16 FT. 3 IN.

THE NUMBER QUESTION IS N = 16 FT. 3 IN. ÷ 5
N = 3 FT. 3 IN.

3 The diagram below provides aid in solving the problem:

1 LB. + 3 LB. = 4 LB.
                    8 LB.   12 LB.   16 LB.   20 LB.   24 LB.
NI. COPPER NI. COPPER NI. COPPER NI. COPPER NI. COPPER NI. COPPER
3   3   3   3   3   3
3 + 3 + 3 + 3 + 3 + 3 = 18   A TOTAL OF 18
                             POUNDS OF COPPER

4 The number line below can be used as an aid in solving the problem:

8   16   24   32   40   48   56
5   5   5   5   5   5   5
FOR A TOTAL OF 35 WINS IN 56 GAMES.

```
        2 HRS.  4 HRS.  6 HRS.  8 HRS.  10 HRS.  12 HRS.
WEST   FROM A  FROM A  FROM A  FROM A  FROM A  FROM A   EAST
   |_____|_____|_____|_____|_____|_____|
ALBERTSVILLE                          BAKERSVILLE
```

At any given hour there will be 7 trains on the track headed east (from Albertsville to Bakersville) and 7 trains headed west. A train going east will have met all these 7 trains by the time it is 6 hours out of Albertsville. During this time 3 other trains have left Bakersville; and by the time the eastbound train arrives at Bakersville, 2 other trains will have left and 1 will be departing. *Answer:* 13.

A forty-eight item, tape-recorded, problem-solving test was developed for the study. The problems were presented orally on tape and were also duplicated on the answer sheets. Analysis of the test revealed a mean item difficulty of .51, a mean index of item discrimination of .57, and a reliability of .90 as determined by the Spearman-Brown formula. The use of the Spearman-Brown formula was deemed appropriate since 95 percent of the students finished each item on the test. This test and the problem-solving section of the Iowa Tests of Basic Skills were used as pretests and final tests to compare the gains of experimental and control groups.

The experimental and control groups were composed of sixth-grade classes from three Iowa cities. Eleven experimental and nine control classes took part in the study. Placement of the classes into experimental or control groups was made on the judgment of the curriculum supervisors, who attempted to distribute experience of the teachers, ability of the students, and teaching ability specifically in arithmetic between the experimental and control groups. A total of 505 sixth-grade children participated in the entire program.

The participating experimental and control teachers were oriented to the program by the writer. Illustrative lessons were provided as an aid to the introduction of each problem-solving procedure.

The tape-recorded test developed for the study and the problem-solving section of the Iowa Tests of Basic Skills were administered to the experimental and control classes. In determining the level of problem-solving materials to be used by a pupil in the experimental groups, the tape-recorded test score was used along with the judgment of the teacher. In the case of borderline students, the teachers were directed to allow such students to attempt some lessons at each level and, with the help of the teacher, to determine the groups with which they would receive maximum benefit.

During the course of the study the experimental classes worked three special problem-solving lessons a week. The problems presented in their textbooks were omitted unless they were an integral part of the introduction of a new topic. The control groups followed the regular problem-solving program presented in the textbooks. Actually the control classes worked a greater number of problems during the study than the experimental classes.

At the end of the ten-week period the posttests on both measures were administered. Questionnaires and essays were used to obtain the subjective judgment of the teachers and pupils concerning the materials.

The tests were corrected and the t statistic was used to compare the mean class gain scores between experimental and control groups. The 10 percent level of significance was used.

Results showed:

1 The difference between mean gains on the tape-recorded test favored the experimental subjects and was significant beyond the .1 percent level.

2 The difference between mean gains on the problem-solving section of the Iowa Tests of Basic Skills favored the experimental subjects and was significant between the 5 and the 10 percent levels.

From the analysis of the questionnaires and essays it can be concluded that:

1 The lessons are of benefit to above-average, average, and below-average problem solvers.
2 The materials are of appropriate difficulty and reading level.
3 The specific procedures are effective in improvement of pupil problem-solving achievement.
4 Children find such materials worthwhile and enjoyable.
5 Children are receptive to "How's Your P.Q.?" type problems.
6 Pupils enjoyed working problems taken from a variety of sources. The pupils in this study revealed a definite preference for problems from foreign and old United States textbooks.

On the basis of the work conducted with pupils and teachers in the course of this study, the following suggestions warrant consideration:

1 Teachers and others who prepare arithmetic materials will enhance the educational value of their teaching materials by developing problem-solving lessons of multilevel difficulty.
2 Greater use should be made of such specific procedures as writing the number question, use of drawings and diagrams, pupil formulation of problems, orally presented problems, and using problems without numerals.
3 Greater use should be made of the tape recorder in test presentation.
4 In preparing materials for increasing pupil interest in arithmetic, curriculum workers will find it profitable to include problems from foreign textbooks and from old United States textbooks.

Problem solving with enthusiasm—the mathematics laboratory

ANNE W. SCHAEFER

ALBERT H. MAUTHE

A promising approach to the teaching of problem solving is the use of the mathematics laboratory. The purpose of the laboratory is to provide children with opportunities to discover mathematical concepts through their active involvement in solving problems. The emphasis is on learning by doing.

The British primary schools, working with the Nuffield Teaching Project,[1] have had much experience with mathematics laboratories. The project has produced a fourteen-minute film about the mathematics laboratory; the title of the film, "I Do—and I Understand," is taken from the Chinese proverb, "I hear, and I forget; I see, and I remember; I do, and I understand."

Our enthusiasm for this method led us to begin a mathematics laboratory with some of the fifth-grade classes in Scarsdale. However, we had two major concerns: (1) Would the problems we devised prove to be meaningful learning experiences for the children? (2) How would the children react to the laboratory? How interested and involved would they become?

The purpose of this article is to share

our experiences regarding these concerns, with a view toward encouraging other teachers to introduce the laboratory approach in their classrooms. Included are ten problems we have found valuable to children in the intermediate grades, the children's reactions to the mathematics laboratory, and our evaluation of it.

Procedures and problems

A regular classroom is used for the laboratory, and common, readily available materials are used in solving the problems. These problems are typed on individual cards and worded in such a way as to allow the children to work on them with a minimum of teacher assistance. The children, working in groups of three or four, record their work in laboratory books. They are encouraged to keep a record of their approach to the problem, including the steps they used to obtain the solution and the purpose behind these steps. During the laboratory period, the teacher makes frequent visits to each group, listening to the discussion, and asks the minimum of questions, all well phrased and based on the comments made by the students, which are needed to guide the group in the desired direction.

Perhaps the greatest difficulty experi-

---

[1] The Nuffield Foundation, Mathematics Teaching Project, 12 Upper Belgrade Street, London, S.W. 1, England.

Reprinted from *The Arithmetic Teacher* (January 1970), 7–14 by permission of the authors and the publisher.

enced in setting up the mathematics laboratory was the development of a workable set of problems that would provide the children with opportunities to think for themselves. The following ten problems have been successfully used in the laboratory in Scarsdale. Each problem is divided into three parts: *a*, *b*, and *c*. The first (*a*) lists the materials needed; the second (*b*) contains the instructions to the students as they would appear on the assignment card; and the third (*c*) supplies additional information to assist the teacher.

TEN SUCCESSFUL PROBLEMS FOR THE MATHEMATICS LABORATORY

1. Polygons and area[2]
    a) Materials needed
    (1) 3 rectangles labeled *A*, *B*, and *C*, with dimensions, respectively, of 6″ × 8″, 6″ × 8½″, and 6½″ × 8½″
    (2) A parallelogram, a triangle, and a trapezoid, labeled *D*, *E*, and *F*, respectively, each with base 6″ and height 4″.
    (3) Scissors
    (4) Ruler
    b) Assignment card
    (1) Measure the length and width of rectangle *A* and record these measurements. Beginning at the left, measure and mark each inch along the top and bottom of the rectangle; beginning at the bottom, mark each inch along the other two sides. Using these marks as guides, draw straight lines from side to side and from bottom to top so that the rectangle is divided into 1 in. squares. By counting the squares, determine the area of the rectangle. Can you find a shorter way to find the area?
    Repeat the same steps you have just used, first with rectangle *B* and then with rectangle *C*.
    (2) By using the information you

have learned above, determine the area of parallelogram *D*. (A parallelogram is a four-sided figure whose opposite sides are parallel.) Hint: You may cut the parallelogram into two parts and rearrange the pieces. Record your answer and explain how you arrived at this solution.

(3) Now find the area of triangle *E*. You may find it helpful to cut out another triangle *congruent,* or of the same size and shape, to the original triangle and work with both triangles. The results of your work with the parallelogram should help you solve this problem. Again, record and explain your solution.

(4) By using what you have just learned, can you find the area of trapezoid *F*? (A trapezoid is a four-sided polygon with two parallel sides and the other two sides not parallel.) You may cut the trapezoid in any manner that might be helpful. Write your answer and explain the steps you used in solving the problem.

c) The objective of this problem is to have the students visualize area as the number of square units a figure contains and learn the standard area formulas. They should perform each of the tasks until these concepts are understood for each type of figure.

The students are encouraged to cut or to duplicate the models of the various figures since this enables them to form a figure with which they are familiar. For example, by cutting the model of the parallelogram along a line perpendicular to the base they can form a rectangle, thereby discovering that the area of a parallelogram equals the area of a rectangle with the same height and base. Also, they see that by properly arranging the two triangles *E* a parallelogram is formed; hence, the area of their original triangle is one-half the area of the parallelogram. And by cutting the model of the trapezoid along a diagonal, thereby forming two triangles, they find that the area of the trapezoid is the sum of the areas of the two triangles.

---

[2] The teacher will recognize that the paper cutouts used in problem 1 are *models* of the polygons rather than the polygons themselves; some teachers may wish to make this distinction on the assignment card.

2. Using equations to record solutions to water-jar problems[3]

a) Materials needed
   (1) Water
   (2) 3 jars or plastic containers of various sizes, the largest having a 76 oz. capacity
   (3) Graduated quart measuring jar
   (4) Tape
   (5) Crayon or a felt-tip pen

b) Assignment card

For each exercise (table 1), pour the

Table 1

Exercise	A	B	C	D
1	24	2	5	17
2	18	43	10	5
3	9	42	6	21
4	20	59	4	31
5	23	49	3	20
6	15	39	3	18
7	28	76	3	25
8	18	48	4	22
9	14	36	8	6

amount of water indicated in columns A, B, and C into three jars. Put a piece of tape at the water level and mark it to show the height of the water. Using only these marked levels as measuring units, you are to obtain the amount of water indicated in column D. To do so you may pour water in or out of the three jars and discard the water that you do not need. Check your result by pouring the water into the graduated quart measuring jar. Record your solutions in mathematical terms by writing equations.

Think of as many equations as you can for each exercise in table 1. For example, only some of the ways exercise 1 can be solved are given below.

$$A - B - C = D$$
$$B + 3C = D$$
$$6B + C = D$$
$$5C - 4B = D$$

c) The purpose of this problem is two-fold. First, it is to overcome the blinding

[3] Adapted from Luchins, A. S., "Mechanization in Problem Solving: The Effect of Einstellung," *Psychological Monographs* 54, 1942, whole no. 248.

effect of habitual, or set, ways of solving problems. For example, since exercises 2–4 can be solved by the equation

$$B - A - 2C = D,$$

children tend to apply rigidly this solution to the remaining exercises. However, exercises 5 and 6 are also solvable by a shorter method using only two jars, A and C. Exercise 7 can only be solved by the method $A - C = D$. Exercises 8 and 9 are solvable by either method. It is interesting to note how many students are able to solve exercise 7, in which the solution does not follow the same pattern as in the preceding exercises.

The second objective of this problem is to provide an opportunity for the students to write mathematical sentences as the solutions to problems.

3. Angles of polygons

a) Materials needed
   (1) Protractor
   (2) Scissors
   (3) Paper
   (4) Ruler

b) Assignment card

Note: All polygons (many-sided figures) referred to in this problem are to be *convex;* that is, each angle is less than 180°.

(1) Draw a fairly large triangle and measure its angles with the protractor. Determine the sum of the measures of the three angles and record your answer. Cut out the paper model of the triangle and cut off the three corners. Place the corners beside one another with their vertices (points at corners) together and describe what happens to the sides on the outside. Does this help to check that your sum was correct?

Repeat these steps using a different triangle.

(2) Draw a quadrilateral (four-sided polygon), measure the four angles, and determine the sum of the degrees. Record your results. Again cut out the model of the quadrilateral and cut off the corners.

Can you check your work by arranging the corners in a certain way?

Repeat these steps using a different quadrilateral.

(3) Now draw polygons of five, six, seven, and eight sides, measure the angles, and determine the sum for each figure, completing the following chart:

No. of Sides of Polygon	Sum of Measures (in Degrees) of Angles
3	————
4	————
5	————
6	————
7	————
8	————

(4) Without drawing the polygon, determine the sum of the measures of the angles of a polygon with ten sides. Check your answer by constructing a ten-sided polygon and measuring its angles.

Angles of Polygons

(5) Can you determine the sum of the measures of the angles of a polygon with twenty-five sides? Is there a fixed relationship between the number of sides the polygon has and the sum of the measures of its angles? Can you explain this relationship?

c) In this problem, the students gain experience in using a protractor as well as finding the sum of the measures of the angles of convex polygons. They should clearly see the relationship between the number of sides of the polygon and the

sum of the measures of its angles. By cutting the corners off the paper model of the triangle and placing them beside one another with the vertices of the angles together, they should note that the exterior sides form a straight line, showing that the sum of the measures of the angles for the model is about 180°. By studying the results of their chart, they should also see that the sum of the measures of the angles of a polygon increases by 180° with each additional side. Hence the sum is given by the formula

$$(n - 2) 180°,$$

where n is the number of sides of the polygon. The children should attempt to explain the formula; the formula can be verified in terms of the sum of the measures of the angles of the $(n - 2)$ triangles formed by connecting any vertex of the polygon to each opposite vertex.

4. Circumference and diameter
   a) Materials needed
      (1) Several circular objects
      (2) String
      (3) Yardstick
      (4) Optional: bicycle and a playground circle

   b) Assignment card

   (1) Using string and a ruler, measure the circumference (distance around) and the diameter (distance across) of five circular objects. Record this information and tell how many times as great the circumference is as the diameter of each circle.

   (2) Now measure the diameters of several other circles and *estimate* their circumferences by using the relationship learned above. Check your answers by measuring the actual circumferences.

   Measure the circumferences of several circles, estimate the diameters using this relationship, and check your answers by measuring the diameters.

   (3) Optional. For this part you will need your bicycle, a yardstick, and a piece

of string as long as the circumference of your bicycle wheel. First, using your bicycle, measure the circumference of the large circle on the playground. After you have recorded this information, estimate the diameter of this circle using the above relationship. Check your answer by measuring the diameter.

c) If the students are able to do so, they determine pi by dividing each circumference by its diameter. Those unable to divide fractions can get the desired result by laying off the diameter's length on the string that measures the circumference; a finding of "a little more than three times" would be sufficient.

To determine the circumference of the large playground circle, the students should find the circumference of their bicycle tire and count the number of revolutions it makes as they ride around the circle. (Since considerable skill is required to ride on the line marking the circle, they should repeat this several times and find the average; this helps them gain an appreciation of the need for averaging.)

5. Probability experiments with dice

  a) Materials needed

    (1) Pair of dice

    (2) Graph paper

  b) Assignment card

    (1) Roll one of the dice 120 times and each time record the number that appears. Make a bar graph of your results by listing the numbers 1 to 6 (for the numbers on dice) along the bottom of the graph and showing by the height of the bars the number of times each number appeared. Is there a pattern to your graph?

    (2) Now throw both dice 120 times, record the sum of the two numbers that appear each time, and make a bar graph of these results. This time the numbers at the bottom of your graph will range from 2 to 12. Do you notice any pattern to this graph? Explain why certain bars are longer than others.

  c) When the students roll one die, they see that each face does not appear exactly one-sixth of the time, but that the probability ratio approaches this as the number of throws becomes very large. They also gain experience in correctly constructing and labeling a bar graph.

When they throw two dice, the students will notice an approximation of a bell-shaped curve, which is explained by table 2.

6. Probability experiments with coins

  a) Materials needed

    (1) 6 coins

    (2) Graph paper

  b) Assignment card

    (1) Flip two coins 100 times and record the number of times 0 heads, 1 head, and 2 heads appear. Make a bar graph of these results, writing the numbers

Table 2

Sum on 2 faces	Combinations that produce sum	No. of combinations	Probability
2	1 + 1	1	1/36
3	1 + 2; 2 + 1	2	2/36 = 1/18
4	1 + 3; 2 + 2; 3 + 1	3	3/36 = 1/12
5	1 + 4; 2 + 3; 3 + 2; 4 + 1	4	4/36 = 1/9
6	1 + 5; 2 + 4; 3 + 3; 4 + 2; 5 + 1	5	5/36
7	1 + 6; 2 + 5; 3 + 4; 4 + 3; 5 + 2; 6 + 1	6	6/36 = 1/6
8	2 + 6; 3 + 5; 4 + 4; 5 + 3; 6 + 2	5	5/36
9	3 + 6; 4 + 5; 5 + 4; 6 + 3	4	4/36 = 1/9
10	4 + 6; 5 + 5; 6 + 4	3	3/36 = 1/12
11	5 + 6; 6 + 5	2	2/36 = 1/18
12	6 + 6	1	1/36
	Totals	36	36/36 = 1

Probability Experiments with Coins

0 to 2 (for the number of heads) along the bottom of the graph and showing by the height of the bars the number of times each number of heads appeared.

(2) Repeat the same steps, but now use three, four, five, and then six coins. (The numbers at the bottom of your graphs will change depending on how many coins you use.)

c) The students should obtain results approximating the probabilities listed on table 3, based on Pascal's triangle.

Table 3

Heads:	0	1	2	3	4	5	6
No. of Coins			Probability				
1	1/2	1/2					
2	1/4	2/4	1/4				
3	1/8	3/8	3/8	1/8			
4	1/16	4/16	6/16	4/16	1/16		
5	1/32	5/32	10/32	10/32	5/32	1/32	
6	1/64	6/64	15/64	20/64	15/64	6/64	1/64

7. Finding a point—treasure hunt
   a) Materials needed
   (1) Measuring tape or a yardstick and string
   (2) Ruler
   (3) Treasure (candy, cookies)

b) Assignment card

Draw a map to scale, or write a clear set of directions which, when followed, will lead to your "buried treasure." You may "bury," or hide, your treasure in the playground or in the school building. Your directions should include several of the topics we have studied in mathematics this year.

c) One group of students does the assignment, and the other groups take turns trying to find the treasure, which, if they are successful, they can keep. The group that finds the treasure in the shortest time prepares the next map or set of directions.

8. Ratios—estimating with water drops
   a) Materials needed
   (1) Eyedropper
   (2) Measuring spoons (such as $\frac{1}{4}$, $\frac{1}{2}$, and 1 tsp.; and 1 tbsp.)
   (3) Measuring cup

   b) Assignment card

Using the eyedropper and the measuring spoons, determine about how many drops of water are necessary to fill the measuring cup to the 8 oz. mark. Then estimate the number of drops in a pint, a quart, and a gallon.

Write your answers in your laboratory

220

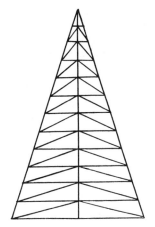

FIG. 1. Triangles and Trapezoids

book and explain clearly the method you used to arrive at these answers.

c) In this problem, students see the value of solving problems in the most efficient way, using the relationships between the various measures.

9. Triangles and trapezoids

    a) Materials needed: a copy of figure 1

    b) Assignment card

      (1) Determine the number of triangles in figure 1. Find a plan of attack which will help you arrive at the solution. Look at the large triangle in several different ways and convince yourself that you have seen all the triangles.

      (2) Determine the number of trapezoids (four-sided polygons with two parallel sides and two nonparallel opposite sides) in figure 1. Note: the only parallel lines are the 12 lines going across (horizontally). Again, have an approach in mind before you begin counting.

    Record your work in your laboratory book.

    c) Although this problem requires that the students recognize triangles and trapezoids, its importance is in the development of a systematic approach to the solution. There are several approaches to each part. For the trapezoid, the children might begin by counting the number of trapezoids that use part or all of the base of the big triangle as the larger base of the trapezoid:

there are five with such a base and a height of 1 unit; three more with such a base and a height of 2 units; three more with a height of 3 units; and so forth. Following this procedure with each of the horizontal lines, they will discover that there are 220 trapezoids.

By another systematic procedure, 113 triangles can be counted.

10. Squares

    a) Materials needed: a copy of figures 2 and 3

    b) Assignment card

      (1) Determine the number of squares in figure 2. Be sure to notice that a square with 2 units on each side consists of five squares: the large square and the four 1-unit squares within it. It's very important that you first plan how to solve the problem before you begin your work. Hint: How many squares are in a square with 2 units on a side? 3 units on a side? 4 units on a side?

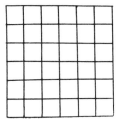

FIG. 2. Squares   FIG. 3. Picture Puzzle

    c) As in problem 9, the development of a systematic approach is important to the solution of this problem. In figure 2, the number of squares is

$$36 + 25 + 16 + 9 + 4 + 1 = 91.$$

In figure 3, there are sixty-eight 1×1, twenty-nine 2×2, seven 3×3, twelve 4×4, nine 5×5, three 6×6, four 7×7, one 8×8, one 11×11, and one 12×12 squares, for a total of 135 squares.

Students' reactions

The students are very enthusiastic about their work in the mathematics laboratory.

In fact, many have requested more frequent laboratory periods than our present one hour per week, while others have asked permission to continue the work during their lunch period. Furthermore, several students have written their own problems at home and are quite eager to have their classmates solve them in the laboratory. One child had a new approach to a problem occur to her just before she went to sleep. Afraid she would forget the solution if she waited until morning, she immediately got up and wrote it down.

The following comments by five students are representative of the classes' enthusiastic reaction to the laboratory.

I think that the laboratory is a great idea . . . I feel that I am learning a lot. I have learned how to work with other people. My group discussed our ideas and opinions until we found a way to solve the problems. We had to think and do research, and yet it was fun.

I think that the mathematics laboratory is a very good thing to have and there are very good points about it. Some of the good points are: (1) it helps you to learn to work well with other people; (2) it increases your ability to solve problems; (3) it helps you to think of many ways to solve one problem.

I know that the laboratory is work, but it's *fun* work!

The idea of people in the fifth grade thinking up new problems, I think, is good.

We get other people's opinions about a problem. I think one thing that is good about the laboratory is that there are many ways to figure out a problem and people all don't go about the problems the same way.

Authors' evaluation

Our experience has convinced us of the value of the mathematics laboratory approach to problem solving in the intermediate grades. We feel that the children benefit from this method in two main areas: first, they are able to achieve a deeper understanding of mathematical concepts through their experiments with concrete objects; second, since they quickly discover that there are many approaches to solving a problem, they come to view each problem with an open mind, eager to think of their own solution. As a result of their experiences in the mathematics laboratory, the children become deeply involved in their work, and learning for them is an active, creative process.

Improving problem–solving skills

JACQUELINE SIMS

In the last few decades, a number of procedures have been developed for improving problem-solving ability. This article presents some examples of interesting approaches to the problem.

Reading, selecting, and evaluating

Careful reading, selecting, and evaluating are all needed to determine the *one* number for which all the statements in the following exercises are true.

1. What number is—
   a) an even number,
   b) divisible by 3,
   c) less than 100,
   d) greater than 9 × 9,
   e) an even number when digits are added?

2. What number is—
   a) between one and one hundred,
   b) smaller than fifty,
   c) not an odd number,
   d) not a multiple of 4,
   e) not smaller than 20,
   f) not a multiple of 5,
   g) not larger than 30,
   h) not the same if the two digits are interchanged?

Using problems without numbers

For each of these problems, tell what you need to know. Then tell what to do to solve the problem.

1. Joe wants to know the average score in his arithmetic tests. In each test 100 was perfect.
   He needs to know: _____
   He solves the problem by: _____

2. Bill wants to know how much it cost his father to fill the gasoline tank of the car.
   He needs to know: _____
   He solves the problem by: _____

3. Sam wants to know how much more it cost for the dessert than for the sandwiches at the party.
   He needs to know: _____
   He solves the problem by: _____

Problems without numbers give practice in the aforementioned skills as well as in determining the order in which conditions should be used.

4. How can Sally find out how much change she should get? She buys some stamps at one price and some more stamps at another price. She knows how much money she started with.

Encourage children to write mathematical sentences that describe the problem.

Locating relevant and irrelevant data

Many students feel obliged to use every numeral in the problem. Some of the more conscientious, but less mathematically inclined, will go so far as to include the number of the problem itself and add in the page number just to be on the safe side. Students need training in the identification of relevant and irrelevant words or needed data in a problem. The following

Reprinted from *The Arithmetic Teacher* (January 1969), 17–20, by permission of the author and the publisher.

examples might serve to provide experiences in selecting relevant data.

1. Joe bought a box of 64 crayons for $.75. He gave the clerk $5.00 How much change should he receive?
   Answer: _____ Extra Number: _____

2. Mike spent 2 hours cutting the lawn and 20 minutes helping his dad plant 3 bushes. How many minutes was this?
   Answer: _____ Extra Number: _____

3. On Tuesday, 230 of the 242 children in Mayfield School were present. The principal said that 15 was the largest number absent any day that week. How many were absent on Tuesday?
   Answer: _____ Extra Number: _____

Finding more than one question in a problem

Problems with more than one question will give additional insight into problem solving. Let the children make a statement such as the following and find as many problems as they can within it. Translate the solutions into number sentences.

1. John has 152 stamps, Phil has 268 stamps, and Ray has 473 stamps.
   a) How many stamps do John and Phil have together?
   b) How many more stamps does Ray have than John and Phil have together?
   c) How many more stamps does Phil need in order to have 500?
   d) How many more than 700 stamps do Phil and Ray have?

Finding more than one way to work a problem

e) How many more stamps do the boys need to make a total of 1,000?

Innumerable questions may be developed from problems of this type.

2. A toy store sign reads:
   Toy cars are 5¢ each.
   Balloons are 3¢ each.
   Limit of 4 each.
   a) What is the least amount of money you could spend?
   b) What is the greatest amount you could spend?
   c) A boy spent 8¢. What did he buy?
   d) Alice gave the clerk a dime and a nickel. She received 2¢ change. What did she buy?
   e) Jimmy spent 17¢. What did he buy?
   f) Could anyone spend exactly 25¢?
   g) What purchase amounts to 21¢?
   Make a chart to show the prices of all the possible combinations of purchases.

Number of Balloons at 3¢ each

	0	1	2	3	4
0	0	3	6	9	12
1	5	8	11	14	17
2	10	13	16	19	22
3	15	18	21	24	27
4	20	23	26	29	32

Number of Toy Cars at 5¢ each

Some of the foregoing activities should help to develop the idea that there may be more than one way to work the same problem. These possibilities should be explored with exercises such as these:

1. Mary bought apples at 4 for 20¢. At this rate, what is the cost of one dozen apples?
   20¢ ÷ 4 = 5¢   One apple costs 5¢.
   12 × 5¢ = 60¢   One dozen apples cost 60¢.

224

Alternative method:

$12 \div 4 = 3$	In 12 apples, there are 3 groups of 4 apples.
$3 \times 20¢ = 60¢$	One dozen apples cost 60¢.

2. The Boy Scouts bought 50 cans of peanuts at $.25 a can. They sold them for $.45 a can. How much did they gain?

$50 \times \$.25 = \$12.50$	Cost of peanuts
$50 \times \$.45 = \$22.50$	Selling price of peanuts
$\$22.50 - \$12.50 = \$10.00$	Amount gained

Alternative method:

$\$.45 - \$.25 = \$.20$	Gained on each can
$50 \times \$.20 = \$10.00$	Gained on 50 cans

3. Joe was earning money by cleaning shoes. He cleaned 4 pairs in 20 minutes. At that rate, how many pairs could he clean in one hour?

$20 \div 4 = 5$	It took 5 minutes to clean 1 pair of shoes.
$60 \div 5 = 12$	Joe could clean 12 pairs in one hour.

Alternative method:

$60 \div 20 = 3$	There are 3 groups of 20 in 60.
$4 \times 3 = 12$	Joe can clean 12 pairs of shoes in one hour.

Developing number sentences

The development of number sentences aids in a clear and logical attack on the problem to be solved. Students should be encouraged to develop this technique.

1. Bill bought 3 baseballs and a bat for $2.46. The bat cost $1.50. What was the price of one baseball if the price was the same for each one?

Solution:

Number of cents for one bat	150
Number of cents for one baseball	n
Number of cents for three baseballs	$3 \times n$
Number of cents for baseballs and bat	$(3 \times n) + 150$
Number of cents for baseballs and bat	246

Sentence: $(3 \times n) + 150 = 246$.

This sentence is equivalent to the sentence $3 \times n = 96$.

Solution: 32

Thus, the price of each baseball was 32¢.

2. Sue saved $3.54. She earned $.75 more by babysitting. How much money does she still need to buy a sweater that costs $7.95?

Number Sentence	*Numbers Involved*
Number of cents Sue saved	354
Number of cents Sue earned	75
Number of cents she needs	$795 - (354 + 75)$
Number of cents she needs	n

Sentence: $795 - (354 + 75) = n$.

Solution: $n = 366$

Statement: She needs $3.66 to buy the sweater.

Using diagrams

The California Mathematics Council in its Strands Report recommends that pupils learn to use various strategies and tactics for general problem solving. The problem may be illustrated by the construction of a diagram or the use of materials.

A special rubber ball is dropped from the top of a wall that is 16 feet high. Each time it bounces up ½ as high as it fell. The ball is caught when it bounces 1 foot high. Find the total distance the ball travels.

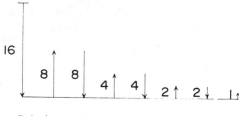

Solution: 16 + 8 + 8 + 4 + 4 + 2 + 2 + 1 = 45 feet.

Estimating a reasonable answer

The practice of estimating a reasonable answer before performing the necessary computation will establish a criterion for plausible responses and help to discourage wild guesses. The estimate can establish the limits within which to work the problem and also can serve as a check on the final solution. Most word problems provide opportunity for practice of this skill.

Conclusion

Word problems are a real challenge to teacher and pupil due to the many factors that must be accurately considered. However, participation in a developmental approach to the logical solution of problems may help to clarify the processes involved. The teacher must provide many opportunities for the children to become actively engaged in these procedures:

Establishing and using guides to good thinking
Following directions
Reading, selecting, and evaluating
Using problems without numbers
Locating relevant and irrelevant data
Finding more than one question in a problem
Finding more than one way to work a problem
Developing number sentences
Using measures
Using diagrams
Estimating a reasonable answer

Promoting problem–solving skills through nonverbal problems

CECIL R. TRUEBLOOD

Solving verbal problems is considered to be a very important part of elementary school mathematics programs. This is based on the assumption that there will be transfer to solving problems faced by pupils in a variety of physical world situations. With textbook problems, the poor reader often cannot abstract the essential elements of a problem situation because of his low level of reading ability. The teacher therefore needs verbal problems that present less interference to the development of problem-solving skills.

Wilson (1939) suggested that "functional problems" or "experience units" be used to present problems of social significance to students. While this approach has merit, it does not consistently provide enough verbal problems based upon the scope and sequence of modern mathematics programs. The type of verbal problems suggested by the following illustrative lesson has this potential. These problems differ from the ordinary since they are non-verbal problems; that is, they are presented without words. Consider this:

Miss Mason has been working on solving verbal problems with her class of 30 sixth graders since the beginning of the school year. Most of the class was doing acceptable work. However, it was becoming increasingly obvious that her pupils represented a wide span of abilities in problem solving. During the last period class, Sue and Janet read books because they quickly solved the assigned textbook problems. José shredded crayons because none of the verbal problems he tried to read made any sense. Sarah and Dwight finished all of the problems but none of the answers were correct. They just used the numerals without paying any attention to the words.

Miss Mason decided to challenge her class with a nonverbal problem. Before class the next day she used a few minutes to sketch the following scene on the blackboard.

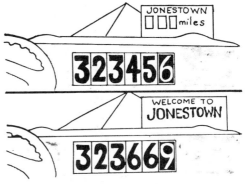

FIGURE 1

She introduced the problem by saying, "Look at this illustration carefully. What problem is suggested?" The teacher then elicited class discussion that defined the problem suggested by her illustration. After several minutes of discussion the class decided the problem could be "How far has the car traveled down the road?" The teacher then said, "When you think you've figured out a method for solving the problem, record your answer. Then check your answer by using a different method." The following represent the various types of solutions placed on the board by Miss Mason's pupils:

Dwight

32345.6	32345.6 (start)
+ ?	+ 21.3 answer
32366.9	32366.9 (finish)

José

32366.9	
−32345.6	
21.3	How far car traveled

Reprinted from *The Arithmetic Teacher* (January 1969), 7–9, by permission of the author and the publisher.

Sue said, "I multiplied 2 × 10 and then added 1.3. The car traveled 21.3 miles."

The teacher then gave the pupils a ditto sheet with several illustrations like those in Figure 2.

FIGURE 2

She said, "Think of a problem suggested by one of these illustrations. If you can show me that your problem can be answered by using the illustration you selected, you may present the problem to the class." The teacher was careful to put illustrations on the ditto sheet that could be interpreted in several ways.

What are the advantages of using such nonverbal problem settings? The major reasons for including some nonverbal problems in your mathematics program are listed below.

1. Nonverbal problems allow pupils to focus quickly on a problem situation without heavy reliance on advanced reading skills. For early primary grade teachers such problems provide another mode for presenting verbal problems at the prereading level. To the intermediate grade teacher they offer a vehicle for presenting more challenging verbal problems to pupils having reading difficulties.

2. The nonverbal problem format is very flexible; that is, many different problems can be generated by pupils from one illustration. To the teacher this flexibility provides another opportunity to meet the varied needs of pupils. He is able to encourage a single or multistep solution to a problem. For the pupils it means an opportunity to exercise some originality and creativity in finding solutions to verbal problems.

3. Perhaps the most potent advantage

is that the teacher can provide the types of problem situation that come closest to the real-life experiences faced by pupils inside and outside the classroom. For the teacher this means more student interest and greater potential for transfer of learning. To the pupils it means solving problems of immediate relevance to them.

4. Nonverbal problems can be tailored to meet the needs of pupils who have special requirements. For example, by exercising judicious choice in selecting illustrations, the teacher can meet the needs and requirements of pupils with culturally different or poor language backgrounds. These pupils are given an opportunity to meet problems that can be stated in pupil language. They then have a chance to listen to the interpretations of their peers.

The following illustrations are included to further focus on the type of illustrations that can be used in your classroom. Why not give them a try and create some of your own? How does it work?

FIGURE 3

FIGURE 4

FIGURE 5   FIGURE 6

Using Flow Charts in Elementary School Mathematics Classes

JAMES J. ROBERGE

TODAY'S ELEMENTARY SCHOOL pupils have been appropriately described as being members of the computer generation. To meet the needs of these pupils, elementary school teachers should provide them with learning activities which will enable them to become knowledgeable about computers while they improve their proficiency in the use of the basic skills. An interesting topic which will facilitate the attainment of these objectives is flowcharting. Stated simply, a *flow chart* is a diagram which contains a list of steps to be performed in sequence.

Reprinted from *Timely Topics* (November/December 1971) by permission of Houghton Mifflin Company.

Pedagogical Value of Flow Charts

The pedagogical value of flow charts is extensive. For example, flow charts can be used as instructional devices with pupils at various ability levels. That is, flow charts can be used as devices for motivating drill and practice as well as for enrichment.

Flow charts also provide a concrete means for illustrating a method of solution for a problem. In addition, flowcharting requires active participation, by the pupils, in the development of a solution for a problem. More precisely, since problem solving is a step-by-step process, in which certain steps have to be performed before others, the use of flow charts encourages pupils to organize their thoughts in a systematic manner, and to explore the various alternatives which arise at each step in the problem-solving process. Thus, when pupils have succeeded in flowcharting a solution for a given problem, they are more likely to understand the underlying concepts.

Finally, the flow charts that pupils construct to solve problems mirror their thought processes. Hence, they provide the teacher with a diagnosis of pupils' weaknesses with specific concepts.

Introducing Flow Chart Symbols

The geometric figures used as symbols in flow charts have specific functions. The basic symbols used include the following:

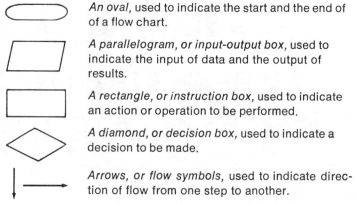

An oval, used to indicate the start and the end of of a flow chart.

A parallelogram, or input-output box, used to indicate the input of data and the output of results.

A rectangle, or instruction box, used to indicate an action or operation to be performed.

A diamond, or decision box, used to indicate a decision to be made.

Arrows, or flow symbols, used to indicate direction of flow from one step to another.

After the symbols have been displayed on the chalkboard, and the pupils have learned to identify each symbol with its function, some simple flow charts for routine activities might be constructed by the class, and discussed. Some sample flow charts are as follows. (see next page)

This activity could be followed by construction of similar flow charts for other routine procedures that are suggested by the pupils. Once they have grasped the basic idea, they are ready to apply flowcharting to mathematical activities.

```
                    ( START )
                        |
              +-------------------+
          +-->|  Take a           |<--+
          |   |  pencil.          |   |
          |   +-------------------+   |
          |           |               |
          |   +-------------------+   |
      +-->|   |  Look at          |   |
      |   |   |  the point.       |   |
      |   |   +-------------------+   |
      |   |           |               |
      |   |        /Is it \    Yes     |
      |   |      <sharp enough?>-------+
      |   |        \      /            |
      |   |           | No             |
      |   |   +-------------------+    |
      |   |   |  Put it in        |    |
      |   |   |  the sharpener.   |    |
      |   |   +-------------------+    |
      |   |           |                |
      |   |   +-------------------+    |
      |   |   |  Turn the         |    |
      |   |   |  handle.          |    |
      |   |   +-------------------+    |
      |   |           |                |
      |   |   +-------------------+    |
      |   +---|  Remove the       |    |
      |       |  pencil.          |    |
      |       +-------------------+    |
      |                                |
      |       +-------------------+    |
      |       |  Put the          |    |
      |       |  pencil on        |    |
      |       |  your desk.       |    |
      |       +-------------------+    |
      |               |                |
      |          /Do you \             |
      |        < have more >---Yes-----+
      |          \ pencils?/
      |               | No
               ( STOP )
```

```
                    ( START )
                        |
              +-------------------+
          +-->|  Select a         |<--+
          |   |  Key.             |   |
          |   +-------------------+   |
          |           |               |
          |   +-------------------+   |
          |   |  Try the Key      |   |
          |   |  in the lock.     |   |
          |   +-------------------+   |
          |           |               |
          |        /Does \     No      |
          |      <  it fit? >---------+
          |        \      /
          |           | Yes
          |   +-------------------+
          |   |  Try to turn      |
          |   |  the Key.         |
          |   +-------------------+
          |           |
          |        /Does \
          +--No---<  it turn? >
                   \      /
                      | Yes
              +-------------------+
              |  Open the         |
              |  door.            |
              +-------------------+
                      |
                 ( STOP )
```

Mathematical Applications of Flow Charts

Constructing a flow chart to test a number for divisibility by 2, 3, 4, 5, 6, 9, and 10 is an activity that could serve to review the topic of divisibility, while giving the pupils added insight into flowcharting in general. You may wish to introduce this activity by a quick review of the tests for divisibility, as follows:

1. A number is divisible by 2 if the digit in the ones' place of its compact numeral is 0, 2, 4, 6, or 8.

2. A number is divisible by 3 if the sum of the digits of its compact numeral is divisible by 3.

3. A number is divisible by 4 if the total value of its tens' and ones' digits is divisible by 4.

4. A number is divisible by 5 if the ones' digit in its compact numeral is either 0 or 5.

5. A number is divisible by 6 if it is divisible by both 2 and 3.

6. A number is divisible by 9 if the sum of the digits of its compact numeral is divisible by 9.

7. A number is divisible by 10 if the ones' digit of its compact numeral is 0.

(*Note:* Since the tests for divisibility by 7 and 8 are more complicated, we have not included them in the flow chart.)

The flow chart for applying the divisibility tests to a given number might be as follows:

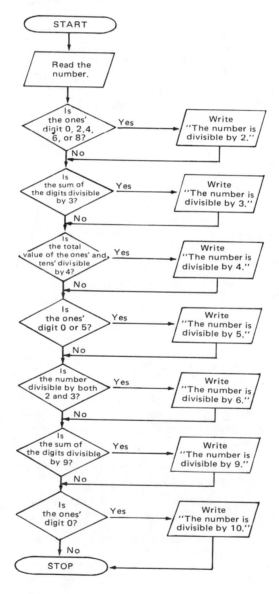

To show how it works, you might have the class test the number 27,330 for divisibility by 2, 3, 4, 5, 6, 9, and 10, as follows:

a. Read the number 27,330.

b. Since the ones' digit is 0, the answer to the question in the first decision box is "Yes," and we write, "The number is divisible by 2."

c. Since $2 + 7 + 3 + 3 + 0 = 15$, and 15 is divisible by 3, the answer to the question in the second decision box is "Yes," and we write, "The number is divisible by 3."

d. Since the total value of the ones' and tens' digits is 30, and 30 is *not* divisible by 4, the answer to the question in the third decision box is "No," and we move on to the next box.

e. Since the ones' digit is 0, the answer to the question in the fourth decision box is "Yes," and we write, "The number is divisible by 5."

f. Since the number is divisible by both 2 and 3, the answer to the question in the fifth decision box is "Yes," and we write, "The number is divisible by 6."

g. Since $2 + 7 + 3 + 3 + 0 = 15$, and 15 is *not* divisible by 9, the answer to the question in the sixth decision box is "No," and we move on to the next box.

h. Since the ones' digit is 0, the answer to the question in the seventh decision box is "Yes." We write, "The number is divisible by 10," and move on to the STOP box.

In summary, the class can write the statement, "The number 27,330 is divisible by 2, 3, 5, 6, and 10; it is *not* divisible by 4 or 9."

The following flow chart is a schematic representation of a sequence of steps which may be followed to determine the quotient and remainder for one-step division problems. This flow chart could be used to introduce pupils to a working form for solving one-step division problems. It could also be used as a motivational device to stimulate drill and review prior to the introduction of more advanced topics in division. In either case, this flow chart will provide you with an opportunity to reinforce pupils' understanding of the terms associated with the concepts of division.

As an example of the use of this flow chart for a classroom learning activity, the class might consider the division problem $6\overline{)25}$ and trace through the steps of the flow chart as follows:

a. Read the dividend 25 and the divisor 6.

b. Let the first estimated quotient be 5.

c. Since $6 \times 5 = 30$, and $30 > 25$, the answer to the question in the first decision box is "Yes."

d. The "Yes" arrow leads to the instruction to make a smaller estimate, so let the second estimated quotient be 3.

233

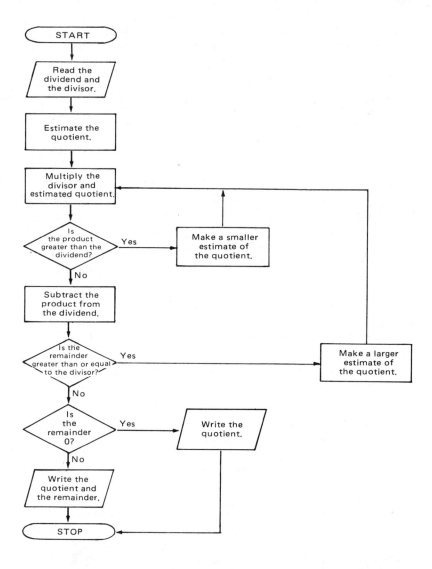

e. The arrow leads back to the instruction to multiply the divisor and the estimated quotient, so we multiply 6 by 3 and get the product 18.

f. Since 18 < 25, the answer to the question in the first decision box is "No."

g. The "No" arrow leads to the instruction to subtract the product from the dividend.

h. Since 25 − 18 = 7, and 7 > 6, the answer to the question in the second decision box is "Yes."

i. This time the "Yes" arrow leads to the instruction to make a larger estimate, so the choice now is 4, and we are led back again to the multiplication instruction box.

j. Multiplying 6 by 4 gives the product 24, which leads to a "No" answer to the question in the first decision box.

k. Since $25 - 24 = 1$, $1 < 25$, and $1 \neq 0$, the answers to the questions in both the second and third decision boxes are also "No."

l. The third "No" arrow leads to the output box, so the quotient 4 and the remainder 1 are written.

Concluding Remarks

The study of flow charts is a fascinating and challenging learning activity which will increase the pupils' awareness of an important means of communication in our computer-oriented era. At the same time, it can serve to sharpen their performance in basic arithmetic skills. In addition, first-hand experience in constructing and using flow charts should help to develop greater ability in problem solving. Thus, a flow chart is an instructional device which an innovative teacher may use in many ways to generate enthusiasm in the study of "modern" mathematics.

References

Duncan, E. R., Capps, L. R., Dolciani, M. P., Quast, W. G., and Zweng, M. *Modern School Mathematics: Structure and Use,* Revised Edition. Boston: Houghton Mifflin, 1972.

Gibney, T. C., and Lengel, J. A. "Utilizing a Flow Chart in Teaching Ninth Grade Mathematics." *School Science and Mathematics,* LXVIII (April, 1968), 292-296.

Kessler, B. M. "A Discovery Approach to the Introduction of Flow-Charting in the Elementary Grades." *The Arithmetic Teacher,* XVII (March, 1970), 220-224.

Lovis, F. B. *Computers 2.* Boston: Houghton Mifflin, 1964.

McQuigg, J. D., and Harness, A. M. *Flowcharting.* Boston: Houghton Mifflin, 1970.

National Council of Teachers of Mathematics. *Computer Oriented Mathematics.* Washington, D. C.: The National Council of Teachers of Mathematics, 1963.

Travers, K. J., and Knaupp, J. E. "The Computer Revolution Needs YOU!" *The Arithmetic Teacher,* XVIII (January, 1971), 11-17.

SELECTED BIBLIOGRAPHY FOR CHAPTER 10
Problem-Solving Techniques and Strategies

Ausubel, David P. "Facilitating Meaningful Verbal Learning in the Classroom." *The Arithmetic Teacher* (February 1968), 126.

Bradfield, Donald L. "Sparking Interest in the Mathematics Classroom." *The Arithmetic Teacher* (March 1970), 239–42.

Eisenberg, Theodore A. and Van Beynen, John G. "Mathematics Through Visual Problems." *The Arithmetic Teacher* (February 1973), 85–90.

Geiselmann, Harrison A. "Have You Tried This?" *New York State Mathematics Teachers' Journal* (June 1972), 123–27.

Grossman, Rose. "Problem-Solving Activities Observed in British Primary Schools." *The Arithmetic Teacher* (January 1969), 34–38.

Henny, Maribeth. "Improving Mathematics Verbal Problem-Solving Ability Through Reading Instruction." *The Arithmetic Teacher* (April 1971), 223–28.

Hess, Marvel. "Second Grade Children Solve Problems." *The Arithmetic Teacher* (April 1966), 317.

Himmon, Dean E. "Problem Solving, Part I." *Science and Children* (April 1966), 22–23.

——. "Problem Solving, Part II." *Science and Children* (May 1966), 16–17.

Kessler, Bernard M. "A Discovery Approach to the Introduction of Flow-Charting in the Elementary Grades." *The Arithmetic Teacher* (March 1970), 220–24.

O'Brien, Thomas and Shapiro, Bernard. "Problem Solving and the Development of Cognitive Structure." *The Arithmetic Teacher* (January 1969), 11.

Orans, Sylvia. "Go Shopping! Problem-Solving Activities for the Primary Grades with Provisions for Individualization." *The Arithmetic Teacher* (November 1970), 613.

Page, Robert L. "Old MacDonald Builds a Fence." *The Arithmetic Teacher* (February 1973), 91–93.

Pottenger, Mary and Leth, Leonard. "Problem Solving." *The Arithmetic Teacher* (January 1969), 21.

Riedesel, C. Alan. "Problem Solving: Some Suggestions for Research." *The Arithmetic Teacher* (January 1969), 54.

Schoenherr, Betty. "Writing Equations for Story Problems." *The Arithmetic Teacher* (October 1968), 562.

Snyder, Barbara B. "Please Give Us More Story Problems." *The Arithmetic Teacher* (February 1973), 96–98.

Steffe, Leslie. "The Relationship of Conservation of Numerousness to Problem-Solving Abilities of First Grade Children." *The Arithmetic Teacher* (January 1968), 47.

Sullivan, John J. "A Few Thoughts on Verbal Problems." *New York State Mathematics Teachers' Journal* (June 1965), 105–07.

——. "Problem Solving Using the Sphere." *The Arithmetic Teacher* (January 1969), 29.

Suydam, Marilyn N. "Using Research: A Key to Elementary School Mathematics— Verbal Problem Solving." Research Bulletin, Pennsylvania State University, University Park, Pa. 16802.

Swenson, Esther J. "How Much Real Problem Solving?" *The Arithmetic Teacher* (October 1965), 426–30.

Weaver, J. Fred. "Applications and Problem Solving." *The Arithmetic Teacher* (October 1965), 412–13.

Wilson, John W. "What Skills Build Problem-Solving Power?" *Instructor* (February 1967), 79–80.

Zweng, Marilyn. "A Reaction to 'The Role of Structure' in Verbal Problem Solving." *The Arithmetic Teacher* (March 1968), 251.

PART V

BEADS, BUTTONS AND THINGS

CHAPTER 11

MACHINES, MANIPULATIVES AND GAMES

Young children are, by nature, active and curious. They want to find out, to discover, to explore, and to operate the things around them and within their own world. Young children do not think in abstractions, but in terms of concrete objects and events. Therefore, they need to be encouraged and guided to explore objects and materials around them. Learners need to interact with one another, with classroom objects, and with other mathematical manipulatives and materials. They need to play mathematical games, pour liquids into containers of different sizes and shapes, measure objects, examine attributes, and explore relationships between objects. Thus, children experience and learn mathematics and mathematical processes.

The first seven articles in this chapter concern the use of manipulative or real-world objects in the teaching of elementary school mathematics. The last eight selections have to do with using mathematical games in the classroom. Perhaps the best way to introduce this chapter is with another short article, "Why Manipulative Materials?" by John Trivett.

Why Manipulative Materials?

John Trivett

Why manipulative materials? Because everyone touches and plays with objects—especially kids. "Don't fidget with the pencil." "Sit still!" "Why are you rolling that piece of chalk down your desk?"

Frequently, manipulation of objects is essential to learning. My daughter can't learn to walk without manipulating her feet, arms and whole body. You can't learn to drive a car without first playing with the machine. It's necessary even if, like me, you agree there's much more to driving! And we can't learn to cook, play the piano or write without moving materials.

Moreover, it's fun to manipulate materials—aesthetically pleasing. That's what pottery, sculpture and yo-yo's are all about. Maybe this even extends to throwing spitballs!

Sometimes one manipulates materials almost unconsciously while one meditates (saying a rosary), concentrates on the end product rather than the manipulation (combing your hair), or thinks (doodling with a pencil, coat button or cigarette). Looking intently through binoculars one plays with the adjustments, and while chatting with friends most people pat arms, toss change in their pockets or scratch their chins.

It is natural and delightful for all of us to manipulate materials. In fact, it's hard to imagine life without doing it. Sometimes these actions are essential to learning. Can such behavior contribute to the learning of mathematics?

Well, clearly, some materials have always been used for this purpose. Fingers and counters, for instance. Unfortunately, children soon get a feeling that they shouldn't depend on such physical aids, but many teachers do not help them develop criteria for learning so that they can internalize a counting method and abandon their fingers.

New important aids have been produced over the past twenty years and are available to all. Colored rods, geometry boards ("geo-boards"), attribute blocks, geometry models, squares and cubes, and a variety of children's

Reprinted by permission of Cuisenaire Company of America from an article in their Spring 1972 newsletter, *The Cuisenaire Reporter*, entitled "Why Manipulative Materials?"

building kits are some of the ones being used. The materials are combined with activities led by teachers who understand their use. Many students learn from an awareness of what they are doing far more richly, joyously and efficiently than from spoken or written words and explanations.

Speech can be a very weak communicator of basic mathematical principles and memory of a second-hand, vicarious experience is a poor substitute for first-hand involvement. Why should children just watch the teacher become involved? Why can't they, too, be turned on by their own involvement? We try to achieve this in art, football, science, or singing. Children make music, throw pots, take part in discussion groups, cook, read for themselves and draw their own pictures. Why shouldn't they also participate in mathematics?

The aids mentioned above have important qualities over and beyond other materials. They are simple, relatively cheap to buy or make and easily used in classrooms by students of all ages.

They lead—under the appropriate guidance—to an immense number of concepts, algebraically, arithmetically, and geometrically. After using them for eighteen years I still find some relevant, fresh use. (Last week's discovery: a display of colored rods arranged in a fan with rods the length of your fingers, and another train of rods representing the span of your hand. Not profound, perhaps, but potentially promising!)

The most important aids are *multivalent* materials. Their whole purpose is to act as bridges of communication. Each student manipulates materials and produces something. He looks carefully at what he has done. He talks about it and listens to accounts by class-mates engaged in similar work. The teacher probably adds useful conventional words and symbols which help students talk about it all and record their findings. But the teacher is almost powerless to add anything else —certainly not concepts. These have to be triggered within each person and do not *come across* from others. Together, students and teachers advance out of a corporate mixed ignorance toward understandings which, have no fear, are in accord with established knowledge and local curriculums. The teacher is not the only authority. A person is *his own authority*—he is the *author* of his own experience. We need children and adults who experience mathematics at first hand. When they do, they love it and they "get it"!

Do we use these aids all the time? Of course not! There is paper and pencil, talk and chalk, books, tables, many other things to use. A blend arising from a proper understanding of processes and facts is the pedagogic objective. Used inappropriately as magical substitutes for ignorance, insensitivity and alternates to other math programs, manipulative materials can be only dead wood. Far better to have left the trees alive and growing and stayed at home ourselves!

What can you do with an egg carton?

MARGERY BAUMGARTNER

I love egg cartons. Doesn't everybody? Whole books have celebrated their potentialities for art projects, and now expensive *objets d'art* based on the lowly egg carton are appearing on the market. But egg cartons also have value for the beginning study of mathematics.

An early learning activity which children pursue for the fun of it is sorting macaroni, buttons, or beans for color, shape, and size; later, also numerals and letters. For this purpose, colored plastic eggboxes are attractive; but pulpy, gray paper ones have the additional beauty of being available, free. When a child can contribute materials instead of having them provided, he has already involved himself; and he learns during the process of preparing his own box.

The beauty of the number 12 is its many factors: 12, 6, 4, 3, 2, 1. Various egg cartons can be cut into sections of these different sizes—i.e., a section to hold 6 eggs, which we call a 6-cup size; a 4-cup size; and so forth. For a carrying tray, the cut-off box top may be used, or even better, another uncut carton with 12 egg cups to fit the sectional ones into, for foolproof comparison of numbers.

To aid in distinguishing the quantities easily at this stage, all the 3's are painted one color, the 4's another, and so on. Both the long boxes (two 6's) and the short ones (three 4's) are used, variously subdivided, to acquaint the child with many possible patterns within the number 12. The children separate and reassemble the components in their containers, and as questions arise regarding combinations

smaller than 12, one egg cup can be fitted into another to show, for instance, that two of the 1-cup size are equivalent to one of the 2-cup size. The child learns how to test his own hypotheses. After a brief demonstration, more visual than verbal, the children manipulate the concrete materials freely.

The child is asked, "How many ways can you make a dozen?" After he has combined 6's, 4's, 3's, 2's, 1's, the teacher may persist, "How else could you make a dozen?" The child may try $6 + 4 + 2$, and go on from there. Those who are using number names may verbalize simple combinations in addition and subtraction as they fill and empty their boxes. The important point is that the material is open-ended; the individual using it may be as primitive or as sophisticated as his stage of development permits.

Many activities suggest themselves. Estimating is one thing children can do well. Fractions come naturally in this game. Counting by 2's gives the idea of multiplication whether or not you call it that. "Three 2's"—the child says what he sees. "How many 3's does it take to make 6?" is in the language of division.

Children use the commutative principle naturally; they don't need the name for it, but the teacher may help to develop the depth of the concept by verbalizing the operation when they notice that three 4's are equivalent to four 3's. The idea of learning in families appeals to them—when you learn one fact, you have learned several related ones.

Reprinted from *The Arithmetic Teacher* (May 1968), 456–458, by permission of the author and the publisher.

A quarter-hour with four-and-one-half-year-olds

Eighteen children, four and one-half years of age, found the egg cartons an exciting way to discover number ideas.

All eighteen had been pretested for ability to count serially and with understanding. Their rational counting seemed adequate for presenting the lesson on dozens to the whole group. The interest held for the entire discussion.

Teacher: "Do you know how many eggs come in this box?"

Tracy: "Sixty-nine."

Susan: "Forty—fifty—sixty."

Andrew: "A dozen."

Teacher: "How many is that?"

Brian: "Twelve."

Teacher: "Andrew says this is a dozen box, and Brian says a dozen is twelve. How can we check whether that's right?"

Children: "Count!" (All count as teacher points.)

Teacher: "Now we know a dozen must always be twelve. But suppose we only wanted to buy half a dozen, like this?"

Marc: "Six!"

Joe: "He thinks there's three and three. Three threes. No, three threes doesn't make that; two threes, I mean."

Teacher: "We can easily find out whether two of the threes will fit into this six. Can you tell that this is a three without counting the cups, just by looking?"

(Teacher held the three-section against a colored background, with the humps up for visibility, to encourage perception of a group without the necessity of counting.)

Children: "Three!"

(Teacher slipped two three-sections into the half-carton they had identified as six.)

Mitchell: "Three and three are six."

Joe: "The box says '12 eggs': 'eggs,' and '12.'"

Teacher: "Oh, so it does, right here— '12 eggs'! But if we buy only half a dozen—"

Children: "Six."

Teacher (holding them up and fitting them into the box): "How many half-dozens in a whole dozen?"

Children: "Two."

Teacher: "So this six plus this six are equal to twelve. But what if we only buy this many—four?"

Mitchell: "Two and two make four."

Teacher: "All right, let's put two twos in—they fit into the four."

Joe: "And how much does three fours make? Twelve."

Teacher: "We'll put in one four, two fours, three fours."

Andrew: "Maybe twelve."

Teacher: "It fits in the dozen box, so it has to be twelve. You can count again if you want to, but if it just fits we already know it's a dozen, twelve. Now anyone who'd like a chance to fit the pieces together might see how many different ways you can find to make a dozen."

The teacher listened to the children's responses and recorded them to reread for clues as to how their minds were working before planning the next step, to avoid teaching either above their comprehension or below their interest level. This time she used the tape recorder and found that it not only lightened the labor of getting the record while doing the teaching but also picked up nuances one might have overlooked during the fast-moving discussion.

What can we learn by studying a divided box? Essentially, we are developing an approach to problem solving. A way of working is what we want the children to learn—not the answer but the attack. As in teaching reading, we do not depend on any one method. The essence of mathematics teaching is this: "How can we find out?"

Let's cultivate intuition—"fill the box" in many ways and see that there may be more than one way to solve a problem. Children are always eager to discover what they can do with materials, until their curiosity becomes stifled. The young child provides his own motivation; his goals may not be the teacher's, but they provide incentive because they are his own.

Sugar-cube mathematics

JON L. HIGGINS

There's a lot of mathematics in a twenty-five cent box of sugar cubes. They can be used to illustrate different numeration systems, the measurement of volume, multiplication of fractions, and even techniques of estimation. A twenty-five cent box of cocktail-sized sugar cubes contains 210 cubes of surprising uniformity. Suppose we begin by having students verify the number of cubes in the box. (You may wish to cover the numbers printed on the outside of the box. It is also advisable to set one box aside for eating purposes only!) The most direct way to determine the number of cubes is to count them one by one as they are removed from the box and piled on the table. Of course, if you're interrupted in this process and forget where you were, you have little choice but to start all over again. With a moderate number of distractions a counting job like this could last all morning for many students!

Organizing the counting process

The first improvement you can make is to organize the counting process. Suppose that instead of forming one big pile, you counted out many piles of ten. If you're interrupted, you have only to repeat the counting process for the current pile of ten. Once the box is emptied, the task is completed by counting the piles of ten. We could apply the same kind of organization to this new counting task—that is, we could form heaps of ten ten-piles. Thus, by the time the box is emptied, we could have as many as three types of organizations in front of us—hundred-heaps, ten-piles, and (possibly) single cubes. Not only have we counted sugar cubes, but we have a physical illustration of the way our numeration system is organized.

Making Dienes-blocks substitutes

With some white glue we can make our illustration into a physical model that can help us investigate addition and subtraction algorithms. Take one of the ten-piles. By gluing faces of the sugar cubes together with a generous dab of white glue, you can transform the pile into a stick one cube wide and ten cubes long. Remembering that the hundred-heap was made by shoving together ten ten-piles should give you a clue about gluing hundreds. You can take. ten of the ten-sticks and glue them together side by side, forming a square of cubes that is ten cubes wide and ten cubes long. What we have formed is an inexpensive set of Dienes blocks.[1] To be sure, the commercial set of wooden Dienes blocks is much more permanent than our sugar-cube substitutes. Nevertheless, the sugar-cube sticks and squares should be durable enough to let you do much exploring to see how students can use this particular model representation.

Consider the addition problem 74 + 58. We can represent the number 74 by laying out seven sticks and four cubes. We can represent 58 by laying out five sticks and eight cubes. Now "add" by sweeping all the cubes together and all the sticks together. Can we simplify these two groups?

---

[1] Commercial Dienes blocks (also known as Multibase Arithmetic Blocks) are available from Herder and Herder, 232 Madison Ave., New York, N.Y. 10016.

Reprinted from *The Arithmetic Teacher* (October 1969), 427–431, by permission of the author and the publisher.

We have a total of twelve single cubes. We could take ten of the cubes and trade them in for a ten-stick. (You may want to have students work in pairs with one acting as a "banker" to facilitate this trade.) After the trade, we have simplified the situation to two single cubes and one ten-stick to be combined with the ten-stick group.

That ten-stick group originally had 12 sticks in it. With the addition of the one stick we obtained by the trade, it should now contain 13 sticks. Can we simplify this group? Yes, we can take ten sticks and trade them in for one square. We will have simplified our groupings to one square, three sticks, and two cubes. Our answer is one hundred thirty-two.

The volume of a cardboard box is measured by filling it with sugar cubes. Each layer of cubes can be considered a multiplication array.

Of course, what we have accomplished by all this trading is a carrying process. Encourage the child to keep a written record of both problems and answers. Hopefully, after a few sessions of this manipulation he will decide that he can save time by evolving his own algorithm for carrying a part of the written problem.

The model also works particularly well for subtracting. Here the process is to represent the original number in squares, sticks, and cubes and to remove enough pieces to form a representation of the number that is subtracted. The pieces that remain represent our answer. Consider the problem 31 − 17. We represent the first number by laying out three sticks and two cubes. To subtract 17, we must remove one stick and seven cubes, which requires some ingenuity. We can change our representation of 32 by trading in one of the sticks for ten cubes. We then have two sticks and twelve cubes. Now we can easily remove seven cubes and a stick, leaving us with the remainder, one stick and five cubes, or 15. With enough repetition most students should begin to evolve this (different) trading process into a written regrouping algorithm.

Looking at other number bases

When we began counting the sugar cubes in the box, we chose to organize our counting process in terms of ten-piles and hundred-heaps because we realized that this would ultimately illustrate our numeration system. There is no inherent reason why we could not choose other basic sizes for our piles and heaps. For example, we could have organized our counting around three's. We would have counted out three-piles, clustered three three-piles into a nine-heap, three nine-heaps into a twenty-seven mound, and three twenty-seven mounds into an eighty-one hill, etc. In this base three numeration system, our box of sugar cubes would count out to two hills, one mound, two heaps, one pile, and zero cubes. (Try it and see.)

Working addition and subtraction problems in base three numeration can be beneficial for older students if it serves to deepen their insight into our commonly used base ten system. To do this, we must be careful to continually connect the work in unfamiliar bases to the analogous situations in base ten. Sugar-cube Dienes blocks can help make this connection, for we can make exactly analogous physical representations for other numeration bases. Consider base three. Three sugar cubes glued together

These blocks represent the problem in **base three**—"112 + 1122." The task of regrouping will be accomplished **by trading** cubes for sticks, sticks for squares, and squares for super-cubes.

will make a stick for this system. Three of these sticks can be glued side by side to form a square. Three squares can be stacked and glued together to form the twenty-seven unit—perhaps a workable designation for it would be a "super-cube." Super-cubes can be combined into super-sticks, which can in turn form super-squares, which can form super-duper-cubes, etc. These base three representations can then be used to develop algorithms for addition and subtraction in base three notation. Because this development involves the same kind of trading procedures we have already discussed for base ten, it should be relatively easy to point out analogous procedures in different base numeration systems. Challenge your students to make base two, base five, and base seven sugar-cube sets of blocks as well.

Determining volume

All this sugar-cube gluing started with the problem of determining how many sugar cubes were contained in our original box. We could have tackled the problem in a different way. Our box that contains the cubes occupies a volume. So does our cube. If we knew the volume of the box, and of a single cube, we should be able to figure out how many cubes could be put together to form a volume equal to that of the box. Thus we could indirectly count the cubes by measuring volumes.

The process of directly measuring lengths, areas, and volumes is basically a process of counting. Suppose we want to measure the length of a line segment; that is, to compare the length of the line segment to the length of a unit segment. We can make the comparison directly by covering the line segment with unit segments laid end to end until the unit segments form a new line segment congruent to the first. We then count the number of unit segments we have used to form this congruent line segment. That number is the measure of the original line. Suppose we take the length of a toothpick as a unit line segment. To measure another line segment like that formed by the edge of a desk top, we simply lay toothpicks end to end along the edge until the edge is covered. The number of toothpicks we use is the measure of the edge of the desk—and that number is obtained by counting. A ruler improves upon this process in two ways. Its edge represents several units already laid end to end and this helps speed up the measuring process. It also has subdivisions of these units to help improve the accuracy of the approximate congruence we wish to form.

Just as the unit of length must itself be a length, so must the unit of area be an area and the unit of volume be a volume. Any shape that occupies space could be used as a unit volume. However, the nature of this shape does affect the ease with which we can carry out the measuring process. As we shall soon see, a cube is a very judicious choice for the shape of a unit volume because it allows us to short-cut the basic counting involved in the measuring process. Suppose we take our sugar cube as one unit of volume. How would we use it to measure the volume of a rectangular solid, such as a box? Just as we covered the line with toothpick units, we must "cover" the volume of the box with sugar-cube units. We then count the number of sugar cubes required to form our congruent volume, and this number is our measure.

You may wish to have students try this procedure with some simple boxes constructed so that the problem of fractional units is temporarily avoided. Small cocktail-size sugar cubes are one-half inch long on each edge. Thus we want to construct boxes that are whole multiples of half-inches in each of their dimensions. You can easily make such boxes from poster board or light cardboard by using the basic pattern shown in figure 1.[2]

Score the cardboard along the dotted lines so that the four side tabs can be bent upwards and taped together to form the box. It's fun to give students lots of these

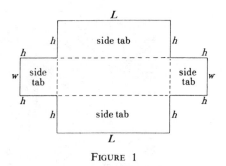

FIGURE 1

boxes and let them guess which will hold the most cubes before they actually begin to fill them up. An interesting set of boxes can be made by using the following dimensions.

Height	Width	Length
1	1	36
1	6	6
2	1	18
2	2	9
3	1	12
3	2	6
4	1	9

The interesting thing about this set of boxes is that they *all* hold 36 sugar cubes. Yet their appearance is quite different, and most children will believe that some will hold more than others.

The process of counting the sugar cubes which fill these boxes can be shortened considerably. Can you find out how many sugar cubes there are in the top layer without counting every cube? You can by realizing that this layer of cubes forms a rectangular array, and that rectangular arrays are one way to define multiplication. Thus you can find the number of cubes in the top layer by multiplying the number along one dimension (length) by the number along the other dimension (width). And the total number of cubes in the box is the number of cubes in one layer times the number of layers (i.e., the number of cubes along the third dimension, height). If we take the length of any one edge of a sugar cube as a length unit, the number of cubes along the width is the same as the measure of the width-line in these units.

Similarly the numbers of cubes along the length and height are just the measures of

[2] Sealing tabs will help to make the box stronger. They are not pictured in the illustration since the basic pattern is sharper without the tabs.

these lines in our edge-units. Thus, instead of counting, we can find the number of cubes in the box by multiplying $\ell \times w \times h$, where ℓ, w, and h are the numbers obtained by measuring the length, width, and height lines in cube-edge units. Since the number of cube units contained in the box is what we mean by the measure of the volume of the box, we have

$$V = \ell \times w \times h.$$

Thus there is really no mystery involved in the formula for the volume of a rectangular solid. *It is simply the result of short-cutting the fundamental counting process.* After some practice with this short-cut, students should be able to return to the original box of sugar cubes and determine the number of cubes it holds by counting only those cubes along each of the three dimensions.

Working with fractions

Taking a sugar cube as a single (but whole) unit of volume avoids the problem of working with fractions. For older students you may want to pose the problem of determining the number of sugar cubes as a problem in multiplying and dividing fractions. Have them measure each edge of a sugar cube with a ruler marked in at least half-inches and compute the volume of the cube. Each edge of the cube measures one-half of an inch, yet upon multiplication they will find the volume of the cube to be much less than one-half of a cubic inch. You may want to have them stack enough cubes to form a cubic inch in order to explain this apparent paradox.

The volume of the box can be computed after measuring its edges. Of course, these measurements will give the volume of the outside of the box, and we are really interested in finding the volume of the inside of the box. Thus the box measurements should be made a little short, especially along the length, since there are two thicknesses of cardboard forming each end. Dividing the volume of the box by the volume of a sugar cube will give the number of sugar cubes in the box.

Estimating

If you look very closely, you can see some of the grains of sugar which make up the sugar cube. Would you estimate the number of grains in a single cube to be in the tens, hundreds, thousands, or millions? By looking closely (a magnifying glass helps) you can get an approximate count of the number of grains along one edge of the cube. Assuming the number of grains to be the same along the other edges, and the distribution of grains to be uniform throughout the cube, you can multiply to get a more accurate estimate of the number of grains in the entire cube. Try changing the approximation of the number of grains along one edge and see what change this makes in your final answer.

We have seen several ways in which the lowly sugar cube can illustrate basic mathematical ideas. Perhaps you can think of others. Do try sweetening up your mathematics class. You may find your discoveries in sugar-cube mathematics just beginning.

A number line without numerals

ESTHER MILNE

An assignment for an in-service class was to think of ways to help first graders become aware of the ways of numbers without writing any numerals. The assignment was instigated by some who feel that the writing of Hindu-Arabic numerals in the first grade hinders a program dedicated to understanding numbers. A feeling for "numberness" must be a part of the child's being before the act of writing the abstract symbols is required of him. Writing symbols involves manual dexterity and internalization of an abstraction, not fortes of most first graders.

To achieve a deep feeling for numberness the students will have to participate in activities involving one-to-one correspondence, counting, joining disjoint sets, grouping to make ten, and so forth. These activities could take all or practically all of the first grade school year.

A number line without numerals can provide opportunities for such participation by students. (Directions for making the number line follow the discussion.)

Counting the cards at each position is one activity. Children can set up the number line by counting the cards in a strip and setting them in order.

Just as basic as counting is the realization that each neighboring strip contains one more card (or one less card) than the strip next to it. How could the number of cards be figured for a position if one strip of cards is left out? (See fig. 1.) How could the number of cards be figured for the next position to the right of the last one pictured? And the next one? And the following one?

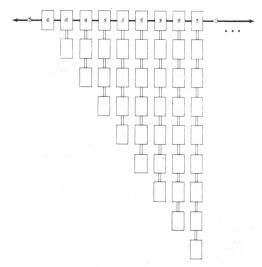

Fig. 1

If two colors (or designs) on cards are used alternately in a strip of cards, "odd" and "even" can be discerned by observation. (See fig. 2.)

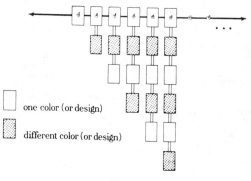

☐ one color (or design)

▨ different color (or design)

Fig. 2

A first look may reveal to some of the young students that every alternate number strip ends with a card of a particular color (or design). The even number will show a one-to-one correspondence be-

Reprinted from *The Arithmetic Teacher* (March 1971), 189–191, by permission of the author and the publisher.

tween the two colors (or designs) in a strip. The real meaning of the concept of "even" can be understood this way.

Adding (joining disjoint sets) to make ten can be done easily by using the cards from the number line. Have half a dozen hooks nearby. Have a student place the ten-card strip on one hook. (It's nice for this activity to have the ten-set of cards a special color that is not used for the one-set through the nine-set.) Next to the ten-set have a student place a set of cards, say the eight-set, and to it hook on the set needed to represent the missing addend, in this case the two-set. Continue with another pair to make ten. (See fig. 3.) Five presents an interesting problem.

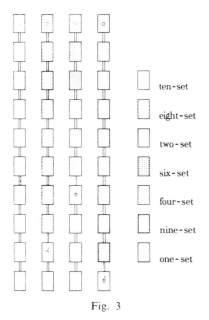

ten-set

eight-set

two-set

six-set

four-set

nine-set

one-set

Fig. 3

Figure 4 illustrates the directions for making the number line.

The backs of the old deck of cards can show for the counting activities.

The other side of the cards can be covered with "contact" paper. Before this side is covered, you should connect the number of cards you want together with mask-

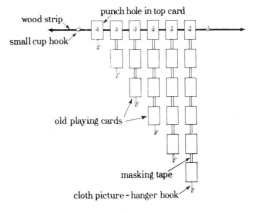

Fig. 4

ing tape and put the cloth picture-hanger hook on the bottom of each strip. Use this side for sums-of-ten activities, as was illustrated by figure 3.

The colored (or design-covered) cards used for the "odd-even" exercises might be stapled together more quickly than taped together. Each top card will still need a hole punched in it.

EDITOR'S NOTE. These ideas are well worth trying with young children. Can you extend these suggestions to counting with multiples? To counting from 10 to 20? And so forth? CHARLOTTE W. JUNGE

Graph paper: a versatile visual aid

ROBERT PARKER

In considering the mathematics program for elementary children, it is important for teachers to examine not only mathematical ideas but also the contexts within which such ideas are used. This usually means an examination of visual aids or tools of learning, since instruction in elementary grades depends upon the concrete to explain the abstract. With this in mind, one such tool that deserves special attention is graph, or squared, paper. Until one looks at the many possible uses of this visual aid, it may be taken for granted, and its applications remain sporadic and disconnected. This article is presented for the purpose of showing that graph paper is an educational device having few peers, especially from the points of view of economy, availability, and ease of use. An ordinary compilation of graph paper uses would be helpful to some degree, but what follows is a sequential development from the primary level through the upper elementary grades.

To begin with, graph paper allows for pupil construction of a number ray. Since the linear coordinates are equally spaced, there exist equal units along both the vertical and horizontal axes. Regular quarter-inch squared paper is suitable in the upper grades, while larger squares are preferred for primary children. Paper having one-half inch squares (which can be teacher-made on a mimeograph stencil) compensates for the lack of muscular and visual

coordination in many primary pupils. This first function of graph paper used as a number ray brings out one-to-one correspondence, reenforces counting numbers, and presents sets having fewer, more, and equivalent members. Many school systems

1	2	3	4	5	6	7	8	9	10
	12			15				19	
21			24						30
		33			36			39	
41	42			45					50
	52	53			56	57			
	62		64			67			70
71	72			75			78	79	80
		83		85	86			89	
	92		94			97			100

FIG. 1.—The One Hundred Square.

provide primary teachers with a form known as the One Hundred Square (Fig. 1). This ten-by-ten array is useful in presenting the concept of ten and multiples of ten.

The One Hundred Square can be thought of as a number line on which children develop somewhat rote numeration at first, followed by the concepts of grouping with tens and one hundred.

Reprinted from *The Arithmetic Teacher* (February 1969), 144–148, by permission of the author and the publisher.

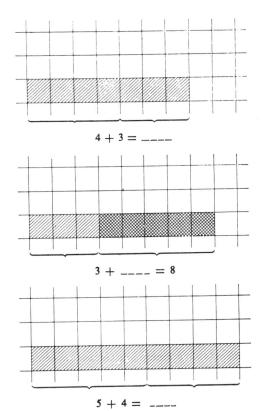

$$4 + 3 = \underline{\hphantom{0000}}$$

$$3 + \underline{\hphantom{0000}} = 8$$

$$5 + 4 = \underline{\hphantom{0000}}$$

FIG. 2.—Sets and equations on graph paper.

+	0	1	2	3	4	5
0	0	1	2	3	4	5
1	1	2	3	4	5	6
2	2	3	4	5	6	7
3	3	4	5	6	7	8
4	4	5	6	7	8	9
5	5	6	7	8	9	10

FIG. 3.—Initial table of addition facts.

A number line or ray along both axes prepares the child for addition facts, found many times in a table (Fig. 2). Listing all numerals up to ten need not be done until children are ready for such combinations. One might begin a table of addition facts by including only the numerals through five and extending these as mastery is gained (Fig. 3). The term "commutative" would not be used at this level, although the con-

cept of 3 + 2 being equivalent to 2 + 3 would be developed.

As a preliminary to even this beginning table, substitute x's or circles for the numerals along the axes. Thus there would be x for one, xx for two, etc. The child would then draw the proper number of x's at the coordinates (Fig. 4).

+	x	xx	x xx
x	xx	x xx	xx xx
xx	x xx	xx xx	x xx xx
x xx	xx xx	x xx xx	xxx xxx

FIG. 4.—Initial table of addition facts substituting x's for numerals.

Beginning work in multiplication is facilitated by having children construct arrays showing rectangular or square arrangements. (See Fig. 5.) This visual approach builds upon the addition skills and shows the child how many two's or three's make up a number. Such arrays are also useful for the inverse operation of division in that a child can use the subtractive approach to remove two's or three's from an array. In

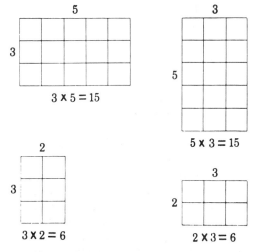

FIG. 5.—Multiplication arrays showing commutative principle.

251

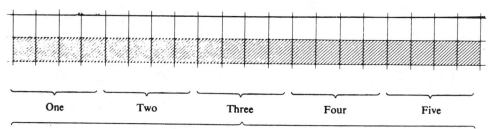

One   Two   Three   Four   Five

Number of times four is found as a factor

$5 \times 4 = 20$

FIG. 6.—Collecting factors along a number line or ray.

addition to the array method of introducing multiplication, the simple collecting of factors along the number ray may also be used (Fig. 6).

Many of the whole-number uses of graph paper apply also to the rational numbers, as illustrated by the following diagrams of equivalent fractions and multiplication of rational numbers.

The study of many geometric forms and the finding of area within many regions can be done visually with graph paper. The larger one-half or one-inch squares

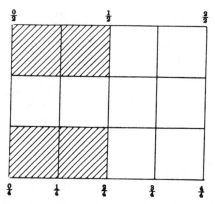

FIG. 7.—Equivalent fractions showing $\frac{2}{4} = \frac{1}{2}$.

should be used, since these concepts call for children to trace an outline with crayon or heavy pencil and to cut such an outline with scissors. While the study of shapes is started in the primary grades, continuous review and presentation should enable the child in the upper elementary grades to see the relations between a formula for area and the particular form of a region.

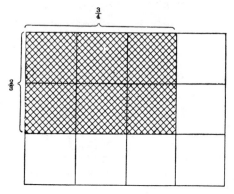

FIG. 8.—Multiplication of rational numbers showing $\frac{2}{3} \times \frac{3}{4} = \frac{6}{12}$.

In conclusion, the following is a random sampling of graph paper use. Pupil readiness and a background of previous graph skills, rather than strict grade placement, determine when children should receive instruction in many of these concepts. For example, the recreational game of Battleship is not just entertainment, but rather an exercise in coordinates. It has been placed in Book 2 of the sixth-grade School

6 units

5 units

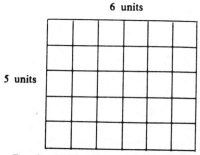

FIG. 9.—Rectangular array showing that length times width is the proper formula for this figure. $5 \times 6 = 30$ square units.

252

Mathematics Study Group program. This same game may be used with fourth graders, since these children have studied intersections when making tables of addition and multiplication facts.

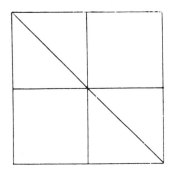

FIG. 10.—Visual relation between area of a square or rectangle ($A = bh$) and that of a triangle ($A = \frac{1}{2}bh$). Cutting and rearranging a square or rectangle allows the child to make two symmetrical triangles, each being one-half the original square or rectangle.

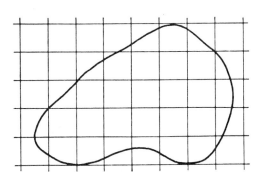

FIG. 11.—Graph paper used for approximation of area of irregular figures. The maximum coverage of square units is thirty-three, with total coverage in fourteen square units and partial coverage in nineteen square units.

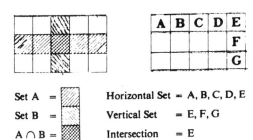

Set A =

Set B =

A ∩ B =

Horizontal Set = A, B, C, D, E

Vertical Set = E, F, G

Intersection = E

FIG. 12.—Set intersection.

8	3	10
9	7	5
4	11	6

FIG. 13.—Magic square

	1	2	3	4	5	6	7	8
A		B						
B								
C			B					
D								
E							B	
F		B						
G						B		
H								

FIG. 14.—Game of Battleship.

To play the game of Battleship, each player prepares a grid with numerical and alphabetical axes. "Battleships" are inserted in five locations. Opponent attempts to guess the location of battleships by naming grid coordinates. Variations of this game use a treasure cache or a jungle safari with animals to be captured.

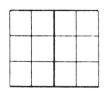 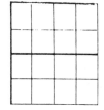

FIG. 15.—Perpendicular lines.

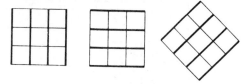

FIG. 16.—Parallel lines.

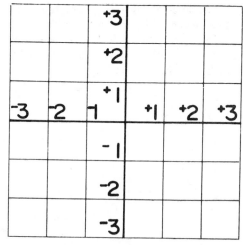

FIG. 17.—Studying integers on graph paper.

Cooking teaches numbers in kindergarten

Sara S. Shipley

Since the first Sputnik went up, science and mathematics have increased in importance at all levels of education. At the highly active kindergarten age, cooking projects offer an opportunity to introduce many fundamental concepts of both. With very simple equipment an imaginative teacher can devise an assortment of projects appropriate for any season of the year.

Making applesauce is a good fall project. If you have a friend with an apple tree, your group may experience the delight of gathering the apples. But, no matter how you acquire the apples, the group will have to do considerable tasting all along. Contrast the crunchy, firm texture of raw apples and gritty sugar with the smooth, soft feel of the cooked applesauce.

No cooking project should be a teacher demonstration. Let them do it. Table knives are sharp enough to cut apples without cutting children. Wash the apples in a pan deep enough for the fruit to float. At our school we use a plastic dishpan. Why do the apples float? While the apples are cooking, try some other things around the room to see if they float. Will a block float? The screwdriver? A pan?

Don't throw out the apple cores right away. Motivate some thinking with them. How many seeds are in your apple core, Anne? Several children will soon be counting apple seeds. Our hamster wants her fair share of the apple seeds. They are one of her favorite foods. The birds like them, too. Let's put a few in the geranium pots in the window. Per-

haps they will grow. If they do, we can lift the plants out into individual pots to take home—or, if there is a place available in the school yard, plant some outdoors. We must note on the calendar when we plant the seeds. When they come up we can count the days to see how long they took to germinate or sprout. Kindergartners are delighted with big words. Never hesitate to use them along with what may be the more familiar term.

Be sure that the applesauce does not burn while all this is going on. We must remember to stir it. Shall we serve it with cinnamon? Take a whiff and see if you like the way it smells. Cinnamon is the bark from a tree. These in this box look like little sticks, but they are really rolled up bits of the bark, hollow in the middle. The cinnamon in the shaker box is the same kind of bark that has been very finely ground. James says his mother makes cinnamon toast. That is a good idea! We might make cinnamon toast at school some day.

Now that all the members of the class have had a good chance to smell the cinnamon, how many want some in the applesauce? Twelve do, eight do not. Is twelve more than eight? They are not sure; some say yes, some say no! We might make a line. The "cinnamon" folks line up in front of my left hand —the "no cinnamon" folks line up in front of my right hand. Can you see which line has more children in it? Of course we can see there are four more who want cinnamon than do not want cinnamon. At this point the applesauce must be measured and divided so that ⅔ is served first and ⅓ is left in the pan to have cinnamon added.

See how many possibilities this simple apple project gives for using numbers? We add — subtract — divide. Any group of children can learn to say 1, 2, 3, 4, 5, 6, 7, 8, 9, 10 in a merry sing-song. But do the numbers have meaning for the child, or are they merely words as in a nursery rhyme?

Throughout the day the effective teacher uses numbers. A good way is to label things. One day Alan made several pictures. Teacher asked him if he would like to have them numbered in the order that he had made them before he took them home. Unfortunately, Alan was so fascinated with the numbering process that his objective came to be to making *more and more* pictures each day. So it took some time to *unlearn* that lesson and establish the fact that quality and not quantity in art is desirable.

Any kindergarten group in an area cold enough to have freezing weather has perhaps watched ice form as a science project. Do you carry this on through the melting of the ice and watching it all drift away as steam? This must be done with care and adequate warning about the dangers of steam. Keep in mind the short attention span of this age, use a small amount of wa-

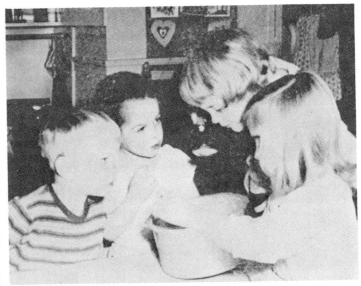

They put a certain number of eggs in the pan and mixed them together with a big mixing spoon and this is how they smelled to small noses...

ter—not more than ¼ inch deep in the pan.

Another project has each child put water in a little cup or dish from the housekeeping corner, then put it out to freeze. Not all the children will want to boil away their ice, but some will want to put it in the big pan and watch it go from solid to liquid to vapor. In this essentially scientific project there are numerous possibilities for using numbers and mathematical concepts. Put a thermometer on the window ledge with the dishes of water. Watch how it goes down. Use the clock for numbers. Mark the position of the minute hand when the dishes are put out by placing a bit of masking tape on the rim of the clock. How far did the minute hand move before someone noticed ice beginning to form? Before the first dish was solid? Before all were solid? (Mark each of these stages on the clock rim for further discussion.) Why did the water in the saucer freeze before that in the deep pitcher? How can Julie get her block of ice out of the sugar bowl which tapers toward the top? If you have a candy or deepfat thermometer, put it in the pan with the melting ice and watch it go up. **What did it register when steam began to appear? Watch the ther-**

mometer and be sure to remove it before it reaches the top limit or the pan boils dry.

It is great fun to watch the ice forming but there are values in the ice making project even if you have to do it in the refrigerator. Fill a plastic or paper container completely full. Why did the ice rise above the top? What happens if we let it melt? Will it run over the edge of the container?

Another variation on the popular ice making project is to place a small block or plastic toy in the water and watch it reappear as the ice melts from around it.

Occasionally we have popcorn at school with our fruit juice instead of the usual crackers. We put in ⅛ cup of popcorn then use that same ⅛ cup measure to serve the popcorn and count how many ⅛ cup servings came out after popping.

The accompanying picture shows the kindergarten group at the Lookout Mountain Methodist Week-Day School making pancakes. (The teacher brought an electric skillet from home for this project.) A mix would have been quick and easy. But, education-wise, the measuring, seeing, smelling, and tasting most of the ingredients that go into making pancakes is important. The recipe did not call for corn meal, but we used some so that we could contrast the feel of smooth, soft flour with the rougher texture of corn meal in the pancakes.

There are other food projects that take no cooking. Fresh cream (½ pint) shaken in a tightly sealed pint jar will make butter by the time everyone has had a turn at shaking it. There are quite simple activities like hot chocolate or rather involved two-day projects like finishing off the Halloween pumpkin as a pie. Cottage cheese making is a project that takes several days, but watching the liquid milk turn first to semi-solid clabber, then with a little heat separate into solid *curds and whey* is most rewarding.

One saucepan and a hot plate are all the equipment needed to start cooking in kindergarten. Many projects need no more than that. To enrich the program, measuring cups and spoons help, preferably metal sets of ¼, ⅓, ½, and 1—as well as individual spoons. These should be available to the children for measuring water, or sand at the sand table. They might also be used in comparing how much sawdust was created at the work bench by various children, or at different times.

Before we are ready for mathematics we must have a firm grasp of the large-small, big-little, greater-than, less-than, equal to, high-low, near-far, etc. relationships. At no time will a kindergarten teacher gather her group together for a numbers or arithmetic lesson, but throughout the busy program she uses numbers and stimulates interest in these basic relationships and mathematical concepts by good questions related to what the children are interested in doing.

Using functional bulletin boards in elementary mathematics

WILLIAM E. SCHALL

Visual aids—films, still pictures, models, bulletin boards, and so on—are among the most useful tools in education, but they do not teach without intelligent planning and use (Glenn O. Blough and Albert J. Hugget, *Elementary School Science and How to Teach It* [New York: Dryden Press, 1957], pp. 33–34). Bulletin boards can play an important role in today's mathematics program. However, a bulletin board, if it is to be successful in achieving its purpose, must gain and be *worthy* of the class's attention.

The writer wishes to extend his appreciation to Mary Jane Koepfle and Deborah Lewis, students in his mathematics methods course at the University of Cincinnati, for their help with this article.

A good bulletin board should also be supportive of and adaptable to classroom activities in a particular subject area, elementary mathematics in this case. Since children like activities or games in which they participate or are actively involved, a game approach is suggested for use here.

The rest of this paper describes several bulletin boards in various areas of elementary school mathematics. Each suggested bulletin board includes a short discussion of the bulletin board's purpose, the appropriate grade level, concepts and objectives that the bulletin board is intended to develop or reinforce, suggested questions that the teacher might use in connection with the bulletin board to stimulate additional thought and discovery, and a short description of a class activity that could

Reprinted from *The Arithmetic Teacher* (October 1972), 467–471, by permission of the author and the publisher.

involve pupils with the bulletin board.

The first bulletin board is shown in figure 1.

A. PURPOSE: To review and reinforce the basic mathematical operations as well as the concept of renaming numbers. The code or the message can be changed frequently, depending on the class, activities, season, and so on.

B. GRADE LEVEL: Fourth or higher depending on the code used. For higher grades a rational-number code can be used.

C. BEHAVIORAL OBJECTIVES:
1. The children will be able to work the problems and read the message.
2. The children will be able to recognize that a number can be renamed in many ways.
3. The children will be able to rename numbers in different ways.

D. DISCUSSION QUESTIONS:
1. What is a code?
2. Why do people sometimes write in codes?
3. How do you read a code?

E. DESCRIPTION OF THE ACTIVITY:
1. The children work the problems to read the message.
2. For additional practice in renaming, each child can write his own message in code and rename the letters as

different mathematical problems. Besides reviewing renaming, this will also review the basic mathematical operations.

Another bulletin board is shown in figure 2. It is used as follows:

A. PURPOSE:
1. To use the written numerals as a daily practice in counting
2. To serve as a daily practice to match a set with the appropriate cardinal number
3. To use the displayed objects as a basis of comparing sets

B. GRADE LEVEL: This bulletin board could be used in the kindergarten and first grade. Counting, the introduction of sets, and the comparison of sets are usually introduced in the kindergarten and could be used for review purposes in the first grade. (However, the specific time it is used depends on the children's progress.)

C. BEHAVIORAL OBJECTIVES:
1. The child will be able to count to 10.
2. The child will be able to identify each numeral; this means knowing a "5" or a "7" when he sees it.
3. The child will be able to name the geometric figures that are the elements of the sets.
4. The child will be able to match the

CAN YOU READ THIS MESSAGE??

$3 + (4 + 3)$ $(5 + 5) - 3$ $(8 + 2) + 0$ $(5 + 8) + 13$

$3 + 20$ $(5 + 5 + 5) - 10$ $(7 + 7) - 13$ $(25 + 4) - 26$

$(1 + 1) + 0$ $2 + 2 + 2 + 2$ $(6 + 6) - 7$ $(4 + 4) + 0$

$(17 + 10) - 6$ $(24 - 8) + 5$ $(1 + 1) - 1$ $(5 + 6 + 6) - 12$

$(1 + 2) + 0$ $3 + (28 - 27)$ $79 - 53$ $(18 - 7) - 2$

$(3 + 1) + (1 + 3)$ $(8 - 8) + 1$

a = 8	f = 25	k = 20	p = 21	u = 2
b = 14	g = 12	l = 4	q = 6	v = 24
c = 22	h = 5	m = 19	r = 15	w = 10
d = 9	i = 26	n = 17	s = 23	x = 16
e = 7	j = 13	o = 3	t = 11	y = 1
		z = 18		

Fig. 1

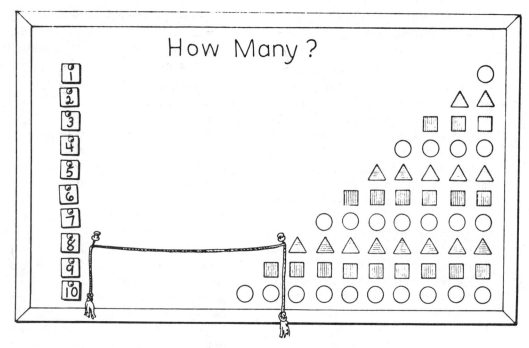

How Many ?

Fig. 2

sets with the correct cardinal number. Rope mounted on the bulletin board can be used to indicate the correspondence.

5. He will be able to compare the sets, thus using the ideas of more or less.

D. QUESTIONS TO STIMULATE THOUGHT:

1. Can you name the numerals written on the bulletin board?
2. Can you show me where the "5" is (similarly for other numerals up to ten)?
3. Can you show me the set of four objects? Or, which group has the four circles?
4. Which set of objects has the largest cardinal number?
5. Which set of objects has the smallest cardinal number?
6. Are there more circles (○) than squares (□)? How can you tell?

E. DESCRIPTION OF THE GAME:

1. A child can select a number and point to the numeral that represents the number.

2. Another child can match the correct set with the cardinal number.
3. If the second child gets the correct answer, he gets to select a number.
4. The activity can be varied. The teacher can point to a numeral and ask children to identify the correct set.

The next bulletin board is shown in figure 3, and its use is outlined below.

A. PURPOSE:

This bulletin board is designed to stimulate thinking about geometric shapes—how they are made and what they are called. There is to be transfer of learning from the geometric shapes illustrated on the bulletin board to geometric shapes in the everyday environment.

B. GRADE LEVEL:

This particular bulletin board is designed for the primary grades; however, the basic idea of the bulletin board (the geo-board) can be used for all grade levels in the elementary school.

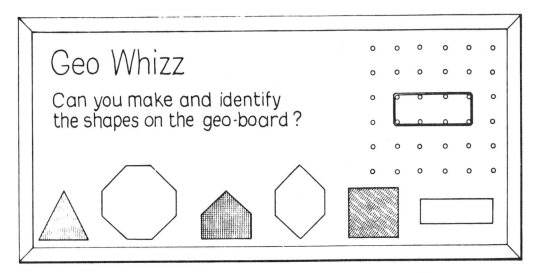

Fig. 3

C. BEHAVIORAL OBJECTIVES:
1. The child will be able to identify the six geometric shapes pictured on the bulletin board.
2. The child will be able to copy the indicated shapes on the geo-board.
3. The child will be able to recognize the number of points connected in each shape.
4. The child will be able to distinguish between the geometric shape and its region.
5. The child will be able to recognize a wide variety of geometric shapes in the classroom.

D. MATERIALS:
1. Individual geo-boards
2. Rubber bands
3. Yarn (for use on the bulletin board)

E. QUESTIONS TO STIMULATE THINKING:
1. The same geometric shapes that you made on the geo-board can be found within the classroom. Can you find some examples?
2. Can you make a given shape on the geo-board?
3. If given the number of points or sides in a geometric form, can you create the corresponding form?

4. How many different geometric shapes did you see on the way to school this morning?

The last bulletin board to be discussed here is shown in figure 4; its use is outlined below.

A. PURPOSE:
To motivate children as they work with the fundamental operations of arithmetic; also, to increase the learners' skill in computation in the fundamental operations

B. GRADE LEVEL:
Most intermediate grade levels—depending on the difficulty of the computation and skills involved

C. BEHAVIORAL OBJECTIVES:
1. The learners will work examples using the four fundamental operations of arithmetic with increased accuracy and speed.
2. The learners will demonstrate increased interest in arithmetic through participation in the "Grand Prix" activity and other mathematical activities.
3. The learners will demonstrate co-operative learning and working skills

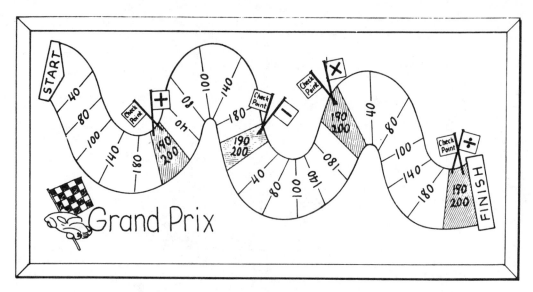

Developed by Mrs. Beth McCracken, Westwood Elementary School, Cincinnati, Ohio.

Fig. 4

through participation in the Grand Prix activities.

D. DESCRIPTION OF THE ACTIVITY:

.Grand Prix is designed to motivate children to attain a higher level of proficiency with the four fundamental operations of arithmetic—addition, subtraction, multiplication, and division. Children within each class are divided into teams of two. These teams

1. Are composed of one student from more skilled or more able groups and one from the less skilled or slower-moving groups to maintain a balance;
2. Are in twos so they can help each other with flash cards and other team activities;
3. Provide good "working together" experience.

These activities are done, of course, after good, meaningful experiences have been provided for the learners with the operation; the emphasis is on refinement of the skill, that is, accuracy, speed, retention, and operational ease. To begin the activities, the addition facts are given for the qualification day. The learners are then given a chance to have a "trial run." They have a week to practice with their partners before "race day," when their combined scores (these scores can be predetermined values) determine where their racer moves on the track. The same procedure is repeated for substraction, multiplication, and division. Trophies, certificates, and so on, are presented to teams that reach the "checkpoint." The checkpoints are team scores of 200 or whatever categories are chosen.

One corner of the room has letters suspended over it spelling "THE PIT," where children may go when other work is finished to get "tuned up." There is a box labeled "Mechanics' Tools" that contains flash cards and various other devices for practice with the basic facts.

EDITOR'S NOTE. Bulletin boards should "teach" as well as decorate! Mr. Schall indicates how this may be done. Particularly helpful are the statements of purpose and the specific suggestions for the use of each bulletin board in teaching. Have you other good ideas to share with us?—ARNOLD M. CHANDLER

Papy's Minicomputer

FRÉDÉRIQUE PAPY

The method we have used to introduce children to mechanical and mental arithmetic employs the distinct advantages of the binary system over other positional systems, while at the same time taking account of the decimal environment in which we are immersed. It is thanks to the *Minicomputer Papy* that we have been able to achieve this.

Inspired by some work of Lemaitre, this variety of two-dimensional abacus uses the binary system on boards which are arranged according to the decimal system.

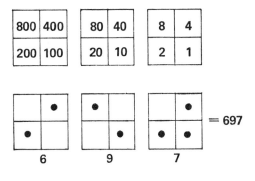

Fig. 1

The colours* recall the red family of the set of Cuisenaire rods and help to make the four rules of the machine easily accessible.

2nd October, 1967

Frédérique hangs one of the display boards on the blackboard

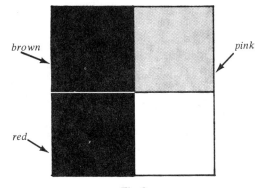

Fig. 2

– Oh, they're our colours! said Jean-Jacques, showing a white, red, pink and brown rod.
– *Yes, indeed!*
  We will play a game together; you will use your rods and I will use the board and these counters.
Frédérique puts two black counters on the white square.
– They stay up by themselves! one child marvelled.
– *Couple two white coaches together.*
– It's the same as the red rod!
– A white-white train is the same length as a red coach.
Frédérique takes two counters from the white square and puts one on the red square.

---

*In this book, the colors used in the original article could not be reproduced. A dot inside a circle, ⊙, or a number in a circle represents *red*. A star, ★, represents a *green* counter. As noted on page 265, the black dots in Figs. 18–20 represent *blue*.

This article is reproduced from *Mathematics Teaching* (Spring 1970), 40–45, the Journal of the Association of Teachers of Mathematics, by permission of the author and the editor.

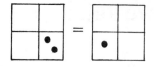

Fig. 3

– *Two counters on the white square are equal to one on the red.*

The children are quick to adopt the shorthand 'a red' for 'a counter on the red square' and state the rule 'two whites are equal to a red'.

– *Let us go on with the game!*

Frédérique puts two counters on the red square and the children put two red rods together.

– It is the same as the pink rod!

Frédérique takes the two counters from the red square and puts one on the pink square.

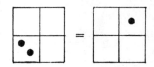

Fig. 4

– Two reds are equal to a pink.

Jean-Jacques is on fire with desire to move the counters and begs:

– Madame! Can I play?

Frédérique agrees and Jean-Jacques lifts the black counter from the pink square and replaces it with two green counters.

– It is beautiful!

– *Couple two pink coaches together.*

– It is the same as a brown rod.

Frédérique removes the green counters and puts one on the brown square.

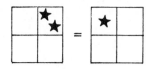

Fig. 5

– Two pinks are equal to a brown.

– *A new game!*

Frédérique places a counter on the white square

– 1

She places a second counter on the white square

– 2

– I can play another way! says Carine. She lifts off the 2 counters from the white square and puts one on the red.

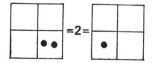

Fig. 6

– *Add 1 to the number 2*, Frédérique suggests, putting a new counter on the white square.

– It is 3.

– Like with rods, red-white

– or light green.

– *Add 1 to the number 3*, Frédérique continues, putting a new counter on the white square.

– It is 4.

– I can play another way, says Sylvie, who takes the two counters from the white square and puts one on the red.

– And another way, states Jean-Jacques, replacing the two counters on red by one on the pink.

– *Let us go on adding 1 . . .*

The recital continues and the numbers 5, 6, 7, 8, 9 appear on the Minicomputer.

Fig. 7

– *Add 1 to the number 9.*

A child puts a counter on the white square.

– It is 10.

– I play in another way, says Anita, replacing the two counters on white by one counter on the red.

– It's the red-brown train.

– The orange rod.

– It's 10.

Frédérique hangs a second Minicomputer board on the blackboard, to the left of the first. She takes the two counters from the red and brown squares and puts a counter on the white square of the new board, saying simply

– *And this is still* 10.

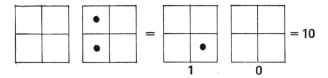

Fig. 8

– Jump! On to the second board, remarks Jean-Jacques, not in the least surprised.

Thus armed, our pupils are able to represent on the Minicomputer the number of elements in any set of counters put on square 1. The application (in reverse order) of the four fundamental rules on a small scale lets us keep a concrete link between a number in its written form and as represented on the machine.

13th October, 1967

Two small Minicomputer boards and a box of counters are on the desks in front of each child.
– *How many counters do you have in your box?*
– 29 . . . 32 . . . 18 . . . 26 . . . 35
The children's boxes are not all filled with the same number!
– *Put the box of counters on the white square of the first board. Play . . . and then write down the result.*

The first extended piece of individual work on the Minicomputer: concentration, precise and quick movements: some children arrive at the result without a mistake.

Clumsy techniques, counters spilled, false moves; the others must be helped.

The second part of the lesson is played on the wall Minicomputer.

We start with the number 25 on the machine. We work all the counters back to the white square on the first board and count them: confirmation!

After a fortnight the demands of the class forced Frédérique to introduce a third board and to accept numbers over 100.

– If I put a counter on each square of the machine, will I make the number 100? asks Didier, who still has a machine with only two boards.

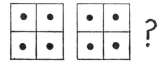

Fig. 9

Frédérique does not answer. She expects the problem to arise later in a different form and prefers to let the child's thinking follow its own course.

24th October, 1967

Didier returns to the attack.
– I want to make 100 on the machine.
– *All right . . . let us have your method!*
– 100 is twice 50, Didier goes on, showing:

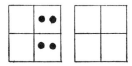

Fig. 10

– I can play!
He replaces the two counters on the white square by one counter on the red square and the two counters on the pink square by one on the brown.

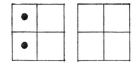

Fig. 11

– I want to play again! he says, taking the counters from the red and brown squares and putting a counter on the left of the second board.

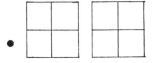

Fig. 12

And Didier demands his rights:
– I must have another board, Madame!
Frédérique gives him one. Didier triumphantly forms the number.

263

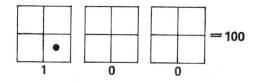

Fig. 13

Overexcited, he shouts out the numbers as he shows them on the machine.

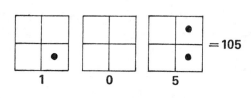

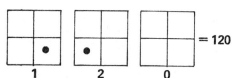

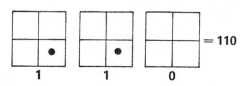

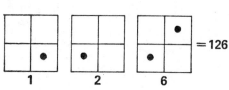

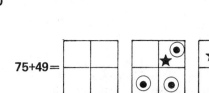

Fig. 14

Impressed, the class has shared in this discovery.

The addition of whole numbers is carried out automatically by 'playing' the machine. The same is true for *doubling* which is a primitive experience for children and fundamental to the Mini-computer.

The part played by arbitrary memorisations, so often a distasteful part of learning to compute, is reduced to a minimum. The addition of small numbers is carried out by means of intelligible rules: a purely binary system when the sum is less than 9 and a mixed decimal-binary system in other cases. From the beginning, therefore, the children are initiated into a positional system of numeration.

– *In the car park I counted 75 Volkswagens and 49 Mercedes. How many cars are there altogether?*
– Show 75 in red!

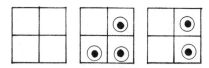

Fig. 15

– And 49 in green!

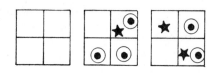

Fig. 16

– Can I play, Madame?

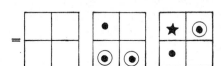

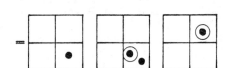

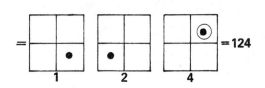

Fig. 17

264

The children make a note of the temperature in degrees centigrade (Celsius) each morning. At the start of the school year, in Brussels, these are always natural numbers, but the winter months force the use of negative numbers. From the *sixth month*, negative integers are part of the common knowledge of the children, having authentic status as numbers 'measuring' a 'size' which has been experienced.

The results of a sequence of two-person games are written with red and blue numbers which are mutually destructive, since each point captured by one of the players 'kills' a point captured by the other. The addition of red and blue numbers follows. From the given notation the children arrive at the additive group of integers.

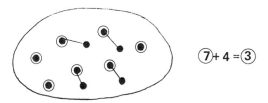

$$\overline{7} + 4 = \overline{3}$$

Fig. 18

Eventually, the writing is simplified and all the numbers are written in black, those originally blue having a bar across the top. In effect this time we have the group Z, $+$, except for a very minor difference, $\overline{3}$ being used instead of -3. We have known the advantages of the $\overline{3}$ notation for beginners for a long time since we have used it to simplify logarithmic calculations.

At the beginning we show the result of each game by putting red and blue counters on a flat surface. A battle to the death gives the final score. The children spontaneously transfer the procedure to the Minicomputer.

– *Let us calculate this to find how big it is:*

$$\overline{100} + 23$$

Great excitement at all the desks!
– *Do it on the machine!*

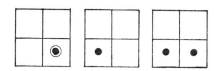

Fig. 19

☐ The black dots in Figs. 18, 19 and 20 should be interpreted as being blue.

Fig. 20 (opposite)

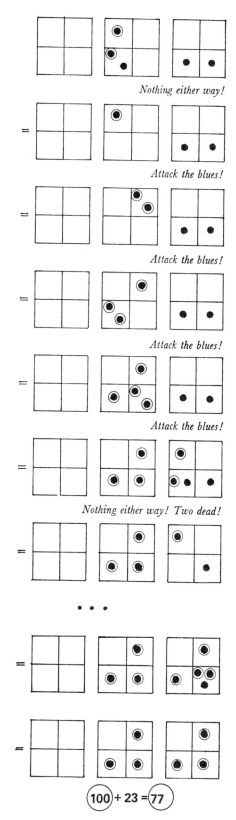

Nothing either way!

Attack the blues!

Attack the blues!

Attack the blues!

Attack the blues!

Nothing either way! Two dead!

• • •

$$\overline{100} + 23 = \overline{77}$$

– Who will win?

– Red!

– Red soldiers! Attack the blues!

In our teaching, the function 'a half of' appears as the reciprocal of the function 'twice' which is a transformation of **Z**; that is, a function mapping **Z** into **Z**. But some numbers exist which are not twice an integer. They do not have integral halves.

A bar of chocolate can be broken fairly into two pieces. A 100 franc note can be changed into two 50 franc notes. 100 is certainly an even number. But one franc can be changed into two 50 centime coins, and fortunately this always appeals to six-year-old children. Consequently it seems very natural to them to try to find a half of 1 on the machine.

It was because they wanted to write 100, got by doubling 50, that the pupils literally obliged me to give them the third board. We can now put 100 on the board and watch the way in which we find a half of it; we can start with 10 in the same way.

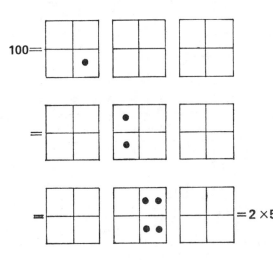

100=

=

= 2 × 50

Fig. 21

How can the same rule be applied to finding a half of 1? The children ask for a new board, to the right of the others, and at once call it the 'tiny

numbers board'. How shall we remember that it is the 'tiny numbers board'? By putting a green line between the boards which will later become the decimal point. So we can calculate a half of 1 and we write 0·5.

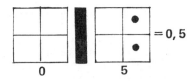

= 0, 5

0   5

Fig. 22

The problem of dividing 100 francs fairly between three children introduces an exceptionally interesting situation. The first glimpse of a non-terminating decimal comes through in this remark of one child: 'We will still be here tomorrow morning . . .'

The interest which six-year-old children show in quite large numbers forces us to respond by giving them unmotivated calculations which offer various kinds of challenge. Thanks to the Mini-computer, our pupils can add and subtract three-figure numbers and multiply them by simple fractions. The work is a good foundation because it requires considerable concentration to find the best strategies each time. These exercises help to produce a harmonious cooperation between the intelligent human being and a true machine, which is what the Minicomputer is in the eyes of the pupils.

Bibliography

G. Lemaitre, 'Comment calculer' in *Bulletin de l'Academie royale de Belgique: Classe des Sciences*, Brussels, 1954.

G. Lemaitre, 'Pourquoi de nouveaux chiffres?' in *Revue des Questions Scientifiques*, July, 1955.

G. Lemaitre, 'Le calcul élémentaire' in *Bulletin de l'Academie royale de Belgique: Classe des Sciences*, Brussels, 1956.

G. Lemaitre, *Calculons sans fatigue*. E. Nauwelaerts, Louvain, 1954.

Papy, *Minicomputer*, Ivac, Brussels, 1968. English version: Collier-Macmillan, New York (in press).

Papy, *Mathématique moderne* 1. Didier. Brussels-Paris-Montreal, 1963. (English version: Collier-Macmillan, New York, 1969.)

Frédérique and Papy, *L'enfant et les graphes*. Didier, 1968. English version: Algonquin Publishing, Montreal, Canada, 1970.

Frédérique, *Les enfants et la mathematique* 1. Didier, 1970. English version: Algonquin Publishing, Montreal, Canada (in preparation).

The accounts of lessons described in this article are extracted from the last reference and are reproduced here by kind permission of the editor.

"Fradécent"—a game using equivalent fractions, decimals, and percents

CHARLES ARMSTRONG

Fradécent is a card game that is useful in helping sixth graders learn equivalent fractions, decimals, and percents as well as aiding them in adding and subtracting fractions. The word *fradécent* contains parts of the words fraction, decimal, and percent.

The game is played with a deck of eighty-one cards. The following fractions are used: 1/2, 2/4, 4/8, 5/10, 1/4, 2/8, 4/16, 3/4, 6/8, 12/16, 1/5, 2/10, 3/15, 2/5, 4/10, 6/15, 4/5, 8/10, 12/15, 1/8, 3/8, 5/8, 7/8, 1/10, 3/10, 7/10, and 9/10. Cards with the corresponding equivalent decimals and percents are used with these fraction cards.

Play begins with the dealer giving each player seven cards and placing the remaining cards in the center of the table face down. The player to the left of the dealer begins by drawing the top card from the deck. Each player in turn tries to make a scoring set. There are two possible scoring sets: (1) a set of three cards consisting of an equivalent fraction, a decimal, and a percent (one of each), and (2) a set of three cards consisting of three equivalent fractions. When a player is able to make a scoring set, he places the cards face up in front of himself. Whether or not he is able to make a scoring set, each player must discard (place a card face up beside the deck of cards in the center of the table) to complete his turn. A player may not go out without discarding. Play continues clockwise.

Additional scoring is possible by adding on to existing scoring sets. This may be done by any player during his turn. He may

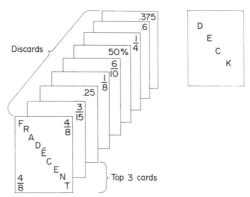

Fig. 1. Drawing from the discards

build on his own sets or on the sets of his opponents. A player builds on his opponents' sets by placing the scoring card in front of himself and declaring the set on which he is building.

During his turn, a player may draw a card from the deck in the center of the table or any number of cards from the pile of discards. If he draws from the discards, he must use the last card he takes in a scoring set during that turn and keep all the other discards in his hand. In figure 1, for example, if the player takes the top three cards, he must use the .25 card in a scoring set and add the 3/15 and 4/8 cards to his hand.

When a player gets rid of all his cards, the hand is over and the points are added. The face values of the cards are counted for points. The total of the cards remaining in a player's hand is deducted from the total of his scoring cards. The winner of the game is the player who gets five or ten points first. Before play starts, players should agree on whether they will play a five-point game or a ten-point game.

Reprinted from *The Arithmetic Teacher* (March 1972), 222–223, by permission of the author and the publisher.

Placo—a number-place game

ROBERT C. CALVO

This game emphasizes fun *and* the learning of positional and place-value aspects of our number system. *Placo* (pronounced play-so) is a number game that boys and girls can play to good advantage in the classroom. The children blindfold each other, one at a time, and then attempt to fit the rings on the highest place-value pole (see diagram). One value of the game is the manipulative aspect. The children enjoy handling the rings, fitting them on the dowels, and then computing to find the amount they "win."

While Placo is chiefly a **place-value** game, in actual classroom practice it contributes to other operations in arithmetic as well. It provides stimulating practice in computation and enables children to have fun while they play the game. It can be used with success in Grades 2 through 6 and can be refined to challenge even the brightest students. It may be played by as few as two children or as many as the whole class.

Three colored dowels are glued on a base, as shown. These dowels are stationary. The dowel to the left is highest in value, the one to the right is least in value. They may be painted blue, red, and white in that order. They are labeled 100's, 10's, and 1's. Thus 100's (in a whole-number game) are blue, 10's are red, and 1's white.

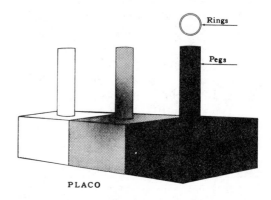

PLACO

One player is **blindfolded** and given some rings (9–18) of which he may put not more than nine on any one **dowel**. This **blindf**olded player then tries to **place** the **rings** on the dowels. The other players (split in teams) **watch** in glee. The object of the game **is to get** the largest score by putting the **rings** on the highest-value dowel. After all **rings** have been placed, the blindfold is removed, and the whole class or a small group computes the amount. The other players then take their turn. Players can play with a goal of 10,000 or with a time limit, the highest score winning. In playing decimal-fraction Placo, the blue dowel represents ones; the red dowel, tenths; and the white dowel, hundredths.

Placo cards may be developed to use with the game. These cards are used as follows: After each turn, a player selects

Reprinted from *The Arithmetic Teacher* (May 1968), 465–466, by permission of the author and the publisher.

one of the cards and follows the directions given. They say, "Double your score" or "Cut your score in half" or "Subtract three tens," etc. This serves to provide variety in the game. These cards can be developed to reinforce specific skills needed by the class or to provide the class with needed drill. Cards might give directions to add, subtract, multiply, or divide. In addition to dramatizing and creating added interest, they serve to equalize scores of fast and slow learners without penalizing the fast ones.

A word here about the use of colored dowels. While some experts say that place-value instruction should not depend on color, we can't rule it out completely at early levels of learning. In place value, it is important that the child knows that "this ring is in the place to the right" or "that peg is in the units column," but that doesn't mean color can't be brought in to augment these instructions. It merely means color should not be used as a sole means of identifying number-place. If color can help the children gain a concept or can make a game more attractive, then its use is defensible.

Other uses can be found for this game. One possibility is to use it with sight-saving classes by cutting niches in the base to identify the corresponding number-value. Computation could be done mentally. Another potential use would be to roll special dice and compute the score in whatever number base the dice show when rolled. The dice could be made so only certain number bases would result.

Materials list for Placo

1. Piece of wood 12 inches long, 4 inches high, and 6 inches wide

2. Doweling (also known as "closet poling"), $1\frac{3}{8}$ inches in diameter, three pieces of 7 inches each

3. Drapery "cafe curtain" rings, $1\frac{1}{2}$ inches inside diameter, 12 to 20

4. Cards and tacks

5. Sandpaper

6. Paint—white, red, and blue high gloss

7. Carpentry tools, including drill and saw

8. For variation noted, 3-by-5 cards

9. Large handkerchief or clean rag for blinder

10. Oak tag for power chart

Building the game

1. Drill three holes equally distant from each other in the 12-by-6-by-4 inch piece of wood, making these holes 1 inch deep and $1\frac{7}{16}$ inches in diameter.

2. Sand one end of each 7-inch piece of doweling smooth. Put some white glue on the bottom inch of the dowels and the inside of the drilled holes. Slip the dowels in and allow to dry overnight.

3. Purchase cafe curtain rings of the size specified in the materials list.

4. Paint the game three different colors, as mentioned previously, making the dowels and the base of matching colors.

5. You may want to put two small nails about 3 inches from the top of each dowel (exact placement would depend on the thickness of the curtain rings). In this way, players could put only 9 rings on each peg—a 10th would have nothing to keep it from falling off. This style of counting uses only 9 units. When 1 more is added, it becomes the number on the left.

Mathematical Games

DAVID S. FIELKER

A seminar group at the 1971 ATM Conference spent four happy sessions on mathematical games. Most of the time was spent just playing, but we did manage to tear ourselves away from *Squares, Dr Nim, Avalanche* and the others, to analyse our own mental processes and discuss the purposes of playing games as a mathematical activity. This is a personal, ill-remembered, brief summary!

Some games are specifically designed to teach certain topics or encourage certain concepts. Often this is a limitation, but sometimes the material or the rules can be modified to widen the scope. *Ten-up*, for instance, is designed to practice certain number-facts, but other properties of numbers can be explored and a re-numbering of the pieces can provide experience of modular arithmetic.

A ready-made game entails the discipline of obeying an arbitrary set of rules. This is perhaps a mathematical activity. It is possible to modify the rules, or make up your own rules as you go, checking consistency and feasibility. This is perhaps more of a mathematical activity.

Some games are based on mathematical structures, and though they are implicit rather than explicit, playing the game involves an experience of the structure. Sometimes in order to play the game well, one must analyse the structure. The *Hex* colour-matching game is based on a tessellation of hexagons. *Think-a-dot* games can be analysed in terms of group structure.

Even when a game does not involve identifiable mathematics, it may *feel* mathematical. This is perhaps because minute strategies are developed, involving ideas of cause and effect (implication), or systematic ways of exploring a variety of possibilities (programming), or assessing which line of action is better (probability). From these an overall strategy may be developed, or such an overall strategy may be possible anyway. *Chess*, for example, involves the minute strategies of 'if I take the pawn he will take my bishop', and the overall strategy of building up an attacking or defensive position on the board.

Inventing your own games involves a variety of mathematical activities. Invention from scratch—an entirely new kind of game—is rare, as is an entirely new piece of mathematics. Most mathematics arises from generalisation of existing mathematics, or from a variation of certain aspects of the situation. So it was with the games that the group were invited to invent. These now follow, and the reader is invited to make his own analysis, invent his own games, play those that already exist, and consider the relevance of all this to his teaching. (The games have not been exhaustively tested—there wasn't time! They may need further modification.)

There was some discussion about where this fits in, physically, to the teaching of mathematics. Nobody suggested tucking away such activities 'at the end of term, after exams', but some had maths clubs or games clubs in the lunch hour or after school. However, games are obviously activities that fit into any maths lesson, especially into a workshop situation where all the class are working individually or in groups at a variety of mainly practical activities. But we all have a certain amount of prejudice and although we might permit *Think-a-dot* in such a situation, would we draw the line at chess? And there was a long thoughtful pause when someone suggested that in maths lesson one could play bridge!

This article is reproduced from *Mathematics Teaching* (Autumn 1971), 11–13, the Journal of the Association of Teachers of Mathematics, by permission of the author and the editor.

Exago (S. Costello)

Instead of the usual grid for noughts and crosses, use this layout of points on an isometric pegboard. Coloured pegs are used to show when a move has been made. Each of two players takes it in turn to insert a peg of his own colour on the board. The first to have three pegs in a row wins.

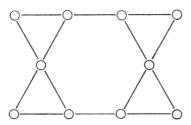

Chinese Draughts   (Julia Bradwell, F. H. C. Oates, Ron Carter)

A game for two or three players played on a regular hexagon marked out on isometric pegboard so that there are five holes along each side. Each player begins with eleven pegs of his own colour, arranged thus:

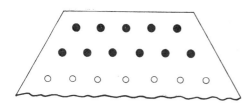

For two players arrange opposing pegs on opposite sides. For three players arrange them on alternate sides.

A peg may be moved into any adjacent hole (i.e. at most to any one of six). A peg may be moved by jumping over one peg of his own or his opponent's; any number of such jumps may be made at one time. Any jump over an opponent's peg results in a capture and the consequent removal of the peg.

Each peg captured counts one point to the capturer. Each peg reaching 'home'—the row of five holes on the opposite edge of the board—is immediately removed from play, and counts three points to that player.

The game ends when all possible pegs have reached home. The winner is the person with the highest score on one game. Alternatively, it is the first person to reach 50 in a series of games.

Jump-ball (Raymond Mercier, Lenon Beeson)

This game can be played with pegs on a pegboard, or with counters on squared paper. Blue

---

Y	Y	Y		///
Y	Y		///	
Y		Ball		B
	///		B	B
///		B	B	B

and yellow counters are placed as shown in the diagram, and a 'ball' is placed in the centre square. Moves are along lines of squares, not diagonally and not backwards. The object is to move the ball into one's own territory, i.e. anything other than shaded squares.

The ball is moved by being 'jumped over'; the ball is then moved back one square in the direction from which the jumper came. A jump must start in a square adjacent to the ball or an opponent's piece, and can continue in a straight line into the next empty square, over any number of the *opponent's* pieces. Pieces jumped over are removed.

When the ball is at the edge of the oppponent's ground, the ball may be forced from the edge by jumping a piece over the ball and over the edge, losing the piece.

In the event of the ball finishing on neutral territory, the winner is the player with more pieces left.

(The board may be better as 7 × 7 rather than 5 × 5, but with the same number of pieces.)

The Queen's Game (R. W. Hewins, B. Overton, H. C. Williams)

Two sets of distinguishable counters each containing 15 of the same kind, *Queens*, and one

Q	Q	Q	K	Q	Q	Q	Q
Q	Q	Q	Q	Q	Q	Q	Q
q	q	q	q	q	q	q	q
q	q	q	k	q	q	q	q

271

different, *King*, are placed as shown on an 8 × 8 square board. Note that the *Kings* are in the same column.

Moves are as for the King and Queen in chess. A King may move into any adjacent square; a Queen may move through any number of empty squares in a straight line parallel to the sides of the squares or diagonally. A move may not finish on a square occupied by one of the player's own pieces, but it may finish on a square occupied by an opponent's piece, which is then 'taken' and removed from the board.

The winner is the player to take his opponent's King.

(*Variations*: replace the *Queens* by some other chess piece).

Lancaster Checkers (*G. Austwick, R. Clarke*)

This is a game for two players played on an $n \times n$ pegboard. Each player has $2(n-2)$ pegs of his own colour which are set out along two adjacent sides of the square, using all the holes except the corners. The object of the game is to interchange the two sets of pegs.

The two players move alternatively, one space at a time, horizontally or vertically. The player getting all his pegs to the opposite side first wins

(An alternative is to move pegs diagonally only.)

A Pegboard Game (*A. Papendick, F. O. Smart*)

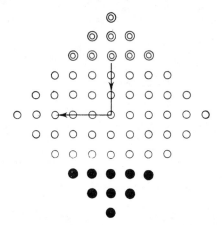

The game is played on an oblique pegboard square as shown, each player having nine pegs. The object is to get all one's own pegs into the opponent's corner.

Moves are 'L-shaped': initially pegs move forward, followed by the same number of holes to either side, provided the way is clear. Once pegs are out of the corner, backward moves are also allowed. Diagonal moves are not allowed. An opponent's peg which is adjacent to one's own may be jumped to get out of the home corner.

Colour-prime (*Marion Walter, Prof. A. Engel, J. A. Dixon*)

(This is a variation of a commercial game called *Ten-up* produced by School Utilities, Romford, Essex, which consists of triangular wooden pieces, each segmented and in three different colours, as shown. The segments in the original game are numbered from 0 to 10 and the pieces may be fitted together so that the colours match and the sum of adjacent numbers is 10. There are other ways of numbering the forty-eight pieces.)

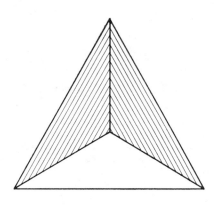

Each of any number of players up to five receives eight pieces. (If more than five wish to play the number is reduced to leave at least eight pieces 'sleeping' or out of play.)

The player with the piece having the lowest face-total will play that piece first. The second player will play so as to match the colour of his piece with one of the colour edges of the first piece, and at the same time ensure that the combined total of the two matching edges is a prime number. Each subsequent player must lay his piece to match an edge of the *last* piece played. If a player cannot follow he takes a piece from the 'sleeping' pile and misses that turn.

The game ends when a player successfully plays his final piece and wins; *or* when no player can follow and all the sleeping pieces are in hands, the player with the lowest total of values left in hand wins; *or*, since the shape of the pieces lends itself to the construction of hexagons, any player who successfully fills in a completely enclosed triangular 'hole' within a hexagon, by so doing obviously stops the game and is the winner irrespective of the points totals in hands.

(Alternatively, one could award points for certain primes—more points for the primes that occur less often in the game—with a bonus for finishing. The winner is the player with the highest total.)

Three games

BRUCE F. GODSAVE

When René Descartes sat up in bed one day and created coordinate geometry, he did a fantastic thing for mathematics. Many students and teachers believe that much of mathematics came to us in a manner similar to that of the Ten Commandments. In fact, something like the coordinate system was made up by a mortal man, in much the same way that man created Monopoly. It would be difficult, if not impossible, to devise an intuitive way of learning the rules of Monopoly. We are told the rules, and after we learn them, we use our skill and intelligence to send our friends to the "poorhouse." The same is true with coordinate geometry and most other mathematical concepts: we need to learn the rules.

The idea of a point, how to name a point, where to begin numbering, and the direction the numbers go are all part of the rules of a coordinate system. Assume now that we have a class that has been given the rules; the children recognize the axis and have been given a method for finding a point. We need to add "experience."

The first two games described below are geoboard activities that provide such experience in naming points, checking points, writing the names of points, and organizing data. This will help children become familiar with coordinate systems before we complicate the idea with lines, circles, graphs, and things like that. The activities will also be fun for the class.

The third game described disguises general review as a game of tag.

Treasure Hunt

This is like the game "Battleship," except it is nonviolent! Use 4 × 4 or 5 × 5 geoboards, balls of clay, and two teams. Each team will put five balls of "gold" (clay) on the pins or nails of the geoboard without letting the other team know where the gold is (see fig. 1). The teams

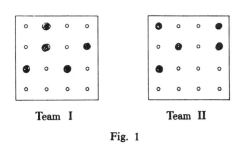

Team I   Team II

Fig. 1

take turns naming points. When one team names a point that has gold on it, this team gets the gold. The first team to lose its five pieces of clay loses the game.

The rules of the game are simple, but the skills required to win efficiently involve practice with using ordered pairs to name points and organizing data so that students know whether or not a point has already been named. While one team gets practice by naming points, the other team gets practice by checking each point to see if it has any gold on it.

For the first game or so, after naming ten or more points the students will forget whether they have already called a particular point or not. This should lead to writing down the ordered pairs as they are named so that the players don't have

Reprinted from *The Arithmetic Teacher* (May 1971), 327–329, by permission of the author and the publisher.

273

to waste time checking the same point over and over. If children don't do this on their own, suggest it. Here again is the practice they need. Later, their list will be rather long and they may spend a long time checking the list. This should lead to organizing the list as shown in figure 2. Again, if children don't do this on their own, suggest it.

First List

(0,0)	(2,2)
(3,4)	(4,4)
(2,1)	(0,1)
(1,3)	(1,1)
(4,2)	
(1,2)	
(0,3)	
(4,1)	
(2,3)	

Organized List

(0,0)	(1,3)	(2,1)	(3,4)	(4,2)
(0,3)	(1,2)	(2,3)		(4,1)
(0,1)	(1,1)	(2,2)		(4,4)

Fig. 2

Through the Maze

This is also a team game involving group versus group, although it may be played by one student against another. Each team constructs a maze, using rubber bands and a geoboard. By calling off ordered pairs, one team is to get through the other team's maze without seeing the maze. Figure 3 shows two mazes with their solutions.

In this game, no rubber bands are to be crossed, nor may anyone land on a peg with a rubber band on it. The beginning is at (0,0), and the goal is to get to (n,n).

To play, someone on team 1 names a point on the maze. If there is a straight-line path between where he is and where he wants to go, he is told he can move. For example, on Maze II he could make one of two possible first moves, (2,1) or (3,1). All other choices are blocked.

This game requires a different recording method from that used in Treasure Hunt. Points that are blocked while a player is in one position will not be blocked if he is in another position. Here are some suggestions for this game:

1. Use a marker to show the position of the team in the maze.

2. The first time the game is played, make up a maze and play while the class looks at the maze. Explain why various moves can or can't be made. Be sure there is a path through the maze.

3. For the next game, make up a maze and show it only briefly to the class. Then have the class work together to choose points. Once they understand the method of playing, they can divide into groups or pairs to play.

Children find this a very interesting game, although it can't be played fast. It can be played during the time before or after school, or during lunchtime. It's a quiet game that provides needed practice in finding points on a plane.

Another game that could be used to provide experience in naming points is "Four-in-a-Row," a game used in the Madison Project. It is similar to tic-tac-toe, except that there is an extra column and row.

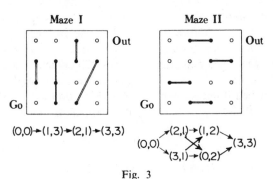

Maze I

Maze II

(0,0)→(1,3)→(2,1)→(3,3)

(0,0) →(2,1)→(1,2)→ (3,3)
→(3,1)→(0,2)→

Fig. 3

The teams take turns naming points where an $\times$ or an $\bigcirc$ is to go.

Mathematical Tag

Sometimes in the teaching of mathematics we feel compelled to do a little review. I'm sure most of you have tried to find different ways to review that don't seem like review. I was faced with a review and also needed to keep a promise that we could have mathematics outside someday. My solution to both problems was a variation on the game of tag. I called it "Mathematical Tag," since that described very well what was going on.

Mathematical Tag is based on the rules of the tag you played (or still play). Here is how it works:

Everyone in class chooses a number from 1 through the number of students in the class. No two pupils can have the same number. Each child will wear his number so it can be seen from the front or the back. Wide tape does a good job. You will have already prepared a pack of 3×5 cards on which are written expressions like the following:

$\square$ is not equal to 5.
3 is greater than $\square$.
$\square + 3$ is an odd number.
$\square$ is a member of {5, 10, 15, 20}.
$\square + 4 = \square + 4$.

You will need to keep a tally of your solution sets to check that each number is used about the same number of times. If after checking you find you need another 1, 16, and 19, you can make up a card with the expression

$\square$ is a member of {1, 16, 19}

—and you're all set, so to speak. Try to avoid cards whose solution set will have only one or two members. It's hard on whoever is "it" to go after only two people.

Now comes the hard part—deciding who will be "it." Let me suggest that you be "it." This shows that you're a real sport, and it will show the children how the game is played.

Now "it" draws a card, reads it to himself, and figures out the solution set. Without telling anyone what the solution set is,

he must tag a person whose number is a member of the solution set. The person tagged is then "it," and the new "it" draws a card. This can continue until the desire to hold class outside is eliminated from everyone, or until the kids are tired enough to do seatwork.

What do you do when "it" tags someone whose number is not a member of the solution set? You could impose a penalty, but that is not necessary. The truth (set) will out.

Just a few more comments. The level of difficulty of the cards depends on the abilities of the students. For example, compare ($\square$ is not equal to 5) with ($x \neq 5$) and ($\square$ is between -3 and 3) with ($|y| < 3$).

After the game is played once, the class can be asked to make up a pack of cards for the next time. They will see how hard it is to get the number of occurrences of each number the same.

Try to work out a way of using negative numbers.

Try to keep the playing area restricted to a small space.

This game can be played indoors; but it is a noisy game, so choose a place accordingly.

I have played this with my classes, and it is a lot of fun. If your class size is about twenty-five, you could use the following sentences for your deck of cards. Copy each sentence on a card by itself. If I figured right, each number will appear in seven different solution sets.

$\square$ is even.
$\square$ is odd.
$\square$ is not even.
$\square$ is not odd.
$\square$ is a multiple of 2.
$\square$ is a multiple of 3.
$\square + 1$ is a multiple of 2.
$\square$ is a prime.
$\square$ is not a prime.
$\square$ is less than 11.
$\square$ is greater than 17.
$\square$ is between 10 and 18.
$\square$ is a multiple of 5.
$\square$ is a member of {1, 2, 4, 7, 8, 11, 13}.
$\square$ is a member of {14, 16, 17, 19, 22, 23, 25}.
$\square$ is not 15.

Have fun with mathematics!

Number Games
with Young Children

Ida Mae Heard

Many number notions can be developed through the use of games. The author found that a group of kindergarten children in the Laboratory School at North Texas State University were very fond of finger plays, using finger puppets. These children especially liked the intriguing finger play about Two Mother Pigs:

Two mother pigs lived in a pen. Each had four babies,
and that made ten.

The rhyme continues:
These four babies were black as night.
These four babies were black-and-white.
But all eight babies loved to play
And they rolled and rolled in the mud all day.
At night, with their mother, they curled up in a heap
And squealed and squealed until they went to sleep.

Children can be most creative as they make up their own finger plays to illustrate this story. Try this game with a five- or six-year-old youngster and you may be amazed at his ability to dramatize this situation. In a spirit of free play the child is introduced to the idea of set union (joining the members of two groups). Out of this experience evolves the addition facts: $1+1=2$; $2+(4+4)=10$; and $4+4=8$.

A Listening Game

Another game to play with young children is one that trains them to listen carefully. Take a small match box and put one or more coins inside. Do not let the child see the kind nor the number of coins placed into the box but tell him how many cents worth of money you have inside. The child is then invited to shake the box and to listen to the sound of the coins as they slide back and forth. Suppose you tell the child there is five cents worth of money in the box. He will be able to distinguish the rattle made by one nickel from that made by five pennies. This game can be gradually made more complex as the child guesses the number of each coin in the box to make 10c, 15c, 20c, or 25c.

A Spinner Game*

Then there are games you can devise to give the young learner practice on mathematical concepts which he already knows. Spinner games are always fun. Use a spinner on a circular shape with eight different colored sectors. Have pockets made of corresponding colors of construction paper attached to a corrugated board. In each pocket have several cards with numerals from 0 to 9 and several cards with any set of pictures

---

*This game was designed by one of the author's students, Mrs. Charlotte Peters.

from 0 to 9. Place the cards in the pockets to correspond with colors which are directly opposite on the spinner so that one end of the spinner will always stop on a color associated with a numeral card and the other end will indicate a number card. See diagrams below.

Invite the child to make a spin. He is then asked to select one card from each pocket with colors that match those on which the ends of the spinner stop. If this game is played by a couple of children, one child can identify the symbol on his numeral card, and the other child can tell the number of pictures on the set card. Then the two children can decide whether the numeral represents a number equal to, less than, or more than the number of drawings on the set picture card. This activity can lead to the writing of number sentences that express inequalities such as: $3<4$ or $4>3$.

A Fishing Game

It's always exciting to "go fishing." Whales with paper clips for mouths can be "caught" with small magnets tied

Colors of number card pockets: *purple, brown, pink, red.*

Colors of numeral card pockets: *orange, yellow, blue, green.*

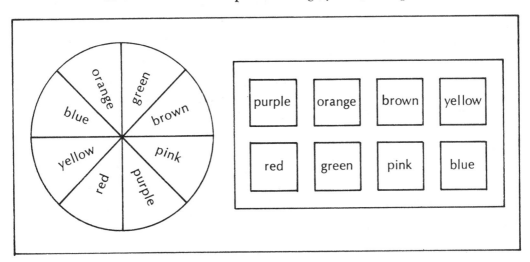

on to the ends of the fishing lines.

Each fish can be named according to the numeral, the number of objects in the set picture, the coin, the geometric shape, the equality sign $(=)$, or the inequality signs such as is not equal $(\neq)$, is less than $(<)$, or is greater than $(>)$. An open number sentence such as the following can also be used on a fish:

$$2+2=\square \qquad \square+\square+\square=9$$
$$6-1=\square \qquad \square+\square=10$$
$$2\times5=\square \qquad 4-\square=2$$
$$12\div6=\square \qquad \square-3=3$$

The child could be asked to tell the number for the box that will make the sentence true. This type of open sentence, oftentimes referred to as frame arithmetic, introduces the child to a kind of informal algebra.

The Rewards

There is a tremendous amount of mathematics a young child can learn and review through game-like procedures. If the child finds the experience enjoyable, his attitude toward the study of mathematics will be more favorable; and if he can feel success in his number work, he will have a more positive concept of himself. Nothing succeeds like success. An informal approach to the study of mathematics through games can alleviate much fear *for* the subject and a sense of failure *in* the subject that many adults experienced as children.

LET'S PLAY CARDS!

Stephen Krulik

With the current emphasis in mathematics education on individualizing the learning experiences of students in the classroom, teachers are constantly looking for devices which are easy to make, easy to use, and good for small-group activities. Card games are probably the simplest device to satisfy all of these requirements and still teach mathematics. Almost every card game that is played by adults at home is adaptable for use in the mathematics classroom; the variety of games is virtually endless.

The only equipment needed is a stack of small blank cards and a set of felt-tipped pens in assorted colors. A handy size of card to use is approximately two and one half inches by three inches. These can easily be made by cutting ordinary $3'' \times 5''$ index cards in half. These cards are light cardboard making them durable; at the same time, their size makes them easy to handle and easy to store.

Reprinted with permission from the *New York State Mathematics Teachers' Journal* (June 1971), 115–117.

Fraction War. In the middle school grades, this game can be used to help students gain practice in comparing the value of fractions. The game can be played by two, three, or four players.

The deck for this game contains sixty-six cards. On each card write one fraction. *Do not reduce these fractions to lowest terms.* Some examples are 1/2, 1/3, 4/8, 2/9, 3/9, 6/9, 5/7, 8/12. Provide each player with scrap paper and a pencil.

The deck is shuffled and dealt one at a time to the players in turn until all the cards have been distributed. Each player keeps his own pack in front of him, face down, and then turns his top card face up. The pupil whose card shows the fraction with the greatest value wins that round. He takes all of the exposed cards and places them face down underneath his own pack. If there is a tie (for example, one player turns up 1/2; a second player turns up 4/8), these two players each turn over a second card. The player whose card now shows the higher value of these two wins the entire round: he takes all the exposed cards. Play continues in this manner until one player loses all his cards or time is called. The player with the most cards in his pack is the winner.

Factoring Casino. At the early high-school level, pupils need a great deal of practice in factoring algebraic expressions. This game should help with this sometimes boring drill. It can be played by two, three, or four pupils.

The deck contains forty-five cards. On twenty of these cards, write in black ink twenty expressions to be factored. These can range from the simplest expressions $(x^2 + 2x)$ to the more complex $(x^3 - y^3)$. On another twenty cards write, in red, the same expressions but in factored form. Thus for each expression on a card in black, there is a card containing the factors of this expression in red. For example:

20 Black Cards	20 Red Cards
$x^2 + 2x$	$x(x + 2)$
$x^2 - 4$	$(x + 2)(x - 2)$
.	.
.	.
.	.
$x^3 - y^3$	$(x - y)(x^2 + xy + y^2)$

On the remaining five cards, in red, write incorrectly factored expressions. Try to use errors commonly made in class, e.g., $(x - 1)(x - 1)$ mistaken as the factors of $(x^2 + 1)$.

The forty-five cards are mixed. Deal five cards to each player in the game. Turn an additional five cards face up in the center of the table. Pupils take turns in consecutive order. Each pupil takes one card from his hand and places it face up in the middle. He tries to match an expression on the table with its factored form from his hand (or vice versa). He replaces the card by drawing one from the pack of unused cards, keeping five cards in his hand throughout the game. If he is successful in matching any exposed card with the correct mate, he takes the matched pair and places it in front of him. The player at the end of the game with the most matched pairs is the winner. Cards left in the center do not count. The game ends when time is called, or when all 45 cards have been used up.

Georummy. Teachers often find it difficult to stimulate a class of geometry students to review the basic properties of some simple geometric figures. In some cases, the teacher simply names a figure and the class lists the properties. This game should make the review a bit more enjoyable. It can be played by two, three, or four players.

The deck consists of fifty-two cards, thirteen in each of four "suits"; colors can distinguish the suits — red, blue, green, and black, for example. Each suit should contain one each of the following:

Figure Cards	Property Cards
triangle	right angle
rectangle	acute angle
square	obtuse angle
pentagon	median of a triangle
parrallelogram	angle bisector
	altitude
	equilateral
	diagonal

DEFINITION. A *group* consists of three or four of the same card, each from a different suit.

DEFINITION: A *run* consists of three or more cards from the same suit. The run must contain exactly one figure card, together with two or more property cards. The property cards must be properties of the figure card that heads the run.

Thus, a run might contain Rectangle (figure card), right angle, diagonal, angle bisector, altitude. Notice that the run need not have all of these property cards; any two are sufficient. This run, however, headed by the "rectangle" card, may *not* contain the acute-angle, equilateral, or obtuse-angle cards.

Play is as in ordinary rummy. Each player is dealt a hand containing seven cards. The rest of the cards are placed face down in the center; this is the "pack." Each player in turn draws one card and discards one card from his hand face up in the center. The player may draw either the top exposed card that has been discarded by the previous player, or the top unexposed card from the pack. The game ends when one player has exactly two runs, two groups, one run and one group, or one long run in his hand using all seven cards. If the pack runs out before this happens, the discard pack may be turned over and play continued.

Comments. Notice that these card games may be simple or complex. They can easily be adapted to any level. Be aware, however, that card games do have some drawbacks:

1. The number of people playing in one game is small. Thus the teacher either must make several decks for the same game, or have different groups of pupils doing different things.

2. Noise is a factor — but constructive noise is good.

3. The teacher must serve as a rule arbitrator and move from game to game. If problems arise that the group cannot solve, only then should the teacher offer an opinion.

4. Pupils will often become bored if the game continues for more than fifteen or twenty minutes.

Enrichment Games Get Pupils to Think

Here's a refreshing change of pace from daily drill

LOLA J. MAY

EXCITEMENT in mathematics comes not from learning the basic skills, but in extending these learnings into many varied directions. The basic tools must be learned, and learned early, but they should not be the "be all and end all" of our mathematics program.

Children need opportunities for enrichment in math in which the emphasis can be placed more on the principles and processes of reasoning than on mere rote drill. Here are two exercises that help children to think imaginatively about math and that, at the same time, make enjoyable classroom activities.

Fun with magic squares

In the array of numbers shown below, the students should find the

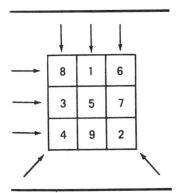

sum of each row, each column, and each major diagonal. When all the sums are the same, as they are in this case, the array is a "magic square."

Have the children multiply each of the numbers in the array by 5 and then repeat the addition. They

will discover that, again, all of the sums are the same, as shown below.

After a number of magic squares have been tried, have the children

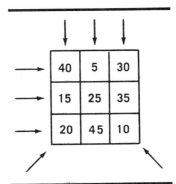

look at the number in the center and compare it to one of the sums. Then have them do the same with several other magic squares. They will discover that in this type of 3 x 3 magic square—three numbers in each direction—the center number is always one-third of the individual sums.

Now have them study the magic square shown below. The numbers 3

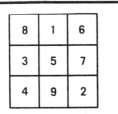

and 1 form what is known as the short diagonal. Other short diagonals in this square are 1 and 7, 9 and 7, and 9 and 3. Have the pupils

look at the number in the opposite corner from each short diagonal.

Ask the students to find the relationship between the sum of the numbers in each short diagonal and the number in the opposite corner. They will see that the sum of the short diagonal is always twice the number in the opposite corner.

Next, direct the students' attention to the numbers that lie on either side of the center number—in this case, 3 and 7 in the horizontal row and 1 and 9 in the vertical column. Note that the sum of each pair is the same—10. The center number is 5. Thus, the sum of the numbers on either side of the center number is always twice the center number.

Knowing these rules, pupils can make a magic square when only three of the nine numbers are given.

In the array below, 8, 10, and 5 are given. Study the numbers and

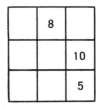

you'll see how the rest of the squares can be filled in. The numbers on the short diagonal—8 and 10—make a sum of 18. The number in the opposite corner, therefore, must be 9, half that amount. Now, the other corner number is 5. Since the sum of the short diagonal is twice the corner number, the sum of the diagonal opposite 5 must be 10. Since one of the two numbers is 8, the other is 2, and we can fill in the missing number, as at right below.

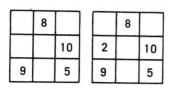

The numbers 2 and 10 are on either side of the center number. Since their sum is 12, the center number is 6, as at left below. If the center number is 6, the sum of the numbers on either side of it must be 12. Since the top number is 8, the bottom one must be 4, as at right below.

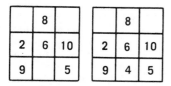

Now, the short diagonal of 4 and 10 gives a sum of 14, so the opposite corner number is 7, as at left below. The short diagonal of 2 and 4 gives the sum of six, so the other corner number is 3, and we can now fill in the last remaining square, as at right below.

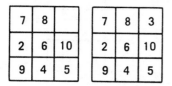

Students find great satisfaction in developing their own magic squares by following the rules they have discovered. Those who become interested in magic squares can go on to 4 x 4 squares—4 numbers in each direction — which produce many interesting patterns. They can also study some of the ancient Chinese magic squares.

The diagram shown below is a "five clock." Start at zero and move seven spaces clockwise. Where do you land? (On 2.) Make 10 moves starting at zero and you land on zero again. Take four moves and you stop on 4. This type of activity should be continued for a few minutes until the children become fa-

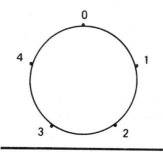

miliar with the procedure. All the moves that land on zero are multiples of five.

To add 3+4 on the clock, you start at 3 and move 4 spaces, landing on 2. In adding 3+3, you land on 1. In adding 4+4, you land on 3. The only digits used in this system are 0, 1, 2, 3, and 4. A finite mathematical system uses only a limited number of digits.

Students should practice with the clock by doing all of the possible additions and filling in the sums, as shown in the chart below. The addends in the lefthand column are added to the addends in the top row. When you add any numbers in the

+	0	1	2	3	4
0	0	1	2	3	4
1	1	2	3	4	0
2	2	3	4	0	1
3	3	4	0	1	2
4	4	0	1	2	3

five system, the sum is always one of the numbers in the system.

When you can perform an opera-

tion such as addition on the members of a set and not get any numbers outside the set, it is said the set is *closed* for that operation. When you add 3+4, the sum is 2. When you add 4+3, the sum is 2. What about the sum of 2+3 and 3+2? The sum is the same. Is the operation of addition commutative in this system? Yes. Add (3+4)+2 in this system. The sum of 3+4 is 2. The sum of 2+2 is 4. Now add 3+(4+2). The sum of 4+2 is 1. The sum of 1 and 3 is 4. So the operation of addition is associative. What number can be added to any other number to obtain that number? It's zero. Thus zero is the identity element of addition in this system.

Next, the students should fill in the multiplication table for this system. The product of 3×4 is found by adding 4 three times. Then the operation should be tested to see if it is commutative and associative. Also, the students should see what the identity element is in multiplication for this system.

The same procedure should be carried out on a six clock. This is when students find out that some number systems are not the same as ours. Working in finite systems will give children an opportunity to test other systems to see if the structural properties they have learned in our number system always are true.

Magic squares and finite mathematical systems are only two ideas for enrichment. Teachers can use many other ideas, some of which are suggested as follows:

1. Study other numeration systems, such as Egyptian, Babylonian, and Roman.

2. Find many different ways of adding, subtracting, multiplying, and dividing.

3. Study the history of measurement and the origin of our standard foot, inch, yard, etc.

4. Work in various number bases to stress our concept of place value.

Schools need to buy mathematics books for each grade level to allow children to learn more about the essentials of mathematics than is presented in the textbook. Mathematics can be exciting if children are allowed to study some phases of the subject in depth.

Disguised practice for multiplication and addition of directed numbers

ESTHER MILNE

Practicing computation can be drudgery. But practicing computation with integers is necessary. Games aren't drudgery. Let's combine games with practice. Then students can have fun as they improve the accuracy of their computation.

Try this game.

INTEGER GAME

How many can play? Two, three, or a whole class divided into two teams.

What equipment is needed? Two spinners* and paper on which to write scores. (When a large group is playing as two teams, it is best to have replicas of the spinner dials on an overhead projector transparency.)

Mark some numerals + (positive) on the first spinner. Mark the rest of the numerals − (negative). On the second spinner label a couple of the numerals "power" and the rest positive and negative. (There might be a zero on a dial.)

How do you play? Spin the first spinner. The number on which the pointer stops is the first factor. Spin the second spinner. Where the pointer stops gives the second factor, or tells us the power to which the first factor is to be raised. (If a large group

---

* If you don't have your own favorite source for purchasing spinners, these are available from Vroman's, 367 S. Pasadena Ave., Pasadena, Calif. 91105.

is playing, place a marker on each dial on the overhead projector to show where the pointer stopped. A coin could be used as a marker.) Record the product. Keep a running score.

Who wins? Decide before the game starts **how many turns will constitute a game. After the last turn the player or team whose score is farthest from zero is the winner. Score is kept by all players. Each one keeps his own score and that of his opponent(s).**

If 10 turns is to be a game, the score sheet looks something like this:

ME

	Factors	Product	Running Score
1.	$^-5 \cdot {^+4}$	$^-20$	$^-20$
2.	$^+4 \cdot {^+4}$	$^+16$	$^-4$
3.	$(^+3)^2$	$^+27$	$^+23$
4.	$0 \cdot {^-6}$	0	$^+23$
5.	$0 \cdot {^+1}$	0	$^+23$
6.	$^-8 \cdot {^-6}$	$^+48$	$^+71$
7.	$(^+1)^2$	$^+1$	$^+72$
8.	$(^+3)^2$	$^+9$	$^+81$
9.	$0 \cdot {^-5}$	0	$^+81$
10.	$(^-7)^2$	$^+49$	$^+130$

THEE

	Factors	Product	Running Score
1.	$^-5 \cdot {^-6}$	$^+30$	$^+30$
2.	$^+3 \cdot {^+4}$	$^+12$	$^+42$
3.	$(^-5)^2$	$^+25$	$^+67$
4.	$0 \cdot {^-6}$	0	$^+67$
5.	$(1)^2$	$^+1$	$^+68$
6.	$^-6 \cdot {^+1}$	$^-6$	$^+62$
7.	$^+2 \cdot {^-5}$	$^-10$	$^+52$
8.	$(^-6)^3$	-216	-164
9.	$^-8 \cdot {^+1}$	$^-8$	-172
10.	$(^+2)^3$	$^+8$	-164

Thee is the winner!

Reprinted from *The Arithmetic Teacher* (May 1969), 397–398, by permission of the author and the publisher.

SELECTED BIBLIOGRAPHY FOR CHAPTER 11
Machines, Manipulatives and Games

Machines and Manipulatives

Ackerman, Judy. "Computers Teach Math." *The Arithmetic Teacher* (May 1968), 467.

Bartel, Elaine V. "Let's Play Computer." *The Arithmetic Teacher* (March 1974), 176–77.

Bernstein, Allen L. "Use of Manipulative Devices in Teaching Mathematics." *The Arithmetic Teacher* (May 1963), 280–83.

Brong, Tedi. "Fun With Pegs and Pegboards." *The Arithmetic Teacher* (April 1971), 234–35.

Clary, Robert. "Teaching Aids for Elementary School Arithmetic." *The Arithmetic Teacher* (February 1966), 135.

Fejfar, James L. "A Note Concerning Teaching Machines and Theories of Learning." *The Mathematics Teacher* (March 1966), 258.

——. "A Teaching Program for Experimentation with Computer-Assisted Instruction." *The Arithmetic Teacher* (March 1969), 185.

Garner, R. C. "Manipulative Material, Geometric Interpretation, and Discovery." *The Arithmetic Teacher* (May 1969), 401.

Gennema, Elizabeth. "Manipulatives in the Classroom." *The Arithmetic Teacher* (May 1973), 350–52.

Gosman, Howard Y. "Mastering the Basic Facts With Dice." *The Arithmetic Teacher* (May 1973), 330–31.

Hawthorne, Frank. "Hand-Held Calculators: Help or Hindrance?" *The Arithmetic Teacher* (December 1973), 671–72.

Holder, Lois. "Of Numberlines and Regions." *The Arithmetic Teacher* (April 1969), 322.

Holz, Alan W. "A Slide Rule for Elementary School." *The Arithmetic Teacher* (May 1973), 353–59.

Immerzeel, George and Wiederanders, Don. "Ideas for Teachers." *The Arithmetic Teacher* (October 1971), 391.

Jones, Margaret Hervey and Litwiller, Bonnie H. "Practice and Discovery: Starting With the Hundred Board." *The Arithmetic Teacher* (May 1973), 360–64.

Leeseberg, Norbert H. "Evaluation Scale for a Teaching Aid in Modern Mathematics." *The Arithmetic Teacher* (December 1971), 592–94.

Livingstone, Isobel. "Live Models in Arithmetic?" *The Arithmetic Teacher* (January 1970), 81–82.

May, Lola. "Teaching Tools—How to Use Them in Math." *Grade Teacher* (February 1970), 126–30.

McDermott, John. "Sample Computer." *The Arithmetic Teacher* (March 1969), 177.

Milne, Esther. "Number Line: Versatility." *The Arithmetic Teacher* (December 1968), 738.

Osborn, Roger. "The Use of Models in the Teaching of Mathematics." *The Arithmetic Teacher* (January 1961), 22–24.

Patterson, Katherine. "A Picture Line Can Be Fun!" *The Arithmetic Teacher* (December 1969), 603.

Patterson, William, Jr. "A Device for Indirect Measurements: An Entertaining Individual Project." *The Arithmetic Teacher* (February 1973), 124–27.

Phillips, Jo. "Mathematics—Maneuvers on a Calendar." *Instructor* (August/September 1969), 110–11.

——. "Math Models." *Instructor* (January 1968), 102–03.

Randolph, Winifred and Jeffers, Verne G. "A New Look for the Hundreds Chart." *The Arithmetic Teacher* (March 1974), 203–08.

Reys, Robert E. "Considerations for Teachers Using Manipulative Materials." *The Arithmetic Teacher* (December 1971), 551–58.

——, (Ed.) "Mathematics, Multiple Embodiment and Elementary Teachers." *The Arithmetic Teacher* (October 1972), 489–93.

Scott, Joseph. "With Sticks and Rubber Bands." *The Arithmetic Teacher* (February 1970), 147–50.

Sowell, Evelyn. "Another Look at Materials in Elementary School Mathematics."

School Science and Mathematics (March 1974), 207–11.

Steinberg, Esther R. and Anderson, Bonnie C. "Teaching Tens to Timmy, Or a Caution in Teaching With Physical Models." *The Arithmetic Teacher* (December 1973), 620–25.

Suydam, Marilyn N. "Teachers, Pupils and Computer-Assisted Instruction." *The Arithmetic Teacher* (March 1969), 173.

Swadener, Marc. "Activity Board–The Board of Many Uses." *The Arithmetic Teacher* (February 1972), 141–44.

Van Arsdel, Jean and Lasky, Joanne. "A Two-Dimensional Abacus–The Papy Minicomputer." *The Arithmetic Teacher* (October 1972), 445–51.

Van Atta, Frank. "Calculators in the Classroom." *The Arithmetic Teacher* (December 1967), 650.

Whitney, Hassler. "A Mini-Computer for Primary Schools." *Mathematics Teaching* (Winter 1970), 14–18.

Games

Aman, George. "Another Discovery in the Peg Game." *New York State Mathematics Teachers' Journal* (April 1973), 63–65.

Coltharp, Forrest L. "Mathematical Aspects of the Attribute Games." *The Arithmetic Teacher* (March 1974), 246–51.

Cook, Nancy. "Fraction Bingo." *The Arithmetic Teacher* (March 1970), 237–39.

Denman, Theresa. "Mathematics–The Attribute Game." *Instructor* (November 1973), 51–52.

Dilley, Clyde and Rucker, Walter. "Arithmetic Games." *The Arithmetic Teacher* (February 1972), 157–58.

Frye, Helen. "Mathematics Throughout the Curriculum." *The Arithmetic Teacher* (December 1969), 647.

Golden, Sarah R. "Fostering Enthusiasm Through Child-Created Games." *The Arithmetic Teacher* (February 1970), 111–15.

Haggerty, John B. "KALAH–An Ancient Game of Mathematical Skill." *The Arithmetic Teacher* (May 1964), 326–30.

Hawthorne, Frank. "A Simple Even Game, With a Winning Strategy." *New York State Mathematics Teachers' Journal* (January 1973), 6.

Homan, Doris. "Television Games Adapted for Use in Junior High Mathematics Classes." *The Arithmetic Teacher* (March 1973), 219–22.

Hunt, Martin H. "Arithmetic Card Games." *The Arithmetic Teacher* (December 1968), 736–38.

Karraker, John. "Battle of Mathematica." *Instructor* (May 1970), 87–88.

Kautz, John, "Games and Activities for All Grades." *Instructor* (January 1968), 44.

Kerr, Donald R., Jr. "Mathematics Games in the Classroom." *The Arithmetic Teacher* (March 1974), 172–75.

Lutz, Marie. "A Place-Value Game." *New York State Mathematics Teachers' Journal* (June 1972), 142–45.

Mathison, Sally. "Mathematicalosterms." *The Arithmetic Teacher* (January 1969), 64.

Mauthe, Albert. "Climb the Ladder." *The Arithmetic Teacher* (May 1969), 354.

Niman, John. "A Game Introduction to the Binary Numeration System." *The Arithmetic Teacher* (December 1971), 600–01.

Porlier, Corinna. "Don't Miss the Train." *The Arithmetic Teacher* (February 1973), 139–42.

Ranucci, Ernest. "The Numbers Games." *New York State Mathematics Teachers' Journal* (April 1973), 88.

Rowland, Rowena. " 'Fraction Rummy'–A Game." *The Arithmetic Teacher* (May 1972), 387–88.

Ruderman, Harry D. "The Greatest–A Game." *The Arithmetic Teacher* (January 1970), 80.

Tucker, Benny F. " 'Parallelograms': A Simple Answer to Drill Motivation and Individualized Instruction." *The Arithmetic Teacher* (November 1971), 489–93.

Williams, Russell L. "Bingtac." *The Arithmetic Teacher* (April 1969), 310.

Wills, Herbert. "Diffy." *The Arithmetic Teacher* (October 1971), 402–05.

Winick, David, "Arithmecode Puzzle." *The Arithmetic Teacher* (February 1968), 178.

Zytkowski, Richard. "A Game With Fraction Numbers." *The Arithmetic Teacher* (January 1970), 82–83.

CHAPTER 12

MATH LABS

The idea of the laboratory in teaching is certainly not new. Elementary school classes have used beads, blocks, and other multi-sensory aids for many, many years.* The term "mathematics laboratory" as it is used today by educators, teachers, and administrators is not easy to define. To some it means a cabinet and some materials in a small corner of the classroom; to others it may mean a special, separate room supplied with equipment from A to Z. For most, the mathematics laboratory is something in between these two extremes.

Among the many reasons for having a mathematics laboratory or a laboratory approach to teaching mathematics, the following are most often cited:

(1) It provides independent experimentation and investigation for individuals and groups (the experimental approach that requires the student's participation).

(2) The teacher is freer to give help to individual learners when needed.

(3) Learners discover mathematical facts and relationships through the manipulation of objects, application of theory, and so on.

(4) The lab is conducted on a less formal level than the classroom, thus providing an opportunity for more interaction among learners and learners and teacher.

(5) It adds variety to the usual classroom routine.

Of course, as with other methods and procedures, the success of the laboratory approach in mathematics depends upon the teacher. The readings offered in this chapter were selected to help the elementary teacher who is laboratory oriented do a better job in mathematics. For instance, in "Mathematics Laboratories and Teachers' Centres—The Mathematics Revolution in Britain" Edith Biggs gives an excellent description of some of the changes taking place in England, and also provides ideas for classroom (laboratory) use. We should all remember Miss Biggs's closing remarks:

> Planning for discovery learning makes heavy demands on teachers, but it provides immediate and invigorating results. As is emphasized in a film produced for the Nuffield Project, "I hear, and I forget. I see, and I remember. *I do, and I understand.*"

---

*See Kristina Leeb-Lundberg, "Kindergarten Mathematics Laboratory—Nineteenth-Century Fashion," *The Arithmetic Teacher* (May 1970), 372–86, or "References for Mathematics Teachers," *The Mathematics Teacher*, Vol. 44 (1951), 422–25, 428.

W. G. Cathcart, W. Liedtke, and L. D. Nelson, in "Laboratory Activities in Elementary School Mathematics," offer some excellent illustrated activities for laboratory use in the classroom. *Read this article!* You will surely want to try some of these ideas.

John Ginther and Barbara Buccos, in "Mathematics Laboratory Activities for Primary School," show examples of the assignment cards they have written and how they are used. After reading this article, try writing some assignment cards of your own.

Of course, no section on math labs would be complete without some articles or suggestions from Lola May. Be sure to read "Math Lab-II," "Math Lab-III," and "Math Lab-IV." You should find many activities here to get your lab approach moving.

Mathematics laboratories and teachers' centres—the mathematics revolution in Britain

EDITH E. BIGGS

In Britain one of the educators quoted most often is Professor A. N. Whitehead. "Every child should experience the joy of discovery" has been our slogan for more than twenty years. In this respect, America has been at least partly responsible for the revolution which is taking place in British education. Although the changes did not start with mathematics, it is in mathematics that the greatest and perhaps most exciting changes are taking place today.

But the most exciting changes are to be found in the classrooms themselves. My first example is taken from a group of six-year-olds in England. The teacher had asked the children which box of a collection they had made (nine boxes of different shapes) would hold the most sand. Discussion followed on what a container was, and one child reported, "We had lots of fun deciding what was a container and what was not. Philip said a house. Simon said a piece of paper, but a flat piece of paper won't do. You would have to fold it into a box shape, and that would be right. A pocket is a container; so is a rocket. Raymond said a leg is because it holds blood and bones, but thinking of blood and bones put me off my tea."

The children then decided to fill each box with sand and to weigh the boxes. This helped the children to arrange the boxes in order—but the sand ran out at the corners, and the children decided to look for something better. It proved very difficult to find objects which would fill the boxes to the children's satisfaction, and it was some time before a boy found that a set of beads shaped like cubes would fill all the boxes. The contents of each box were then placed in columns above each box. But there were more than forty children in this busy classroom, and soon the cubes were knocked on the floor. A boy then decided to mount the boxes in order on a piece of paper with half-inch squares and to mark "one square for each cube." (The edges of the cubes were a little larger than half an inch.) From the resulting block graph, the children discovered that two of the boxes contained the same number of cubes but that "they are very different shapes."

A class of six- and seven-year-olds in Ontario was also introduced to ideas of volume through containers of quite varied shapes, which the children collected themselves. They made several discoveries,

Reprinted from *The Arithmetic Teacher* (May 1968), 400–408, by permission of the author and the publisher.

using a variety of cereals or water to fill the containers. For example, they were surprised to discover that three containers of very different shapes held the same quantity of rice. They were then asked if they could find out how many pieces of macaroni there were in a large bag, without counting every piece. The first suggestions were, "I'd count in twos" and ". . . in tens." "In twenties," said others. They were then shown a small container and asked if they could use this to solve the problem. A boy said that he would fill the cup with macaroni, count the contents, then find how many cups he could fill with the rest of the macaroni—and work out the answer from this. Another boy said that he would tip the contents into a jar with straight sides and mark the level. He would then tip the entire contents of the bag of macaroni into the jar and mark cup levels up the jar. A girl said she would halve the contents of the bag, count one half and double that. (We left her thinking how she could halve the contents.) Here were three good suggestions from the children themselves, which the teacher allowed them to try out. The cup contained 110 pieces, and it took 6½ cups to fill the bag. This final count presented some difficulty to these young children until a boy paired the cups and counted 220, 440, 660. Another boy calculated the number of pieces in half a cup to be 55 and explained that half a hundred was 50 and half of ten was 5, giving 55 in all.

The children's real experience with volume had given them the ability to solve a problem of some difficulty.

Problem situations

When children learn mathematics by active methods, they are constantly learning through problem situations—planned by the teacher or posed by the children themselves.

Teachers usually find it difficult to see how arithmetic can be taught by discovery methods. Of course, children need to know the basic number relationships, but pattern is a great help in this and other respects. All too often, as the following example shows, we teach methods before the children have realized the need for these.

Some "disadvantaged" children of nine and ten had estimated the number of grains of rice in a jar and were making a "count" in various ways. A group of girls counted the number of grains in a small cup (57) and found that it took 40 cups to empty the jar. They had no idea of how to find the total until one girl decided to use the blackboard. She began to write 57's, one beneath the other; there were thirteen 57's in the first column and twelve in a second when I moved over, hoping to discuss the problem—but, unfortunately, the bell sounded for recess. On the following day, the teacher gave this group eight examples of work to be done in multiplication, and each girl succeeded.

A five-year-old boy makes shapes.

This seems to me to be putting the cart before the horse. First we should put the children in a situation demanding multiplication and allow them to arrive at their own solution, however crude. Then we can help them to refine their own method. Finally, we give them as much practice as they need to become efficient (but not *too* much practice).

Time is another topic which children find difficult. A group of seven-year-olds had recorded their bedtimes (from 7:00 P.M. to 11:00 P.M.!) and their times of getting up. They drew pictures of themselves going to bed and getting up and arranged these in the appropriate sets. The teacher then asked the children to find out who was in bed the longest time. "Those who go to bed at 7:00 P.M." was the immediate and unanimous reply.

Vainly the teacher pointed out that one of the children in the 7:00 P.M. set was up at 6:00 A.M. The children were firmly convinced that those in bed first were in bed longest. So the teacher helped each child to work out how long he was in bed. Then *she* drew a graph in which she took one inch to represent one hour and made an entry for every child. Figure 1 pictures this on a reduced scale.

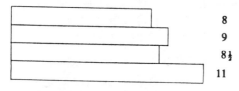

FIGURE 1

The children's prompt comment on seeing the graph was, "But that's not right. You've made us all go to bed at the same time." Wisely, the teacher abandoned any further attempt to press an idea which was too sophisticated for these young children.

Some nine-year-olds were far more successful because the ideas were their own. They captured (and later released) various insects and observed the time each insect took to travel 1 foot. They found, for example, a spider that covered the distance in 10 seconds and a snail that covered it in 7 minutes.

Gail's brother collected caterpillars, so she wrote about the caterpillar. She found that it covered 1 foot in 45 seconds and therefore it covered 1 yard in 135 seconds. She calculated that at this rate it would cover a mile in 66 hours. She reported, "It is a mile to the Guildhall. So if the caterpillar starts at noon on Monday, it would not reach the Guildhall until six

o'clock on Thursday morning. This does not allow for meals, traffic, or sleeping."

But the work in elementary classrooms is no longer limited to arithmetic; many ideas in geometry and algebra (particularly in graphical representation) are included. The outstanding capacity children have for thinking mathematically is illustrated by the following example of a group of eight-year-olds in a class in Ontario, very recently introduced to discovery methods in mathematics.

They were working with squares and were collecting examples of squares in the classroom. I gave each of the children a sheet of paper and asked if they would make the largest square they could without using a ruler. One girl folded a piece over at one end until she thought the shape looked like a square. I asked her to check that the shape was a square, and, after some thought, she matched two adjacent sides and adjusted her square. A second girl who matched adjacent sides immediately said that her mother had told her how to do this when she was paper-folding one day. By now the six children in the group had made similar squares, cut off the odd piece at one end, and folded the square along the diagonal to check. I asked them what shape they now had. "A right-angled triangle with two equal sides," they said. "How do you know they are equal?" I asked.

They matched the two sides of the triangle to show me. One said, "Because we started with a square." They found that they had made another isosceles triangle. Excitedly they continued the sequence, making one isosceles triangle after another, each one smaller than the preceding one.

"What will happen?" I asked.

"We shall come to a very wee triangle, but we shall never get right to the middle point," was the reply.

"As the triangle gets smaller, the paper gets thicker," said another.

I asked them to open the paper again to see what else they could discover. They soon found that each successive triangle was half the preceding one and began to give me the sequence of fractions

$$1, \ \frac{1}{2}, \ \frac{1}{4}, \ \frac{1}{8}, \ \frac{1}{16}, \ \frac{1}{32}$$

as we refolded our paper.

Then I asked them if they could make the original square into a triangle twice the size of the first triangle. Once one girl thought of cutting the square along the diagonal, this was soon done. By now we were all working on the floor. I asked them if they could make a triangle four times the original.

"Me and Scott will," said Margaret, immediately piecing the triangles together and jumping up and down in her excitement. (I was as excited as she was!) Here is the sequence—but the larger examples are not included, in case the reader should like to try this for himself. There are many other discoveries which can be made from this sequence, shown in Figure 2. (The lines in the square mark successive folds.)

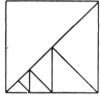

FIGURE 2

These eight-year-olds had met their first example of a limit, had discovered some patterns (sequences) in arithmetic and geometry, and, above all, they had become intensely interested in mathematics.

Materials used

The materials used in teachers' workshops and in the classroom are mostly inexpensive (although certain items such as good scales for weighing are a necessity), but they should be as attractive as possible.

A teacher was working with some ten-year-olds who had cut out a sequence of colored squares with edges of 1 inch, 2 inches, 3 inches, etc. She asked them to make patterns with these and then to write down all the sequences they could find. The data they reported is shown in Figure 3.

Sides	0	1	2	3	4	5	⟶
Perimeter	0	4	8	12	16	20	⟶
Area	0	1	4	9	16	25	⟶

FIGURE 3

The teacher asked the group what the graph of the perimeter sequence would be. "A straight line, because it goes up in 4's," was the reply. Discussion followed about intermediate squares (edges measuring between 2 and 3 inches, between 3 and 4 inches, etc.), and the graph was drawn.

The teacher then asked them what the graph of the areas would look like. They worked out the sequence of differences

$$1, \ 3, \ 5, \ 7, \ 9$$

and were excited by the pattern. "It can't be a straight line because the differences are not equal," they said.

"But it can't go up and down because there is a pattern," said a boy. Eventually they decided that the graph must be a curve and drew their first curved graph, seen in Figure 4.

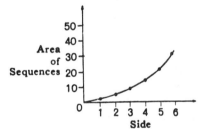

FIGURE 4

By now the children had developed an intense interest in problems in mathematics, and their next investigations concerned rectangles with the same perimeter. They made an exciting discovery when they cut out and fitted together a sequence of rectangles that had a perimeter of 2 feet. When they turned their attention to the areas of this sequence of rectangles, they recognised the pattern of differences (the reader is invited to try this for himself) and once more predicted the graph they would get by plotting areas and widths.

This led one boy to investigate the perimeters of rectangles with the same area. He could not spot the pattern this time, and the graph he made of perimeters and widths caused much discussion.

The last example of classroom work that we shall give is that of work done by a group of three boys led by Peter, aged ten (of average intelligence). Peter and the forty other children in his class had had a traditional course in arithmetic until the last year in the elementary school. His new teacher had experimented in discovery

A ten-year-old boy, using a trolley on a slope, studies variable loading and weight relationships.

methods in mathematics for some years and had planned an exciting programme and made attractive materials available in the classroom. But the children were very cautious in their response, and only gradually did they begin to experiment with the materials. A toy truck on a track running down an adjustable slope attracted them most, and this led to some interesting work on gradients.

One day Peter came to the teacher with a graph of $y = x^2$ which the class had done as a group project. He said he had thought of a problem which he would like to try to solve. He pointed out that the gradients of the curve were different at $x = 1$, $x = 2$,

etc., and said he wanted to find the pattern. Before the teacher could answer, Peter told her of a second problem. He showed her that the areas under the curve were changing, too, and said he wanted to find this pattern also. Astonished, the teacher asked, "Peter, what makes you think there *is* a pattern?" Peter replied, with great confidence, "But in mathematics there's always a pattern; you've only got to look for it." And Peter did look for the patterns—and discovered them both after six months of work.

Neither the teacher nor Peter realised that Peter's problem was one Archimedes set himself twenty-two centuries earlier which led him to discover the calculus. Peter's methods resembled those of Archimedes. For example, to measure area Peter counted squares and approximated. At one stage Peter was held up because he could not see a pattern in his sequence of areas:

$$\frac{1}{3} \qquad 2\frac{2}{3} \qquad 9 \qquad 21\frac{1}{3}$$

(Again, this problem is left for the reader to solve.)

Peter's excitement at his discovery was probably as intense as that of Archimedes. At the end, he read his work through from the beginning and commented on the many mistakes he had made.

Peter's work has very important implications for those of us who are working to improve the teaching of mathematics. It suggests that there are children who do not appear to have a high IQ and yet possess creative abilities and powers of persistence which can lead them to make important discoveries in mathematics.

Teachers' centres

In order to understand the changes of the past twenty years, it is important to know something of the English educational system. Education is compulsory for ages five to fifteen years (some time to be raised to sixteen). An ever increasing number of students remain at school for one, two, or three years beyond the age of fifteen. Many of the students continue in some form of higher education, for example, at colleges of advanced technology, colleges of education (teachers' colleges) and universities. The first two now provide degree courses, although a degree course is not compulsory, for example, for a teacher.

There are some variations, but the large majority of children transfer from elementary to secondary education at the age of eleven years. About ten years ago, doubts were felt about the written examination at eleven years for all children in the state education system. The examination normally consisted of an intelligence test, an arithmetic test in which the emphasis was on speedy and accurate computation, and an English test which comprised comprehension, formal exercises, and sometimes composition. Today many local education authorities have abolished the "11+ examination" and give more emphasis to the schools' reports and IQ ratings.

At the secondary stage there are two "public" examinations. The Certificate of Secondary Education was introduced about five years ago to cater for less academic students. Teachers have taken a major part in planning area syllabuses for this and in devising the type of examination to be given. Grade I in this examination is equivalent to an Ordinary Level pass in the other examination: The General Certificate of Education. This examination can be taken at two levels: Ordinary Level (usually taken at fifteen or sixteen years) and Advanced Level (usually taken at seventeen or eighteen years). The General Certificate of Education is frequently used as a qualifying examination. For example, the minimum requirement for entrance to a college of education is normally five subjects at Ordinary Level. (Many colleges now also require one or two subjects at Advanced Level).

It is important to mention these external or public examinations because, in England, neither the Department of Education and Science nor the local education author-ities require schools to follow specific courses of work. Each principal (called a headmaster or a headmistress) is responsible for the courses of study in his own school. Usually the headmaster discusses the courses and his aims with his staff. Moreover, headmasters and headmistresses are appointed because they are good teachers, and they are expected to take some active part in the teaching programme. (In large schools, where there is great mobility of staff, the head may spend his time helping teachers fresh from college and on probation for one year.)

In these circumstances, it can be seen that the examination syllabus could and sometimes does restrict the scope of the course of study, even though each school enjoys freedom to devise its own courses; but it is interesting to note that changes in the teaching of English in elementary schools were effected despite the "11+ examination," and today many children write fluently and with a vivid vocabulary by the age of seven or eight. As they progress through the elementary school, many of their teachers base their language work on each child's creative writing, and the result is that by the age of ten many children write with accuracy and precision. The idea also holds true for mathematics experiments in elementary schools; this began about ten years ago in schools where numbers in classes were large (over 40) and in areas where there was an "11+ examination."

What are the characteristics of these changes which now affect every aspect of the curriculum in English elementary schools? As Dr. Dienes indicates, there is need for us to "shift the emphasis from teaching to learning, from our world to the children's world." Above all, we provide opportunities for the children to think for themselves, so that learning for them is an active, creative process. Our aims in the teaching of mathematics at *all* levels are, in summary, to give our students (1) the opportunity to think for themselves, (2) the opportunity to appreciate the order

Teachers enjoy discovering, too! Investigation with shapes.

and pattern which is the essence of mathematics, not only in the man-made world but in the natural world as well, and (3) the needed skills. In brief, we provide our students with an environment containing the best materials for learning. "Teachers observe and plan. Children experiment and discover. This is what makes maths alive."[1]

But if teachers are to provide opportunities for their students to learn mathematics through firsthand experience, they must first be convinced that this is possible. After we had experimented in the classroom ourselves, we started on a campaign of in-service training with teachers which involved them immediately in learning mathematics through experience and discovery. The first requests for help came from elementary teachers, but from the outset we invited secondary teachers to take part. At first, secondary teachers of mathematics were reluctant to come to the in-service courses for elementary teachers, but gradually we extended the age range covered, first to five to thirteen years and now to

five to sixteen years. Today it has become an established practice that elementary and secondary teachers work together at these courses. The materials provided are quite varied and are those we hope to see available in classrooms for use by the students.

In the beginning, these practical mathematics courses lasted two or three days and were organised on a local basis. But, despite our efforts, there was rarely a sustained follow-up, and even the most enthusiastic teacher lost interest when no one was at hand to encourage him so that most of the stimulus of the early work was lost. Early in 1963, however, the first teachers' centres were set up in the rural county of Dorset. These first centres, used as classrooms by day, became rooms where teachers could meet at regular intervals during the evening to discuss difficulties and successes in their experiments in mathematics in the classroom—and to learn more mathematics by discovery methods themselves. This idea of a centre where teachers could meet regularly spread slowly, too slowly to meet the growing need. Late in 1963, I visited the Nuffield Foundation to urge the directors to sponsor a mathematics teaching project (to

---

[1] *Maths Alive*, 16mm. film, available for sale from the National Council for Audio Visual Aids in Education, 33 Queen Anne Street, London, W.1.

294

cover age range five to thirteen years) which would require the local education authorities participating to provide a teachers' centre.

A scheme was drawn up; in-service courses for "leader" teachers from each of thirteen pilot areas were provided, and the Nuffield Foundation Mathematics Teaching Project (organiser, Dr. Geoffrey Matthews) was launched in 1965. These leader teachers (twelve to fifteen from each area) formed the nucleus at each centre to provide in-service training in their own areas, first for teachers in schools within the Nuffield Project and later for other schools anxious to use the facilities of the centres. By 1966, nearly a hundred centres had been set up in mathematics, in science, or in both, as part of the second phase. The leader teachers for the second phase received in-service courses from teams of teachers from consultative areas (Devon, Dorset, Nottingham, Sheffield), areas in which teachers had been experimenting with the teaching of mathematics for some years. Today the value of teachers' centres has been accepted in Britain, and there are, in all, some three hundred of these. Most of these are open to teachers from all schools, and most include subjects other than mathematics (although mathematics and/or science was usually the first subject studied). The leader teachers have done a magnificent job at these centres. Small groups of teachers (twenty to twenty-four) meet regularly—e.g., one afternoon and one evening a week for eight weeks—and sometimes have a second series of weekly sessions after an interval of time, under the guidance of a leader.

Many of the centres are not part of the Nuffield Mathematics Project but make use of the guides as these are published. Most centres require some help from time to time and call on lecturers from colleges of education, on other centres, or on Her Majesty's Inspectors of Schools. The Department of Education and Science has organised an increasing number of courses in mathematics each year, not only to bring together all those concerned in the learning of mathematics, but also to appraise methods and content and to encourage work at new centres as these are set up. Groups of teachers, elementary and secondary, come from each area for these courses.

Promising developments

It has been estimated that about 25 percent of our elementary schools have substantially changed the teaching of mathematics (and of other subjects), shifting the emphasis to active learning. At the secondary stage there are now eleven mathematics projects (including the Scottish project which will eventually involve 80 percent of the academic schools). All eleven projects have been concerned with teaching presentation as well as with new content; but in secondary schools, and particularly in academic streams, teachers have been more cautious about a change which will mean students' working in small groups whenever new content is attempted. There are an increasing number of secondary schools experimenting in this way, but the large majority still prefer teaching the entire class as one large group. At ten national regional courses this year, every effort will be made to consider those topics which secondary teachers find difficult to present by discovery methods, in order to encourage and help these teachers to plan for active learning in small groups.

What of the in-service courses themselves? These became more effective from the time teachers took an active part in the planning. Teachers explore the possibility of materials for themselves (materials from the environment as well as materials structured for mathematics). They also prepare topics for the students they teach, to cover the age range five to thirteen or sixteen years, and write questions intended to give their students opportunities to think and to plan their own investigations. Teachers try out each other's questions and rewrite the questions if they want to criticize them. They also learn more mathematics themselves—by guided discovery—and take a new look at arithmetic. They are given detailed notes when they leave, to remind them of all the workshop activities.

What about the children? How do they profit from this active, creative way of learning mathematics?

In Ontario, we organised two series of workshops in six regions. Each workshop lasted for three days, and during the six weeks intervening between the first and second workshops the teachers undertook to experiment in their own classrooms. We spent the day before the second workshop began in four or five classrooms.

This has been a most encouraging experience. Most of the eighty teachers at each workshop had provided some opportunities for small groups of their students to learn mathematics by discovery methods. A few made a beginning with one group at a time, gradually increasing the number of children involved. Others organised the whole class in groups and gave each group materials and a new topic to investigate and discuss. It was very interesting to find that even though some of the teachers' work was highly structured, the children were able to launch out on much freer investigations on their own when encouraged to do so. They had been set free to think for themselves. Many teachers were eager to talk of the effect of these methods on the children: of their lively interest, of their increased sense of responsibility, and of their versatility. A large and varied quantity of work was exhibited by the teachers at the second workshops, and some of this already showed great variety and gave evidence of the students' willingness and ability to think.

Two other most promising developments deserve mention. Teachers were chosen to participate in the workshop on the basis of their ability to give future help to their colleagues in their own regions. One programme consultant has already collected material for subsequent workshops. A teacher in her first year of teaching conducted a workshop one evening from 4:00

P.M. to 9:00 P.M. for the nineteen other members of her school. The five members of the workshop team from another region

A nine-year-old girl makes her own drawing for a cube.

planned and conducted four workshop sessions in school time with thirty-five volunteers (teachers of Grades K–13) during the six weeks between the workshops. So in-service training in Ontario by leader-teacher teams has already begun, and there have been many requests to the Ontario Department of Education for workshops in regions not included in the initial stage.

Three aims

Let me finish by mentioning my three aims in the teaching of mathematics, whether in the classroom, at teachers' colleges, at universities, or at in-service workshops for teachers:

1. Let the children think for themselves.

2. Let the children discover for themselves the mathematical patterns which are to be found everywhere in the man-made and natural environment.

3. Give the children the skills they need.

Planning for discovery learning makes heavy demands on teachers, but it produces immediate and invigorating results. As is emphasized in a film produced for the Nuffield Project, "I hear, and I forget. I see, and I remember. *I do, and I understand.*"[2]

[2] "I Do and I Understand," photographed for the Nuffield Mathematics Teaching Project, can be purchased from Sound Services Ltd., Wilton Crescent, Merton Park, London, S.W. 19.

Laboratory Activities in Elementary School Mathematics

W. G. Cathcart, W. Liedtke, L. D. Nelson

THE LABORATORY PROCEDURE is gaining wider acceptance in Alberta schools as a method for teaching mathematics in the elementary grades. In this setting pupils are usually divided into small groups and certain tasks are suggested to the groups on activity sheets. The purpose of the tasks is to lead each child to a deeper understanding of a particular mathematical concept, principle, relationship, or process. He learns by becoming involved with materials and by interacting with other pupils. The laboratory procedure contrasts sharply with procedures usually observed in mathematics lessons. Pupils move around more and the noise level is higher. Teachers soon find, however, that this situation stimulates effective learning. They also find this to be an ideal way of meeting the problem of individual differences.

Topics and ideas for activities are easy to find. Almost any idea or procedure in the textbook can be presented in a laboratory setting. Many modern publications contain suggestions for such activities.

Concrete and manipulative materials used in connection with laboratory activities are generally easy to collect or to construct. There are commercial materials available as well, of course. Teachers should give considerable thought to building an adequate collection of manipulative and other materials for laboratory use.

Parts of some activity sheets are shown on the following pages. There are also photographs of pupils in Grades II and VI in Campbelltown and McKernan Schools engaged in the activities. Campbelltown School is in Sherwood Park and McKernan School is in the City of Edmonton.

SUGAR LUMP FORECASTING

1. Choose a partner and take a sugar lump with one colored face. Now start tossing the lump as people toss dice. What fraction of the time would you guess it would land on its colored face if you tossed it many times? Record your guess.

2. Now toss the lump 24 times. Keep a record of your tosses and of the number of times it landed on its colored face. How close was your forecast to what happened. Revise your guess if you wish.

3. Start again and toss the lump fifty-four times. What fraction of the times did it land on its colored face. How close is this to your forecast?

From *The ATA Magazine* (March/April 1969), 14–17, published by the Alberta Teachers' Association, by permission of the publisher.

4. Take a lump that has two colored faces. Forecast the fraction of times it would land on a colored face. Experiment as in numbers 2 and 3 to see how close your forecast is to what happens.

5. Take two lumps with one colored face each. Forecast what fraction of the times they would *both* fall colored face down when tossed at the same time. Experiment to check your forecast.

6. Suppose you took three lumps of sugar and colored three faces of each of them. Forecast the fraction of times you would expect them *all* to land on a colored face if tossed together.

TESSELLATIONS

An interesting problem that has come down to us from ancient times is the problem of tiling a surface without leaving any spaces. We shall call such tiling a *tessellation*.

To the right is an example of a tessellation using a triangle.

You can see that the area enclosed by the dotted line can be completely covered with triangles.

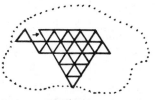

1. Cut out a four-sided figure from a piece of cardboard. Trace around it to see if it can be used to make a tessellation. (Do not turn the figure over.)

2. Use the 5-sided (pentagon), 6-sided (hexagon) and 8-sided (octagon) figures to make a tessellation. Will all of them work?

3. Cut out any polygon you like. Trace around it to see if it can be used to make a tessellation.

4. Suppose you were to use a 5-sided figure (pentagon) and one other figure to make a tessellation. Show by tracing what shape the other figure would have to be.

PASCAL'S TRIANGLE

1. Betty and Joyce were about to bake a cake. After the cake was mixed, they could add any or all of the following ingredients to the cake (or they might decide not to add any of them).

Nuts
Fruit
Coconut
Chocolate

Try to find out how many varieties of cake they could make with these four ingredients. Use the cards to help you. Enter your results in the table. (Part of it has been completed for you.)

INGREDIENTS ADDED

	None	One	Two	Three	Four
No. of Varieties	1		6		1

2. Suppose there were only 3 ingredients:

Fruit
Coconut
Chocolate

Fill in the table below to show the number of varieties of cake that can be made with these three ingredients.

INGREDIENTS ADDED

	None	One	Two	Three
No. of Varieties				

3. Suppose there were 5 ingredients:

Nuts   Coconut
Fruit   Chocolate
Marshmallow

LET'S MEASURE

Use the string and stick on the table to measure.

Look at each of these lines. First write down how many sticks or strings long you think the line is. Then measure it and write down your answer.

My guess: .... sticks
Measured length: .... sticks

My guess: .... sticks
Measured length: .... sticks

Here are some other measuring units you could use.

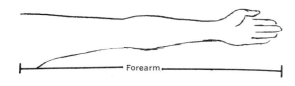

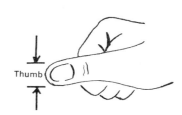

Thumb

Use these units to measure the following. First guess and then measure.

                      GUESS   MEASURE

a) Length of the blackboard in
      forearms   ....

b) Width of this paper in thumbs ....

299

c) Width of the door in spans ....

d) Width of a book in thumbs ....

e) Length of your pencil in
   thumbs ....

f) Height of the door in forearms ....

g) Width of a window in spans ....

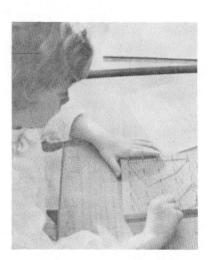

FIND THE SOLIDS

Find the solids shown in the diagrams. Count the number of edges, faces, and corners.

Write your answer in the table.

	Cube	Cone	Cylinder	Prism	Pyramid
Edges					
Faces					
Corners					

FIGURES AND SPACES

1. Use the rubber-band (elastic) and make a figure with three sides on the board with the nails.

   Do you know a name for this figure?

2. Make a figure with four sides.

   Do you know a name for the figure you have made?

3. Now make a figure that has five sides.

   Does this figure remind you of something familiar?

4. Can you make a figure with six sides?

   What does this figure remind you of?

BUILDING NUMBERS

1. Open the envelope and arrange the cards in order from smallest to largest.

2. We can use these cards to show different numbers. Here is how we can show the number 247.

200 +	40 +	7 +

Find these cards and make the number 247.

3. Now make each of these numbers and keep them in front of you.

(a)	329	(c)	86	(e)	504
(b)	661	(d)	400	(f)	110

4. Find a flat block that you would use to show 100.

5. Choose one of the numbers you made with the cards in Question 3. Use the blocks to build that number.

6. Look at the number your friends are building. Can you guess their number?

7. What name would you give to the largest block in the box?

8. Now see if you can make the number 1252 with the blocks.

Mathematics Laboratory Activities For Primary School

Barbara J. Buccos and John L. Ginther

AN ASPECT of the school mathematics program which is now beginning to receive attention in this country is the mathematics laboratory.

What is the "math lab," and how does it differ from the usual textbook-oriented mathematics program? One view of the mathematics laboratory typically involves the following elements:

1. A description of an investigation which students could carry out, a situation with mathematics built into it. This description often is written on a file card (sometimes called an "assignment card"); a working mathematics laboratory includes a large set of such cards.

2. Physical apparatus. Each assignment card will note what equipment is needed for the investigation described on the card.

3. Teams. Many teachers who have successfully inaugurated mathematics laboratories find that having students work together in pairs on the various investigations is ideal. The two students can discuss what they are doing, and each is important to the team.

4. Evaluation. The evaluation of student work in the mathematics laboratory usually stems from a combination of written and oral reporting by the team of students upon completion of their assignment card.

Reprinted from *Contemporary Education* (May 1971), 298–299, with permission of the authors and Indiana State University.

The authors became interested in the possibilities of this type of mathematics laboratory applied at the second grade level. The remainder of this article briefly reports the procedures followed and the results observed.

The first task was the writing of assignment cards appropriate to second graders, considering their interests and mathematical competencies. Examples of the cards written are these:

Card No. 103　　　　Apparatus: mirror
geometry — symmetry, reflection

1) Use the mirror to help you find what these code words are:
Hint: Below your eyes! ⊔∩⫎⫎⋀
Hint: A little goat. ⋀⊥⋃

2) Can you write letters or words in code like this?

3) Make some designs on a sheet of paper and see what happens when you stand the mirror on the pattern.

Card No. 118　　　　Apparatus: ruler
measurement
geometry

1) Pick one wall of this room and find out how many cement blocks are in it.

2) How big is one block?

3) What else can you tell about the wall?

4) Draw a picture of the wall. Try to make things in your picture the right size!

Next, the apparatus to accompany the cards had to be obtained — and on a very limited budget! Nearly all of this "apparatus" consisted of common items, readily available — such as rulers, string, scissors, a thermometer, cardboard, graph paper. Toy stores had some equipment that proved useful in the mathematics laboratory. Games and puzzles with mathematical aspects, e.g., checkers, Soma, Qubic, etc., were used and interested a number of the children.

Then we introduced the mathematics laboratory to the children. We did this by spreading the cards and apparatus on a large table, and having each pair of children choose the assignment card that interested them. We found that the students "played" with the apparatus — ignoring the cards — for the first few lab periods. Upon reflection, we saw that we might have expected this — the play stage usually precedes the more goal-directed activity.

The mathematics laboratory was used for about one hour per week. On several occasions the second graders were joined by a first-grade class for the math lab. The results were quite encouraging. Both classes did quite well, and the second graders were able to help the first graders with the reading of the cards.

Much of the evaluation was done orally as the teacher discussed the results discovered by the little "investigators." To follow the progress of the class, a graph chart system was employed, with the children's names on one axis and the assignment cards listed on an adjacent axis. Completed cards were "commemorated" with a gold star in the appropriate boxes on the chart.

As a beginning, the mathematics laboratory as we organized it seemed to us a worthwhile addition to the primary mathematics program. Student interest was high, and opportunities abounded for finding mathematics in the environment. We also liked the creative aspect of the children's work.

Of the many areas for improvement, these are of chief interest to us now:

1. Coordinating the mathematics laboratory activities with the remainder of the mathematics program so that specific concepts can be reinforced.

2. Sharpening the techniques for evaluating a team's work on an experiment.

3. Designing a "bookkeeping" system to chart a student's long-term progress in his total mathematics program.

We would welcome comments or questions from anyone interested in starting a mathematics laboratory in his classroom.

Math lab—II

Your math lab is now in full operation. So it's time
to introduce some new activities to reinforce
and enliven the daily routine

LOLA J. MAY

BY NOW YOUR math lab is off and
winging in your classroom. To follow up the review activities given
last month (see "Math Lab!", *GT
Sept. '71*, p. 102), here are some activities to challenge your youngsters.

For the primaries

1. Make a ten. Prepare a group of
cards each with 10 squares or rectangles. Write each of the following
number facts on an index card: $8 +
5 =$ ____, $9 + 6 =$ ____, $7 + 8
=$ ____, $8 + 9 =$ ____, $6 + 7 =$
____, $8 + 7 =$ ____, $6 + 8 =$
____, $7 + 9 =$ ____, $5 + 9 =$ ____
Have a pile of counters available (each child will need 18).

The child takes one of the fact
cards, say, $6 + 7 =$ ____. He
should then take six counters, then
seven more and put the unused
counters aside. He places one
counter on each square until all the
squares are filled. The remaining
counters—three in this case—are put
beside the 10-square cards. Then,
the child fills in the sum, i.e.,
$10 + 3$ (see Figure 1, below).

Two children can play in pairs.
One acts as a checker as the other
works the problem. If a player gets
the right answer, he receives five
points. If he makes an error and is
corrected by the other player, the
challenger earns five points. The
children alternate cards. The first
child to gain 30 points is the winner.

2. Sorting game. Write a different
addition fact on each of 54 index
cards. Make sure you have some
that total less than 10, exactly 10
and more than 10. Then, take three
sheets of paper. On the top of one
write, " < 10" with the wording
"less than 10" below it; on the second, " > 10" and "greater than
10"; and on the third, "$= 10$" and
"equal to 10."

There are two players for the
game. The 54 cards are placed faced
down in a pile. The first player picks
three cards and places them on the
sheets according to whether the
sums are less than 10, greater than
10 or equal to ten. He receives two
points for each correct answer. If an
error is made and is challenged by
the second player, the challenger receives five points. The second player
then draws three cards and plays on
the sheets. The game continues until
all the fact cards are used. The
player with the highest score wins.

The same game can be played
with 54 subtraction cards. In this instance, the three sheets of paper will
read, "less than 6," "greater than 6"
and "equal to 6."

3. Make a shape. Make up a card
showing four geometric shapes and
the worth of each in cents. See Figure 2 for examples.

Challenge the children to make a
picture using the four shapes and
then add up the "worth" of his picture. The child that makes the picture worth the most money wins
(see examples in Figures 3 and 4).

For the intermediates

1. Number sentence game. Prepare index cards as follows:

☐ Six cards with " $>$ " on one side
and "$=$" on the other side.
☐ Six cards with "$\times$" on one side
and "$\div$" on the other side.
☐ Fifteen cards with one of the following numerals on each of them:
18, 21, 24, 27, 35, 36, 40, 42, 45, 48,
54, 56, 63, 72 and 81.
☐ Twenty-four cards, three each, of
numerals 2, 3, 4, 5, 6, 7, 8 and 9.

Two to six youngsters can play
the game. The 39 numeral cards are
placed face down in a pile. The 12
symbol cards are spread on the
table. Each player takes one of each
symbol card and then draws three
numeral cards from the pile. With
the three cards and either side of
each symbol card, each player must
make a number sentence (sentences
such as $6 \times 4 > 18$, $24 \times 35 >
8$ or $54 \div 6 = 9$).

A player receives five points for
an inequality sentence and 10 points
for an equality sentence. If a player
makes a mistake, another can challenge and receive 10 points. The
player making the error receives no

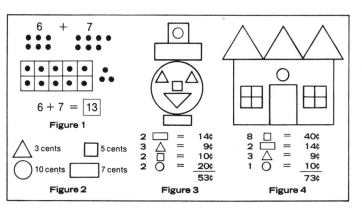

6 + 7

6 + 7 = 13

Figure 1

△ 3 cents   ☐ 5 cents
○ 10 cents ▭ 7 cents

Figure 2

2 ☐	=	14¢
3 △	=	9¢
2 ▱	=	10¢
2 ○	=	20¢
		53¢

Figure 3

8 ☐	=	40¢
2 ▱	=	14¢
3 △	=	9¢
1 ○	=	10¢
		73¢

Figure 4

points. After each play, the numeral cards return to the bottom of the pile. The first player to net 50 points wins.

You can vary the play by allowing the youngsters to invert the " > " symbol.

2. *Remainder game.* Make up a number board similar to the one on page 65 (or you make a transparency of the page itself). You'll also need four cards of each numeral from 0 to 9. Two to four can play the game. Each player will need a counter.

The first player draws a card from the numeral pile where the cards have been placed face down. He divides the first number on board by the number drawn. For example, if he draws a 5, then 18 ÷ 5 is the problem. The player names only the remainder. In this case the remainder is 3. He moves his counter forward three places and lands on 34. (All players start on 18.) The next player draws a card. Say his card is a 9; this divides 18 by 9. The remainder is 0 so he cannot move and must stay on 18. Play continues until all players have had their first turn. Then the first player takes a second turn. He draws a number from the pile. To get "home" a player must come in on the exact number of squares. For example, if a player is on the 31 square, he must have a remainder of 2 to win the game, if on 19 he must have remainder of 1. Players can call for time to work out problems with paper or counters that are there for that purpose. Poker chips can be placed out for the children to use. If an error is made and corrected by another player, then the player that made the error must take his counter back to his previous position. The player that gets home first is the winner of the game.

Rules for playing *The Remainder Game:* The first player draws a card from the numeral pile. Divide the first number on the board (18) by the numeral drawn. (For example, if you draw a 5, then 18 ÷ 5 is the problem.) Name only the remainder (3, in this example) and move your counter the number of places coinciding with the remainder. The next player follows the same procedure. Say his card is a 9. Since 18 ÷ 9 leaves no remainder, this player would not move from the 18 square. To get "home," a player must come in on the exact number of squares. (For example, if you are on the 31 square, you must have a remainder of 2 to win.) The first player to get "home" is the winner.

Math lab—III

After two months' experience in the math lab, your youngsters
are ready for more sophisticated activities. Here are
some that will guarantee fun while learning

LOLA J. MAY

PREVIOUS INSTALLMENTS of this series of articles on setting up a math lab in your classroom have provided a number of activities which you can make available to your pupils. Those activities, plus a number of your own favorites which you have by now added, should have made your math lab a going, daily part of your math program.

This month we'll be adding only one new activity (i.e., one for the primary level and one for intermediates). However, there are many variations to each.

For the primaries

In the primary grades, place value needs continual reinforcement. The game of *Dominosums* fills the bill. All your youngsters need are a set of dominoes (or a set of simulated dominoes made from tagboard or poster board) and a game board, which is easy to make (see Figure 1 on opposite page). The width of the game board should be the same as the long side of a domino.

The object of the game, which can be played by one or two students, is to place the dominoes on the game board so that they add up to, but do not exceed, a chosen sum. Figure 2 shows a game where the goal sum of 100 has been reached.

Here's how the game is played: If there is one player, he draws five dominoes from the pile which is face down and tries to reach, but not exceed, 100. If two are playing, each draws three dominoes. Each player looks at his own dominoes, but shields them from his opponent.

The players take turns placing a domino on the game board trying to reach the desired sum (100, in this example). The game ends when the sum is reached or exceeded. If a player reaches exactly 100, he wins and gets 10 points. Each player must take his turn; no passing is allowed. If the domino he plays exceeds 100, he loses, the game ends and his opponent gets five points. If, after all six pieces are played, the sum is less than the goal (100), more dominoes are drawn one at a time and played until the sum is reached or exceeded.

The players then start all over, drawing three new dominoes each. When all the dominoes are used up, the player with the most points wins.

You may have to remind the children that each domino can be read two ways; that is, a domino that has a 3 and a 6 can be either 36 or 63.

VARIATIONS

1. In the lower primaries, the dominoes can be stacked lengthwise and the single digits added or subtracted.

2. Prepare a set of index cards, each with a three-digit sum written on it. The pile is placed face down, and the top card is turned over. Each youngster draws a specific number of dominoes (from four to six) and tries to get a sum as close as possible to the number on the card.

3. Try this one on your mathematical whizzes: Each child draws from four to six dominoes and by using both addition and subtraction

tries to get a sum as close to 0 as possible.

4. Same as above, only this time each child tries to get as close to the sum on an index card.

5. By using the two numbers on a domino as factors, you can create similar multiplication games.

(Note: Variations 3, 4 and 5 can also be used with intermediate-graders.)

For the intermediates

This month's activities for the intermediate-graders center on line designs. Looking for patterns and making designs will attract the youngsters and is also good motivation for more work in geometry. Here are a series of parallel-line, angle and circle activities.

PARALLEL LINES

Duplicate sheets of paper with a pair of parallel lines marked with equal segments; on the left line, number each point from 1 on and number the right line in reverse order (see Figure 3). Here is the sequence of activities for the youngsters:

Activity 1. Draw line segments from 1 to 1, 2 to 2, and so on, until all points are connected.

Activity 2. On another sheet, the youngster follows the rule of three; i.e., go from 1 on the left line to 4 on the right line, 2 on the left line to 5, etc., continuing until the last dot on the right line is reached.

Activity 3. Make up new rules: "Add 2," "Add 5," "Add 7," etc. Start with 1 on the left line and go to the number on the right line according to the rule. The student continues until he has gone as far as he can on the right line.

Question 1. What pattern is observed about the crossing of lines as you change the rule? (Answer: The point moves down according to the rule.)

Activity 4. The rule is "Subtract 3." This time the youngster starts with the highest number on the right line and goes to the number on the left line according to the rule. For example, if 8 is the highest number on the right line, then 8 – 3 = 5, so he goes to 5 on the left line. Then he starts at 7 on the right line and goes to 4 on the left line, etc.

Activity 5. Make up another rule for subtraction, such as 2, 4, 5 or 6. The same procedure as for Activity 4 is followed.

Question 2. What pattern is observed about the crossing of the lines when the rule involves subtraction? (Answer: Point moves up according to the rule.)

Activity 6. Give the rule "Multiply by 2." The child starts at 1 on the left line, says "one times two" and goes to 2 on the right line. Next he points to 2 on the left line, says "two times two" and goes to 4 on the right line, etc. He continues until he reaches the end of the right line.

Activity 7. Make up another rule for multiplication, such as 3, 4 or 5. The same procedure as for Activity 6 is followed.

Question 3. What is observed about the pattern of the crossing of the lines? (Answer: Point moves to the left.)

ANGLES

Duplicate sheets of paper with angles of various sizes, with equal segments marked off on each ray. Number by ones in opposite directions (see Figure 4).

Activity 1. The student draws lines from 1 to 1, 2 to 2, etc. Ask: "Do you see a curve?"

Activity 2. On another sheet of paper, the student draws a line from 1 to 3, 2 to 5, 3 to 7, etc. (from consecutive numbers to consecutive *odd* numbers). Ask: "Do you see a curve? Is it different from the curve in Activity 1?"

Activity 3. Make up a rule, such as "Add 3," and have the student draw the lines and see what kind of curve he gets.

CIRCLE

Duplicate sheets of paper with a circle; mark off and label equal segments (see Figure 5).

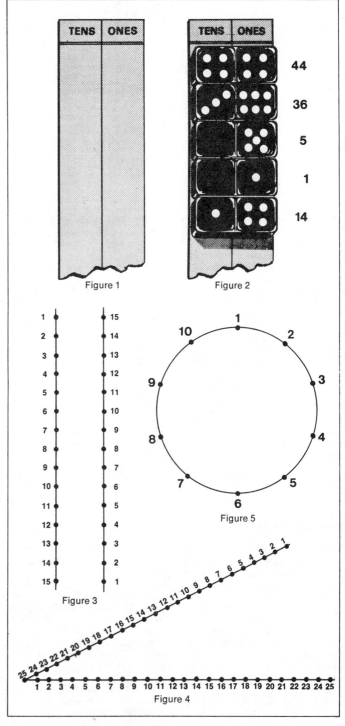

Figure 1

Figure 2

		44
		36
		5
		1
		14

Figure 3

Figure 5

Figure 4

Activity 1. Have the youngster draw lines from 1 to 3, 2 to 4, 3 to 5, etc.

Activity 2. Make up rules and have the youngsters try them out on other circles to see what kind of designs they get.

FOLLOWUP ACTIVITIES

Some youngsters will want to make the line designs on tagboard and use colored string for the line segments, thus making interesting, colorful designs. By all means, let their creativity go wild!

Math lab-IV

There's more to math than computation. "Checker Challenge"
can help your intermediate-graders improve
their problem-solving skills

LOLA J. MAY

A WELL-ROUNDED math lab should include activities that do not require computation skills. Children who have difficulty using numbers also should have fun with mathematics in your math corner.

"Checker Challenge" will fill the bill for intermediate-grade youngsters in this fix. (Indeed, any intermediate-grader will enjoy the game.) The activity requires game strategy—that is, involvement in experimenting and using trial and error techniques. As the children become absorbed in the activities described in this article, they will be building skills that will improve both their problem-solving abilities and power of concentration. And success in these may win them back to computation.

The only materials needed for "Checker Challenge" are checkerboards (or squares of paper marked off in 64 squares) and some markers (eight per board).

Make up a sample checkerboard sheet (see Figure 1 below) which illustrates that a *row* runs horizontally and a column goes vertically; also, that there are two *main diagonals* in a square. The impor-

tance of the child knowing these basics is clear in the rules of "Checker Challenge." Write the following on a card:

"Place one marker in each row of the checkerboard, with these two conditions: (1) No row, column or main diagonal may contain more than one checker, and (2) Each row, column and main diagonal must contain one checker. Now, using eight markers, try to find a solution. It will not happen in a few minutes. Keep moving the markers around."

Figures 2 and 3, below, show incorrect solutions. (Figure 2 has two checkers in a main diagonal, and Figure 3 contains two checkers in the same row.) Clip out the two illustrations and paste them on the instruction card, pointing out that they are incorrect.

There are many correct solutions. Figure 4 shows one of them.

When a child finds one correct array, the next step is to record it. It is important to record each solution, so that the child can see if he is actually coming up with different solutions as he continues to play the game.

For recording, the columns have

Roman numerals and the rows have decimal numerals (see Figure 4). The solution is recorded from left to right, with the column numeral first for each marking. Thus, the record for the solution in Figure 4 reads: I-1, II-3, III-4, IV-6, V-2, VI-8, VII-5, VIII-7. (Recording can be simplified if everyone understands that he will record from left to right in order and then just omit the Roman numerals for the columns. Thus, the simplified recorded solution for Figure 4 would be: 1, 3, 4, 6, 2, 8, 5 and 7.)

Once a child has found one solution and has recorded it properly, challenge him to find nine other solutions, recording each one.

New challenges

Your successful problem-solvers are now ready for some new checker challenges. Put the following instructions on separate cards:

"Put down the markers for the first four positions from one of your recorded solutions. Say, the first four positions are I-1, II-3, III-2, IV-5 (or 1, 3, 2, 5 if you are using the simplified system). Put markers on the board in those four squares.

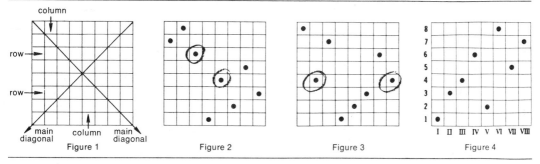

main diagonal	column	main diagonal	
Figure 1		Figure 2	Figure 3	Figure 4

Now, find at least four correct solutions that have the same markers in the first four columns."

"Pick any one of the solutions you have found and place the eight markers on the board. Using a different type of marker, find a second solution which utilizes eight different squares—in other words, no two markers can be in the same square.

When you have a correct solution, you will have two markers (no more, no less) in each column, each row and each main diagonal. After you get one double solution, try others. Don't forget to record each solution."

"Now you're ready for the big leagues. First, find three solutions on the same board that do not over-

lap (that is, only one marker in any square). Then, go on to find four, then five, six, seven and finally eight."

If anyone finds a solution for all 64 squares (that is, eight solutions on one board), by all means make sure it's recorded for posterity.

SELECTED BIBLIOGRAPHY FOR CHAPTER 12
Math Labs

Barson, Alan. "The Mathematics Laboratory for the Elementary and Middle School." *The Arithmetic Teacher* (December 1971), 565–67.

Biggs, Edith. "The Role of Experience in the Learning of Mathematics." *The Arithmetic Teacher* (May 1971), 278.

Brumbaugh, Douglas K. and Hynes, Michael C. "Math Lab Activities." *School Science and Mathematics* (October 1974), 541–42.

Clark, H. Clifford. "Before You Individualize Your Elementary Math . . ." *School Science and Mathematics* (November 1971), 676–80.

Clarkson, David M. "A Mathematics Laboratory for Prospective Teachers." *The Arithmetic Teacher* (January 1970), 75–78.

D'Augustine, Charles H. "Preservice Laboratory Experiences for Mathematics Methods Courses." *The Arithmetic Teacher* (April 1974), 324–28.

Davidson, Patricia S. and Fair, Arlene W. "A Mathematics Laboratory—From Dream to Reality." *The Arithmetic Teacher* (February 1970), 105–10.

Ewbank, William A. "The Mathematics Laboratory: What? Why? When? How?" *The Arithmetic Teacher* (December 1971), 559–64.

Fitzgerald, William M. "A Mathematics Laboratory for Prospective Elementary School Teachers." *The Arithmetic Teacher* (October 1968), 547.

Kieren, T. W. and Vance, J. H. "The Theory of Active Learning: Its Application in a Mathematics Workshop." *The Manitoba Journal of Education* (November 1968), 33–40.

Kluttz, Marguerite. "The Mathematics Laboratory—A Meaningful Approach to Mathematics Instruction." *The Mathematics Teacher* (March 1963), 141–45.

Matthews, Geoffrey and Comber, Julia. "Mathematics Laboratories." *The Arithmetic Teacher* (December 1971), 547–50.

May, Lola J. "Individualized Instruction in a Learning Laboratory Setting." *The Arithmetic Teacher* (February 1966), 110.

——. "Learning Laboratories in Elementary Schools in Winnetka." *The Arithmetic Teacher* (October 1968), 501.

Schnell, Jacqueline H. and Klein, Lane Krasno. "Development of a Mathematics Laboratory." *The Arithmetic Teacher* (October 1974), 492–96.

Silverman, Helene. "Where Are the Children?" *The Arithmetic Teacher* (December 1971), 596–97.

Swan, Malcolm and Jones, Orville. "Preservice Teachers Clarify Mathematical Percepts Through Field Experiences." *The Arithmetic Teacher* (December 1969), 643.

Sweet, Raymond. "Organizing a Mathematics Laboratory." *The Mathematics Teacher* (February 1967), 117–20.

Vance, James H. and Kieren, Thomas E. "Laboratory Settings in Mathematics: What Does Research Say to the Teacher?" *The Arithmetic Teacher* (December 1971), 585–89.

PART VI

CORRELATIONS AND INSPIRATIONS

CHAPTER 13

CORRELATION OF MATH-SCIENCE ACTIVITIES

The previous chapter on math labs indicated that mathematics is *not* a spectator sport—you need to be involved if you are to profit from its study. What better way to involve children than capitalize on their interest in scientific phenomena and show them the essential role of mathematics in some real-world science activities?

Mathematics and science seem to be terms that go together in our everyday language almost as much as bread and butter. In fact, science and mathematics have grown together. Kusch* has stated it this way:

> . . . a great deal of mathematics has been devised for the explicit purpose of serving the needs of an evolving science. A striking example is that of Sir Isaac Newton who invented the calculus to aid him in investigating the motions of the planets. The calculus was not invented by someone who would, at the present time, be described as a pure mathematician; rather it was invented by someone who desperately needed more powerful mathematical tools than were currently available.

The readings in this chapter were selected to encourage and facilitate the essential complementary nature of science and mathematics. For example, Charles D'Augustine develops some interesting ideas in his article, "Reflections on the Courtship of Mathematics and Science," that could be used in elementary and junior high classrooms. The alert teacher will find many useful applications here.

James Hogan and William Schall in "Coordinating Science and Mathematics" provide two complete math-science lessons, "The Use of Space" and "Where the Wind Doth Blow." Investigations such as these provide many opportunities to utilize and elaborate on mathematical concepts.

"Developing Concrete Experiences in Elementary School Science and Mathematics" by Hy Ruchlis offers a strong rationale for learning as a sensory experience. He claims that concrete experiences are needed and follows up with some excellent activities, such as "Developing Concepts of Electricity," "Using Mirrors to Develop Ideas about Angles," "Approaching Angles Through Bent Tubes," and "Measurement in Science and Mathematics."

Nature abounds in suggestions for patterns in mathematics and science. Nathan Ainsworth, in "An Introduction to Sequence: Elementary School Mathematics and Science Enrichment," gives a fine discussion of the Fibonacci sequence with an

---

*Polykarp Kusch, "Mathematics in Science Education," *The Science Teacher* (May 1966), 20–21.

ample supply of illustrations from the learners' real world. This article is a good example of the correlation of mathematics, science, geometry, and number theory.

Our environment is better understood and becomes more meaningful to students through measurement and quantitative experiences. Robert Lemmon, in "Quantitative Descriptions in Science," provides two lessons which give children opportunities to use scientific procedures in developing quantitative descriptions of their environment.

What better way to apply measurement skills than outdoors? "The Slope of the Land" by Clifford Knapp provides a very complete lesson for having children make their own examinations of the slope of the land. This lesson is an excellent beginning for a file or unit on "outdoor mathematics."

Elton Beougher's article, "Blast-Off Mathematics," is another one that you may wish to put in your "outdoor mathematics" file. Beougher provides some excellent ideas that teachers might use to give variety to their lessons. In fact, you might have your learners build some model rockets* and have your own "blast-off."

William Schall, in "Discovering Centigrade and Fahrenheit Relationships," provides a discovery lesson in which an overnight science probe provided an opportunity for some good learning—an understanding of the centigrade and Fahrenheit thermometers and the relationships between the two scales. The development of a topic in this way is certainly much more meaningful and interesting than the isolated rote memorization of traditionally introduced formulas.

Verne Rockcastle, in "Upper Winds for Upper Grades," provides another interesting application for mathematics and another lesson for an "outdoor mathematics" file. This exercise will increase your students' knowledge of measurement and recording data, as well as lead to a greater understanding of the behavior of the winds.

Opportunities for using the mutually supportive nature of mathematics and science class activities are enormous. Many possibilities exist for the teacher who is alert and willing to implement them.

*For additional information, write to Estes Industries, Inc., Penrose, Colorado.

Reflections on the courtship of mathematics and science

CHARLES H. D'AUGUSTINE

It has been said that mathematics is the queen of the sciences. However, if one wants a truly dynamic elementary school mathematics program, then one must promote the marriage of mathematics and science curriculums. We have only to study the current elementary school science and mathematics programs in order to realize that not only are mathematics and science unmarried, but neither curriculum reflects a very serious awareness of the other's existence.

Let us examine and illustrate how the mathematical skills utilized by a scientist might also be skills we would wish to develop in a mathematics curriculum.

One of the tasks of a scientist is that of finding patterns that exist in nature. In his quest for these patterns he often lists his quantitative data in a table or translates it into a graph. After he has collected and organized his data, he attempts to construct a generalized mathematical model that will facilitate predictions about future events. He then tests the reasonableness of his generalized model by making further observations. Let us examine some simplified examples which depict sequences of activities a scientist might engage in when searching for these patterns.

FIGURE 1

Suppose a scientist is concerned with the relationship between the stress applied to a wire of one square unit of cross-section area and the amount of elongation produced on the wire by the stress. (For simplicity we will assume the very crudest tools at the scientist's disposal.) He attaches one end of the wire to a ceiling beam; to the other end of the beam he attaches an eyelet that will allow him to hang various weights on the wire. (See Fig. 1.)

After each weight is added, he notes how much the wire has elongated with respect to each unit of length. He collects his data and records them in a table (see Table 1).

Reprinted from *The Arithmetic Teacher* (December 1967), 645–649, by permission of the author and the publisher.

312

Table 1

Pounds per square unit	Unit elongation (per unit of wire)
1,000	.001*
2,000	.002
3,000	.003
4,000	.004

* Recorded to the nearest thousandth of a unit.

He then constructs a graph.

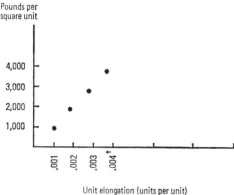

Pounds per square unit

Unit elongation (units per unit)

† Each number has been rounded to the nearest thousandth.

It seems to him that a straight line would pass through each of the points on his graph. (This is called a linear relation.)

He hypothesizes that the ratio of elongation per unit to pounds per square unit is equal to

$$\frac{.001}{1,000}, \text{ or } \frac{E/\text{unit}}{P/\text{unit}^2} = \frac{.001}{1,000}.$$

Thus, he predicts that if he applies 6,000 pounds per square unit, then the elongation can be derived from the equation

$$\frac{E}{6,000} = \frac{.001}{1,000}.$$

He proceeds to place 6,000 pounds on the wire with one square unit of area in cross section and verifies that the mathematical model does appear to fit the pattern he has noted. However, upon continued experimentation he finds a point (called the yield point) where his mathematical model will no longer serve as a predictor of elong-

ation. However, for all practical purposes his model will serve as a useful tool because he will wish to use his model for instances below the yield point.

Many experimental situations yield data that, when graphed, can be fitted to a straight line. Elementary school teachers may wish to familiarize themselves with several of the relationships so that they can coordinate their efforts in science and mathematics. A few of the more common linear relationships (which will lend themselves to a ratio-type model) are as follows:

a) Temperature of a gas versus pressure (assuming a constant volume)

b) Weight of a floating body versus the weight of the water displaced

c) Velocity of a body projected upward versus the time it takes to reach the greatest height

d) Elongation of a spring versus the weight attached to the spring

e) Reading of an ammeter versus voltage (resistance held constant)

f) Relationship of distance traveled versus time (velocity held constant)

These relationships represent the simplest of linear relations in that a zero with respect to one of the variables will correspond to a zero of the other variable. In other words, the idealized line produced by the experimental data passes through the zero-zero coordinate.

Let us observe another type of activity in which a scientist might engage. This activity requires the scientist to make a frequency count (counting the number of times a particular event occurs).

He breeds a certain type of water bug and makes the following observations. When he breeds two short-legged bugs together, he always gets short-legged bugs. When he breeds two long-legged bugs together, one of two things happens: either some of the offspring have short legs and some long, or all of the offspring have long legs. He records his data in a table such as Table 2.

313

Table 2

Parents	Number of short-legged offspring	Number of long-legged offspring
Both short-legged		
Pair A	42	0
Pair B	75	0
Pair C	84	0
Pair D	67	0
Pair E	39	0
Both long-legged		
Pair F	0	73
Pair G	19	60
Pair H	0	48
Pair I	0	64
Pair J	13	42
Pair K	15	39

He now notices three distinct patterns emerging. When both parents are short-legged, the offspring are short-legged; and when both parents are long-legged, either the offspring are all long-legged, or the offspring have a chance of being short-legged in the ratio of about one to four. He proceeds to test his new hypothesis by further experimentation. In time he will arrive at a generalized model for predicting the following, assuming no previous knowledge of the heredity of the parents:

1. The probability that when he breeds two long-legged bugs at least some of the offspring will be short-legged

2. The probability that when **two long-legged** bugs both having short-legged siblings are mated, they will have only long-legged offspring

Can you think of some other problems for which he might derive a generalized model? None of the above models are easily derived. The scientist could either attempt to derive the mathematical model by further experimentation or seek some model already in existence that could be modified to fit his experimental data.

If he chooses the latter course, he might select the mathematical model that fits the data obtained when flipping a coin. For example, if he designates one side of a coin

heads (H) and the other side tails (T), then if he flips a coin twice, the following are possibilities for what he would see:

Both heads	HH
Both tails	TT
First toss heads; second tails	HT
First toss tails; second heads	TH

The scientist knows from the literature that each parent contributes to the offspring one gene from each pair of genes that he has. Therefore, if he designates L as a gene for long-leggedness and S for short-leggedness, the possible gene combination for any parent is either LL, or LS, or SL, or SS. Since the order of the genes will not make any difference, he winds up with the possibilities that any given bug will have either two long-legged genes, two short-legged genes, or one of each type of gene.

The scientist reasons that since some of the long-legged bugs have short-legged offspring, **at** least one of the parents has a **short-legged** gene; but by logic he can quickly establish that if some short-legged offspring are to be produced by the mating of two long-legged bugs, then both of the long-legged parents need to carry one of the **short-legged** genes. Study the two models (**Fig. 2**) and see if you can use logic to establish why it must be Condition B and not Condition A. (Hint: What did the LS parents look like?)

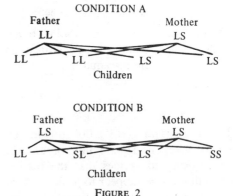

FIGURE 2

How was the scientist able to reason that when both mates possessed LS pairs, one could reasonably expect one out of every

314

four offspring to have short legs? Can you figure out the expected ratio of short-legged to long-legged offspring when one parent is SS and the other is LS?

Another scientist is interested in predicting the temperature in his area. He studies data sampled from a great many points at the surface of the United States and notices a very unusual pattern. It appears to him that when temperatures are recorded at the same elevation and in close proximity to each other, they do not change abruptly, and that any given temperature appears to represent the average of temperatures on each side of it. For example, suppose he knows the temperatures of points B, C, D, and E. (See Figure 3.) By taking the sum of these temperatures and dividing by four, he finds that this gives him a fair estimate of the temperature at A.

$$B$$
$$E \quad A \quad C$$
$$D$$

FIGURE 3

He notices a variation in the temperature pattern when a front passes through his area. By studying the temperature drop with respect to the passage of the front, he is able to make predictions about the temperature in his area after the front passes.

As he studies more and more data under various conditions, he finds many conditions that affect his prediction of temperature (such as cloud cover, winds, terrain, etc.)

The goal of this scientist is to keep refining his mathematical model as the influence of more and more variables is determined.

While the mathematical skills used by these aforementioned scientists are not inclusive, they do represent mathematical skills that are not consistently found in the elementary mathematics curriculum. The first scientist utilized the skill of fitting a curve to data and then used this curve to make further predictions. While the author is not advocating the introduction of non-linear curve fitting, there exist many realistic types of problems from the fields of economics, science, business, and the social sciences that involve linear relationships that could be developed at the elementary school level. Not only would these situations represent realistic problems to the child and help him gain insight into his world, but they would also serve as a medium for developing a very useful tool for future problem solving.

The second scientist utilized both the skills of statistics and logic. Problems from economics, science, business, and the social sciences can give the child real insight into the role that statistics and logic play in scientific problem solving.

From the third scientist we gain some insight into the dynamic nature of problem solving and the mathematical skills that play a role in this scientist's search for patterns. The scientist searches for the effects on the pattern brought about by varying the factors. This would be analogous, in mathematics, with studying the effect on the area of a rectangle if the length is increased uniformly, if the width is increased uniformly, or if both are increased uniformly.

The skills relating to "dynamic" models are skills overlooked in the elementary mathematics curriculum. One does not have to look very far to find many places in our society where the skills relating to a dynamic model play a far more significant role than the skills relating to a static model. (Most problems in the elementary school mathematics curriculums are static in nature in that the effects of only one set of conditions are calculated.) For example, the scientific farmer often chooses a crop based on the cost of production, marketing cost, expected market price, and other variables; or the scientist concerned with traffic patterns studies the effects of traffic density, industrial density, residential density, and climatic conditions in arriving at a mathematical model which permits him to recommend various timing cycles for the traffic lights; or the engineer takes into account shear, bending moment, fiber compression, deflection, expansion due to temperature, dead weight, and live weight in designing a beam for a bridge. The farmer, the scientist and the engineer each used a dynamic model.

In summary, if the elementary mathematics curriculum is to be an effective tool in a technological society, it must take into account in its construction those aspects of a science curriculum that may lend themselves to the construction of mathematical models.

The author has not attempted to make an all-inclusive list of the science skills one might incorporate into the mathematics curriculum. As a matter of fact, the completion of a list of potential skills and concepts from science that could be included in the mathematics curriculum would be a valuable addition to mathematics education. However, the very nature of such a list would necessitate either convening a national symposium, attended by representatives of the many sciences; or having a committee designated to make such a survey, which would involve not only questionnaires but observations by mathematicians of scientists at work, followed up by personal interviews.

While the problem is complex, its solution becomes more and more urgent as more and more aspects of our society reflect the effects of the technological revolution.

Coordinating Science and Mathematics

JAMES R. HOGAN and WILLIAM E. SCHALL

SO YOU have been looking for a way to integrate the curriculum? After all, that is what is supposed to be done in the current thrust for relevance in school. Spelling words are taken from social studies and science, language arts now includes the study of communication in societies, and writing and reading put it all together in an experience we call the modern elementary school. This move toward relevancy and integration of the curriculum does have the potential for adding sparks of vigor in the emerging vision for elementary school experiences of the 70's. One aspect of the integration of the traditional subject matter areas often overlooked is the essential complementary nature of science and mathematics. When you probe into the nature of science as a method investigation it becomes apparent that scientific knowledge rests upon measurement or quantification. What better way is there to capitalize upon children's interest in science phenomena than to show them the essential role of mathematics in the things children like to read about and investigate on their own or under the guidance of an adult?

Opportunities for using the mutually supportive nature of mathematics and science class activities are enormous. For example, virtually all children have had some study of weather phenomena. This has traditionally been regarded as a science subject. The study of weather is more valuable, more productive of hypotheses and testing if numerical data are kept. The measurement of time in minutes, hours, days, weeks, months, and even years can be part of weather study and yet these measurements are generally reserved for a unit in the mathematics text. Graphs of temperature changes with determination of highest, lowest, and average temperatures over a span of time may appear to be science subject matter concerned with seasonal changes. The mathematics in this science includes addition, subtraction, division, and multiple addition (multiplication). This is mathematics!

We have all tried to teach children how to graph data. Why not use the same unit in science to obtain the data children will represent on the graph? Day to day or hour by hour records of atmospheric pressure and temperature are easily obtained. The data can be graphically represented and requires review of the procedures for reading science instruments and for marking the data or record sheets. Should mathematicians keep records of

Reproduced with permission from *Science and Children* (May 1973), 25–27. Copyright 1973 by the National Science Teachers Association, 1742 Connecticut Avenue, Washington, D.C. 20009.

316

their work? Should scientists describe the phenomena they observe? Of course! Graphing data is a means both a mathematician and a scientist use to keep records of what they have done. When the graph is complete more information is available from the record itself. Scientists would study the graph to help predict future occurrences of a phenomenon. Hypotheses can be made from the data and experiments can be devised to test them.

Have you tried the "experiment" where candles are burned under inverted glass containers? The purpose of this demonstration is to show students that a constant supply of air containing oxygen is needed for combustion to continue. Such a demonstration could be correlated with a theme such as Fire Prevention Week. In health or social studies the lesson might be included under a study of home safety. The opportunities to teach mathematics concepts are also plentiful. The measurement of time could be useful in gaining potential science data about combustion. Again, records must be kept. Does the size of the inverted container make a difference in the amount of time a lighted candle will continue to burn? If two lighted candles are used under the same container in this investigation, will the amount of time the candles burn be changed?

Investigations like this one in particular involve many opportunities to elaborate on mathematical concepts. Measurement of the volume of the containers is an important aspect of the investigation. Yet, the determination of cubic or liquid volumes is often divorced from the science curriculum, being isolated as a separate unit of study in mathematics. Does the number of candles used in each step of the investigation change the results obtained? Record keeping, graphing data, and number operations using fractions should be used as part of this common science textbook lesson.

Science texts suggest many demonstrations and investigations which can be made a part of the science lesson. Often the possibilities for teaching or reinforcing mathematics concepts are suggested by the authors. More possibilities usually exist if the teacher is alert to the opportunities and will implement them. Basic mathematical operations should not be overlooked as possible mathematics-science activities. Elementary science data usually involves simple mathematical operations. When these operations are accepted by the student they not only reinforce elementary mathematics but also reinforce the value of numbers as a means for obtaining significant science data. Interpretation of the mathematical (or numerical) data is not simply a mathematical operation but an exercise in logic.

CLASSROOM ACTIVITIES

The first of the following two activities, "The Use of Space," is less complicated and could even be used in the primary grades. The second activity, "Where the Wind Doth Blow," is a much more extensive activity and develops many of the relationships discussed in this article.

The Use of Space

One of the concepts children learn is that the same space is shared by many different living things. This activity gives practice in counting and observation. Acquire several metal hangers and bend each of them to form a rough square—a "study space." The children can trace the inside of the square on a sheet of paper which becomes the record sheet for student observations about the number, kind, location, and paths of living things. A specified period of time for the observations should be determined; generally fifteen to thirty minutes will be ample.

The children should be encouraged to find different areas around the school grounds to study. Sunny and shady areas, areas in open grass or under trees and bushes, even areas of the sidewalk should be investigated. Children should keep records of the kinds and numbers of living things they observe which are in or come to the area in the coat hanger squares.

Children should find answers to such questions as:

1. How many plants are in the square?
2. How many different kinds of plants can you find?
3. How high was the tallest plant?
4. How many flowers were in the square you observed?
5. How many animals did you see in the area?
6. What animals did you see?
7. How many different kinds of animals did you see? How many of each were in your square?
8. Did you see any animals eating? What did they eat?
9. Could you find traces of dead animals or plants? What did you see?

Children can become aware of the relationships between plants and animals and grow more skillful in the use of numbers as a tool of science in this investigation.

Where the Wind Doth Blow

Children are generally fascinated with weather vanes and wind socks. They can make some observations about wind on their own by trying their skill at launching and flying kites in brisk breezes. Many good science and mathematics lessons can be part of kite making and kite flying contests. What is the direction of the wind? Are the children able to describe the wind according to the compass points? Can they predict which kite flying sites are apt to be good selections because of a prevailing wind direction in their geographic area? A lesson about safety and kite flying would make a valuable corollary to this investigation.

Meteorologists describe the wind according to the direction *from* which the wind blows. Keeping a record of wind direction and the weather encountered over an extended period of time (two or more weeks) may lead children to make some tentative suggestions about the weather and the kinds of winds they have experienced. The students can also find out if the direction of the wind has anything to do with the temperature of the air and the kind of weather to expect.

To make a weather vane, the following materials are needed. For assembly, see Figure 1:

wooden dowel, ¼-inch diameter and 2 feet long
wooden dowel, ½ to 1 inch diameter and 2 feet long
2 six-penny finishing nails
heavyweight tagboard or cardboard, 2-foot-square
tagboard, 2 x 6 inches
magnetic compass
piece of ¼-inch plywood 2 x 12

317

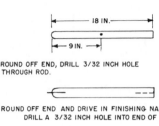

ROUND OFF END, DRILL 3/32 INCH HOLE
THROUGH ROD.

ROUND OFF END AND DRIVE IN FINISHING NAIL.
DRILL A 3/32 INCH HOLE INTO END OF ROD.

(TAGBOARD 2 INCHES
X 6 INCHES)

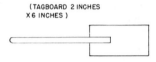

GLUE THE SMALL PIECE OF TAGBOARD TO THE
END OF THE THIN ROD.

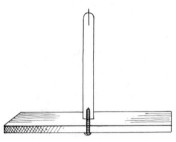

FASTEN THE LARGE ROD TO THE PLYWOOD BASE
WITH THE SCREW. THE ROUNDED END MUST BE UP.

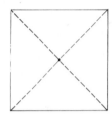

ON THE 2X2 FOOT TAGBOARD DRAW A SET OF
RAYS 10° APART. USE THE CENTER OF THE TAG-
BOARD AS THE FOCAL POINT OF THE RAYS. AFTER
THE RAYS HAVE BEEN DRAWN CUT A HOLE IN
THE CENTER OF THE TAGBOARD LARGE ENOUGH
SO THAT THE SHAFT OF THE WIND VANE CAN PASS
THROUGH THE HOLE.

USE YOUR MAGNETIC COMPASS TO DECIDE WHERE
NORTH IS LOCATED. MARK NORTH AS 0° (OR 360°),
EAST AS 90°, SOUTH AS 180°, AND WEST AS 270°.
PUT THE SHAFT OF THE WIND VANE THROUGH THE
HOLE IN THE TAGBOARD AND THEN PLACE THE
ARROW-LIKE VANE ON THE FINISHING NAIL SO
THAT THE VANE CAN TURN EASILY ON THE SHAFT.
REMEMBER TO KEEP THE 0° DEGREE LINE POINT-
ING TO MAGNETIC NORTH.

FIGURE 1.

inches with a 3/32-inch hole drilled in the center of the piece
one-inch wood screw
thermometer

Find a spot outdoors where the wind does not have to blow around a building or a large bush and set up the wind vane according to the directions in Figure 1.

The direction the vane points is the direction of the wind and is also the name of the wind. A wind from 180° is a southerly wind and a wind from 270° is a westerly wind. What are the names of other winds, from the directions from which they blow?

Winds may also be named for the direction from which they blow in terms of the degree of the direction. For example: 160°, 235°, 320°, 25°, etc. During this investigation, have the children keep their records in terms of the degrees from which the wind blows rather than the names of the compass points like North or South. They should keep a record of the date, time, temperature, and wind direction; there should also be a record of the kind of weather there is each time. Keep a new record sheet each day.

A sample report form is shown below.

After children have kept records for a while, have them check the data for relationships, similarities, etc. Does the wind seem to come from one predominant direction? When the weather changes to rain, warmer temperatures, cooler temperatures, or cloudy or clear days, what happens to the wind direction? When the temperature is warmest, what direction is the wind blowing?

REPORT FORM			
Recorder's Name _____ Date _____			
Time	Temperature	Wind Direction	Weather Description

DEVELOPING CONCRETE EXPERIENCES IN ELEMENTARY SCHOOL SCIENCE AND MATHEMATICS

Hy Ruchlis

The problems of the elementary school teacher are substantially different from that of the high school teacher. Whereas the high school teacher is expected to be a specialist in one or two subject areas, the teacher of elementary grades is expected to be reasonably competent in geography, history, arithmetic, art, music, and science, and is also expected to know something about the psychology of children and their parents, baby care, and housekeeping in the classroom. Quite frequently, the elementary school teacher is a bookkeeper charged with handling funds. And all this must go on in a relatively small room in which 30 to 35 wiggling young children are to be kept simultaneously interested in teacher-directed activities for a period of 5 hours a day. When you stop to think of it, it's a wonder that the teacher actually accomplishes so much.

That is not to say that the life of a high school teacher is a bed of roses. It isn't. He has his own problems—but of a different kind. However, when we view the actual teaching situation in the elementary schools we no longer wonder about why science is not taught, but rather begin to think about how we can help the teacher do the kind of job that needs to be done in science.

Concrete Versus Abstract Experiences

Many, or perhaps most, elementary school teachers fear teaching science. And many do not teach science at all because of this fear, which is heightened by unreasonable expectations of what is to be taught.

In the elementary grades, the main objective of teaching science is to provide a wide variety of simple *experiences* which establish the foundation upon which science concepts may be built. Conceptual development should, of course, take place in the elementary grades and even in pre-school training. But a properly planned set of experiences can also play a vital role in preparing the child for concepts that he will encounter beyond the elementary school – in high school, college, and adult life.

The base of all learning is sensory experience. We read with our eyes. We listen to a discussion with our ears. We write with our hands in coordination with sensory impressions in our eyes. Thus the basic tools for learning abstract concepts are the senses, and it is well known that defective senses lead to defective learning.

It is rather obvious that if we desire to teach a child to appreciate music we should arrange to have him *listen* to lots of it, and the earlier we do this the greater the probability that the child will be musically inclined. It is also known that having the child *play* music will help develop a sensory base for later appreciation and enjoyment.

Just *talking* or *reading* about music is probably a poor way to learn to appreciate it. Such a procedure – an abstract experience (if there is such a thing) – does not involve the senses *directly* and is therefore inadequate. What is required is a *concrete* experience.

We might establish an index of concrete to abstract experience (C-A ratio) as a guide to teaching procedure. An index of 1.00 would represent all concrete experience and no abstract experience. An index of 0.00 would mean no concrete and all abstract experience. It is apparent that the index for an infant is 1.00, while that for a mathematician would be close to zero. But the mathematician gets to that exalted state from infancy by gradually reducing the C-A ratio as he matures through elementary school, high school, and college. Obviously, the picture will differ for different intellectual pursuits, but the general pattern is clear.

Perhaps some of the difficulty encountered with teaching science and mathematics (and other subjects as well) stems from a too rapid reduction of the C-A ratio in the early grades. One can generally introduce any topic or subject at an early age by making the C-A ratio approach 1.00. One can generally destroy a young child's interest in a subject by making the C-A ratio approach zero. The secret of success for the early introduction of a concept is the development of appropriate concrete experiences that draw from a child's previous experiential base.

Moreover, the subject matter of science – the study of our environment – provides even more reason for involvement of the senses, because the only way man can really know about his environment is through the use of sensory impressions. He operates with his logical mind on this sensory experience and makes deductions that then extend his knowledge. But periodically he must touch the earth, so to speak, and verify his deductions with additional sensory experience.

It is possible, of course, to tell someone what has been discovered about rain, or about volcanoes, or about the stars, and no doubt one's past experiences can often be stretched to some extent to encompass what one has not experienced. But an actual experience with a phenomenon clarifies what is being discussed and provides a high degree of motivation for learning — especially for the younger child.

The Child's Dilemma

Let us consider a problem that might be faced by an explorer who reaches a remote tropical island. He attempts to explain ice skating to the natives and they listen in astonishment as he regales them with tales of water that becomes as hard as a rock, of people walking about on the surface of the water, and of the fun of ice skating. He tells them about people who put on shoes with sharp blades and glide about on the surface of the hardened water. What does the inhabitant of the tropical island think about all this? More likely than not he is skeptical. But one trip to the Arctic and its frozen seas would give that tropical islander a basis for understanding the nature of freezing of water and its relation to temperature. Or, he could substitute for that trip a direct experience with the formation of ice in a refrigerator — if one were available.

The child in the elementary grades is often in a similar predicament when learning science. Much of what he is taught lies beyond his experience, and the things he is told about the world in the course of some of his science lessons must seem as strange — and therefore as difficult to understand — as ice skating is to the tropical islander. Despite the fact that so much can easily be done in the schools to provide appropriate sensory experiences, science is still taught largely through abstract reading without the essential accompaniment of actual experience by children.

Most children can comprehend the *what* and *how* of a simple situation even if they can't understand *why* something happens. But they generally need to know what and how before they can understand why. Consequently, much of elementary science should be experiential, with increasing emphasis, as children develop and their conceptual background expands, on why things happen. Children deprived of such early experiences will probably flounder and be confused. Indeed much of the difference in intellectual development among children stems from this deprivation. Genetic endowment plays an important role, of course, but in many cases is used as an excuse for continuing deprivation of the vital experiences children must obtain to understand science.

The Role of Simple Equipment in the Classroom

Since we are often dealing with phenomena which are not directly part of a child's normal experience, it is essential to bring experiences into the classroom by means of equipment of some kind — not necessarily the complex precision equipment of the science laboratory, but materials that are specifically designed to produce events which encompass the experiences we seek.

Developing concepts of electricity. Consider the subject of electricity, for example. Children generally have wide experience with it, and they are quite able to accept some of the things adults tell them about it. But how do we explain the cause of current electricity, which is said to be due to the motion of tiny, invisible, charged particles in wires?

In this situation, a few simple experiences with static electricity — with attraction of unlike charges and repulsion of like charges — can make a world of difference in the depth of understanding which a child has about the subject of electricity.

For example, we can provide the child with a sheet of smooth plastic which he presses against a sheet of paper and rubs with his hand. Upon attempting to pull the plastic away, he discovers a strong force which makes the plastic stick to the paper. If the plastic is then placed over his head, he feels his hair standing up. When he subsequently brings the plastic close to his ear, he feels and hears tiny sparks. The plastic and paper both cling to clothes, walls, doors. Particles of vermiculite or other light objects jump about erratically or fly off when placed on the plastic and the finger is moved around underneath.

Thus the child experiences electrical force and electric charge, basic to our universe, and also basic to understanding electrical phenomena. Moreover, he produces electrical forces and charges using simple, easy-to-obtain materials, and with nothing more than a simple rubbing action. There is nothing mysterious about *what* and *how*.

These experiences not only make the idea of electrical force meaningful, but the very dramatic nature of the experience arouses curiosity and motivates the child to want to know more and do more. In this way the experience makes the teaching alive and real for the child as well as for his teacher.

The experiences described above may be extended to involve the concept: Like charges repel; unlike charges attract. Two identical dark-colored plastic strips that produce negative charges are rubbed on paper, then suspended from the hand. They strongly repel. The same thing is done with two clear plastic strips that produce positive charges. They also strongly repel. But when this is done with one dark colored plastic strip and one clear plastic strip, they strongly attract. Obviously, the electric charges on the dark and clear strips must be different in some way.

From such experiences it is much easier for children to understand the meaning of like charges and unlike charges, and then the meaning of positive and negative charges. In effect, the children are recapitulating experiences similar to those of early scientists and generating the same kind of enthusiasm at making new discoveries. They are new to the children even if they aren't new to mankind.

Using mirrors to develop ideas about angles. Consider the subject of angles. Normally we would think of this topic as suitable for study in a mathematics lesson, not in science. But neat compartmentalization of subject matter often violates experience, and tends to deprive children of rich experiences which can contribute to conceptual development in different subject areas. Often a properly designed material has its own logic that does not fit into a prescribed cubbyhole. The mirror, for example, usually studied as part of the subject of light, can be used to breach the traditional subject-matter boundaries and reach out into the study of angles with enormous benefit to children.

Suppose we hinge two mirrors as shown below, and stand them on edge on a table. A penny is then placed between the mirrors. The child sees several

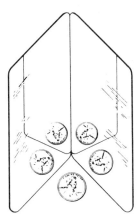

pennies in the mirrors. As he moves the mirrors closer (makes the angle smaller) he observes more pennies. As he moves them farther apart (makes the angle larger) he sees fewer pennies. Without naming it, the activity of moving the mirrors to "make more money" involves a direct experience with angles, in this case the angle between mirrors. The child literally "makes angles" with mirrors instead of drawing abstract angles on paper. Moreover, there is a visual relationship between angle and number. This can be developed by providing special cards with circles divided into 3, 4, 5, or more parts. If the child places the mirrors on an angle that is one third of a circle, he sees three pennies; on one fourth of a circle, he sees four pennies; on one fifth of a circle, five pennies. Thus the idea that an angle can be measured as a portion of a circle is made real. At the proper time, it is a relatively simple step to the idea of a degree as 1/360th of a circle, and to the protractor as an instrument with which we can draw any desired angle.

If the child places the hinged mirrors on a colored photograph from a magazine, a new world of design opens up before him. Beautiful kaleidoscopic images appear, which change as the angle between mirrors is changed. Just as the pair of hinged mirrors jumped the gap between science and mathematics in the realm of angles, it now does the same thing for science and art. The child can now make his own designs by drawing pictures, cutting and pasting magazine photos, and by varying the angle between mirrors. And rather than passively viewing designs through a kaleidoscope,

the child has active control of the variables which produce changes in design.

Approaching angles through bent tubes. A set of plastic tubes and pipe cleaner connectors may be used to provide a different kind of experience with angles. Angles are formed when two rays share a common endpoint, but they can be conveniently indicated by the following procedure. Short lengths of pipe cleaner are inserted into tubes, and pushed together to form a flexible junction at the point at which they join. The child can bend the tubes to make a large (obtuse) angle, a small (acute) angle, a right angle, a straight angle, and a zero angle. He can even rotate one of the tubes from a zero angle through acute angle, right angle, obtuse angle, straight angle, and then on to angles from 180° to 360°. In effect, he is obtaining a functional experience with variable angles that will stand him in good stead when he needs the concept in a future trigonometry lesson.

The child can then extend his experience with angles to triangles of all kinds (equilateral, isosceles, scalene, and right). He can make quadrilaterals, including the square, rectangle, rhombus, and parallelogram. He can, with his own hands, tilt the side of the square to change the right angle to an acute angle and thereby convert the figure into a rhombus. He can observe the lack of rigidity of the four-sided shape, in contrast to the complete rigidity of the triangle, and thereby experience the importance of bracing a structure with triangular frames. This concept will be most helpful at a future time if he should study engineering, if he becomes a carpenter, or if he just builds things for a hobby.

With a slight modification of the relationship of diameters of tubes and connectors, it becomes possible to insert two pipe cleaners into one tube. This, in turn, makes it possible to produce junctions of three or more tubes, opening up a new world of solid shapes. A fourth-grade child can now make a tetrahedron, cube, prism, pyramid, octahedron, icosahedron, and many more shapes. Without being aware of it, the 10-year-old leaps beyond the confines of the normal mathematics curriculum into an area of solid geometry.

Measurement in science and mathematics. Much of modern science is based upon measurement, and proper development of scientific concepts requires that the child obtain measurement experiences at as early an age as possible. Consider the specific problem of measurement of weight. Suppose we construct an inexpensive equal-arm balance of the kind shown below. The traditional instrument of this kind is a heavy, bulky, expensive device with a special bearing to ensure accuracy to one tenth of a gram (about 1/300th of an ounce). Such precision is fine for a chemistry student interested in quantitative analysis, but is not necessary for elementary school children. And the traditional standard accurate weights are obstacles for the young child. We can, however, give the child something he is familiar with, paper clips, for example, to use as standards of weight.

A pencil is put into the pan at the left, and paper clips

into the one at the right. The child proceeds to add paper clips until they balance the pencil. What does the pencil weigh? One child may find that his pencil weighs ten clips; another, twenty clips. Which pencil is heavier? Obviously, the pencil that weighs twenty clips is heavier. How many times heavier? A second grader could see that it is twice as heavy.

A year or two later the same type of procedure can be extended to determine ratios of weights. A penny is put into one pan and a dime into the other. The side with the penny goes down. Another dime is added to the first one. Now the side with two dimes goes down. So a penny is less than twice as heavy as a dime. The child alternately adds pennies and dimes until balance is achieved with four pennies and five dimes. He can easily conclude that one penny weighs 5/4 dimes.

A similar procedure reveals that a nickel is found to equal two dimes in weight. The teacher can pose the problem: How many pennies would balance how many nickels? A table is constructed, as follows:

| Weight | | Weight | |
Pennies	Dimes	Nickels	Dimes
4	5	1	2
8	10	2	4
12	15	3	6
16	20	4	8
		5	10
		6	12

Children observe from the table that ten dimes weigh the same as eight pennies and also the same as five nickels. So it seems reasonable that five nickels would balance eight pennies.

Is that really so? A quick check with real nickels and pennies reveals that the prediction is correct. The child thereby experiences an important aspect of scientific process — the making and testing of a prediction. He also receives a mathematical experience that will later help him understand variables, functions, and equations.

In designing educational material it is often feasible to make the material serve various purposes by making use of properly selected physical features. For example, in the balance just discussed the flat arms can serve as a lever with a fulcrum in the center. The upper surface can be imprinted with a scale showing zero at the center and uniform numbered divisions extending outward toward both ends of the arms. The numbers on the flat arm represent distance from the center.

A whole new range of activities now becomes possible. For example, in the activity illustrated below, on one side of the balance one washer is placed at position 2

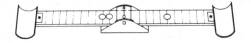

and another washer is placed at position 3, and on the other side of the balance one washer is placed at position 5. When the bar is balanced in this way, does the position of one washer on one side always equal the sum of the positions of the washers on the other side? Like true scientists, the children test the hypothesis and find it to be so. Is the hypothesis true for three or four washers on one side? They test and again confirm the hypothesis.

Suppose we put two washers at position 3 on one side. Where will one washer balance the two? By actual trial the children find that balance occurs at position 6. Will two washers at 4 be balanced by one at 8? A quick test confirms this guess. As a related fact, it is a simple matter to show that this arrangement is a machine because a small weight (one washer) can lift a larger one (two washers).

If we place two washers at position 3, where will three washers on the other side balance the two? The children find that three at 2 balance two at 3. Will four at 3 balance three at 4; or two at 6; or one at 12; or twelve at 1? They find that the product of the number of washers by their position on one side equals the product of the number of washers by their position on the other side. Important commutative and associative mathematical relationships are thereby experienced. Factors may also be demonstrated.

If we place two washers at 3 and one at 4, the children find that one at 10 on the other side balances this combination. Why? Two washers at 3 on one side would be balanced by one at 6 on the other side. One washer at 4 is balanced by one at 4. So the combination on one side is balanced by one washer at 6 + 4, or 10. Without realizing it, the children are solving the following equation:

$$1 X = (2 \times 3) + (1 \times 4)$$
$$X = 6 + 4$$
$$X = 10$$

If the above combination of two washers at position 3 and one at position 4 is balanced by one washer at 10, where will two washers produce balance? Obviously, if one at 10 produces balance, so will two at 5. The children are now solving this equation:

$$2 X = (2 \times 3) + (1 \times 4)$$
$$2 X = 6 + 4$$
$$2 X = 10$$
$$X = 5$$

Do we wish to study the metric system of weights? That's not too difficult with this equipment. The nickel, which weighs very close to five grams, can serve as the standard of weight. A length of copper wire can be cut until its weight balances that of a nickel. The wire can then be cut into five equal parts to produce fairly accurate weights of one gram. The one-gram length of wire can be cut into two to make one-half-gram weights. Multiples are easily formed into two, ten, and more grams. Various objects can then be weighed using these standard weights.

Is this process as accurate as using manufactured precision weights? No, but the experience of making his own weights is much more rewarding to the child because it stresses the process and the doing, rather than high standards of precision that have little meaning in the elementary grades.

By the time the educational potential of this equal-arm balance is exhausted, the child would have received a very wide set of experiences involving the force of gravity, machines, levers, measurement of weight, calibration, scientific process, addition, subtraction, multiplication, division, ratio, functions, variables, equations — and no doubt some others that would be discovered by actual use in the classroom.

What teacher could not do an outstanding job of science and mathematics teaching if she had a supply of such devices for small groups of children? Could most teachers succeed in teaching some of the concepts noted above without materials of this kind? These devices illustrate the logic of simple equipment in the classroom. Properly designed and properly used, such materials transcend the narrow confines of the traditional curriculum, and involve the child and the teacher in an unforgettable sensory experience that is clear and meaningful because it is direct, concrete, and real.

The Need for Special Design in Elementary Grades

With the first shock of Sputnik and the sudden realization that we were neglecting science, funds became available for the purchase of science equipment for elementary schools. It was perhaps natural that science supervisors, usually trained to handle high school equipment, would provide those materials with which they were familiar. As a result, schools bought large amounts of equipment such as precision scales, Centigrade-Fahrenheit thermometers, traditional ring stands and test tube racks, etc. Much of this material is unsuitable for the elementary school because it is costly, bulky, or breaks easily.

The problem of designing appropriate materials for elementary schools is different from that of designing materials for the high schools because the children are different, and so require different kinds of experiences. And the teachers are also different and have different skills and working conditions than in the high schools. In short, materials for the elementary schools must be specifically designed for the purpose and needs at hand.

Thermometers in the lower grades. Let us consider the problem of designing a thermometer for the lower grades. The traditional Fahrenheit thermometer may be used at grades 3 and up, but it is relatively costly and subject to breakage. Furthermore, thermometers for the elementary grades don't have to be as precise as those designed for high school laboratory use.

Suppose we use a simple thermometer with an aluminum back that will not corrode and can therefore be immersed in water. We can use a double thickness of glass, which responds more slowly to changes in temperature than does a single thickness but is less subject to breakage. Although such a thermometer is less precise than conventional thermometers, it is much less costly. The children can then have a direct experience with measurement of temperature rather than depending upon second-hand sources.

Can we extend experiences with thermometers down to grade 1 or even kindergarten? We can, if we design a special scale — in this case a picture scale of temperatures. Alongside 10°F we might show an Eskimo; for 30°F, a child skating; for 90°F, a child in a bathing suit; for 110°F, a hot sun. The first-grade pupil can then use the thermometer with temperatures shown by pictures, as well as by numbers. It is then a natural step to shift to the numbered scale as soon as the class reaches understanding of numbers beyond 10. In fact, the teacher can make excellent use of the set of thermometers to help develop concepts of 10, 20, 30, etc. In effect, the thermometer is a number line that can be used for a wide variety of arithmetical exercises.

The static electricity materials mentioned previously also illustrate the problem of special design for elementary schools. The use of plastic strips of different color to illustrate that like charges repel and unlike charges attract greatly facilitates conceptual development. Without such difference in colors the children would tend to confuse the strips and thereby confuse the concept.

Similarly, the mirrors mentioned previously should be specially designed for elementary schools. Traditional mirrors made of glass, while smooth and highly reflective, might break, and teachers would hesitate to use them with little children. The thickness of traditional glass mirrors is also an obstacle for producing multiple images when two mirrors are hinged. Inexpensive metal mirrors would be more suitable. They do not break and they are thin enough to be hinged properly with gummed tape. In addition, they may be distorted with moderate force to produce the "funny-face" type of images produced by concave and convex mirrors.

While such mirrors do not have the clarity of image that a glass mirror produces, they do produce an image that is of sufficient quality for elementary school purposes. Will the children cut themselves on the corners or edges of the metal? The mirrors can be stamped with round corners, and the dies can be made in such a way that the edges are not sharp.

Improving the flower pot. As a final illustration of this point about special design for elementary schools, consider the traditional flower pot. It's fine for growing plants, but school children view it as a mysterious "black box," inside of which things happen and suddenly burst forth into visibility. But what's going on inside? Any child would love to know. A plastic bag with holes punched in the bottom can serve as an effective "see-through" planter, but special durable plastic cups are now available for this purpose. Such a transparent cup mounted on a tumbler which contains the water supply, easily reveals the way the young plant grows underground (see below). The roots grow downward, through the holes at the bottom of the cup, into the water in the tumbler below, and eventually fill the space below the cup.

Do we wish to observe the effect of depriving the plant of minerals? Just lift out the plant, pour out the water (or mineral solution) and replace it with any desired solution. Does the child have difficulty deciding how much water to add each day? Does he over-water the plant? This problem is easily solved with the "planter-cup." The child is instructed to simply keep the level of the water in the tumbler just up to the bottom of the cup. The water then soaks up properly through the holes into the vermiculite (or soil) above and ensures proper watering.

Do we want to show how the root finds a water supply? Pour out some water as soon as a root penetrates one of the holes in the cup. The root will grow downward into the water and soak up an adequate supply, even though the vermiculite becomes dry. Does the child wish to cut off a root and study it with a magnifier or microscope? He lifts out the planter-cup, cuts off a root with scissors, and replaces the intact plant. Does he wish to mark a root tip with India ink and see how it grows? The cup is lifted up, a root is marked, and back it goes without injury to the plant.

The planter-cup is also an effective filter, which can be used to show how to purify debris-laden water. As shown in the drawing below, a piece of cotton stuffed into the bottom filters out the debris.

Organization of the Materials

Handling and storing materials, especially in quantities sufficient for groups of children, is normally quite time consuming. Many schools are not fortunate enough to have laboratory assistants available to prepare and store equipment. The solution to this problem implies a kit or "laboratory" in which the materials are organized in a storage box. However, in view of the good and bad experiences many schools have had with such kits, it may be worthwhile reviewing briefly some of the characteristics of a good "laboratory." Here are some points to consider:

1) Are the materials provided organized around an appropriate conceptual scheme?

2) Is the laboratory equipped with a clearly written teacher's manual that indicates where the materials can be found for each activity, and describes in some detail how the activity is to be performed?

3) Are the activities closely related to the activities described in the classroom textbooks, so that the teacher can fit the laboratory investigations into her curriculum?

4) Is the outer container sturdy enough to withstand daily use? Will it break if dropped by accident? Will it cave in if accidentally bumped?

5) Are the inner storage compartments adequate? Are the parts jumbled into a mass and hard to find? Are fragile items protected in some way?

6) Is there an overall index of parts which enables the teacher to return any part to its proper place in the box?

7) Does every part have an assigned place?

8) Is the location of every part designated in such a way that it can be found in a few seconds?

9) Is there adequate provision for activities by groups of children, rather than for one teacher demonstration?

10) Are all materials readily replaceable so that the laboratory is not rendered useless by normal wear and tear?

11) Are consumable materials provided in sufficient quantity so that replacement occurs rarely, perhaps once in about 10 uses?

Educational Technology

The dramatic aura of esoteric electronic gadgetry hangs over the word "technology," which makes us automatically think of computers, talking typewriters, television in the classroom, and the like.

No doubt these devices are destined to play an educational role, but how deep that role will be remains to be determined. We are just beginning to put educational equipment for teaching concepts in the elementary schools on the same footing as the mass production of refrigerators, typewriters, pencils, and automobiles. This is the area in which the greatest need exists, and this is the area in which educational designers will make the greatest progress in the next decade.

An introduction to sequence: elementary school mathematics and science enrichment

NATHAN AINSWORTH

The nature sequence

Around the beginning of the thirteenth century, Leonardo Fibonacci defined a certain mathematical sequence that proved to have many interesting connections to nature.[1] His sequence has been coined, "The Nature Sequence." Fibonacci's Sequence can be written by starting with two 1's; the next number of the sequence can be formed by adding the two 1's together to get two (2); the next number of the sequence can be found by adding the last two numbers together, which would be $1 + 2 = 3$; other numbers of the sequence can be found in a similar fashion, that is, by adding the last two numbers together. The first few terms of the Fibonacci Sequence are: 1,1,2,3,5,8,13,21,34,55,89, and so on. Let us now look at some of the interesting occurrences of the Fibonacci Sequence.

THE SUNFLOWER

During the months of July and August, the sunflower is very pretty with its bright yellow face. When the fall months come, the flower petals die, the stamens fall off,

FIG. 1. The seeds of the sunflower are arranged with 34 counterclockwise spirals and 55 clockwise spirals.

and the seeds of the sunflower are exposed. To most people, the sunflower would now be quite unattractive, but with close examination, the beautiful and mysterious arrangement of the sunflower seeds is revealed. When we study the seed arrangement, we notice that the seeds are arranged in such a way that spiral patterns are formed—one spiral pattern proceeding in the clockwise direction and one spiral pattern proceeding in the counterclockwise direction. This is an amazing thing that nature has done, but the most mysterious thing is that no matter how large or small the sunflower may be, the number of individual clockwise spirals and the number of individual counterclockwise spirals should always be two consecutive numbers of the Fibonacci Sequence. Experiments that the author has carried out have shown a spiral ratio of 55 clockwise and 34 coun-

1. A sequence is a list of numbers, where any one number is related, by some mathematical way, to the numbers before and after that number.

Reprinted from *The Arithmetic Teacher* (February 1970), 143–145, by permission of the author and the publisher.

terclockwise spirals, and in a large sunflower, a spiral ratio of 144 clockwise and
89 counterclockwise spirals. (Figure 1 will
help in understanding the spiral arrangements.)

CONES

Many other occurences in nature have
spiral arrangements. One example is in the
cones of any evergreen tree. In a cone,
whether a pine cone, hemlock cone, spruce
cone, or other kinds of pine, there will be a
certain number of counterclockwise spirals
and a certain number of clockwise spirals.
For example, the cone of the white pine
tree (fig. 2) has 5 clockwise spirals and
8 counterclockwise spirals. If we compare
these two numbers with the Fibonacci Sequence, we can see that 5 and 8 are two
consecutive numbers in the sequence.

FIG. 2. White pine cone has 5 clockwise spirals
and 8 counterclockwise spirals. The cross hatch-
ing shows 1 clockwise spiral. Two adjacent numbers of Fibonacci's sequence are 5 and 8.

On the cone of the Loblolly Pine,
(Southern Pine) there are 8 counterclockwise and 13 clockwise spirals. The numbers
8 and 13 are again two consecutive numbers of the Fibonacci Sequence.

OTHER THINGS IN NATURE

The sections of a pineapple (fig. 3) are
arranged in this spiral fashion, having an
8 and 13 spiral arrangement. Most daisies
have a 21 and 34 spiral arrangement, again

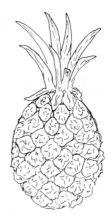

FIG. 3. Look at this pineapple carefully. Notice
the clockwise spirals. The pineapple has an 8 and
a 13 spiral arrangement.

two consecutive numbers of the Fibonacci
Sequence. Also, the leaves or buds on a
twig make their way down the twig in a
spiral fashion. The distance between any
two leaves is equal to the sum of the distance between the previous two leaves. This
is the exact way in which the Fibonacci
Sequence is constructed (see fig. 4). Many
twigs will not show this spiral pattern too
clearly. For best results, look on the new
growth of some bushes that may be near
your home. Bushes seem to show the spiral
pattern and the spacing better than twigs
from a tree.

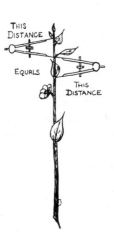

FIG. 4. Some twigs from bushes will show the
spiral descent of the leaves and buds and will
show the most important spacing of the leaves.

Another example of this mysterious occurrence is on the cap of an acorn. The spiral rays are rather difficult to count, but nevertheless, they are usually arranged in 21 counterclockwise spirals and 13 clockwise spirals. (See fig. 5.)

FIG. 5. The acorn has 13 clockwise and 21 counterclockwise spirals. These numbers are two adjacent numbers in the Fibonacci sequence.

The next time you are around trees and flowers, try to notice any spiral patterns that may appear. If you should find some, take several specimens and examine each carefully and count all the spiral rays that go both clockwise and counterclockwise. After recording your findings, compare your results with the Fibonacci Sequence to see if your numbers are, in fact, two consecutive numbers of the Fibonacci Sequence. If you find that your numbers are close to the sequence numbers, then start over and recount the spiral rays.

SHELLS

There are some 80,000 species of shells (snails) or gastropods known today. Gastropods can be identified by their coiling form or their spiral shells. Each of the 80,000 species of shells has a direct connection to the Fibonacci Sequence. The distance between any two spirals is equal to the distance between the next larger spiral distance. For example, if the first spiral distance is 2 cm (centimeters), the second spiral distance is 3 cm, then the third spiral distance, according to the Fibonacci Sequence, would be 5 cm in length. This pattern persists, and the illustrations in figures 6 and 7 show this idea.

For an enrichment program in mathematics and science, the classroom teacher may instruct the class in the structure of the Fibonacci Sequence and then have the class bring in items that may demonstrate either the Fibonacci Sequence or perhaps another sequence. Set up some guideline for experimentation such as a method of counting spirals and a method of measuring distances, using centimeters and millimeters. As a conclusion to this experimentation, a fairly easy and yet excellent display can be made of "The Nature Sequence."

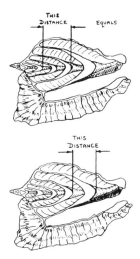

FIG. 6. The shell illustrated is a cross section of a conch shell. This figure shows the spacing between the spiral sections of the shell.

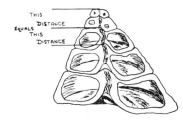

FIG. 7. A cross section of an Indo-Pacific top shell shows that the distance between two spiral units is equal to the next spiral unit. This is the same way the Fibonacci sequence is constructed.

Quantitative Descriptions In Science

ROBERT A. LEMMON

SCIENCE in the primary grades should involve more than learning some of the properties or characteristics of things. Our environment —the universe—is better understood and becomes more meaningful to primary-grade students through measurement and quantitative descriptions. By using certain methods of measuring and collecting data, and then interpreting these data, pupils in the early grades can learn to predict quantitative events. The following two units have been designed to give children opportunities to use scientific procedures in developing quantitative descriptions of their environment.

UNIT I.

Magnetism

The usual study of magnetism in the first or second grade includes a "will and will not" box. The children test many articles and classify them according to whether or not the magnet attracts them.

As an outgrowth of this activity, the teacher can develop the discussion, "How can we measure the strength of a magnet?" The children soon discover they can compare magnets by describing them as a "ten-thumbtack magnet" or a "five-paperclip magnet." This comparison will lead them to ask the question, "Can we tell the strength of a magnet by looking at it?" The develop-

ment of guesses or hypotheses relating that shape, color, size, or weight may serve as a guide in predicting the strength of a magnet, which can be studied by having the children record and interpret data. The pupil's inquiries can produce the type of data shown in Figures 1-4. From such data, one classroom of children drew the following conclusions:

1. The *shape* of magnets does not help us to tell their strength.
2. The *color* of magnets does not help us to tell their strength.
3. The *size* of magnets does not help us to tell their strength.
4. The *weight* of magnets does not help us to tell their strength.

The collecting of original data as well as tabulating and analyzing of the data involved in this activity leads the children through a significant process to a scientifically sound conclusion: The strength of a magnet cannot be measured by its ap-

Figure 1.

SIZE OF MAGNETS

Magnet	Size	Number Of Tacks
1	Largest	21
2	Large	46
3	Middle	6
4	Small	19
5	Smallest	15

Figure 2.

SHAPE OF MAGNETS

Magnet	Shape	Number Of Tacks
1	Horseshoe	10
2	Horseshoe	8
3	Bar	4
4	Bar	8
5	Bar	47
6	Bar	16

Figure 3.

COLOR OF MAGNETS

Magnet	Color	Number Of Tacks
1	Red and Silver	7
2	Black	9
3	Silver	8
4	Silver	19
5	Silver	12
6	Red and Silver	15

Figure 4.

WEIGHT OF MAGNETS

Magnet	Weight	Number Of Tacks
1	Heavy	13
2	Heavy	7
3	Middle	8
4	Middle	19
5	Light	40
6	Light	8
7	Light	15

Figure 5.

pearance. The entire process from the first statement of the problem to the final conclusion is well within the thinking abilities of the primary pupil and is much more challenging than the identification of objects attracted. As an outgrowth of this unit, electromagnetism may be investigated, which will produce some positive results to similar investigations.

UNIT II.

Beam Balance

Construct a beam balance as shown in Figure 5. On a strip of wood, mark equally measured distances from the center. Then obtain several wooden blocks of equal weight to use for the experiments.

The study of balance can be initiated in the lower grades by raising the question: "How can I keep this board balanced?" Discussion will lead to the use of words meaningful to the unit—balance, center support, balance board—and also the use of equivalent weights.

The children can be encouraged to try different arrangements to produce a balance. As they attempt each pattern, have them state what they intend to do and what they expect to occur—their hypothesis or guess. The results of every test should be recorded (whether successful or not!). Every attempt (experiment) is important for complete understanding.

As the collection of "successes" begins to build up, children can be-

gin tabulating the data. The teacher can ask, "How can we measure this arrangement so that we can describe it better?" She can encourage the students to record the data in chart form.

The teacher plays an important role in guiding the children into making worthwhile measurements and recording these measurements on meaningful tables. At times, it may be necessary for her to encourage the rearrangement of the data in order that the children will more readily see the relationships.

All attempts should be recorded on the first chart, but after refining the data only the successes need be used. Identify attempts by recording the number of units of distance away from the center of the board where the stacks of blocks are placed and the number of blocks placed on each side. Figure 6 shows data gathered from a beam balance arrangement like that shown in Figure 5.

From these data, the children can generalize that it is necessary to have equivalent situations on each side (just as in a conditional equation in arithmetic, the two sides must be equivalent):

1. If we have more blocks on Side 2, we will need more distance units on Side 1.

2. When the number of blocks on Side 1 is one and the units on Sides 1 and 2 are equal, then the number of blocks on Side 2 is one.

If the generalizations and interpretations of the data are carefully made they will have predictive value. The children, therefore, should be encouraged to use these generalizations to make predictions.

Conclusions

The purpose of each of these teaching units is to help children discover relationships about the phenomena for which they have collected data. In each lesson, the children have been encouraged to develop guesses (hypotheses) about what they expect to happen; to establish a situation (experiment) to test their guesses; and, as a result of these guesses and the ensuing situations, to collect original data and classify them into tables. After studying and understanding the material in the tables, the children can form generalizations with predictive value. Classroom activities such as these also give children practice in the use of scientific procedures by involving the quantitative measurement of events and the interpretation of data.

Figure 6.

BEAM BALANCE EXPERIMENT

Experimenter	SIDE 1 OF BOARD		SIDE 2 OF BOARD		Results
	Distance In Units	Number Of Blocks	Distance In Units	Number Of Blocks	
Lee	1	1	1	1	Balanced
Elliot	1	1	1	2	Not Balanced
Diane	1	2	1	2	Balanced
Bob	1	3	1	1	Not Balanced

The Slope of the Land

CLIFFORD E. KNAPP

CHILDREN are well acquainted with the hills, valleys, and slopes on the surface of the earth. They know that when it rains, water flows from a higher place to a lower one; a ball that is placed in one spot may roll to another, even though it has not been touched; and pumping a bicycle uphill is much more tiring than pedaling downhill.

Many students are aware that the slope of the land is an important concern to engineers, architects, surveyors, and farmers who must take into consideration factors such as the amount of water that flows downhill in an area, or the amount of soil erosion that takes place from the water flow. Many roads have been constructed higher in the middle than on the sides for drainage and safety in wet weather.

Several interesting observation and measuring activities can be developed in the classroom, and from a short field trip around the school grounds, by having students make their own examinations of the slope of the land with a very simple homemade device called a "sighting level." Students can make many of the observations and measurements that the engineers, architects, surveyors, and builders had to make before they could construct a school, home, or some other building of interest.

Materials:

6-inch piece of cardboard tubing
small glass or plastic pill container
small piece of mirror
masking tape
thread
4 buttons
rubber cement
paper clip
string

Construction:

1. Fill the pill container almost full with water. Seal the cap in place with glue so that the water will not escape. When the water-filled container is tipped, an air space (bubble) will be visible. The bubble will move about freely as the container is moved. See Figure 1.

2. Place the container on a level surface and indicate the position of the bubble with two lines, one on either side of the bubble.

3. Cut a flap in the middle of the 6-inch tube wide enough to hold the piece of mirror which is being used. The sizes of the flap and mirror will vary with the sizes of the other materials used. Glue the mirror to the underside of the flap.

4. Bend and tape a paper clip to the back of the flap so that the position of the mirror can be adjusted.

5. Attach pieces of thread through each end of the tubing so that they are parallel. These threads will serve as sighting lines. The buttons can be used to secure the threads in

position. The instrument is now ready for use.

Operation:

Two or three students can work together and use the sighting level to determine how much higher one place is than another. See Figure 2. One student will need to prepare a stick that is as long as the distance from his eye level to the ground when he is standing up straight. Another pupil can prepare a 100-inch piece of string.* (He should allow a few extra inches of string for attachment to the stick.) The pupil can make sure that the string is perfectly level, by attaching a carpenter's string level or a homemade string level to it. See Figure 3.

Finding an area to measure is the next assignment. In an area where there is a slope for about 10 feet, the pupil with the measured

---

* A 100-inch length of string was chosen because slope is measured in percent. By using 100 units, the number of inches measured on the stick will be equal to the percentage of slope; i.e., if the measurement is 25 inches, the area has a 25 percent slope. However, for finding differences in height, any length string may be used.

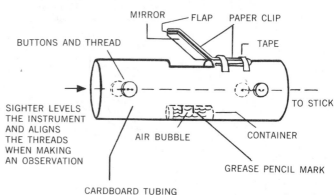

FIGURE 1: The Sighting Level.

MIRROR — FLAP — PAPER CLIP
BUTTONS AND THREAD — TAPE
SIGHTER LEVELS THE INSTRUMENT AND ALIGNS THE THREADS WHEN MAKING AN OBSERVATION
TO STICK
AIR BUBBLE — CONTAINER
GREASE PENCIL MARK
CARDBOARD TUBING

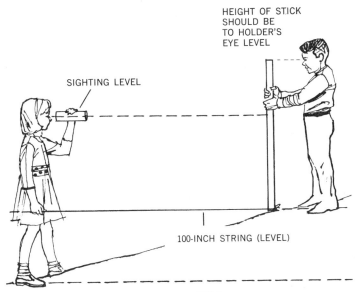

HEIGHT OF STICK
SHOULD BE
TO HOLDER'S
EYE LEVEL

SIGHTING LEVEL

100-INCH STRING (LEVEL)

FIGURE 2.

stick should position himself and the stick up the slope from the position of the pupil acting as the sighter. The stick (with the string tied to the bottom end of it) should be placed straight up and down. Then the string should be stretched out level from the stick to the sighter 100 inches away.

While looking through the sighting level, the sighter should instruct the stick holder to move a finger up or down on the stick until the finger lines up with the two strings inside the sighting level. When the proper spot is found, the position is marked

on the stick. Measurement from this point to the top of the stick will indicate how many inches higher the ground is at the stick holder's position than the ground at the sighter's position. This measurement in inches is also a measure of the slope of that area. Comparison measurements can be taken of several sites in the same area.

Additional Activities

1. On bare hilly ground, have pupils compare the amount of erosion on various slopes. Can they

find more soil erosion on a steeper slope?

2. Students can compare the speed of water flowing over different slopes to see how slope affects the speed. Will swiftly flowing water carry more soil?

3. With a sighting level, pupils can make a simple contour map of a hill.

4. Students can mark a contour line on a hillside by standing in one place and taking a few sightings along the hillside. They can mark each position with a stake and connect the stakes with a line to mark the contour.

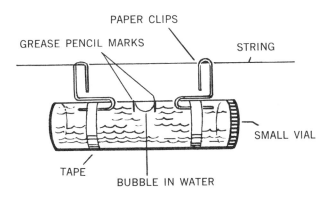

PAPER CLIPS

GREASE PENCIL MARKS

STRING

SMALL VIAL

TAPE

BUBBLE IN WATER

FIGURE 3.

331

Blast-off mathematics

ELTON E. BEOUGHER

A more appropriate and informative title for this discussion might be "Astronomy, Space Travel, and Arithmetic." An attempt will be made here to support the claim that the study of arithmetic and of topics from space science can be coordinated in the elementary school program. Further, it is claimed that this coordination can enhance both the mathematics and science areas and that each can promote learning in the other. The author hopes to provide some ideas that teachers might use to give variety to their lessons.

A number of specific gains can be made through this endeavor. A few that will be illustrated are:

1. Practice using large numbers
2. Practice in estimation and rounding of numbers
3. Provision of numerous examples of the use of ratios
4. Development of skill in measuring and using angles
5. Provision of examples of scale models and development of skill in building such models
6. Provision of examples of many geometric figures, both plane and solid
7. Practice in use of the function concept, that is, relationships between varying quantities
8. Development of a better understanding of our system of time measurement
9. Drill in the fundamental arithmetic operations
10. Provision of many examples for graphing
11. (Trite as it may sound, but still valid) preparation of students to be intelligent citizens of the space age

The reader will note that the advantages claimed generally deal with mathematics understandings and skills. This reflects the fact that the author's main interest is in mathematics education.

At this point the reader may say, "Stop! Put the countdown on HOLD. These claims could be made for many areas of science or life, and such advantages could accrue in many other areas." This is correct. However, there are distinct advantages in favor of the approach through astronomy and space travel.

One advantage is that students already have words in their vocabulary to prepare them for this study: g-force, weightlessness, LM, satellite, zero g, orbital path, apogee and perigee, solar orbit, extraterrestrial, and so on. (This may not be true of the teacher's vocabulary!)

A second advantage is that a great deal of motivation is already present for this study. These are truly children of the space age with whom we deal. Space travel is as real and acceptable to them as a drive to the next town. If the reader doubts the motivation possibilities, he should take a look at the Saturday morning and after-school shows on television which cater to children. Science fiction and space travel have long been the strong suit of these television hours. Consider as added evidence the attraction of children to space toys and games.

A third, major advantage is that there are a multitude of aids, booklets, and films available on the subjects of astronomy and space travel, many of which are free in class quantities. At the end of this article is a short, annotated list of such resource materials and sources of aids.

Reprinted from *The Arithmetic Teacher* (April 1971), 215–221, by permission of the author and the publisher.

Specific topics

In order to add credence to the previous claims, some specific examples of how this study might be accomplished in the elementary school are in order. A more extensive listing of topics and projects is appended at the end of the article. Each topic might be one day's lesson, or, if enough interest developed, could be expanded into a unit.

1. *Connection between astronomy and time measurement.* Few people realize the strong connection between our position in space and our measurement of time. A historical look at time measurement might be revealing in this aspect. It is recorded fact that ancient peoples told time by observance of heavenly bodies—the sun, moon, and stars. Our present-day units of time—years, months, days, hours, minutes, and seconds—all developed out of these early methods. To show the arbitrariness of our time system and its structure, a study of the length of "days" and "years," and possibly "months" (if they had moons), of other planets would be very revealing. For example, Mercury's "day" is about 59 Earth days and its "year" is about 88 Earth days. In view of these facts, a lively discussion could ensue from the question, "If we lived on Mercury, how could we measure time?"

Pupils of early and middle elementary grades would be fascinated by the calculation of their ages in Mercury "years" or Jupiter "years." They would be surprised to find they were less than one Jupiter "year" old. (Jupiter's "year" is about 12 Earth years.)

2. *Study of astronomical navigation.* There are two facets of this—earthbound navigation and navigation in space. A teacher and his students could study the relationship between earthbound navigation and the angle of elevation of the sun at a given spot on the earth. The measuring of this angle could be accomplished by using a simple instrument whose construc-

tion is discussed later. The study of the connection between this angle and the ideas of longitude and latitude could then lead to an interesting study of navigation.

Navigation in space is seen to be a much different and more complex task. This results from at least three facts. No reference system such as latitude and longitude on earth exists in space. On earth, points on land and in the ocean stay fixed. The poles and equator do not move relative to the earth. In space, no such fixed points exist. Any reference system is constantly moving, since all bodies in space are in constant motion. Secondly, two numbers suffice to fix a position on the earth (latitude and longitude), but in space three or more numbers are needed. Thus space navigation necessarily would be based in a three-dimensional geometry (or maybe **four**-dimensional, if time needs to be **considered**) rather than in a two-dimensional **one**. This is a fascinating study in itself. A third inherent difficulty of space navigation is that much more planning and plotting of the course must be done before the flight begins. This was evident in the **recent** moon flights. Only minor corrections **can** be made after such a trip is in progress. Essentially, the flight must be completely preplanned down to the most minute detail.

A study of navigation with your students could **lead** to fruitful results for later work in **algebra**, since there is a strong connection between navigation and graphing and graphing plays a vital role in algebra.

3. *Measuring distance in space.* The distances between natural objects in space are so great that ordinary units of measure and notation for numbers become unwieldy. For example, the distance to the star nearest to us (besides the sun) is approximately 25,800,000,000,000 miles. Generally, astronomers and other scientists write such large distances in scientific notation. For example, the number mentioned would be written as 2.58×10^{13} miles in scientific notation. Another example would be the

333

mean distance of the planet Pluto from the sun, 3.67×10^9 miles. This would be 3,670,000,000 miles in usual notation. (Note the relationship between the power of 10 and the number of places that the decimal is moved in changing from one notation to the other.) Astronomers also use another unit of measurement in naming such large distances. This is the light-year. One light-year is the distance that light travels in one year at the fantastic rate of 186,000 miles per second (approximately). To provide an idea of the enormity of this, there are 86,400 seconds in one day. In each of those seconds light would travel 186,000 miles. Think now of the 365¼ days per year to get an idea of the size of a light-year. To place it in the familiar framework, this is approximately 6,000,000,-000,000 miles. In this system, the closest star is 4.3 light-years away. (This means it takes 4.3 years for light to reach us from that star.) As mentioned before, this is 25,800,000,000,000, or 2.58×10^{13}, miles.

Another fascinating study is that of methods of measuring distances in space. There are a number of sources on this subject of "indirect" measurement, and many of these are within the grasp of elementary school students. Later we will discuss the applications of some of these methods which your students could make.

4. *Variation of gravity on the planets.* A fruitful study of ratios could accrue from this topic. Pupils would be highly motivated to find what their weight would be on other planets and the moon. They would be surprised to learn that if they were fairly good high jumpers, on the moon they could probably jump over their family's car with ease. Furthermore, they could probably pick up their own mother and carry her around, assuming she weighed about 100 pounds. This is because the moon's gravity is ⅙ the earth's, and objects would weigh ⅙ their earth weight on the moon. Your students could find the weight of familiar objects and multiply by ⅙ to find moon weight so as to test their "moon strength."

However, your students would be astonished to learn that on Jupiter they might not even be able to lift a medium-sized dog, since the gravity of Jupiter is about 2½ times that of Earth. The gravity of the sun is about 28 times that of the earth. In such a situation, one of your students could not even lift his own arithmetic book! The possibilities for practice in multiplication and in the use of ratios are almost limitless.

5. *Shapes of orbits.* Most of your students would be aware from diagrams on television and in the newspaper that orbiting objects, whether man-made satellites or planets, do not travel in perfect circles. The path is an ellipse. You (or your students) can easily construct an ellipse by using two pins stuck into a board, a pencil, and a piece of string with its ends tied together to form a closed loop (see fig. 1).

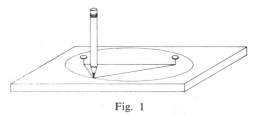

Fig. 1

The **pencil** point is placed within the closed loop **and** then the loop is pulled taut by the point. A closed figure is then drawn on the board by **moving** the pencil around, keeping the string taut. Viewed from above, the **result** is as in figure 2. The closed figure **drawn is** the ellipse.

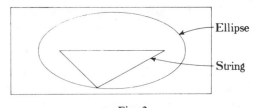

Fig. 2

The two points where the pins are placed are called foci (singular, focus). When a planet orbits the sun, the sun is at one of these foci. In general, when a small body orbits a much larger one under gravitational influence, the large body is at a focus of the ellipse. There are precise mathematical laws governing this motion which would be out of reach of most students' minds, even at the upper levels of elementary school. However, the motion paths are simple enough to understand.

The other topics on the list at the end of the article offer many possibilities for study. With a little bit of thought and preparation and some reading on the part of the teacher, each topic could be developed into an interesting lesson for students. What should be kept in mind is that the purpose of the lessons is to use the subject matter of astronomy and space travel to illuminate and expand arithmetic lessons and to motivate students in arithmetic. The sources listed at the end of the article would prove valuable in this light. Some of these would be useful strictly for their content of facts. Others list aids that may serve to make the teacher's task more appealing both to him and to his students.

Models and instruments

One possibility for an interesting activity is the construction of simple measuring instruments and devices for demonstrating various concepts. One such device works nicely to illustrate number 18 on the list of topics. The device is made using a spool, two weights (large fishing weights from the sporting-goods store will do nicely—experiment to find best sizes), and a piece of string about 3½ feet long. Assemble these materials as in figure 3. Weight 1 is whirled around in a circular path by grasping the spool. Weight 2 will hold weight 1 in a path of a certain radius depending on the relative weights of 1 and 2. Weight 2 acts as the force of gravity does. If we increase the "force" of gravity by increasing the weight of 2, then we note that weight 1 "orbits" closer to the spool. Experiments will show that it will also orbit faster in this new position. Planets in orbit around the sun behave in this way. The planets close to the sun—Mercury, Venus, Earth, and

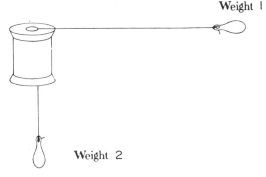

Weight 1

Weight 2

Fig. 3

Mars—move faster in their orbits than the ones farther out—Jupiter, Saturn, Uranus, Neptune, and Pluto. This is because gravity is stronger close to the sun than it is farther from the sun.

In addition to the mental practice to be gained from these studies are the skills to be learned in planning and carrying out projects involving mathematics and space science. Some such projects are suggested in the list at the conclusion of the article. This is only a small sample of the possibilities.

There is much to be gained from the building of scale models. The topics of ratio, multiplication, division, and scales are just a few concepts that are involved. Also, pride in the completed project can contribute much to its value as an educational tool. Depending on the age of the child, the teacher might have to offer more or less guidance in the construction.

Building of a scale model of the solar system

A word of caution should be offered here. It would be impossible to construct a scale model of the solar system all on one scale, that is, with the size of planets scaled the same as the interplanetary distances, and have it fit in a classroom. If this were attempted, in order for Mercury to be large enough to see (say ⅛ inch in diameter) the model of Pluto would have to be about 25 miles away from the model of the sun. (This gives some idea of astronomical distances). So unless you want your "scale" model of the solar system to be 50 miles across, you'd better use two separate scales! A convenient choice might be ⅛ inch = 1,000 miles for the size of planets and 1 inch = 100,000,000 miles for the size of the solar system. This would make the solar system model about 6 feet across. A model of the sun could not be made on the same scale as the planets and have it fit in your model solar system (it would be necessary to make it 9 feet across). Thus it would be best just to indicate its position at the center of the solar system and not have a scale model of it in your solar system.

Projects involving graphing

Projects that could certainly involve graphing are numbers 3, 5, and 8 on list B. An easily made instrument could be used for measuring the angles in numbers 3 and 8. The instrument is constructed by using a nail, a piece of string, a small weight, and a piece of scrap wood about a foot square (see fig. 4). The nail is placed at the center of the arc on which the degree scale is indicated (i.e., at the center of the circle of which the arc is a part). The weight and string are attached to the nail and are used as a plumb line. In use, the instrument is held so that the string passes through the 90-degree mark on the scale.

To measure the angle of elevation of the sun, the device is held plumb so that the nail casts a shadow down across the scale, and the angle is read from where the shadow intersects the scale. The angle measured in figure 4 is approximately 35 degrees. In carrying out number 3 of list B, it would be important to measure the angle at the same time every day (preferably at noon) and to face the instrument in the same direction. A record of these measurements over a length of time would contribute to an understanding by students of the relationship between the changing angle of the sun's rays striking the earth and the change of seasons. A point of caution that should be stressed to your students is that they should **never look directly at the sun** in carrying out any of these measurements.

One source indicated that nine-year-old students carried out project 6 with use of a kit available for less than two dollars. The telescope was powerful enough for one to see many details of the moon's surface and possibly to observe four of Jupiter's moons. They could be seen with the naked eye were they not so close to Jupiter as to be blotted out by its bright light.

A number of the sources listed under C describe how to construct sundials of various types. This would be instructive again in relation to our system of time measurement as well as valuable as a project to be carefully completed.

Indirect measurement and triangulation

To get a feel for the methods of indirect measurement used by astronomers, students could carry out "earthbound" measurements indirectly. Either trigonometry or ratio and proportion could be applied. Use of the latter would involve measuring the length of the shadow cast by the sun of an object of known height (a student) and, at the same time, measuring the shadow of an object of unknown height (a tree). These two are then compared as follows:

$$\frac{\text{boy's height}}{\text{boy's shadow length}} = \frac{\text{tree's height}}{\text{tree's shadow length}}$$

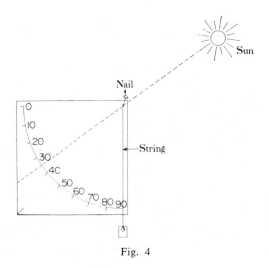

Fig. 4

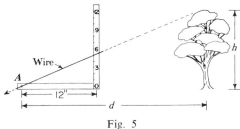

Fig. 5

The use of ratio is obvious. This is somewhat simpler than the method astronomers use, but the basic principle is the same—"triangulation."

An easily constructed instrument for finding measurements indirectly is based on the principles of triangulation. Materials needed are two pieces of scrap lumber about two inches wide and one foot long. These are fastened at right angles and a one-foot scale is drawn on one as in figure 5. A sighting wire is attached at A by use of a small bolt or screw so that the wire can be rotated around A. To use the instrument, one sights along the wire at the top of the object whose height, h, is to be measured (fig. 5). Care must be taken so that the lower edge of the horizontal board is kept horizontal (a plumb line along the vertical board would facilitate this). The distance to the object, d, is then measured and a ratio is set up as follows:

$$\frac{\text{reading on scale}}{12} = \frac{h}{d}$$

Again the use of ratios is obvious. Pupils in the later grades would have had experience with ratios, proportion, and similar triangles and could easily comprehend the mathematics of the instrument.

Conclusion

The versatile teacher can make many uses of the instruments described previously. Furthermore, the topics, projects, and sources given in the appended lists would provide a wealth of material for the same teacher.

A. Some suggested topics for study
1. Connection between astronomy and time measurement
2. Navigation and astronomy
3. Units of measure and methods of measuring distance in space
4. Variation of gravity on planets
5. Relative size of planets
6. Relative distance of planets from the sun
7. Length of "years" of planets
8. Shapes of orbits of natural and artificial satellites
9. Number of moons of planets as related to size of planet
10. Weightlessness and orbiting
11. g-forces
12. Shapes of spaceships and reasons for these shapes
13. Size of rockets as compared to man
14. Size of payloads compared to size of rockets
15. How rockets work
16. Decrease in pull of gravity as distance from earth increases
17. Why planets are spherical
18. Speed of orbiting object as related to distance from object it is orbiting

B. Some suggested projects
1. Construct scale models of the sun and the planets.
2. Construct a scale model of the solar system.
3. Keep a record of change of the angle of elevation of the sun over time and study the use of such a record by ancients to determine seasons.
4. Observe and time an eclipse of the moon.
5. Keep a record of the change in the length of days with the seasons.
6. Build a simple telescope and calculate its power.
7. Observe the moon's craters and compare their size with the size of land masses on the earth.
8. Determine the relationship between the angle of the tilt of the earth's axis with the plane of its orbit and the seasons.
9. Construct a sundial.
10. Measure distances and heights indirectly by triangulation.
11. Taking the conditions of amount of light, temperature, gravity, and so on, on a given planet, determine the physical characteristics of beings that could live there. Perhaps even write a story about their lives.

C. Sources of content, ideas, and aids
Anthony, James K. "Events That Led to the Discovery of Pluto." *School Science and Mathematics* 53 (January 1953): 316–18.

Arey, Charles. *Science Experiences for Elementary Schools.* New York: Bureau of Publications, Teachers College, Columbia University, 1961.

Bernardo, James V. "The Space Age: Its Impact on Education." *School Science and Mathematics* 63 (January 1963): 5–19.

A short history of the space age and NASA. Discusses the role of NASA in elementary and secondary education and details the aids and services available to educators through NASA.

Blough, Glenn O., and Marjorie H. Campbell. *Making and Using Classroom Science Materials*. New York: Dryden Press, 1954.

The Book of Popular Science. New York: Grolier.

Deason, Hillary J., and Ruth N. Fey. *Science Book List for Children*. Washington, D.C.: American Association for the Advancement of Science, 1961.

Gleason, Walter P. "Real Astronomy." *School Science and Mathematics* 54 (January 1954): 31–38.

Discusses construction and use of some easily made instruments that demonstrate astronomical concepts and principles: (1) instruments for measuring the altitude of the sun, (2) sizes for a scale model of the solar system, (3) a sundial, and (4) a model to demonstrate how gravity keeps a satellite in orbit.

Johnson, LeRoy D. "A Physical Science Demonstration: The Planets." *School Science and Mathematics* 53 (October 1953): 569.

Jones, Louise M. "A Trip to the Moon." *School Science and Mathematics* 60 (June 1951): 486–87.

Discusses an excellent and promising interdisciplinary idea: Have your class plan a trip to the moon—planning food, clothing, jobs they would take in transit, and so on.

Kambly, Paul, and Winifred Iadley. "Elementary School Science Library for 1961," *School Science and Mathematics* 62 (June 1962): 419–38. "Elementary School Science Library for 1962," *School Science and Mathematics* 63 (May 1963): 387–414. "Elementary School Science Library for 1963," *School Science and Mathematics* 64 (June 1964): 467–82.

Good selection of astronomy, mathematics, and general science books and books on rockets, jets, and space travel. Each book is annotated and the grades 1–6 are given for which the material would be most appropriate.

Mallison, George C., and Jacqueline V. Buck. *A Bibliography of Reference Books for Elementary Science*. Washington, D.C.: National Education Association, National Science Teachers Association, 1960.

National Aerospace Education Council. *Aerospace Bibliography*. Washington, D.C.: Government Printing Office, 1970.

———. *Free and Inexpensive Pictures, Pamphlets and Packets for Air/Space Education*. Washington, D.C.: National Aerospace Education Council.

A list of many free and inexpensive materials, including sources from which these may be obtained by teachers.

Nelson, Leslie W., and George C. Lorbeer. *Science Activities for Elementary Children*. Dubuque, Iowa: Wm. C. Brown Co., 1955.

Norris, Theodore R. "An Inverse Square Relationship in Science." ARITHMETIC TEACHER 15 (December 1968): 707–12.

Report of a sixth grader's experience in experimenting and drawing conclusions based on data and observations.

Pettorf, H. Ronald. "Fundamental Ideas of Space Travel." *School Science and Mathematics* 63 (May 1963): 405–10.

Background information for the teacher (to catch you up with your students!).

Ray, William J. "Just for Fun: From Arc to Time and Time to Arc." ARITHMETIC TEACHER 14 (December 1967): 671–73.

Reuter, Kathleen. "Sixth Graders Compose Space Problems." ARITHMETIC TEACHER 11 (March 1964): 201–4.

Report of fruitful results of discussions in class which followed some of the first U.S. orbital flights. A host of student-developed problems based on recorded facts about the flights are given.

Ruchlis, Hy, and Jack Engelhardt. *The Story of Mathematics: Geometry for the Young Scientist*. Irvington-on-Hudson, N.Y.: Harvey House, 1958.

Excellent resource for teacher. Practical problems in many areas, including space travel.

Smith, Bernice. "Sunpaths That Lead to Understanding." ARITHMETIC TEACHER 14 (December 1967): 674–77.

Deals with the problem of helping children understand the summer and winter solstices and the equinoxes.

"Space and Oceanography Bibliography." *Instructor* 79 (January 1970): 68.

A fairly extensive bibliography of audio-visual materials, pamphlets, and books suitable for elementary school students.

Discovering centigrade and Fahrenheit relationships

WILLIAM E. SCHALL

The following approach has been used successfully with elementary school children in developing an understanding of centigrade and Fahrenheit thermometers.

It was first developed with a group of sixth graders when we planned an overnight science probe into the nearby mountains. One group was going to analyze a soil sample, and thermometers were needed for soil temperatures in different locations and at different depths.

The group of children were taken to the science room to check the equipment needed. One boy suddenly said, "Hey, these thermometers aren't the same! Which one do we use?" Others said, "What's the difference?" and "Why are they different?" It was realized that this was a golden opportunity to instigate some purposeful learning. The teacher said, "We have two days before our camping trip; let's see if we can discover any relationships between these two thermometers. Let's start in tomorrow's arithmetic class."

Well, that's how the lesson had its start. (This lesson was developed with a small group of five, but is flexible enough that it could be used with an individual student or as an introduction for an entire class.)

Presentation

A representative image of the two thermometers is placed on a transparency to be used with the overhead projector, or it could just as easily be drawn on the chalkboard or placed on ditto (Fig. 1).

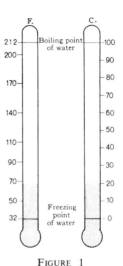

FIGURE 1

Explore. Compare. Look them over carefully. What similarities—what "knowns" —can be found here?

KNOWNS

	F.	C.
Boiling point	212°	100°
Freezing point	32°	0°
	??	50°

If the boiling point and freezing point are known, how can we determine what the Fahrenheit scale will register if the temperature is 50° on the centigrade scale? (Time should be allowed to explore and to try to discover the answer, even if the children only engage in trial-and-error techniques.)

When guidance or assistance is needed, usually the broken arrow line shown in

Reprinted from *The Arithmetic Teacher* (October 1968), 556–559, by permission of the author and the publisher.

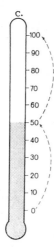

C.

FIGURE 2

Figure 2 is enough to get the group started, and the discussion might proceed as follows:

"Yes, 50° is halfway or midpoint of the *range* on the centigrade scale." (This concept may need to be defined for some of the students.)

"How did we find this?" (Often there is a need to reinforce what has been done.)

$$\begin{array}{r} 100° \text{ C.} \\ - \quad 0° \text{ C.} \\ \hline 100° \text{ C.} \quad \text{Range} \end{array}$$

$\frac{1}{2} \times 100°$ C. $= 50°$ C.   Midpoint

The next step is to have the students transfer this thinking to the Fahrenheit scale. They are asked to find the midpoint on the Fahrenheit scale, and follow the same procedure:

$$\begin{array}{r} 212° \text{ F.} \\ - \quad 32° \text{ F.} \\ \hline 180° \text{ F.} \quad \text{Range} \end{array}$$

$\frac{1}{2} \times 180°$ F. $= 90°$ F.   Midpoint

"Look at our own Fahrenheit scale. Does 90° appear to be the midpoint? No, it does not. What needs to be done? We must add 32° to 90°." (This should be made meaningful to the students. It can be shown simply by moving up the thermometer 90° from 32°, or moving down the thermometer 90° from 212°, thus arriving at the midpoint each time.)

$90°$ F. $+ 32°$ F. $= 122°$ F.   Midpoint

Record this on the transparency or on the chalkboard, as shown in Table 1.

Table 1

C.	0	10	20	30	40	50	60	70	80	90	100
F.	32					122					212

Investigate at this point how the centigrade scale increases. It increases by 10— or each division in Table 1 is a tenth of the range, i.e.,

$$\frac{3}{10} \times 100° = 30°.$$

"Can we find ten corresponding divisions for the Fahrenheit scale? Recall what the range was? 180°. Each portion will be how much? One-tenth."

$$180° \times \frac{1}{10} = 18°.$$
$$32° + 18° = 50°.$$

Now, place this on the transparency or on the chalkboard.

Continue to discover or develop the relationship: as the centigrade scale increases by 10°, the Fahrenheit increases by 18°. The children should see this pattern rather quickly.

Now we can complete both the scales on the transparency, as shown in Table 2.

Table 2

C.	0	10	20	30	40	50	60	70	80	90	100
F.	32	50	68	86	104	122	140	158	176	194	212

Suppose we had a reading of 35° C. How would we find the corresponding Fahrenheit reading? 35° is 5/10 or 1/2 of the interval between 30° and 40°, therefore:

$\frac{1}{2} \times 18° = 9°$

$$\begin{array}{ll} 86° \text{ F.} & \text{Corresponding F. reading for } 30° \text{ C.} \\ +9° & \\ \hline 95° \text{ F.} & \text{Corresponding F. reading for } 35° \text{ C.} \end{array}$$

Many corresponding readings could be developed at this point for additional understanding. If desired, one could attempt to have the children discover the formula for this relationship, although this is not a necessary consequence at this point. One could proceed as follows:

Look at the completed transparency. What observations can you make?

1. Fahrenheit increases 18° for every 10° increase in centigrade (or 1.8° to 1°).

2. We multiplied 18/10 times the centigrade, then added this to 32° to find the corresponding Fahrenheit scale. (Children may need considerable guiding to arrive at this observation.)

Now the traditional formula,

$$F. = 9/5 \ C. + 32,$$

should be meaningful. To reinforce this, one additional step should be helpful. This linear equation can be shown on a graph, again using the overhead projector (Fig. 3).

Conclusion

The development of a topic such as this can be of much more interest and can add more to understanding than the isolated rote memorization of traditionally introduced formulas.

The author does not wish to indicate that this could be developed in one lesson. The length of time for this lesson would certainly depend on the learners and the ease with which they grasp these relationships—the lesson could conceivably require several class periods.

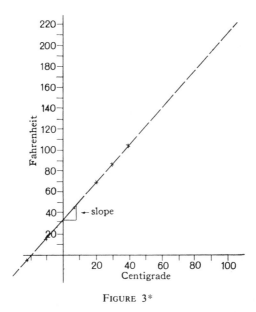

FIGURE 3*

---

*Ward Bouswma, *Algebra for Teachers*, p. 68. Unpublished.

Upper Winds For Upper Grades

VERNE N. ROCKCASTLE

THE study of the winds plays an important part in any science unit on the weather. As part of their study, children in the upper elementary grades can calculate the velocity of the winds using a homemade weather instrument and a simple procedure which require easily-obtainable materials and a minimum of time and energy. This exercise will increase the students' knowledge of measurement and recording skills and lead to a greater understanding of the behavior of the winds.

Materials:

Balloon filled with "lighter-than-air" gas, two protractors, metric ruler, stick ½ x ½ x 12 inches, fish sinker, three carpet tacks, watch with second hand, and spool of white thread.

Procedure:

Tack a protractor to the side of the wooden stick as shown in Figure 1, keeping the straight edge of the protractor parallel to the top of the stick. Hang the fish sinker on a piece of thread from the center of the protractor to serve as a plumb bob. (A soda straw taped to the top of the stick will improve the sighting.) You now have a simple hand transit.

When the plumb bob indicates 90° on the protractor, the transit is horizontal. When the plumb bob indicates 80°, the transit is inclined 10°. An indicated angle on the protractor must be subtracted from 90° to find the inclination of the transit.

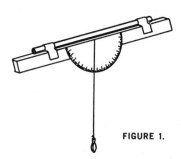

FIGURE 1.

Let a child stand a measured distance (10 or 20 feet) from the wall of the classroom, and from that point find the angle that his line-of-sight to the top of the wall makes with the horizontal. He can do this by finding how many degrees above the horizontal the soda straw had to be elevated in order to sight the top of the wall. On a graph paper, the child should measure off horizontally the number of units equivalent to his distance from the wall. At the end of this horizontal distance, he should copy the angle of elevation indicated by his transit. His scale drawing of his distance from the wall, and his angle of elevation, will indicate the height of the ceiling *above his eye level.*

Suppose a child standing 20 feet from the wall found the top of the wall to be 30° above the horizontal (a protractor reading of 60°): The ceiling would be nearly 12 feet above his eye level. To find the height of the room, add the child's eye-height to the 12 feet. (See Figure 2.)

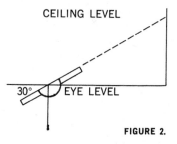

FIGURE 2.

Now tie a long thread to the gas-filled balloon so it can be pulled down if released in the high-ceiling room. Pull the balloon to the floor, release it and time the balloon until it strikes the ceiling. Do this several times to get the average rate of ascent from floor to ceiling. Divide the ceiling height by the time of rise of the balloon to find its rate of ascent.

Now the children can take the balloon out of doors to measure the upper winds. One child should be assigned to each of the following tasks:

1. Keep the balloon in sight through the soda straw.
2. Read the angle of the plumb bob every 30 seconds.
3. Keep time, calling off each 30 seconds to the angle-reader.
4. Write down the elapsed time and the angle of sight at the end of each time interval.

When the data for a few minutes' sighting has been recorded, the position of the balloon at the end of each 30-second interval can be plotted. When the balloon's position at the end of each time interval has been plotted, its horizontal movement can be measured using the same scale for both vertical and horizontal distance.

A group of sixth grade children did this exercise, using a balloon that was found to rise at the rate of 6 feet per second. (It would then be 180 feet high in 30 seconds, 360 feet high in 60 seconds, and 540

Students calculate wind velocity with a homemade weather instrument.

feet high in 90 seconds.) Their figures, and the resulting plotting of the balloon's position are as follows:

Time elapsed	Protractor reading	Inclination of transit	Height of balloon
0 second	90°	0°	0 feet
30	45	45	180
60	60	30	360
90	70	20	540
120	75	15	720

The children, with the help of a supervising adult, used the above figures to plot the position of the released balloon. Figure 3 shows how they diagrammed their data. After the data were plotted, the children were able to measure the horizontal distance between successive positions of the balloon. Knowing that each position represented a time lapse of 30 seconds, they were able to compute the speed of the balloon (average speed) between successive sightings.

An interesting supplementary activity is to tie a self-addressed postal card to the balloon to see how far it really does travel after being lost to sight. One class recently released several balloons in central western New York State. One of the cards was returned from Portsmouth, New Hampshire—about 325 airline miles away!

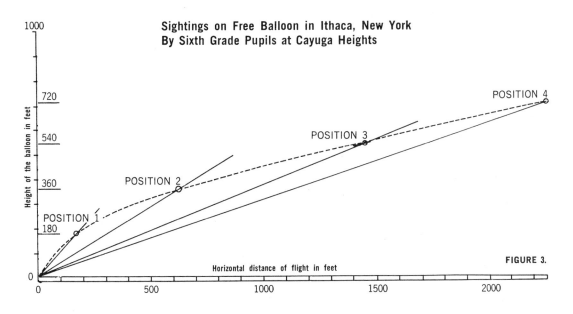

Sightings on Free Balloon in Ithaca, New York
By Sixth Grade Pupils at Cayuga Heights

FIGURE 3.

Height of the balloon in feet

Horizontal distance of flight in feet

SELECTED BIBLIOGRAPHY FOR CHAPTER 13
Correlation of Math-Science Activities

Banks, Harvey W. "Kepler's Laws." *Science and Children* (December 1964), 11–12.

Blanc, Sam S. "Mathematics in Elementary Science." *The Arithmetic Teacher* (December 1967), 636.

Brehm, Shirley A. "Investigations Afield." *Science and Children* (September 1966), 9–10.

Brennan, Matthew. "Science Out of Doors." *Science and Children* (December 1963), 29.

Cordier, Mary Hurlbut. "Let's Take a Field Trip to the Woods." *Science and Children* xmxmxmxmxmxmxmxmxmxmx

"The Fibonacci Rabbits." *New York State Mathematics Teachers' Journal* (April 1971), 59–61.

Gleason, Andrew. "The Interface of Science and Mathematics." *NASSP Bulletin* (April 1968), 118–28.

Hawkins, Robert C. and Bonney, Catharine Y. "An Outdoor Laboratory Program for the Elementary School." *The National Elementary Principal* (February 1968), 20–24.

Hope, John A. and Aikenhead, Glen S. "Theory Into Practice for Elementary Mathematics and Science Methods Students." *School Science and Mathematics* (April 1974), 280–92.

Johnson, Donovan A. "Mathematics Outside the Classroom." *School Science and Mathematics* (February 1974), 129–34.

Joseph, Joseph Maron and Lippincott, Sarah Lee. "Let's Count the Stars!" *Science and Children* (January/February 1969), 40–41.

Kidd, Kenneth P. "A Project Correlating Mathematics and Science." *School Science and Mathematics* (October 1967), 593–97.

Kullman, David E. "Correlation of Mathematics and Science Teaching." *School Science and Mathematics* (October 1966), 645–49.

Kusch, Polykarp. "Mathematics in Science Education." *The Science Teacher* (May 1966), 19–23.

Leonard, Edward H. "A Modified Meter Stick." *Science and Children* (March 1966), 7, 9–10.

Mayor, John R. "Science and Mathematics in the Elementary School." *The Arithmetic Teacher* (December 1967), 629–35.

Morgan, Martha A. and Gilbert, Carol M. "Gnomon Measures." *Science and Children* (December 1964), 23–25.

Newbury, N. F. "Quantitative Aspects of Science at the Primary Age." *The Arithmetic Teacher* (December 1967), 641–44.

Oakley, John S. "Reflections on Integration of Mathematics and Science." *School Science and Mathematics* (April 1974), 293–96.

Rains, Sylvester L. "Can You Prove It?" *Science and Children* (March 1966), 20–21.

Ramsey, Irvin L. "Science and Mathematics in Today's Schools." *School Science and Mathematics* (December 1968), 773–80.

Rettaliata, J. T. "Mathematics and Science: Partners in Progress." *School Science and Mathematics* (March 1964), 173–79.

Rising, Gerald R. "Research and Development in Mathematics and Science Education at the Minnesota School Mathematics and Science Center and the Minnesota National Laboratory." *School Science and Mathematics* (December 1965), 811–20.

——. "Developing Ancillary Skills: Mathematics Deficiencies in the Science Classroom." *The Science Teacher* (November 1967), 29–31.

Ruchlis, Hy. "Multiple Images in Mirrors." *Science and Children* (April 1968), 22–23.

——. "Integrating Science and Mathematics." *Science and Children* (December 1963), 23–25.

Samples, Robert E. "Backyard Universe." *Science and Children* (January/February 1969), 35–39.

Shugrue, Sylvia. "Where Am I?" *Science and Children* (December 1964), 18–20.

CHAPTER 14

OTHER AREAS OF CORRELATION

Have you been looking for a way to integrate the curriculum? This is, after all, a major objective in the current thrust for openness and relevance in school. Spelling words are taken from social studies and science, language arts includes the study of communication in societies, and writing and reading put it all together. This move toward relevancy and integration of the curriculum has the potential for adding vigor to the emerging elementary school experiences of the 70s.

Opportunities for using the mutually supportive nature of the different content and skill areas are enormous. For example, Frayda Cooper's article, "Math As a Second Language," is a must. The reading of mathematics is a natural part of language development. Frayda Cooper makes us aware that children must read word problems, numerals, mathematical symbols, mathematical sentences, directions, and pictorial diagrams, and gives some good practical ideas on how to develop this "language of mathematics."

Jo Phillips's article, "Mathematics: Reading Math Content," fits perfectly with the Cooper article. Every teacher, including math teachers, must pay constant attention to reading.

Evelyn Swartz, in "Interrelationships Between Mathematics and Art for the Kindergarten," stresses the many components common to mathematics and art. Both areas contribute to the conceptual development and visual perception of children. Read this article for an indication of some of the relationships to be found in the objectives, content, and teaching procedures.

" 'Stock-Market' Unit" by Beatrice Thomashow is another good example of the kinds of correlation that are possible. This article directly or indirectly illustrates self-motivation, self-direction, group interaction, the interrelationship of mathematics and other subject areas, creative problem-solving, etc. Be sure to read it; the "stock market" might prove to be an exciting experience for your classroom.

William Ray, in "Just for Fun: From Arc to Time and Time to Arc," presents an interesting article which includes the areas of arithmetic, social studies, and science. Learning about time zones can be a frustrating experience and Mr. Ray suggest ways to help children develop understanding of these ideas.

A field trip in mathematics? Of course! Read Robert Fernhoff's article, "Making the Most of Your Field Trip," and you will understand why. As Fernhoff says, "This is education at work. Learning is fun, not a daily chore. The task is in getting the kids involved; they learn by doing!" So do teachers—try this field trip idea with your class.

MATH AS A SECOND LANGUAGE

FRAYDA F. COOPER

WHEN we associate reading with mathematics, we are usually thinking of word problems. But children must also read numerals, mathematical symbols, mathematical sentences, directions, and pictorial diagrams. *The Random House Dictionary of the English Language* gives, as one definition of language, "any system of formalized symbols, signs, gestures or the like, used or conceived as a means of communicating thought, emotion, and so on." The language of mathematics must be taught from that point of view.

Children enter first grade with several years of oral language experience—the basic tool for learning to read, that is, to understand the meaning of something written or printed. But children's oral mathematics language is very limited. Usually a first grader knows little more than rote counting. Before he can begin reading mathematical language intelligently, he must learn to speak it. Help children build this oral mathematical language by giving them experiences with manipulative and concrete materials and pictures and number lines, along with word descriptions of what they are doing.

Acquiring a basic oral vocabulary in mathematics may well take an entire school year. A child must learn to count to ten meaningfully—matching objects to numbers. He should be able to recognize one object, two objects, three objects, four objects, or five objects without counting each time. He should be able to explain orally the addition and separation of one object to or from five or six objects. Just as reading is talk written down, an equation is mathematical talk written down.

"One and two is three" becomes $1 + 2 = 3$.

"Three is less than four" becomes $3 < 4$.

"Four is greater than three" becomes $4 > 3$.

Substituting a symbol for a missing number is the next stage in this development.

$4 + 1 = \square$ Four and one is what number?

$5 - 2 = \square$ Five take away two is what number?

Encourage students to translate such sentences into words. As they begin to find the missing numbers, many children will still need manipulative materials.

When a child develops a need for larger numbers, place value and reading and writing of larger numbers must be introduced together. Fourth grade is a good time to explain the place value chart extended to the million period, and to assist children in reading numbers as large as a million. Bundles of sticks help stress the meaning of place value.

But as the child becomes more skillful, it's difficult to illustrate quantity with concrete materials. I find it useful to let students make their own place value charts, and then write in the numbers. Next we insert the commas. Show children how they can read each group.

For an interesting activity while learning this concept, I ask students to bring in newspaper or magazine articles using large numbers. After displaying them on the bulletin board, children take turns reading the numbers. Many times I have students use sticks or counters to illustrate the value of a large number. They soon realize they can't show all the sticks, but they can represent bundles of 100, 1000, 10,000, 100,000.

We also make use of number lines to order numbers, locate points, and name points. Consider the numbers between 1,000 and 2,000. (See bottom of page.) Ask students to locate 1,225, 1,675, 1,800, 1,350. Or hang

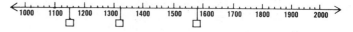

Reprinted with permission from the *Instructor* (October 1971), 76–77, © 1971 by The Instructor Publications, Inc.

lettered tags from different points to be identified.

Use the same techniques to explain rational numbers and integers. When a rational number is introduced, give both the fraction and the decimal name.

A ruler is the best number line. Many now have ten marks to the inch. Again, concrete experiences are necessary. Don't forget that art, physical education, science, and social studies provide good opportunities.

As students consider scale drawings, construct projects, conduct experiments, or set up a game area such as a badminton court, have them carefully describe the process. This should include accurate measurements. Encourage students to write up their descriptions; then make descriptions available to the class.

Even before they can read number symbols, students learn that there are many names for a number. Some names are compound names. Five, for example, can be called one and four, three and two, six take away one, and so on. But children need some of the mathematical symbols to write these sentences. (*Plus,* +; *minus,* —; *is equal to,* =; *is not equal to* ≠; *is less than,* <; and *is greater than,* > appear in many first-grade books.)

First-grade children need to write number stories the same way they write experience stories. As the occasion arises, it's even a good idea to incorporate numbers into experience stories.

Second and third graders are taught the multiplication signs (× or •) and the division signs (÷, ⌐, or ∓). By the time they reach the middle grades, they are introduced to many more symbols: *percent (%)*, *pi* (π), *ratio sign (:)*, *square root sign* (√), *is approximately equal* (≈), *number pair designation* (1, 2), various geometric symbols; use of parentheses; and frames or letters to represent missing numbers.

A teacher should introduce each of these symbols simultaneously with its word meaning. It's important to give children many experiences in reading number sentences and illustrating the meanings with manipulative materials, flannel-board pieces, or pictures. Consider:

4 + 3 = ☐ Four plus three equals what number? Or, 4 and 3 is what number?

4 — ☐ = 2. Four minus what number equals two? Or, 4 take away what number is 2?

Initially, I don't combine the the reading activity with finding the solution set.

Early in the year, I encourage my middle-grade students to start a mathematics notebook. The first pages are reserved for symbols and terms. As we learn a new symbol, each student writes sentences using the symbol; then writes the sentence in words under the mathematical sentence. Sometimes pictures and diagrams are helpful, too.

(5 • 6) + 3 = ☐ Five times six plus 3 is what number?

☐ × 8 > 17 Some numbers times 8 are greater than 17.

a + b = 17 Some number *a* plus some number *b* is 17.

By asking the child to write the mathematical sentence in words, you can see whether a mistake is a reading error or a computational error.

Another reading skill that is sometimes overlooked is the ability to follow the steps of an algorithm. Letter lines to help students focus on each step.

Consider: 226   A
          x 74   B
           904   C
          1582   D
         16,724   E

Can a child state the problem? (Multiply 226 by 74.) Can he explain the derivation of lines C, D, and E?

(line C 226 x 4)
(line D 226 x 70)
(line E add 904 + 15820)

What numeral could be placed in the blank space in line D? (D

means 1582 tens because you are multiplying 226 by 7 tens)

Many students have difficulty because they are unable to follow their textbook directions. Read directions with the class; then do the first example together. At the same time clarify special mathematical terms and have students add them to their notebooks. Students may need help with *pattern, number pair, ordered pair, sequence, table, graph,* and *estimate,* for instance.

Moreover, directions may be inexplicit. Consider the following:

1. Find the patterns:
   a. 4, 7, 10, ☐, ☐, ☐,
   b. 17, 13, 9, ☐, ☐,
   c. 1, 3, 4, 6, 7, 9, ☐, ☐, ☐,
2. What is my rule?
   a. 4, 7, 10, ☐, ☐, ☐,
   b. 17, 13, 9, ☐, ☐,
   c. 1, 3, 4, 6, 7, 9, ☐, ☐, ☐,

In the first instance, is the child being asked to find the missing numbers, or to explain the rule for the pattern? The second example appears to ask for only the rule. Should a child show the missing numbers too, or just think about them?

If a child isn't sure of a meaning, will his junior dictionary clarify it? Can he find help in his textbook glossary? A teacher must help a student check reference books. If possible, have a larger dictionary (college edition) available.

Ask questions that will help students clarify a word's meaning. For the word *pattern,* you might ask: Do the numbers count in a special way? Are they arranged in a special way? Is there the same "jump" between each of the given numbers and the number that follows it? Are the numbers going up or down? (If up, think about adding or multiplying; if down, think about subtracting or dividing). Can a number-line picture help show the pattern?

Number pairs are sometimes associated with a graph. They can also be associated with number facts. What are the number pairs for the product 48? (8, 6); (4, 12); (2, 24); (1, 48); (3, 16); (6, 8); (12, 4); (24, 2); (48, 1); (16, 3).

What are the number pairs for the sum of 12?

Instruct students making a graph to place the first number of an ordered pair on the horizontal axis (x axis) and the second number on the vertical axis (y axis). They must learn to write the ordered pair in the form (x, y).

The table, graph, and its mathematical sentence are related. The relationship may not be completely appreciated unless students have many experiences reading the three together.

They also need help with abbreviations that are used in measurement problems, such as *m* (meter), *cm* (centimeter), *ml* (milliliter), *lb.* (pound), *oz.* (ounce), and so on.

Can a student read a picture, diagram, graph, or cartoon appearing on a page? First and second graders are likely to use picture clues, but older pupils often ignore them as they depend more and more on the written words. We must encourage students to study pictures, diagrams, and graphs carefully.

Early in the year I ask my students to look at a picture with many unrelated objects or activities for thirty seconds; then tell or write about all the things they remember in the picture. They are always surprised at how little they remember. When we view the picture again, we organize it by classifying the objects and making a list. How are things the same? How are they different? Are two or more sets related in any way? Then we look at the picture a second time to list all the things we remember. This activity is always enjoyed, and simultaneously gives students many experiences in reading pictures and remembering just what they read.

To study a diagram, map, or graph it's often helpful to have students point to the key and identify the labeled parts of the picture.

Consider a picture with several triangles. You might have students point to triangles RST, XYZ, and ABC as you spot-check answers. Have them point to angle STR, angle CBA, and so on. Have students tell something about each triangle.

Students also need instruction in reading number lines, rulers (both standard and metric), thermometers, protractors, and slide rules (if used). I encourage students to ask themselves: How many marks are there between 0 and 1? What is the value between two marks?

I emphasize measurement as an approximation and encourage the use of the symbol ($\approx$). It's most important that students be given many opportunities to use instruments of measure once they've been introduced.

If students learn to read mathematical sentences in these suggested ways, written word problems should be a natural outgrowth. Of course a teacher has used oral problems throughout his program, but children should be encouraged to draw a diagram to show action, make a table or a graph when appropriate, write down the information that is given, find the question or questions in the problem, write a mathematical sentence. Have students consider these points with many problems before solving word problems.

Children can be guided to research information, follow directions, and make up rules (generalize). Encourage students to ask themselves, How does this math fit into what I have learned before?

The reading of mathematics is a natural product of language development. As we encourage a more meaningful approach we can raise the mathematical fluency level of students.

Reading math content

Jo McKeeby Phillips

DO YOU ever think of math books as literature? Probably not, although if you look hard, you will find a few that have some literary merit. Do you think of your math books as reading books, in any sense of the term? I doubt it, but in my opinion, you should.

There are many complaints that children cannot read their math books and have a lot of trouble following the directions on activity cards and such. The usual response appears to be a search for books and activities that have a minimum amount of material for children to read. No attempt is made to help the children develop the special skills needed in reading directions and explanations. A wiser response, it seems to me, would be to look for material that is clearly written at a level suitable for children and teach the children to read it. This is not as hard a job as it may sound.

Readability formulas are useful in a limited way for assessing the difficulty of a given piece of material, but they take no account of the organization of the material, a crucial factor in understanding it. (Take anything you like, the Gettysburg Address, for instance, and find its readability index by whatever formula you choose. Rewrite the piece, last word first, next to last word second, and so on, and use the same formula again. You'll come out with the same score for an incomprehensible mess.)

Again, such formulas take no account of idea density, or of the presence or absence of summaries or diagrams or other illustrative materials, all of which surely have a profound effect on comprehension. Perhaps most important of all, they take no account of the interest level. You know what happens when you give a child something he does not want to read. He tells you he can't read it, and he knows where that leaves you—and him.

As far as interest in mathematics is concerned, nothing succeeds like success. I am continually astonished when I see little youngsters absolutely thrilled with something that seems to me to be deadly dull. Why are they thrilled? Because they have learned to do something they could not do before. So I think it's dangerous to prejudge material. "My children couldn't read this." You cannot be sure without trying it, especially if your own attitude exudes confidence that they can read it, and may even find it fun. "My class would not be interested in this." The only thing predictable about children is their unpredictability, isn't it? The strong advice I can give without reservations is: Pick something worth doing and *expect success.* Bolster your expectations by giving instruction and practice in reading mathematical materials of appropriate kinds.

One thing you have to teach in mathematical context is a modest amount of special vocabulary. For this purpose, use the methods you use successfully in teaching vocabulary of any kind. Break the words up into syllables. Sound them out. Show what they mean. Use pictures where that is possible. Teach prefixes and suffixes and word roots. Discuss other words that contain these same prefixes or suffixes or word roots. Clarify the meaning of each term by pointing out what it does not mean; for example: These pairs of segments are *perpendicular:*

These pairs of segments are *not perpendicular:*

Very rarely, if ever, ask children to memorize a definition. What they need is operating knowledge of a term, not an erudite statement about it. Recognize that definitions, within limits, are subject to choice. No one should be distressed by two apparently different definitions of the same term. Almost always, they are different ways of singling out the same class of objects.

Point out the tremendous economy, the

Reprinted with permission from the *Instructor* (November 1971), 64–65, © 1971 by The Instructor Publications, Inc.

elegance even, of the technical vocabulary. Take the simple word square. Used in its geometrical sense, it should call to mind a picture of a plane figure with four congruent sides and four congruent angles—all that in one word. There's more to it than that, too. As one boy put it, when you have four congruent sides and four congruent angles, it forces you to have the opposite sides parallel and the adjacent sides perpendicular and the diagonals congruent and the diagonals both congruent and perpendicular.

One major difficulty with mathematical vocabulary comes with words having meanings in common everyday contexts that are sometimes different from those they have in mathematical context. One such word is *half*. A half moon is not half of the moon, in the mathematical sense. And where in mathematics would you find the larger half of something? And when is a thing half-baked? Surely it is not the same as something that has been in the oven for half of the time needed to cook it properly. Another example is *diagonal*. In mathematics, *diagonal* is not synonymous with *oblique*. When a square is "standing on one corner," one of its diagonals is vertical and the other is horizontal. When a square is "sitting level," both of its diagonals are oblique. For another example, think what a child might do if asked to find the *area* of the playground *area* in the park. You might find it useful to make a list of these "devil words" as you come across them. (Would you include *square?*)

An absolutely essential skill that everyone needs if he is to read a math book with understanding is that of referring to an example or a diagram or a graph or a picture that goes with an explanatory paragraph. This is a skill to be developed gradually over a long period of time, with much practice using appropriate material.

I tell my pupils that they need both hands, as well as their eyes and their brains, when they read a math book. They use one hand to hold their place in the paragraph and the other hand to point to the part of the example or illustration they are reading about. They use their hands, too, to dramatize whatever motions may be involved in the description. Think about it. Don't you do that yourself when you are reading directions or explanations for something brand-new to you? I do. Have your pupils do this under careful supervision for a while, and you'll be gratified at how soon they can figure things out for themselves.

Finally, believe what's true: Not everyone finds a collection of symbols, a chart, a table, a graph, or a diagram easier to comprehend than a paragraph. Find out what your pupils are "reading" from these graphic devices. Some of your slowest, average, and fastest learners may be getting no useful message. Sometimes part of their trouble lies in not reading captions and legends. Sometimes it lies in not seeing the requisite relationships among the parts of the device, which is harder to remedy. But probe until you find out what the difficulty is, and try everything you can think of to clear it up. Often the most productive technique is to call on other children who do see what they are supposed to see in a given situation. Be sure to let them explain it in *their* way.

Especially with certain kinds of individualized instruction, success depends in large measure on the learner's ability to read content material—to read in the broadest sense, that of getting a correct message from printed and graphic material of many kinds. Surely every teacher must pay constant attention to reading, regardless of what subject may be listed on the schedule at a given time.

Interrelationships between mathematics and art for the kindergarten

EVELYN SWARTZ

When some of the current mathematics programs and proposed art programs for the kindergarten are compared, many components common to both subjects are found. An understanding of the interrelationships which exist between the two makes it possible for the kindergarten teacher to capitalize on them and to reinforce and extend children's learning in both areas. Both can contribute to the conceptual development and the enhancement of visual perception of children so that they can learn to live comfortably in a world of technology and a world of art.

Some of the most important relationships are to be found in objectives, content, and teaching procedures. Both mathematics and art are concerned with the objective of helping children develop visual perception skills so that they can recognize and identify shapes, sizes, and colors. Those who teach mathematics and art know that children use shapes, sizes, and colors in identifying, classifying, and categorizing objects in the environment. As children explore their world tactually and visually, they learn to recognize circles, rectangles, and triangles as they are used in many objects and to compare these on the basis of size. Not only do children identify and classify according to shapes, sizes, and color; they also express themselves through the use of these concepts. Kindergarten teachers are familiar with children's use of shapes of many sizes as they develop representative symbols in their artwork, and the use of color to enhance these symbols.

Shape and size are basic concepts in the content of mathematics and art, and are to be formed and enriched by children. Although color is used in both areas, its importance differs in each. In mathematics color is regarded as a property which is used to help children learn to identify and classify shapes and to understand concepts of regions of shapes. In art, however, color may be considered as a concept basic to understanding art, and color theory is an integral part of kindergarten art. Although the emphasis differs, color plays an important role in the learnings acquired in each area.

The content of both subjects can be organized logically and sequentially in keeping with the disciplines themselves and the psychology of growing minds. Teachers are quite familiar with the "concrete—semiconcrete—abstract" continuum used in mathematics, but fewer, perhaps, are aware that this is common to art as well. Art learning experiences can be and, in some cases, are organized along the same continuum: initial concrete learning experiences eventually lead into the visual abstraction of these in confrontations with works of art and nature. After children have had direct, manipulative ex-

Reprinted from *The Arithmetic Teacher* (May 1968), 420–421, by permission of the author and the publisher.

351

periences with shapes of various sizes, they use what they have learned visually as they view their environment.

Relationships between mathematics and art can also be found in the teaching procedures used with children. Since objectives in both areas include helping children form and enrich concepts and develop extended conceptual frames of reference, both attempt to use those procedures which will most effectively help children in the conceptual process. For young children those procedures are ones that include the use of multisensory learning devices. In mathematics, kindergarten children are given shapes of various sizes and colors to help them learn to identify shapes and sizes. They are given countless opportunities to manipulate a variety of objects which will enhance their perception of the shape, size, color. Such concrete experiences help children form and enrich mathematics concepts.

In art programs which have been developed to promote art learnings, similar procedures are used. To aid in their conceptual development, children are provided with multisensory aids—square, circular, triangular shapes in primary colors —and have directed procedures comparable to those used in mathematics so that they will learn to form and enrich art concepts and enhance their visual perception.

The procedures used also include problem-solving activities which give children an opportunity to apply their understandings. And it is in such activities that the differences between the two subjects may be seen. For, although there are relationships between objectives, content, and procedures, the end products for mathematics and art are not the same. The problem-solving activities in mathematics that utilize the concepts of shape and size are designed to help children build a foundation for geometry. In art, on the other hand, problem-solving activities are concerned with helping children use these concepts in their own artwork as they solve problems of expressive intent. Problem-solving activities may also include the use of the conceptual frame of reference as children see the ways artists have solved problems using shapes, sizes, and colors.

Since the reader may be familiar with mathematics learning experiences that are designed to help children learn shapes and comparisons of sizes and to use these understandings, an illustration of comparable art learning experiences will show the interrelatedness and the different end products of the two areas.

In a series of art learning experiences children could have lessons which—

1. Start with line.

2. Proceed with the closure of line to make shapes—circle and square.

3. Use primary colors to enhance the shapes.

4. Allow for comparisons and discriminations of different sizes of circles and squares.

5. Develop the concept of repeat pattern through the arrangement of shapes.

6. Provide opportunities for the visual abstraction of the preceding learnings in a reproduction of a work of art, such as Mondriaan's *Broadway Boogie-Woogie*. (Reproductions are available from the Museum of Modern Art, New York, New York.)

From the preceding sample, the similarities between mathematics and art can be discerned, and the difference between end products can also be noted. For, although the interrelationships should be understood, neither subject should become a vehicle for the other—neither should become a "handmaiden" to the other. But recognition of the common components can help kindergarten teachers bring children, content, and learning experiences into such a combination that the foundations for the development of mathematically and aesthetically knowledgeable and sensitive adults can be laid.

"Stock-market" unit

BEATRICE E. THOMASHOW

Wouldn't teaching and learning any concepts and skills be more enjoyable if we could do it by being able to utilize all the above techniques and goals? Is not our most important mission as teachers to be able to instill in our children a "joy in learning"?

Permit me to share with you, through the use of this stock-market unit, how I have attempted to use the techniques and goals listed at the beginning of this article, to help an advanced fifth-sixth grade mathematics class experience the joy of learning.

The "stock market" proved to be an exciting experience, something the children looked forward to doing each day. In fact, they asked to be permitted to work on the stock market the first thing each morning! It was a way of bringing the outside, live, financial world into the classroom, and into the arithmetic program daily!

Through the stock-market unit, in which each student has a principal role, the class is learning and reinforcing the skills and understandings, through daily practical application, of adding, subtracting, and multiplying fractions and mixed numbers; converting fractions into decimals; percentage;

concept of negative numbers; checking and double-checking their work; the use of graphs; and some aspects of economics! Learning how to compile statistics could also be added.

The basic idea for this unit emanated from two sixth-grade students, one of whom had received a birthday gift of several shares of stock. As it was taught to succeeding classes, evaluating and reevaluating as we went along, improvements were made, but the original idea that evolved from the students' own interests has been retained.

How did we introduce this unit? There are various ways to do this. Much depends upon the sophistication of the group. If it is a young group, it is advisable not to get too involved with a detailed account of the economic aspect of the stock market, but rather to keep it down to a very simple level.

To start off, have several students bring the financial section of the newspaper to class. Small groups of students can work together on one newspaper (or you could use the opaque projector) as you explain that the price quotation next to each given stock is the closing price of that particular stock.

Reprinted from *The Arithmetic Teacher* (October 1968), 552–556, by permission of the author and the publisher.

Social studies are interrelated here, as students discuss the various industries that could be considered the backbone of a nation, and the industries' interdependence upon each other. As a result of these discussions in class, each student will be motivated to select a stock which he will pretend to purchase, and on which he will keep a daily record for the rest of the term.

In addition to each student's privately owned stock, we also have a class joint-ownership stock, on which each student also maintains a record. Large replicas of the work sheet and the work space sheet are made on tagboard, and for the first week the transactions are worked out daily with the class on this jointly owned stock. After the first week, each student individually attempts to work out the daily gain or loss on the class joint-ownership stock. Then the teacher works it out on the tagboard chart, which is clipped on the chalkboard, and the students have an opportunity to check their work, discuss it, and question it with each other and the teacher. After a few days of this group work, and after they have attempted to work out the transactions themselves, the students take turns dictating to the teacher the results of their own computations. They explain why and how they arrived at their results. This procedure always provokes much group interaction and discussion.

By the third week the students are ready to start work on their own privately owned stock. As we proceed in this unit, each student will receive individualized instruction and will profit from the group interaction in large-group instruction. Self-motivation is very strong as they indulge in this simulated situation where each of them is actually involved in "ownership" and is thus an integral part of the unit.

Inasmuch as the unit is set up in a manner that involves checks and double-checks on the child's daily work, the student makes progress in becoming less dependent upon the teacher, and more dependent upon his own learning. The teacher's role will gradually become that of a consultant!

Each student has a folder for each of his stocks. The worksheet form is kept on the right-hand side, and the work-space paper on the opposite side of the folder. (Several sample transactions, which should be self-explanatory, are on the work sheet and work-space sheets at the end of this article.)

The manner in which the work-space sheet is used is very important. The student *must* label his computation with the line and column number of the work sheet to which it refers. There are several reasons for this:

1. The student can check his work quickly and find errors with greater ease.

2. The teacher is provided thereby with a situation in which weaknesses can be quickly diagnosed. The student's strengths in certain areas can also be determined, and this can be communicated to him.

If the diagnosis indicates that the student has been building a toppling structure on a weak foundation, then he must be taken out of the stock market unit temporarily, and an individualized mathematics program devised for him to follow until he has reached the point where he can return to the unit. The desire to "go back" to the stock market is usually sufficiently great to secure the student's cooperation with remedial instruction.

The words "to date" in Columns 6 through 9 of the work sheet can introduce the use of negative numbers. Note that Line 4, Column 7, and accompanying labeled work space indicate that the daily gain of $1.00 merely reduced the "loss to date" from 1⅛ to ⅛. Note, also, the check and double-check computations in the work-space area for Line 5: We came up with the figure of 9⅜ as the "total gain to date" on 12/18 for 15 shares in Column 8, and ⅝ "gain to date" per share on the same date in Column 6. To check these figures you will note that we sub-

CLASS JOINT-OWNERSHIP STOCK

Room _____

Teacher: Mrs. Thomashow

Purchaser's name ___ John Doe ___

Name of stock _____

NOTE: As you insert information on this, your individual record, start making a GRAPH from each of these transactions as they take place.

12/11/67 Orig. Purchase Price: __15__ shares @ (10¼) $10.25 per share = (153 − ¾) $153.75

	Col. 1 Date	Col. 2 Daily Closing Price	Col. 3 Daily Gain per Share	Col. 4 Daily Loss per Share	Col. 5 Daily Gain or Loss (+ −) for 15 shares	Col. 6 Gain per Share to Date	Col. 7 Loss per Share to Date	Col. 8 Total Gain to Date (15 shares)	Col. 9 Total Loss to Date (___ shares)
1	12/12	12½	(2¼) $2.25		+(33¾) 33.75	(2¼) $2.25		(33¾) $33.75	
2	12/13	13⅜	(⅞) .875		+(13⅛) 13.125	(3⅛) 3.125		(46⅞) 46.875	
3	12/14	9⅛		(4¼) $4.25	−(63¾) 63.75		(1⅛) $1.125		(16⅞) $16.875
4	12/15	10⅛	(1) 1.00		+(15) 15.00		(⅛) .125		(1⅞) 1.875
5	12/18	10⅞	(¾) .75		+(11¼) 11.25	(⅝) .625		(9⅜) 9.375	
6									
7									
8									
9									

tracted the 12/18 closing price of 10⅞ from the original purchase price of 10¼, and the remainder was ⅝, which checked out with the figure in Column 6, Line 5, "gain per share to date," of ⅝. However, to double-check, we added this ⅝ to the original purchase price per share (10¼), which equaled 10⅞; then we multiplied by 15, which was 163⅛ for the fifteen shares, representing their value as of 12/18. Next, we subtracted the original purchase price for fifteen shares (153¾) from 163⅛, which left a remainder of 9⅜, and this figure checked with the figure for "total gain to date for fifteen shares" in Line 5, Column 8.

Note also that the price quotations in the newspapers are listed as whole numbers and fractions, which we do also on our work sheet, placing these in parentheses. Immediately next to these figures, we convert these mixed numbers or fractions into dollars and cents, thereby giving students constant practice in converting fractions into decimals. Also, in a very short time, through daily use of this unit, students immediately know at a glance that ⅞ of a dollar is $.875, or that ⅜ of a dollar is $.375, etc.

By choosing a number other than from 1 to 10 for the total number of shares purchased, through daily use students become quite adept at multiplying by that number. After two months or so, you might wish to "purchase" a few more shares, or "sell" a few, in order to have a different figure for multiplying regularly. Or from the outset, the total number of shares purchased for the class-owned stock could be different from the total purchased for privately owned stock.

There are many ways of handling different details of this unit. The teacher and the students should work out the implementation of the unit to best satisfy the needs of each class.

LINE 1, COL. 3

$12\frac{1}{4} = 12\frac{2}{4}$

$-10\frac{1}{4} = 10\frac{1}{4}$

$\overline{\qquad 2\frac{1}{4}}$ GAIN

LINE 1, COL. 5

$2\frac{1}{4} \times 15 = \frac{9}{4} \times \frac{15}{1} =$

$\frac{135}{4} = 33\frac{3}{4}$

LINE 2, COL. 3

$13\frac{3}{8} = 13\frac{3}{8} = 12\frac{11}{8}$

$-12\frac{1}{2} = 12\frac{4}{8} = 12\frac{4}{8}$

$\overline{\qquad \frac{7}{8}}$

LINE 2, COL. 5

$15 \times \frac{7}{8} = \frac{105}{8} = 13\frac{1}{8}$

LINE 2, COL. 6

$2\frac{1}{4} = 2\frac{2}{8}$

$+\frac{7}{8} = +\frac{7}{8}$

$\overline{\quad 2\frac{9}{8} = 3\frac{1}{8}}$

LINE 2, COL. 8

$3\frac{1}{8} \times 15 = (3 \times 15) + (\frac{1}{8} \times 15)$

$\qquad 45 + 1\frac{7}{8} = 46\frac{7}{8}$

CHECK: ADD LINES 1+2, COL. 5:

$33\frac{3}{4} = 33\frac{6}{8}$

$+13\frac{1}{8} + 13\frac{1}{8}$

$\overline{\qquad 46\frac{7}{8}}$

LINE 3, COL. 4

$13\frac{3}{8}$

$-9\frac{1}{8}$

$\overline{\quad 4\frac{2}{8}} = 4\frac{1}{4}$

LINE 3, COL. 5

$4\frac{1}{4} \times 15 = (4 \times 15) + (\frac{1}{4} \times 15)$

$\qquad 60 + 3\frac{3}{4} =$

$\qquad\qquad = 63\frac{3}{4}$

LINE 3, COL. 7

$4\frac{1}{4} = 4\frac{2}{8}$

$-3\frac{1}{8} = -3\frac{1}{8}$

$\overline{\qquad 1\frac{1}{8}}$

LINE 3, COL. 9

$1\frac{1}{8} \times 15 = (1 \times 15) + (\frac{1}{8} \times 15)$

$\qquad 15 + \frac{15}{8}$ or $1\frac{7}{8}$

$\qquad\qquad = 16\frac{7}{8}$

CHECK:

COL. 5 LINE 3: $-63\frac{3}{4} = 62\frac{14}{8}$

COL. 2 LINE 2: $-46\frac{7}{8} = 46\frac{7}{8}$

$\overline{\qquad\qquad 16\frac{7}{8}}$

DBLE CHECK:

ORIG. PURCH. PRICE: $10\frac{1}{4} = 10\frac{2}{8}$

12/14 CLOS. PRICE: $9\frac{1}{8} = 9\frac{1}{8}$

$\overline{\qquad\qquad 1\frac{1}{8}}$

LINE 4, COL. 7

$1\frac{1}{8}$ LOSS TO DATE

-1 TODAY'S (12/15) GAIN

$\overline{\frac{1}{8}}$ LOSS TO DATE

LINE 5, COL. 3

$10\frac{7}{8}$

$-10\frac{1}{8}$

$\overline{\quad \frac{6}{8}} = \frac{3}{4}$

LINE 5, COL. 5

$15 \times \frac{3}{4} = \frac{45}{4} = 11\frac{1}{4}$

LINE 5, COL. 6

$\frac{3}{4} = \frac{6}{8}$

$-\frac{1}{8} = -\frac{1}{8}$

$\overline{\quad \frac{5}{8}}$

LINE 5, COL. 8

$15 \times \frac{5}{8} = \frac{75}{8} = 9\frac{3}{8}$

CHECK:

12/18 CLOS. PRICE: $10\frac{7}{8} = 10\frac{7}{8}$

ORIG. PRICE: $10\frac{1}{4} = -10\frac{2}{8}$

$\overline{\qquad\qquad \frac{5}{8}}$

DBLE CHECK:

ORIG. PRICE: $-10\frac{1}{4}$

$+\frac{5}{8}$

$\overline{\qquad 10\frac{7}{8}}$

$10\frac{7}{8} \times 15 =$

$(10 \times 15) + (\frac{7}{8} \times 15)$

$150 + (\frac{105}{8} = 13\frac{1}{8}) =$

$163\frac{1}{8} = 162\frac{9}{8}$

$-153\frac{3}{4} = -153\frac{6}{8}$

$\overline{\qquad 9\frac{3}{8}}$

356

Just for fun: from arc to time and time to arc

WILLIAM J. RAY

Some units in the arithmetic, social studies, and science curriculum areas present concepts of meridians and parallels, rotation and revolution, and differences in time zones around the globe. Learning about time zones can be a frustrating experience both for teachers and children unless concepts of arc and time are thoroughly understood.

We know that the motion of the sun and the stars around the earth is only apparent—an illusion created by the rotation of the earth itself. In twenty-four hours the earth rotates through the 360 degrees of a full circle. The direction of rotation is counterclockwise, or in a west to east direction.

A set of *imaginary lines,* called meridians, are drawn from the north to south pole. The east-west position of any place on the globe is located by its relationship to these meridians. The *prime* meridian was chosen as 0° longitude and passes through Greenwich, England.

Try to imagine a directed beam of light from a light in the ceiling. (This might represent the sun.) Suspended below it, and rotating in an easterly direction, is a globe. The beam of light, focused on the globe, will at some point in the rotation focus directly on each meridian. When the light falls directly on meridian it is noon at that location.

Every celestial observation is timed according to the time at Greenwich, England (0° longitude). To clarify the relationship between time and arc, we'll assume it is exactly noon at your meridian. When the sun is directly overhead, it is noon as far as that meridian is concerned. Imagine you are standing at 90° west with the sun directly overhead. It would be noon where you are, consequently the sun is at 90° west. Since leaving Greenwich, the earth has rotated 90° or has traveled through 90° of arc. It was 12:00 noon when the sun was at 0°. The time at Greenwich, therefore, must be 12:00 noon plus the time it took the earth to rotate through 90° of arc.

If it takes the earth 24 hours to rotate 360°, in 1 hour the earth travels 15° of arc ($360° \div 24 = 15°$). If the earth rotates 15° in one hour, it must take 4 minutes to go one degree ($60 \div 15 = 4°$). To travel an arc of 90°, it takes 90×4 (min.) or 360 minutes, which is equal to 6 hours. Six hours ago it was 12:00 noon Greenwich time, therefore placing Greenwich time at 6:00 p.m. and your 90° west-longitude time at 12:00.

What you have actually done is to convert 90° of arc to 6 hours of time. In doing so, you have stated the basic relationship between arc and time, which is:

15° of longitude (arc) equal 1 hr. of time.

Before proceeding further, it is necessary

Reprinted from *The Arithmetic Teacher* (December 1967), 671–673, by permission of the author and the publisher.

to know how distances are measured along the circumference of a circle. Measurement along a meridian is expressed in *degrees of arc*. A good example of circular measurement in degrees of arc is the compass card. Whatever the size of the card, its circumference always contains 360°, each degree contains 60 minutes (′), and each minute contains 60 seconds (″). Therefore meridians are measured in degrees (°), minutes (′) and seconds (″), enabling us to determine the following values:

Equivalents — Arc and Time

Time to Arc		Arc to Time	
24 hr.	= 360°	360°	= 24 hr.
1 hr.	= 15°	15°	= 1 hr.
1 min.	= 15′	1°	= 4 min.
1 sec.	= 15″	1′	= 4 sec.

Interpretation of the arc to time would appear this way: It takes the earth 24 hours to travel 360°; one hour to travel 15°; four minutes to travel one degree and four seconds to travel one minute. Understanding this relationship would help upper-grade elementary children understand the importance of establishing a uniform or standard time belt.

Possible enrichment activities

By knowing the relationship between arc to time and time to arc, some children will be able to locate various countries and cities by using two easy methods of conversion.

Let us assume a child wishes to change 5 hours, 19 minutes, and 10 seconds of time to its equivalent in arc. Arrange the figures in a column using the degree, minute and second symbols of arc measurement. Suggestion: Referral to the *Time to Arc and Arc to Time* value table will assist in computation and understanding.

Time	Arc		
5 hours	°	′	″
19 minutes	—	—	—
10 seconds	—	—	—

There are 15° in 1 hour, so multiply the number of hours (5) by 15.

	Arc		
5 hours	75°	′	″
19 minutes	—	—	—
10 seconds	—	—	—

Next, divide the minutes (19) by 4. Reason: It takes the earth 4 minutes to travel one degree; therefore, for 19 minutes of time, the earth rotated 4° with a remainder of 3 minutes. Now multiply the remainder 3 by 15. Since 1 minute of time equals 15′ of arc, 15′ × 3 = 45′.

Write 4 under degrees and 45′ under the minutes category.

5 hours	75°	′	″
19 minutes	4	45	—
10 seconds			

Using the same approach, compute the seconds. Divide seconds of time (10) by 4. It takes the earth 4 seconds to travel one minute of arc time; therefore 10 ÷ 4 = 2 with the remainder of 2. Write the first 2 under the minutes and multiply the remainder 2 by 15. This gives you 30 seconds. Now you have:

5 hours	75°	′	″
19 minutes	4	45	—
10 seconds		2	30
	79°	47′	30″

Consequently in 5 hours, 19 minutes, 10 seconds of time, the earth traveled 79° 47′ 30″. Now we can compute the operation in reverse, determining time when arc is given.

Set up the degrees, minutes and seconds of arc as you did the hours, minutes and seconds of time.

Arc	Time		
	Hours	Minutes	Seconds
79°	—	—	—
47′	—	—	—
30″	—	—	—

There are 15° in one hour, so divide the degrees by 15; in this case 79° ÷ 15° = 5 with a remainder of 4. Place the 5 under the hours column and then multiply the remainder 4 by 4 and place the resulting 16 in the minutes column.

	Hours	Minutes	Seconds
79°	5	16	—
47′	—	—	—
30″	—	—	—

Now divide the minutes by 15; in this case 47′ ÷ 15′ = 3 with a remainder of 2. Write the 3 under the minutes and multiply the remainder of 2 by 4, placing the result under the seconds.

	Hours	Minutes	Seconds
79°	5	16	
47′	—	3	8
30″		—	—

Divide the seconds by 15″; this will give exactly 2 without a remainder. Write the 2 under the seconds column.

	Hours	Minutes	Seconds
79°	5	16	
47′	—	3	8
30″	—	—	2
	5	19	10

Understanding this conversion would en-able children to determine time difference from a meridian west of Greenwich.

Perhaps an activity of this type could encourage children to gain deeper insights into the role mathematics performs in other content areas. If not informative or practical, maybe—just for fun.

EDITOR'S NOTE.—How often have you asked, as you place a long distance telephone call, "What time is it in — — — ?" Confusion regarding time is not uncommon among adults, and many children find longitude and time difficult to comprehend. Mr. Ray suggests one way children can develop understanding of these ideas.—CHARLOTTE W. JUNGE.

Making the most of your field trip

ROBERT FERNHOFF

John, stop running around! No, Susan, you may not go to the bathroom. Franklin, don't touch that machine!"

A teacher was conducting her class on a tour of a bank when I overheard these typical admonishments. Later I asked the children whether they enjoyed their recent trip.

"Yes," they answered in unison. "It was a lot of fun; we didn't have any school this morning." John and Susan added, "But our feet got so tired!"

It was apparent that little or no educational value was derived from that trip to the bank. The relation to the school and the community never became apparent, and all opportunity for personal understanding was lost entirely.

The bank manager had been a gentleman. He had quickly responded to the teacher's request for a visit. A tour date was arranged. The kids toured every part of the bank.

First, there had been the insurance section, with a talk; then the vault, with a lecture; the hike back upstairs to the checks division, past those complicated and unexplained machines; and, after that, a prepared station offering cookies and milk. A lecture on thrift completed the tour.

This total waste of time, energy, and honest effort produced nothing but tired feet.

There seems to be a trend toward walking more and learning less. How do we reverse this? Given the same materials, how do we produce *learning children* who are motivated towards banking and wish to pursue the topic further? How do we promote respect for the banking system and

Reprinted from *The Arithmetic Teacher* (March 1971), 186–189, by permission of the author and the publisher.

the understanding that every quarter spent by each child relates to this system?

This is a problem that, once solved, appears to be simple. My solution is not a mere theory; it works! It has worked for five years.

Premise number one is not to sell the kids short. Secondly, be honest with your students. Lastly, have several plans ready for discussion and adoption by the students.

The students were of junior high school age, the subject was mathematics, and the unit was banking.

I proposed a plan to a group of twenty-nine young people at the introduction of the unit on banking. I told them that when we finished the unit of study we would have experienced banking and really have become involved in it. I asked, "Are you willing to experiment with me?"

"Yes," they answered.

That same day, after school, I visited the County Trust Bank in Croton-on-Hudson and sought out the manager, a Mr. Harring.

"May I bring in my group for a visit. . . ?"

Before I could complete my question he smiled and said, "Sure, when?"

"Perhaps you had better hear me out first. I'm not interested in the fifty-cent tour. I wish to develop firm concepts and a real knowledge and understanding of banking, loans, interest, savings accounts, trust accounts, and Sensimatic 400. I want the kids to be able to answer questions for themselves, like these: Why are you a banker? Are banking people well compensated for their efforts? Do the forty people working here enjoy their work? Where are the bank records? How does the bank serve the community? Is banking an institution?"

I placed my plan of action on his desk. His automatic smile had left him, and he began to perspire. It was then that he sat down at his desk and devoted his full attention to my proposal.

What I proposed was that he provide me with a list of the many jobs that are daily routine, with an outline of the responsibilities of each position. In class, using the overhead projector as an aid, I would describe each position and its duties. When this had been accomplished, each student would have time to think about what position he would most like to tackle. After assignments were concluded on a voluntary basis, each student would prepare a list of questions concerning "his" position at the bank. He would secure the answers to these questions and later report the answers to the entire class.

This would, of course, mean an orientation for the bank personnel: Why are the children coming? What do they expect to learn? How can the bank employees contribute to the students' education and experience?

Reluctantly and somewhat cautiously, Mr. Harring agreed to receive my class in two weeks' time. He would allow the students to work in assigned sections in the bank during a two-hour period. (It turned out to be for three hours.) Mr. Harring requested that I orient his employees so that they would know what to expect.

The next day I whistled all the way to school; I was really excited!

The class was informed of Mr. Harring's willingness to go along with our experiment. We believed that this would be the first time a bank would be staffed by junior high school students under the guidance of experienced personnel.

The class pulse quickened, and quips were passed. Already the students had begun to think of themselves as Town Bankers.

They studied hard; the units on interest, principal, and loans were gone over in class, with many differences of opinion. Questions were listed which neither students nor teacher could resolve. Some of them follow:

1. What are the essential differences between savings, commercial, and industrial banks?

2. What is the true interest paid by people when they borrow money?

3. Who, or what, determines whether or

360

not a person is granted a personal loan?

4. How much money does a bank have on hand?

5. From what source does a bank get its money?

6. Why do people use checks? What happens to checks after they are processed?

Two weeks never passed so quickly before, as they completed the theoretical aspects of the study and became masters of "banking lingo" (at their own level). They completed the reading and were eager to enlarge and test their information.

I received applications for the various positions offered in the bank. They were:

1. Bank manager
2. Assistant bank manager
3. Secretary to the bank manager
4. Payout and receiving tellers
5. Head bookkeeper
6. General staff members
7. Vault department staff
8. Checking-account posters
9. Filing clerks
10. Payroll clerk

With permission granted by our administrator, who scratched his head and muttered, "The bank will never be the same after today," we took our leave of school.

Incidentally, I never discussed either dress or discipline. The students did me proud on both counts.

We entered the bank at ten minutes after nine. Each student was directed to his selected station and introduced to the co-worker who would direct him. It took exactly twenty-five minutes for the ice to melt. The staff at the bank acted as catalysts: what these kids learned was greatly enhanced by the teaching of the bank personnel, who were as excited as the kids. All attitudes were positive. The transaction of business this day included learning and an exchange of ideas.

It was a joy to tour the various stations and observe the students in action. The IBM Sensimatics were being operated by thirteen-year-olds. A depositor was greeted by a student with a smile as she posted the results and gave change under careful supervision. A student secretary answered the phone, saying, "Good Morning. County Trust Bank.—Mr. Armstrong is on another line; would you care to wait, or shall I have him return your call?"

Downstairs was a room called the "Coffee Klatch Chamber." On one side of a table sat the bank manager, his assistant, and the head bookkeeper. Opposite them sat their student counterparts. Questions and answers were exchanged and real values established.

The hum of action was felt throughout the bank.

At one point, a clerk and two student assistants toured the village in the company of a police officer. Their job was to collect the coins in the parking meters. The students were surprised to learn that no hands touched the money; the coins fell into a locked dolly that was rolled from meter to meter throughout the village. When this job was completed they all returned to the bank. Another bank official, in the presence of the police officer, opened the sealed dolly and transferred the money to an automatic coin counter. The amount was entered into the village account by the students, who also packaged the coins in individual wrappers and stored them for later use.

The time passed. Suddenly it was ten after twelve, time to return to school. Never before had I heard so many groans and protests about going to lunch.

We spent the next three periods of our mathematics class reporting on our experiences. First, the students in each group explained their duties and told the answers to the questions they had asked. Discussion followed.

All the conclusions made were related to everyday life. Concepts for daily living emerged. Indebtedness was understood for the first time, and the idea of living beyond one's means was discussed and understood.

Ways and means of resisting the Madison Avenue want-makers were suggested. Responsibility, community affairs, and a pledge to pay became more than mere words.

The children saw examples of good grooming, dress, and personal pride in appearance—concepts hammered at them from kindergarten on.

This unit is only one small segment of a year's work. For additional mathematics units I propose other alternatives. For a plane geometry unit, I suggest work with architects and flight controllers; for a unit on fractions and decimals, working in a supermarket. (See *The Instructor*, April 1959.)

This is education at work. Learning is fun, not a daily chore. The task is in getting the kids involved; they learn by doing!

Just imagine that evening at home: "Hey, Dad! You know whose picture is on a thousand-dollar bill? You know, I had eight of them in my hand today. Oh, boy, did we have fun!"

EDITOR'S NOTE. Not only were the children well prepared for these experiences, but the teacher did much advance planning and later follow-up to ensure maximum involvement and learning. For the bank personnel it was probably an unusual day in which they gained a deeper understanding of modern education. What excellent public relations for the school! CHARLOTTE W. JUNGE

SELECTED BIBLIOGRAPHY FOR CHAPTER 14
Other Areas of Correlation

Barney, Leroy. "Problems Associated With the Reading of Arithmetic." *The Arithmetic Teacher* (February 1972), 131–33.

Earp, N. Wesley. "Procedures for Teaching Reading in Mathematics." *The Arithmetic Teacher* (November 1970), 575.

Freeman, George, "Reading and Mathematics." *The Arithmetic Teacher* (November 1973), 523–29.

Grant, Nicholas and Tobin, Alexander. "Let Them Fold." *The Arithmetic Teacher* (October 1972), 420–25.

Henderson, George L. and Van Beck, Mary. "Mathematics Educators Must Help Face the Environmental Pollution Challenge." *The Arithmetic Teacher* (November 1970), 557.

Kuhn, Doris Young. "The Library and Science and Mathematics," *Theory into Practice* (February 1967), 811–13.

Orans, Sylvia. "Go Shopping! Problem-Solving Activities for the Primary Grades With Provisions for Individualization." *The Arithmetic Teacher* (November 1970), 613.

Rosenberg, Edwin. "Aesthetics in Elementary Mathematics." *The Arithmetic Teacher* (April 1968), 333.

Seal, Joan. "The Listening Post." *The Arithmetic Teacher* (December 1965), 645.

Sommers, Bernice. "Minding a Store Teaches Math in a Hurry." *Grade Teacher* (January 1967), 85, 138–39.

Strain, Lucille. "Children's Literature: An Aid in Mathematics Instruction." *The Arithmetic Teacher* (October 1969), 451.

Sullivan, John J. "Improving the Reading of Elementary School Mathematics." *New York State Mathematics Teachers' Journal* (June 1972), 123–27.

PART VII

PROVIDING FOR INDIVIDUAL DIFFERENCES

CHAPTER 15

SLOW/DISADVANTAGED LEARNERS

Tremendous amounts of time, work, and money have been spent on educational programs for the slow and/or educationally disadvantaged. Although much has been accomplished in this area, a great deal still remains to be done. The articles in this chapter are given with the hope that each classroom teacher will be motivated to start a file or collection of techniques, strategies, and materials to be used with the slow or disadvantaged learner in mathematics.

Billy Paschal in "Mathematical Readiness" claims that the dogmas of the past are inadequate for the "space age" of today. We *can* develop mathematical readiness in disadvantaged youngsters! Read this article for some stimulating thoughts on mathematical readiness, then pursue and apply his suggestions in the classroom.

Jerome D. Kaplan, in "Some Criteria for Selecting Arithmetic Materials for Culturally Disadvantaged Youth," suggests thirteen things to consider when choosing instructional material in arithmetic. If you are involved in selecting or purchasing mathematical materials, be sure to read this article.

"Simple Materials for Teaching Early Number Concepts to Trainable-Level Mentally Retarded Pupils" by Jenny R. Armstrong and Harold Schmidt presents a series of step-by-step aids for teaching numeral-quantity association. They are characterized as being (1) sequential, (2) easily utilized by pupils of low-level motoric, cognitive, and verbal skills, and (3) easily constructed with inexpensive materials that are readily available to most classroom teachers. Be sure to read this article—you'll like these ideas.

Building understanding and skill in computation for slow or disadvantaged learners requires planning, careful development, and much ingenuity on the part of the teacher. "A Color-Coded Method of Teaching Basic Arithmetic Concepts and Procedures" by Roberta Green provides such a lesson. Hopefully this article will stimulate others to search for similar teaching aids which will motivate these children and provide continuity in their learning.

Ruth S. Jacobson in "Fun With Fractions for Special Education" describes a visual aid that she has found effective in helping children with learning disabilities to learn about addition and subtraction of fractional numbers. Elementary classroom teachers will find this visual aid helpful and meaningful with all learners.

Teaching mathematics to slow or disadvantaged children is a crucial area that has its own unique problems and challenges—but also special rewards. Those who wish to pursue this topic more deeply will find an extensive bibliography at the end of the chapter.

MATHEMATICAL READINESS

Billy J. Paschal

No METROPOLITAN AREA, RURAL AREA, OR THE NATION AS A WHOLE can afford a school system which fails to educate a majority of its disadvantaged children and youth. We have not even begun to face up to this problem. Education is the most important factor in determining who is destined to be poor. By 1970, education will be even more important in determining who will be poor than it is today. In his State of the Union Message, President Johnson emphasized the need for child-care centers for the poor in order to prepare youngsters to benefit from first grade activities.

A proposal to extend public education down to the nursery level or even the child-care level for disadvantaged youngsters is not nearly as extreme as it sounds. About sixty-five years ago, Dr. Maria Montessori demonstrated the influence of early childhood education on poverty in Italy. The Montessori approach has the advantage of making learning appear to be fun, and it provides each child an opportunity to find the circumstances which match his own particular stage of development. Dr. Ronald Koegler, a neuro-psychiatrist at UCLA among others, is experimenting with a Montessori nursery program for disadvantaged youngsters.

The Montessori example is by no means the only one. Israel has already extended public education down to the nursery level and is in the process of establishing nurseries for the Oriental Jews. The Oriental youngsters start school at three and their curriculum closely resembles the one Dr. Martin Deutsch is developing in New York City. Deutsch's curriculum is designed to teach primarily verbal and perceptual skills and to bolster the sense of self. Deutsch also works with the parents.

The SMSG Special Curriculum Project,* which is in progress, has already established the fact that disadvantaged youngsters can benefit greatly by being subjected to a structured mathematics program at the kindergarten level. This study also suggests that it is feasible to start a formal program at the nursery level for three- or four-year-old youngsters. Research indicates that youngsters show an early interest in numbers. The preschool child possesses quantitative ability to a degree which needs the attention of educators (pairing the names of the counting numbers with objects, the ability to recognize the cardinal numbers of a given set, etc.). Our schools can meet the needs of these youngsters by providing a carefully controlled environment and manipulative materials (mathematical toys, etc.) in order to develop a program for readiness in mathematics. These lessons need to be informal, but formal planning by teachers is necessary in order that these youngsters may benefit from sequential number activities. The natural thrust of motivation can be clearly seen in these youngsters. They are aware of their limited experiences and are dependent upon the teacher for success in mathematics.

American educators are beginning to view the child as an open system and are concerned about what the child is capable of becoming. This idea is crucial if any program is to be successful, and it is imperative that we recognize I. Q. scores as being meaningless as a guide to a culturally disadvantaged child's potential.

The concepts of time, space and number relationships appear outside the range of knowledge or vision of nursery youngsters. These are abstract concepts, and three-year-olds are only at the gateway to abstract thinking. This fact is what makes the learning of time, space and number concepts an important part of preschool experiences. They are the foundation for the further study of mathematics, geography, geology and other sciences. It is important that these beginning experiences be correct, vivid, real and stem from concrete experiences. Thus, a background for abstract concepts is provided. The teacher can rely on the everyday sensory

*Data are not currently available.

Reprinted by permission from *Journal of Negro Education* (Winter 1967), 78–80.

experiences of the nursery school and the kindergarten because they provide the basis for mathematics in the elementary school.

The chief elements in the world of things in which the child finds himself are time, space, number, form, texture, color and causality. The youngster acquires his command of these elements slowly, first through motor experiences, but these experiences lay the foundation for later judgments and concepts.

In the nursery school, it is important to have in evidence and in use various devices for measuring time. Clocks, sundials, watches, calendars, and any other simple instruments for measuring time are an integral part of life and of the school. As a result of the teacher using these devices, the youngster may accept them as a part of his world.

Direct sensory experiences which utilize space, accompanied by verbal comments which help the preschool child understand the relationships with which he is confronted, enable him to develop concepts of space. The teacher can guide him to make maps, to identify landmarks, to develop contrasting concepts of space (indoors and outdoors, big and little, short and tall), and to develop associations which help him distinguish a particular kind of place.

The teacher can aid the child's mathematical development by providing him with interesting and realistic experiences. The teacher can attempt to guide the child's development by means of a variety of qualitative and very simple quantitative experiences (succession of numbers, counting, less than, more than, joining, etc.). The order in which youngsters may best learn the content of mathematics is to some degree inherent in mathematics itself as a logical organization of ideas and relationships. Mathematics cannot be easily rearranged except for details. A sensible application of topical sequence in mathematics is to take care that each child, so far as possible, has an opportunity to gain understanding and skill with a given topic before he is pushed on to other topics that are built of necessity on those that precede them.

Educationally speaking, the wheel is out of balance in the U. S. as far as disadvantaged youngsters are concerned. The root of the problem educationally is that the slum child has a verbal deficiency and has difficulty in handling abstract concepts. These youngsters can be predicted to be less capable in mathematics, and they will need to depend more upon concrete events rather than being readily able, as we too often assume, to operate at an abstract level.

It is obvious why we have failed since we start much too late, after the damage is already done. The basic problem remains, even with compulsory kindergarten.

The environment in which lower-class youngsters grow up does not provide the "school-relevant" experiences they so desperately need in order to be successful in school. Frequently, the disadvantaged child has not learned how to ask or answer questions, how to study, or how to relate to the teacher.

Hebb[1] makes the assumption that the early childhood period is of decisive importance in determining later intelligence. He believes that disadvantaged youngsters have had insufficient stimulation in their early development, and that this accounts for their lower functioning intelligence at a later age. Others have said that the early challenges of problems to be solved are the preconditions of attaining some measure of our full potentiality. It is believed that the more a child has seen and heard, the more he wants to see and hear.

A compulsory nursery school for three- or four-year-old disadvantaged youngsters is a partial solution to the problem. These youngsters suffer from an overall poverty of environment — visual, verbal and tactile — and they need a formal curriculum of compensatory education. The curriculum should develop verbal and perceptual skills by utilizing the games format of teaching. These youngsters respond exceptionally well to games, and teachers who have worked with them know that one of the best ways to involve them in a learning activity is to make it into a game. The Israeli educators have had tremendous success with Oriental youngsters working toward the government's goal of acculturating their immigrants in one generation.

[1] Donald O. Hebb, *Organization of Behavior* (New York: John Wiley & Sons, 1949).

Social development goes through a series of stages toward social maturity. These stages are not rigidly demarcated from each other. Parten[2] classifies social participation on the part of preschool children into six levels: (1) unoccupied behavior, (2) solitary independent play, (3) on-looker behavior, (4) parallel activity, (5)

associative play, and (6) cooperative play. A youngster is essentially individualistic, and it is not before the end of the third year that cooperative play, as opposed to parallel play, becomes relatively fixed. Learning activities can be designed with this in mind as Piaget demonstrated in his work in the area of social development.

The dogmas of the past are inadequate for the "space age" of the present. We can develop mathematical readiness in disadvantaged youngsters!

[2] M. B. Parten, "Social Participation Among Preschool Children," *Journal of Abnormal and Social Psychology* XXVII (1932), 243-269. 1932.

SOME CRITERIA FOR SELECTING ARITHMETIC MATERIALS FOR CULTURALLY DISADVANTAGED YOUTH

Jerome D. Kaplan

Much attention in the past few years has turned toward the education of culturally disadvantaged youth. While the term "culturally disadvantaged" is rarely defined, most writers probably mean the economically-deprived who have not had the opportunities during childhood to obtain appropriate educational stimuli at home and who consequently meet with little success in the schools.

The Federal Government's Anti-Poverty Program has generated considerable interest in the training of large numbers of such youth. The "cultural gap" that exists between the demands of the labor market and the educational levels of most high school dropouts is being recognized as one factor that has contributed to our inability to eradicate an unemployment rate of five percent or more during the past few years. A major part of the Anti-Poverty legislation passed in the summer of 1964 will attempt to close this cultural gap by training unemployed high school dropouts between the age of 16 and 21. Under the Anti-Poverty bill these youth will be prepared for jobs at Job Corps Centers. These centers are divided into two main types, Rural and Urban. In its Rural Centers, the Job Corps will offer training in basic academic skills such as reading, writing, arithmetic and communication skills. The Urban Center program of the

Reprinted with permission from *The New York State Mathematics Teachers' Journal* (April 1966), 46–50.

Job Corps will reinforce basic academic skills and provide training for specific jobs.

The initial Job Corps Centers opened early in 1965, while the first Urban Center began training youth in mid-February. By the end of 1965, approximately 20,000 trainees were enrolled in Rural and Urban Centers offering remedial education and job training.

The author was a member of a committee that was responsible for recommending instructional material in arithmetic for the Job Corps Rural Centers and developing an instructional system in arithmetic that would teach basic arithmetic skills to these youth.

The arithmetic curriculum of the Rural Centers was determined by the practical goals of the over-all Job Corps program: to train youth to obtain and hold jobs in the labor market. The curriculum is accented toward computational skills such as the basic operations with whole numbers, fractions and decimals; linear, liquid and weight measurement; fundamental descriptive geometry; and the solution to problems using arithmetic and geometry. The specific ingredients of the curriculum were those skills that would best serve the trainees in the future, either on the job or in their personal lives.

Since Job Corps can expect an academically heterogenous group of recruits, the arithmetic materials must meet the demand of young adults who possess a large variance in mastery of reading and mathematics skills. The over-all philosophy that guides instruction in the Job Corps Rural Centers is the offering of the opportunity to trainees to advance as rapidly as possible and to "promote" them whenever they demonstrate competence. Thus the system that was devised to instruct arithmetic is essentially a non-graded, self-instructional, self-checking one that would permit trainees to move at their own pace and to receive instruction mainly in those skills where incompetence has been demonstrated. Because the system is a non-graded, self-instructional one that requires a great deal of flexibility, programmed instruction is, of necessity, playing a prominent role.

To initiate this system, a set of criteria for selecting instructional material was established. All material that was initially chosen came from published material that existed late in 1964. The Job Corps office will determine the value of instructional material so that revisions in the system and changes in the mode of instruction can be made that will improve the system. "Improvement" means a change or changes in the components or design of the system which will facilitate learning in arithmetic over a given period, as compared to what is now possible.

The following is the initial list of criteria for selection of instructional material in arithmetic:

1. *Availability of Data*

   If publishers offer data with their instructional materials, then these data should be one of the major guidelines for selection purposes. One should take into account, of course, the population for which the material has been tested. For the most part, data that are available with most existing (programmed) textbooks are for conventional school populations. Although these populations have some relevance for selection purposes, they do not have the special characteristics of a culturally disadvantaged group. Because most instructional material has not been tested with any sample students, any evidence that demonstrates learning with any population should certainly be considered.

2. *Self-sufficiency*

The degree to which materials are self-sufficient is an important factor because the Job Corps philosophy is one by which instruction is trainee-centered and self-pacing. In addition, trainees in Job Corps will possess varying mathematical abilities and achievements. Hence, it is important to inspect materials with this kind of heterogeneity in mind. This criterion will probably be satisfied by programmed materials, workbooks which break down skills into component sub-skills, films, filmstrips, and parts of some textbooks.

3. *Teachability*

Any material that is selected will be used in a variety of teaching situations with a variety of teachers. Hence, it becomes important that the materials have what might be called teachability. By teachability is meant the ease of use in actual classroom situations, the ability to be used on an individual basis, and ease of handling in terms of format, content organization, instructions to the student, and so forth.

4. *Entry Level*

The arithmetic curriculum for the Job Corps Rural Centers will teach basic arithmetic operations with decimals, fractions and percents. For some trainees there will probably be a need to teach such skills as addition and multiplication facts. For others multiplication and long division will have to be reviewed intensely. In addition, the curriculum will offer training in different measure systems and basic descriptive geometry. The set of material which is selected for for instructional purposes in arithmetic will have to be flexible to provide for the variety of entry levels.

5. *Terminal Level*

The arithmetic materials that are selected will be serving a very pragmatic purpose. Since Job Corps will eventually offer instruction in many different vocational skills, it is important to bear in mind that the "Luxury" mathematical items of a conventional school curriculum will have to be omitted. Attention should be directed toward those materials that teach specific sets of skills that will be useful to the trainee when he leaves the Job Corps program.

6. *Ease of Handling Materials*

Whatever materials may be selected, it is important that they be easy to handle in terms of amount and difficulty of instructions to trainees, format and general layout of problems and exercises.

7. *Readability*

This criterion is self-evident; materials should be easy to read for all who use them. Linked with whatever mathematical problems exist are the problems in reading. It is probably wise to aim at materials that have been written for approximately the third, fourth or fifth-grade reading levels. This criterion creates a new set of problems. The Job Corps trainee is a young adult interested in an adult world about him. He is being trained to assume the responsibilities of an adult world. Most materials with reading levels of the third, fourth or fifth-grade contain too many examples specifically designed for the elementary school child.

8. *Interest Factors*
Whenever possible, materials should be selected which have some relevance to the trainee in terms of his environment, culture and background. Although few materials are available that have been written specifically for young adults from disadvantaged backgrounds, it might be possible to find materials which contain examples and problems that are not overly unrealistic for the goals of the Job Corps program.

9. *Boredom*
In considering this criterion, it is important to bear in mind that what is considered boring to students in the conventional situation may not be boring to Job Corps trainees, and vice versa. Some educators maintain that highly-structured and concrete materials work best with culturally disadvantaged youth. Repetition might produce a sense of accomplishment and security that these trainees need when they come into contact with educational materials. In this realm, the general over-all development of the skills from the beginning to the end should be considered; that is, the extent to which different sequences actually teach skills as opposed to presenting information.

10. *Size of Actual Teaching Unit*
Generally speaking, the size of the unit of instruction should be kept relatively small. Observations made with culturally disadvantaged youth indicate that their attention spans are relatively short. By reducing the size of each individual unit, trainees will be able to obtain a feeling of reward after finishing each unit.

11. *Sequencing*
The sequence of the instructional material or of examples should be such that trainees can understand the logical development and inter-relationship among various skills. To what extent do sequences proceed from the simple to the complex? Do sequences review previously-taught material? Do sequences tie together a number of concepts that have already been taught?

12. *Level of Abstraction*
Materials should be chosen so that the degree of abstraction is kept at a low level. From the best information that is available, it appears that culturally disadvantaged youth learn mathematics best when it is presented via tangible objects or a concrete orientation. These youth have a relatively hard time learning how to generalize, learning abstract concepts and remembering arithmetical rules.

370

13. *Variety*

Probably no single set of materials, regardless of its entry and terminal levels, will be suitable for Job Corps populations. Perhaps the best set of materials for Job Corps use will be a mixed set of materials. Instructors should be given a variety of weapons with which to train youth in the job Corps program. Some approaches will work with some trainees, while other approaches are needed with other trainees.

Simple materials for teaching early number concepts to trainable-level mentally retarded pupils

JENNY R. ARMSTRONG and HAROLD SCHMIDT

As a part of a project designed to teach early number concepts to trainable-level mentally retarded pupils, a step-by-step series of materials was developed. The materials were characterized by being (1) sequential, (2) easily utilized by pupils of low-level motoric, cognitive, and verbal skills, and (3) easily constructed with inexpensive materials that are readily available to most classroom teachers.

Numeral-quantity association taught in the enactive mode

In a research study designed to compare the early number learning of trainable-level mentally retarded when taught in an enactive mode of representation (using manipulative materials) with the learning

acquired through a pictorial mode of representation (using nonmanipulative materials) following the information-processing theory of Bruner (1964), it became apparent that there were very few manipulative materials available for teaching the concepts related to numeral-quantity association (Armstrong 1969). Therefore, as a part of the project a series of simple step-by-step materials was developed.

One of the major criteria of the materials was that they be manipulative in character—that is, in order to use the material the pupil must manipulate objects. Another criterion of the materials was that they be specifically designed to teach a numeral-quantity–association concept. Implicit, therefore, in the use of each material developed was manipulation and focus on a numeral-quantity–association concept.

Materials design with concern for sequential development

A series of materials structured in a very rigorous developmental sequence was then

The authors wish to acknowledge the funding of the project by the Bureau of the Handicapped, USOE, Department of Health, Education, and Welfare, under grant numbers 32-59-0500-1002 and OEG-0-8-080568-4598(032).

Reprinted from *The Arithmetic Teacher* (February 1972), 149–153, by permission of the author and the publisher.

designed. Since the pupils for whom the materials were originally designed were at the very beginning stage of number learning (that is, several had not even learned to count by rote, let alone associate any meaningful concept of quantity with a given numeral), the first set of materials in the sequence was designed to provide a simple manipulative aid to the oral verbalization of the numerals from 1 through 5.

This first set of materials in the sequence consisted of block-chain counters (see fig. 1). Pupils were given the chains of blocks and instructed to count the blocks and verbally tell the number of blocks on each chain. This activity was repeated several times until the pupils could give verbally the numeral that expressed the number of blocks on each of the block chains. The pupils were highly motivated by this ma-

Fig. 1.   Block chains

terial, and in quite a short time they were able to achieve the instructional goal.

A second set of materials that was developed for this same instructional goal was not as successful. This set of materials consisted of glass jelly jars containing varying numbers of lima beans (see fig. 2). Because of the low level of the motoric skills of this group of children, the jars were frequently dropped. This was especially hazardous as the jars were made of

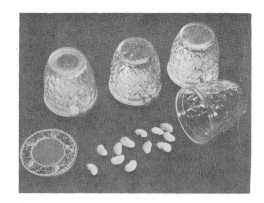

Fig. 2.   Bean jars

glass. Also, the lids were much too difficult to get off the jars.

The next set of materials in the sequence was designed to aid the pupils in associating the number symbols visually and tactilely to quantities of objects. This set of materials consisted of a series of plasticene molds that had insets for different numbers of candles and imprints of the associated numerals, from 0 through 5 (see fig. 3). Pupils were given one "candle counter" at a time and instructed to count the candles in the holder, to verbally give the number that told the number of candles in the holder, and then to use a finger to feel the shape of the number symbol; they were then told to look at the shape of the number symbol, verbalize it, and retrace it with a finger. The pupils enjoyed their work with this set of materials and were aided in getting one step closer to reading the number symbol. This material also

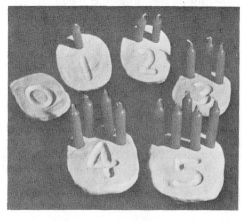

Fig. 3.   Clay candle counters

provided readiness for the actual writing of the number symbol associated with a particular quantity.

The next set of materials in the sequence also allowed for both visual and tactile experiences with the number symbols as well as for quantity association with the symbols (see fig. 4). Each board had two numerals in sequence made from sand-paper to provide a tactile sensation as the pupil ran his finger over the numeral to get a feeling of the shape as well as how it would be to write the numeral. A small well in the board by each numeral provided

well in accomplishing its goal, when con-structing the material for future use one might wish to separate the single numerals and wells by cutting each board in half. In this way there would be only one numeral on each board (that is, 2 and 3 would not be confused with 23).

The next set of materials in the sequence was designed to provide the first experi-ences with writing the numerals with some guidance (see fig. 5). Each board pro-vided a stencil for the numeral concerned and removable blocks on small pegs to illustrate the quantity. Pupils were pro-

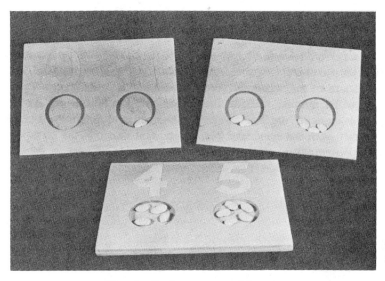

Fig. 4. Sandpaper numeral boards

a place for the pupil to place the number of beans, rocks, or other types of counters that illustrated in quantity what the nu-meral meant.

In using this set of materials, the pupil was instructed to work with each number individually. First he was asked to verbally identify the numeral. Next, he was asked to count out the number of beans that the numeral represented and place them in the well associated with the numeral. Then he was asked to run his finger over the nu-meral several times to get an idea of its shape as he repeated the number name. Although this set of materials worked quite

vided with a sack of blocks, paper, and felt tip marking pens. They were instructed to place the number of blocks the numeral suggested on the board and then using the stencil portion of the board to write the numeral. After writing the numeral using the stencil, pupils were asked to write the numeral without using the stencil, using the stenciled numeral as a pattern.

The next set of materials in the sequence consisted of egg-carton counters (see fig. 6). Since the major focus was on quantity rather than on size or shape, this material served two functions. First, it served its goal as the next step in the developmental

373

Fig. 5. Numeral stencil boards

learning sequence, and second, it served to emphasize the dominant characteristic of concern. The size of the carton did not directly relate to the number of balls the carton held (for example, the one carton and the six carton were the same size but represented quite different numbers of objects). This material was designed to take

the pupils one step further in the sequence. Each pupil was provided with a can of table-tennis balls and a carton. He was instructed to fill the carton with "eggs" and then count to determine whether or not the numeral shown in the lid of the carton told the number of eggs in the carton. Pupils were asked in this case to look at a pat-

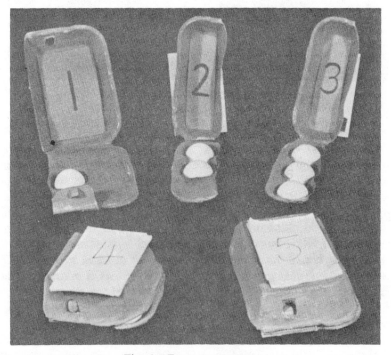

Fig. 6. Egg-carton counters

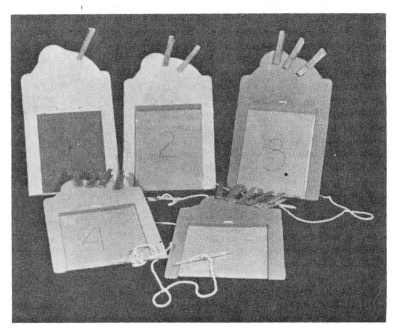

Fig. 7. Magic slate boards

tern of the numeral on the inside of the carton lid at the same time as they looked at the number of eggs (table-tennis balls) the carton would hold. Next, the pupils were instructed to close the lid of the carton and write the numeral on a tablet provided on top of the carton. After writing the numeral, they were instructed to remove from the tablet the sheet of paper on which they had written, open the carton lid and compare the numeral they wrote with the numeral pattern in the lid. This material worked quite well except for the problem of tennis-ball attrition.

The final set of materials in the sequence consisted of magic-slate tablets (see fig. 7). Each pupil was provided with a magic slate with a certain number of clothespins clipped to the top. The pupils were instructed to count to determine the number of clothespins on their slate. Then they were instructed to write on the magic slate the numeral that told the number of clothespins pinned to the slate. This material worked very well in accomplishing the desired goal. It was very flexible in its utility for the provision of individual dif-

ferences. One needed only to add or delete clothespins to provide for pupils at different levels on the numeral scale.

Summary

The purpose of this paper was to describe in detail a series of materials designed to teach numeral-quantity association to pupils at very early stages of cognitive and mathematical growth. The materials were developed to provide very small steps in the learning pattern. In this way, the learning of pupils with learning problems is most particularly facilitated. These materials can be inexpensively reproduced for classroom use and have been shown to be quite successful with at least one sample of pupils with severe learning problems.

References

Armstrong, J. R. "Teaching: An Ongoing Process of Assessing, Selecting, Developing, Generalizing, Applying, and Reassessing," *Education and Training of the Mentally Retarded* 4 (1969): 168–76.

Bruner, J. S. "The course of cognitive growth," *American Psychologist* 19 (1964): 1–15.

A color-coded method of teaching basic arithmetic concepts and procedures

ROBERTA GREEN

The method that is explained below was developed by the author in an effort to aid a fifth-grade boy who was severely retarded in arithmetic achievement. All conventional methods of explaining place value, such as the abacus, pocket charts, Cuisenaire rods, etc., were too abstract for this child, and the use of the materials had not led to his developing understanding of concepts.

The first step in the color-coded method was to provide periods that were designed to elicit an understanding of value equivalency through the use of money. During these periods, a play store was set up and various items were purchased with equivalent amounts of money. That is, ten pencils could be purchased with either ten pennies or one dime, or a pen could be purchased with either ten dimes or one dollar. Sufficient experiences were given until the child thoroughly understood that ten pennies are equal in money value to one dime, and ten dimes are equal in value to one dollar.

Various numerals, written in colored pencil, were then presented to the child. The colors of these numerals corresponded to the colors of dollars, dimes, and pennies. Thus ones were represented by a brown numeral, tens by a silver numeral, and hundreds by a green numeral. (See fig. 1.)

O = **Dollars**

O = **Dimes**

O = Pennies

Note: The color legend used here is adapted to the color requirements of this journal.

635

FIGURE 1

Colored notations of this sort had the effect of making more concrete the abstract concept of notation. It also clearly showed the concept of place value, since it was quite obvious that the number represented by the silver **3** represented greater value than the number that was represented by the brown 5.

The next step was to try and develop the child's understanding of the use of zero as a place holder. Initially one dime and several pennies were used in various combinations, and practice was provided until the child was able to respond quickly and

Reprinted from *The Arithmetic Teacher* (March 1970), 231–233, by permission of the author and the publisher.

$$\begin{array}{cc} 10 & 10 \\ +3 & +7 \\ \hline 13 & 17 \end{array}$$

FIGURE 2

accurately to all sums involving ten plus another number, without resorting to counting the actual money which was before him. Notations of such problems are shown in figure 2. This same idea was extended to other decades (see fig. 3) until the child gained full realization of the function of zero as a place holder and the advantages of grouping numbers by ten.

$$\begin{array}{cc} 40 & 70 \\ +6 & +9 \\ \hline 46 & 79 \end{array}$$

FIGURE 3

Simple two-place and three-place addition was begun, using similar color-coded notations. Various amounts of money were separated into two groups which corresponded to the addends of specific problems. As an initial step, the child would combine the groups of money and find the total. He would then solve the problems on paper using colored pencils that corresponded to the place value of the digits. Examples of such simple addition problems are shown in figure 4. When simple addition had been mastered, subtraction problems were presented in much the same manner.

$$\begin{array}{c} 706 \\ +152 \\ \hline 858 \end{array}$$

FIGURE 4

Preceding the teaching of regrouping in addition and subtraction, a bank (the lid of a shoe box) was established for the purpose of exchanging monies. It was established that only one digit of each color would be allowed in an answer. The procedure for teaching carrying in addition was similar to the procedure used in teaching simple addition, except that a step

was added following the joining of the two separate groups of money. After the monies were joined, the child would proceed to the bank and convert his combined pennies and/or dimes that equaled ten to their equivalents in the next denomination. He would then total his actual money and proceed to solve the paper computations. Thus, to solve the problem shown in figure 5a, six and seven pennies were totaled and ten of the pennies were exchanged at the bank for one dime, which was added to the other dimes. Similarly, to solve the problem shown in figure 5b, ten of the eleven pennies were exchanged for one dime, and ten of the sixteen dimes were exchanged for one dollar.

a)
$$\begin{array}{c} \overset{1}{1}6 \\ +27 \\ \hline 43 \end{array}$$

b)
$$\begin{array}{c} \overset{1}{4}\overset{1}{6}7 \\ + \ 94 \\ \hline 561 \end{array}$$

FIGURE 5

The same procedures were easily extended to the subtraction process and were quite effective in the way they illustrated that "at the subtraction bank, there could be no 'borrowing,' only equal exchanges."

$$\begin{array}{c} 2\,3\,17\,\overset{7}{8}\,\overset{1}{5} \\ - \quad 9\,8 \\ \hline 2\ 8\,7 \end{array}$$

FIGURE 6

The procedure to solve a problem such as the one shown in figure 6 would develop in somewhat the following manner:

The child would be given $3.85.

Teacher: Please tell me how much money you will have left for yourself after you have given me $.98.

Student: I can't give you eight pennies, because I only have five pennies, so I'll go to the bank and exchange one of my dimes for ten pennies. [*Immediately following the bank transaction, the child would make the proper notations on paper with colored pencils.*]
I have fifteen pennies now, and if I give you eight, I'll have seven left. [*The child*

would again make the appropriate notations on paper.]

Teacher: Fine. Now how many dimes do you have to give me?

Student: I have to give you nine dimes, but I only have seven dimes left, so I'll go to the bank and exchange one of my dollars for ten dimes. Now I have seventeen dimes, and if I give you nine I'll have eight left. I don't have to give you any dollars, so I have $2.87 left. [He concluded his problem on paper.]

Eventually the child for whom this method was developed progressed to the point where he was able to do computations accurately without the aid of color-coding. It should be noted that because of the one-to-one situation, it was possible to use real money which undoubtedly had some intrinsic motivational value. Nevertheless, a classroom teacher might easily adapt this method using adequately reproduced play money.

It appeared obvious that the success of this method was primarily due to the manner in which the color-coded digits related to the varying values of monetary denominations. This concrete relationship led at least one fifth-grade child to finally understand and master the basic arithmetic concepts and processes which had for so long eluded him.

EDITOR'S NOTE. Building understanding and competence in computation for the slow learner requires careful planning, step-by-step development, and a great deal of ingenuity on the part of the teacher. A variety of teaching aids is, at times, effective. At other times, use of a *single aid*, such as money in this case, provides continuity in the child's thinking. CHARLOTTE W. JUNGE.

Fun with fractions for special education

R U T H   S .   J A C O B S O N

A device to aid the learning of addition and subtraction of fractional numbers named by like and unlike fractions has been developed in a class for children with special learning disabilities at Boston University School of Education. It is a manipulative device that provides immediate feedback on the correctness of the child's answer to a problem.

This method may be used as preparation or reinforcement for addition and subtraction of fractional numbers when different denominators are involved. Although the example in this article involves halves, fourths, and eighths, the device may also be presented in other fractional values, such as fifths and tenths. This device appears enjoyable to use and, most important, gives the child a method of attack when working with the often difficult concepts of addition and subtraction of fractional numbers.

The aid consists of an eleven-by-eight-inch acetate projectual and corresponding fractional parts (fig. 1). Acetate is used because (1) it can be used on the overhead projector by the instructor and (2) the student can see through the material to the line diagram underneath. The fractional parts are made from a second projectual of another color. This second acetate is cut into pieces that represent fractions. Thus

The author is grateful to John J. Callahan, Boston State College, for assistance in the preparation of this article.

the child's aid has two components, the acetate and the matching pieces, the latter being kept in an envelope.

The procedure for the understanding of addition is as follows:

1. Place the acetate projectual (fig. 1) on the desk.

2. Gather the pieces that represent the fractional numbers to be added, for example, $\frac{1}{2} + \frac{2}{8}$ (fig. 2).

3. Join the pieces and place them on the portion of the acetate marked "1 whole."

4. Slide the pieces down until the right end meets a line—for this example, $\frac{3}{4}$ (fig. 3).

5. Look above the line to the name of the point. This represents the sum of the two fractional numbers.

At this time in the lesson some children might discover that $\frac{1}{2} + \frac{2}{8}$ also equals $\frac{6}{8}$ or $\frac{12}{16}$. This would be a suitable time to discuss the fact that $\frac{3}{4}$, $\frac{6}{8}$, and $\frac{12}{16}$ name the same number. When our class discovered that many seemingly different results are sometimes found by using this procedure, we developed a rule stating that although all answers are correct, the best result is the solution on the line closest to the line labeled "1 whole," the simplest form of the fractional number.

Reprinted from *The Arithmetic Teacher* (October 1971), 417–419, by permission of the author and the publisher.

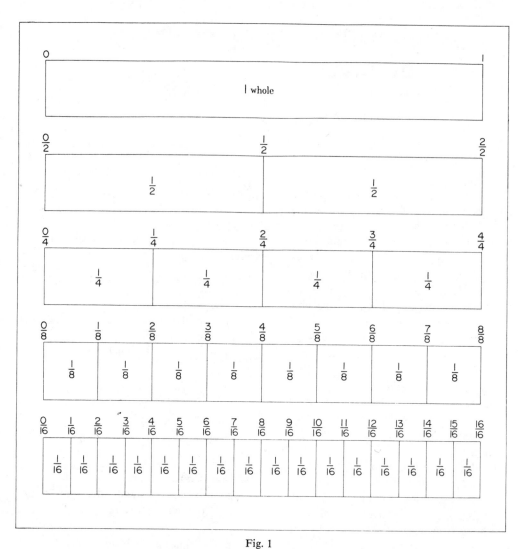

Fig. 1

Fig. 2

For subtraction the first and second steps are the same, but in step three the form is different.

In the example $\frac{1}{2} - \frac{1}{8}$, the piece that represents $\frac{1}{8}$ is placed on top of and on the right hand side of the piece that represents $\frac{1}{2}$. The child starts at the top and moves the pieces down until the left side of the $\frac{1}{8}$ piece is aligned with a point named by a fractional numeral, for example, $\frac{1}{2} - \frac{1}{8} = \frac{3}{8}$ (fig. 4).

Many problems with fractional numbers less than one were pursued in our class and the answers were derived with the help of this aid. Further, after using this device for a short period of time many of the children could solve similar problems with paper and pencil only.

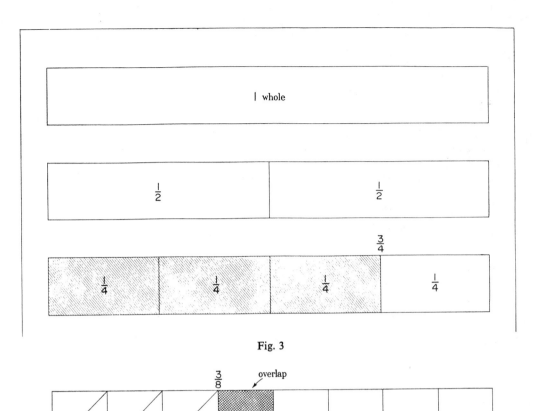

Fig. 3

Fig. 4

Adler, Irving. "Mathematics for the Low Achiever." *NEA Journal* (February 1965), 28.

Beckman, Milton W. "Teaching the Low Achiever in Mathematics." *Mathematics Teacher* (October 1969), 443–46.

Braumfield, Peter and Wolfe, Martin. "Fractions for Low Achievers." *The Arithmetic Teacher* (December 1966), 647–55.

Callahan, John and Jacobson, Ruth. "An Experiment with Retarded Children and Cuisenaire Rods." *The Arithmetic Teacher* (January 1967), 10–13.

Castaneda, Alberta. "A Mathematics Program for Disadvantaged Mexican-American First Grade Children." *The Arithmetic Teacher* (May 1968), 413–19.

Cawley, John and Goodman, John. "Interrelationships Among Mental Abilities, Reading, Language Arts, and Arithmetic with the Mentally Handicapped." *The Arithmetic Teacher* (November 1968), 631–36.

Chandler, Arnold. "Mathematics and the Low Achiever." *The Arithmetic Teacher* (March 1970), 196–98.

Connolly, Austin. "Research in Mathematics Education and the Mentally Retarded." *The Arithmetic Teacher* (October 1973), 491–97.

Davies, Robert A. "Low Achiever Lesson in Primes." *The Arithmetic Teacher* (November 1969), 529–32.

Dickson, Judy. "The Plight of a Child." *The Arithmetic Teacher* (January 1968), 19–22.

Dunkley, M. E. "Some Number Concepts of Disadvantaged Children." *The Arithmetic Teacher* (May 1965), 359–61.

Dutton, Wilbur H. "Teaching Time Concepts to Culturally Disadvantaged Primary-Age Children." *The Arithmetic Teacher* (May 1967), 358–64.

Easterday, Kenneth E. "An Experiment With Low Achievers in Arithmetic." *The Mathematics Teacher* (November 1964), 462–68.

———. "A Technique for Low Achievers." *The Mathematics Teacher* (October 1965), 519–21.

Folson, Mary. "New Math (Too Verbal for the Disadvantaged)." *Instructor* (March 1967), 26–27, 72–74, 166.

Fremont, Herbert. "Some Thoughts on Teaching Mathematics to Disadvantaged Groups." *The Arithmetic Teacher* (May 1964), 319–22.

Fremont, Herbert and Ehrenberg, Neal. "The Hidden Potential of Low Achievers." *The Mathematics Teacher* (October 1966), 551–57.

Greenholz, Sarah. "What's New in Teaching Slow Learners in Junior High School?" *The Mathematics Teacher* (December 1964), 522–28.

Hammitt, Helen. "Evaluating and Teaching Slow Learners." *The Arithmetic Teacher* (January 1967), 40–41.

Henry, Boyd. "Why Can't Johnny Cipher?" *The Arithmetic Teacher* (January 1971), 37–39.

Higgins, Conwell and Rusch, Reuben. "Remedial Teaching of Multiplication and Division." *The Arithmetic Teacher* (January 1965), 32–38.

Homan, Doris Ruth. "The Child With a Learning Disability in Arithmetic." *The Arithmetic Teacher* (March 1970), 199–203.

Howell, Daisy. "Project SOSO (Save Our Slow Ones)." *The Arithmetic Teacher* (January 1972), 29–33.

Jacobson, Ruth. "A Structured Method for Arithmetic Problem-Solving in Special Education." *The Arithmetic Teacher* (January 1969), 25–27.

Kaplan, Jerome D. "An Example of a Mathematics Instructional Program for Disadvantaged Children." *The Arithmetic Teacher* (April 1970), 332–34.

Keiffer, Mildred. "The Development of Teaching Materials for Low-Achieving Pupils in 7th and 8th Grade Mathematics." *The Arithmetic Teacher* (November 1968), 599–604.

Kevra, Barbara, Brey, Rita and Schimmel, Barbara. "Success for Slower Learners, or R$_x$: Relax . . ." *The Arithmetic Teacher* (May 1972), 335–43.

Lerch, Harold and Kelly, Francis. "A Mathematics Program for Slow Learners at the Junior High Level." *The Arithmetic Teacher* (March 1966), 232.

The Low Achiever in Mathematics. Superintendent of Documents, U.S. Government Printing Office, Washington, D.C. 20402 (35¢).

Mintz, Natalie and Freemont, Herbert. "Some Practical Ideas for Teaching Mathematics to Disadvantaged Children." *The Arithmetic Teacher* (April 1965), 258–60.

Paschal, Billy. "A Concerned Teacher Makes the Difference." *The Arithmetic Teacher* (March 1966), 203.

——. "Teaching the Culturally Disadvantaged Child." *The Arithmetic Teacher* (May 1966), 369.

——. "Geometry for the Disadvantaged." *The Arithmetic Teacher* (January 1967), 4–6.

Proctor, Amelia D. "A World of Hope— Helping Slow Learners Enjoy Mathematics." *The Mathematics Teacher* (February 1965), 118–22.

Rebec, Linda. "A Case Against Teaching Mathematical Concepts to Slow-Learning Children." *The Arithmetic Teacher* (May 1972), 333–34.

Richbart, Lynn A. "Remedial Mathematics— A Clinical Approach." *New York State Mathematics Teachers' Journal* (June 1971), 125–29.

Ross, Ramon, "A Description of Twenty Arithmetic Underachievers." *The Arithmetic Teacher* (April 1964), 235–41.

Rosskopf, Myron and Kaplan, Jerome. "Educating Mathematics Specialists to Teach Children from Disadvantaged Areas." *The Arithmetic Teacher* (November 1968), 606–12.

Rouda, Eileen M. "Success for All: An Adventure in Learning." *The Arithmetic Teacher* (January 1972), 35–37.

Snedeker, Noreen D. "Methods, Math, and Mothers, or What Can A Poor Parent Do?" *The Arithmetic Teacher* (February 1968), 156–57.

Stenzel, Jane. "Math for the Low, Slow, and Fidgety." *The Arithmetic Teacher* (January 1968), 30–34.

Weiss, Sol. "What Mathematics Shall We Teach the Low Achiever?" *The Mathematics Teacher* (November 1969), 571–75.

Welsh, Margaret. "The Disabled Child." *Instructor* (January 1968), 155.

CHAPTER 16

REMEDIATION AND ENRICHMENT

Tired of the same old ideas? Looking for something new, different and exciting? All math teachers feel this way on occasion and need to add new stock to their resource files. The articles in this chapter were selected to provide some new (or revised) techniques, strategies, and materials for use in the classroom, or at least point you in that direction.

Lola May, in "Calendar Arithmetic," suggests some interesting and challenging activities that use an ordinary calendar. If you are looking for a functional and interesting exercise with easily obtained materials, read this article and give calendar arithmetic a try.

It isn't much of a leap from "Calendar Arithmetic" to "Clock Arithmetic" by W. G. Quast. This article contains many ideas and activities that can be used with finite mathematics in the elementary classroom. As Quast states, "The study of clock arithmetic will reinforce your students' knowledge of the familiar arithmetic of whole numbers. The students will enjoy discovering familiar properties and new practices in the various clock systems, and at the same time will learn something of the true nature of mathematics."

Magic squares may not be what most people think of as elementary school arithmetic, and perhaps there are few reasons why they should. However, after reading John Cappon's "Easy Construction of Magic Squares for Classroom Use," one can easily see opportunities to fire the imagination of the bright, able student without boring the slower learner.

Have you thought of using *machines* in your elementary math classroom? "They cost too much," you say. Not the "machines" suggested in the article by Marvin Karlin—you can draw these machines. Machines help illuminate basic properties and encourage discovery; children love to work with them and accept drill much more cheerfully when machines are used. You will certainly want to add this idea to your strategies file.

Mental arithmetic—Why not? William Schall, in "Mental Arithmetic: Suggestions for Instruction," provides many examples of mental arithmetic lessons. "Mental arithmetic" encourages the discovery of computational shortcuts and can lead to deeper insights into the number system. Challenge your class with a "follow-me mental exercise."

Need more ideas and suggestions? Then be sure to read some of the articles listed in the bibliography; they all offer many excellent ideas.

Calendar Arithmetic

Lola J. May

A large calendar, preferably one made by the pupils themselves, makes a functional and exciting bulletin board display. Holidays and days of special interest, weather data, the phases of the moon, etc., can be marked on such a calendar, which may also serve as the starting-point of a lively introductory lesson for pupils in the lower and middle grades of elementary school.

SUN.	MON.	TUES.	WED.	THUR.	FRI.	SAT.
	1	2	3	4	5	6
7	8	9	10	11	12	13
14	15	16	17	18	19	20
21	22	23	24	25	26	27
28	29	30	31			

Reprinted from "Mathematics Enrichment Notes," No. 1, by permission of Harcourt Brace Jovanovich, Inc.

	FRI.	SAT.
	5	6
	12	13

Suppose you have a large bulletin board calendar similar to the one shown on the preceding page. Have the pupils select a 2-by-2 square of dates from the calendar. Such a square is shown at the top left. Now have part of the class find the sum of the numbers on one diagonal, say 5 and 13, while the remaining children find the sum of the numbers on the other diagonal, 6 and 12. They will be interested to note that the two sums are the same. Challenge them to choose other 2-by-2 squares on the calendar and test for a similar relationship. The pupils will find this exciting and much addition will be practiced before the search is completed.

THUR.	FRI.	SAT.
4	5	6
11	12	13
18	19	20

Now have the children select a 3-by-3 square, as, for example, the one shown at the middle left. Ask them to test whether a relationship similar to that found for 2-by-2 squares holds for this one. When they have discovered that the sums for the diagonals are the same, tell them to find the sum of the center column and the sum of the center row. Some of them, at least, will be surprised that these sums are the same as those for the diagonals. Then have the pupils consider the center number of this square. If they have studied multiplication, have them multiply this number by 3 and compare the product with the sums for the diagonals and for the center column and center row. Pupils who have not yet learned to multiply may check the relationship by adding the center number three times. Encourage the pupils to find their own 3-by-3 number squares on the calendar and to find the magic sum for the center row and column and the diagonals.

WED.	THUR.	FRI.	SAT.
3	4	5	6
10	11	12	13
17	18	19	20
24	25	26	27

For pupils in the middle grades who may need practice on more advanced addition, try working with a 4-by-4 number square, such as that shown at the bottom left. The pupils will of course note that the sums for the diagonals on this square are the same. There are other relationships worth exploring in such a square. Have the class find the sums for the first 3 columns starting on the left, then the sums for the first 3 rows starting at the top. Now have them note the relationships between the sums for the columns and then predict, without adding, the sum for the last column. In a similar way, have them predict the sum for the last row. You may wish to challenge pupils to explain why the sum for each row is 28 larger than the sum for the row above it, and why the sum for each column is 4 larger than the sum of a column to its right. Some pupils may be able to see that, since the numbers in each row are 7 greater than the corresponding numbers in the row above them, then the sum must be 4×7, or 28, greater. Similarly, since the numbers in each column are 1 greater than the corresponding numbers in the column to the right, the sum of each column will be 4×1, or 4, greater than the sum for the column to its right.

Another interesting relationship may be explored with a 3-by-3 number square. Such a square is shown at the right. Challenge pupils to tell why the sums for the diagonals and for the center column and center row are the same. If no pupil is able to explain this relationship, it may be demonstrated as follows: Place a pencil on the center column (2, 9, 16 in this case), so that it touches the numerals. Now rotate the pencil clockwise one space. The pencil is now on the diagonal 3, 9, 15). Pupils should then note that the numeral under the top of the pencil has changed from 2 to 3, an increase of 1, while that under the bottom of the pencil has changed from 16 to 15, a decrease of 1. The center of the pencil is still, of course, on 9. Thus they can see that the top number for the diagonal is 1 greater than the top number for the center column, and that the bottom number for the diagonal is 1 less than the bottom number for the column. The center number of the diagonal and the center number of the center column are, of course, the same, and consequently the sum for the two sets of numbers will be the same. A similar demonstration will enable pupils to see how the sum for the other diagonal is the same as the sum for the center column. By way of a final example, by rotating the pencil from the center column clockwise until it lies along the center row, pupils will be able to see that the top number of the column has been increased by 8, while the bottom number of the column has changed from 16 to 8, a decrease of 8. Since the center number has remained the same, the sums of the two sets of numbers will also be the same.

MON.	TUES.	WED.
1	2	3
8	9	10
15	16	17

A further interesting relationship on a calendar is seen when you examine the sums of the numbers named by the dates in any one row on the calendar.

If, on the calendar shown at the beginning of this article, you were to take the sum of the numbers represented by the dates from, say, Sunday the 7th to Saturday the 13th, you would note that it is 70. If, further, you divide this sum by the number represented by Wednesday's date, 10, you would find that the sum is 7 times as large as the date for Wednesday. Is this true for another complete row on the calendar? Test the next row. You find that $(14 + 15 + 16 + 17 + 18 + 19 + 20) \div 17 = 7$. Challenge pupils to see if this is true for the remaining complete row on the calendar. Then ask them to. determine if a similar rule holds for weeks in other months. The pupils may develop informal "formulas" concerning the calendar. In the cases discussed above, such a formula would read: "The sum of the dates of the week, Sunday through Saturday, equals seven times Wednesday's date."

387

CLOCK ARITHMETIC

W. G. Quast

We are about to examine some new mathematical systems which use familiar numerals, but whose numbers have properties different from those of the real number system, the numbers we commonly use. In essence, the arithmetic about to be developed deals with the numbers represented on the face of a clock. We shall begin with the familiar clock face which shows twelve hours, but any set of numbers shown on a clock face may be used. Only some of the familiar properties of real numbers will hold true for clock numbers. Those of you who are familiar with the topic of linear congruences will recognize its relationship to clock arithmetic.

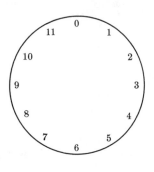

Consider the clock face at the right. The numbers 0 through 11 are associated with particular points on the clock. Zero is more convenient than twelve to indicate the starting point. We shall refer to these numbers as numbers on the "12-clock". The symbol $\stackrel{12}{=}$ shows equality of clock numbers and distinguishes clock equations from other equations involving similar numerals. The symbols $\oplus$, $\ominus$, $\otimes$, and $\oslash$ will indicate the operations we are about to define.

We may define addition of clock numbers as clockwise movement on the face of the clock. Thus on the 12-clock, $3 \oplus 4 \stackrel{12}{=} 7$; that is, four hours after 3 o'clock is 7 o'clock. $5 \oplus 9 \stackrel{12}{=} 2$ since counting 9 units from 5 in a clockwise direction takes us to 2.

You can use the definition of addition stated above to complete this table for addition of 12-clock numbers.

Addition of 12-Clock Numbers

$\oplus$	0	1	2	3	4	5	6	7	8	9	10	11
0	0	1	2	3	4	5	6	7	8	9	10	11
1	1	2	3	4	5	6	7	8	9	10	11	0
2	2	3	4	5	6	7	8	9	10	11	0	1
3	3	4	5									
4	4											
5												
6												
7												
8												
9												
10												
11												

Looking at the completed 12-clock addition table, we can see that the set of 12-clock numbers is closed for addition since each sum is a number from 0 to 11. The first row and the first column of the table show that 0 is the additive identity element for the set of 12-clock numbers. Does $2 \oplus 6 \stackrel{12}{=} 6 \oplus 2$; $8 \oplus 5 \stackrel{12}{=} 5 \oplus 8$? A careful look at the table assures that 12-clock addition is cummutative. If a, b, and c are *any* 12-clock numbers, do you think that $a \oplus (b \oplus c) \stackrel{12}{=} (a \oplus b) \oplus c$? If so, the addition of 12-clock numbers satisfies the associative property. Check some examples to convince yourself that this is the case. Consider the following six equations:

$$4 \oplus 8 \overset{12}{=} a \qquad\qquad 9 \oplus d \overset{12}{=} 0$$

$$7 \oplus 5 \overset{12}{=} b \qquad\qquad 11 \oplus e \overset{12}{=} 0$$

$$2 \oplus 10 \overset{12}{=} c \qquad\qquad 6 \oplus f \overset{12}{=} 0$$

It is clear that for any 12-clock number, a, there exists a 12-clock number, b, such that $a \oplus b \overset{12}{=} 0$; that is, every 12-clock number has an additive inverse.

Since addition of 12-clock numbers was defined as clockwise movement, it would seem logical to think of subtraction as counterclockwise movement. Ten hours before 7 o'clock is 9 o'clock; that is, $7 \ominus 10 \overset{12}{=} 9$. Subtraction exercises with 12-clock numbers may also be solved by using the addition table since subtraction is the inverse (opposite) of addition. For example, to solve the clock equation $4 \ominus 6 \overset{12}{=} x$, we locate the given addend, 6, along the left edge of the table, then read across the row until we come to the sum, 4. The 4 is in the column labeled 10, so the missing addend is 10. Every missing addend is a number from 0 to 11, so the set of 12-clock numbers is closed with respect to subtraction. Unlike addition, subtraction is not cummutative. For example, $5 \ominus 10 \overset{12}{\neq} 10 \ominus 5$. Nor is subtraction associative, since, for example, $5 \ominus (10 \ominus 11) \overset{12}{\neq} (5 \ominus 10) \ominus 11$.

Now, to define multiplication of clock numbers, we use the familiar idea of repeated addition, that is, repeated clockwise movement. $3 \otimes 5$ would be $5 \oplus 5 \oplus 5 \overset{12}{=} 3$. With this definition in mind, you can complete the following table for multiplication of 12-clock numbers.

Multiplication of 12-Clock Numbers

$\otimes$	0	1	2	3	4	5	6	7	8	9	10	11
0	0	0	0	0	0	0	0	0	0	0	0	0
1	0	1	2	3	4	5	6	7	8	9	10	11
2	0	2	4	6	8	10	0	2	4	6	8	10
3	0	3	6	9	0	3						
4	0	4	8	0								
5	0	5	10									
6	0											
7												
8												
9												
10												
11												

When the table is complete, you will see that the set of 12-clock numbers is closed for multiplication. The multiplicative identity element is 1, which you can verify by looking at the row and column labeled 1 in the table. As with addition, multiplication for 12-clock numbers is commutative and associative. Multiplication of 12-clock numbers is also distributive over addition of 12-clock numbers as shown in the following example.

$$8 \otimes (9 \oplus 6) \overset{12}{=} 8 \otimes 3 \overset{12}{=} 0$$

$$8 \otimes (9 \oplus 6) \overset{12}{=} (8 \otimes 9) \oplus (8 \otimes 6) \overset{12}{=} 0 \oplus 0 \overset{12}{=} 0$$

Use the table to verify these examples:

$$10 \otimes 8 \overset{12}{=} 8 \otimes 10 \overset{12}{=} 8 \qquad\qquad 5 \otimes (7 \otimes 2) \overset{12}{=} (5 \otimes 7) \otimes 2 \overset{12}{=} 10$$

389

Solve the following 12-clock equations:

$$5 \otimes 5 \stackrel{12}{=} a \qquad 11 \otimes 11 \stackrel{12}{=} c \qquad 6 \otimes a \stackrel{12}{=} 1 \qquad 2 \otimes a \stackrel{12}{=} 1$$

$$7 \otimes 7 \stackrel{12}{=} b \qquad 4 \otimes a \stackrel{12}{=} 1 \qquad 8 \otimes a \stackrel{12}{=} 1 \qquad 9 \otimes a \stackrel{12}{=} 1$$

From the above equations what can you conclude about the existence of multiplicative inverses for each 12-clock number?

Since division of 12-clock numbers is the inverse operation for multiplication of 12-clock numbers, the multiplication table will be helpful in solving division exercises. For example, $6 \div 7 \stackrel{12}{=} x$ can be solved by locating the given factor, 7, along the left edge of the table and then reading across that row until the product, 6, is located. The missing factor is at the top of the column where the product appears. Thus $6 \div 7 \stackrel{12}{=} 6$. In some clock systems, including the 12-clock system, division is not closed. As a result, a division exercise may have more than one solution or no solution at all, as in the following examples.

$$(1)\ 4 \div 6 \stackrel{12}{=} d \qquad\qquad (2)\ 8 \div 4 \stackrel{12}{=} e$$

No solution   Solutions: 2, 5, 8, or 11

We know that if the product of any two real numbers is zero, then one of the factors must be zero; if $4 \times y = 0$, then $y = 0$. Notice that this property does not always apply to 12-clock numbers. For example, $3 \otimes 4 \stackrel{12}{=} 0$, but neither of the factors is zero. Look at the multiplication table for other pairs of nonzero 12-clock numbers which have a product of zero.

An interesting difference between the set of whole numbers and the set of 12-clock numbers arises when we try to order the set of 12-clock numbers. Given any two whole numbers which are not equal, we can say that one is either greater than or less than the other. In general, for two whole numbers a and b, $a < b$ if and only if there is a whole number c such that $a + c = b$. Let us apply this principle to the numbers of the 12-clock system.

$$5 \oplus 4 \stackrel{12}{=} 9 \text{ implies that } 5 \stackrel{12}{<} 9, \text{ but}$$

$$9 \oplus 8 \stackrel{12}{=} 5 \text{ implies that } 9 \stackrel{12}{<} 5.$$

Since the above statements contradict each other, we must conclude that the numbers on the 12-clock are not ordered. This last statement makes sense because, on a clock, 5 is clockwise from 9 but 9 is also clockwise from 5. The numbers in *any* clock system cannot be ordered.

We have developed a simple arithmetic which involves a finite set of numbers, 0 to 11, rather than the set of numbers of arithmetic, which is infinite. Actually we may form a clock arithmetic with any finite set of numbers. Consider the 7-clock at the right with the numbers 0 to 6. If we define the fundamental operations as we did for the 12-clock (that is, addition as clockwise movement), we can form another clock system.

Complete these tables for addition and multiplication of 7-clock numbers.

Addition of 7-Clock Numbers

⊕	0	1	2	3	4	5	6
0	0	1	2	3	4	5	
1	1	2	3	4			
2	2	3	4				
3	3	4	5				
4	4	5					
5	5						
6							

Multiplication of 7-Clock Numbers

⊗	0	1	2	3	4	5	6
0	0	0	0	0	0	0	0
1	0	1	2	3	4	5	6
2	0	2	4	6	1		
3							
4							
5							
6							

These tables will help you to verify that the following properties will hold for the set of numbers on the 7-clock.

(1) Addition is closed.
(2) Multiplication is closed.
(3) Addition is commutative.
(4) Multiplication is commutative.
(5) Addition is associative.
(6) Multiplication is associative.
(7) Multiplication is distributive over addition.
(8) There is an additive identity element.
(9) There is a multiplicative identity element.
(10) Each number has an additive inverse.
(11) Each nonzero number has a multiplicative inverse.

What is there about the number 7, as compared to 12, that might explain why property (11) holds for the 7-clock system but not for the 12-clock system? Test your conjecture with other clock systems.*

A set of numbers for which two operations are defined and which satisfies the 11 properties listed above is called a *field*. The 7-clock system of numbers forms a field while the 12-clock system does not, since the 12-clock numbers do not all have multiplicative inverses. Under the ordinary operations of addition and multiplication, the set of rational numbers and the set of real numbers, for example, form a field; but the whole numbers, the integers, the odd integers, or the nonnegative rational numbers do not. The set of whole numbers contains no multiplicative or additive inverses, while the set of integers contains no multiplicative inverses. There are other examples of sets of numbers which do not form fields. Not only are there no multiplicative inverses for the set of odd integers, but also the set is not closed with respect to addition, and it contains no additive identity element. It can also be argued that properties 3, 5, and 7 do not work either since they would involve even numbers. The set of nonnegative rational numbers does not contain additive inverses.

Clock arithmetic provides an example of one of the simplest of mathematical structures, a *group*. We will define a group abstractly.

A group is a set S with an operation $*$ defined on the elements of the set such that the following conditions are satisfied:

(1) Closure: For any elements a and b in S, $a * b$ is in S.

(2) Identity: There exists an element of S, which we will denote as i, such that for any a in S, $i * a = a * i = a$.

(3) Inverse: For every element a in S, there is a unique element, called the inverse of a, denoted a^{-1}, such that $a^{-1} * a = a * a^{-1} = i$.

(4) Associativity: For any elements a, b, and c in S, $(a * b) * c = a * (b * c)$.

If in addition to the properties above, the group is commutative, that is, for any a and b in S, $a * b = b * a$, the group is said to be an Abelian group. Addition with the numbers of any clock system forms an Abelian group, but multiplication with the nonzero numbers of a clock system will form an Abelian group only if the number of elements in the system is prime (as, for example, multiplication with the numbers 1 to 6 associated with the 7-clock). As it turns out, a clock system will also form a field if the number of elements on the clock face is prime.

The study of clock arithmetic will reinforce your students' knowledge of the familiar arithmetic of whole numbers. The students will enjoy discovering familiar properties and new properties in the various clock systems, and at the same time will learn something of the true nature of mathematics.

*7 is a prime number.

Easy construction of magic squares for classroom use

JOHN CAPPON, SR.

In the December, 1963, issue of THE ARITHMETIC TEACHER Bryce E. Adkins gave six rules for construction of odd-cell magic squares.[1] His result for a five-by-five magic square is given in Figure 1. It is

17	24	1	8	15
23	5	7	14	16
4	6	13	20	22
10	12	19	21	3
11	18	25	2	9

Figure 1

noticed that all horizontal rows, vertical columns, and diagonals have the total of 65.

Actually, the construction of this scheme can be simplified considerably and is easier to remember when certain numerals are replaced in symmetric positions outside the borders of the square as is customary, for example, in the calculation of determinants of the third rank.

A European method proceeds as follows: Cells are added outside the square by building a pyramid on each border line and placing the digits in the order illustrated in Figure 2 for a five-by-five square. Then, the extra cells of the pyramids are used to fill the empty cells within the

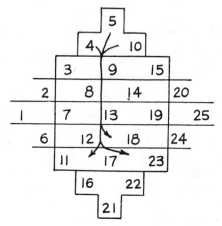

Figure 2

3	16	9	22	15
20	8	21	14	2
7	25	13	1	19
24	12	5	18	6
11	4	17	10	23

Figure 3

square along the opposite border. The result is shown in Figure 3.

Although the integers 1–25 were used in the example, this is not necessary. Any sequence of numbers which is part of an arithmetical progression will suffice. Moreover, as Frances Hewitt[2] has shown for a

[1] Bryce E. Adkins, "Adapting magic squares to classroom use," THE ARITHMETIC TEACHER, X (December, 1963), 498–500.

[2] Frances Hewitt, "4×4 magic squares," THE ARITHMETIC TEACHER, IX (November, 1962), 392–395.

Reprinted from *The Arithmetic Teacher* (February 1965), 100–105, by permission of the author and the publisher.

four-by-four square, there can be gaps in the progression between the parts of the rows. This holds also for all odd-cell magic squares; e.g., the sequence of numbers for a three-by-three square can be chosen as 2, 5, 8 (11, 14), 17, 20, 23 (26, 29), 32, 35, 38, where the numbers in parentheses are omitted (Figs. 4a and 4b).

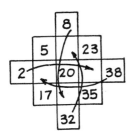

Figure 4a

5	32	23
38	20	2
17	8	35

Figure 4b

Furthermore, the gaps can be taken as any number, positive or negative, if they

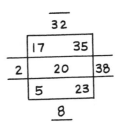

Figure 5a

17	8	35
38	20	2
5	32	23

Figure 5b. *This is the reverse of* 4b!

are all the same, e.g., 2, 17, 32; 5, 20, 35; 8, 23, 38. The gap here is 27 (Figs. 5a and 5b).

All these difficult cases can be summarized in one general rule: The sequence of numbers in the "pyramidally extended square" is formed by two arithmetical progressions, one determining the numbers from the most left-placed cell to the upper cell, and another one determining the numbers from the most left-placed cell to

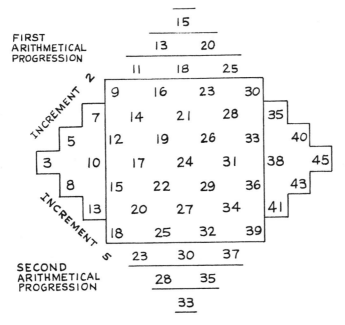

Figure 6a

9	23	16	30	23	37	30
35	14	28	21	35	28	7
12	40	19	33	26	5	33
38	17	45	24	3	31	10
15	43	22	15	29	8	36
41	20	13	27	20	34	13
18	11	25	18	32	25	39

Figure 6b. *The magic square formed by Figure* 6a.

the bottom cell. Both arithmetical progressions can be taken arbitrarily. (See Figs. 6a and 6b.)

The rule given by Frances Hewitt, that a new magic square can be formed from a given square by adding or multiplying the digits with a certain number or fraction, follows immediately from our general rule.

Adkins gave an easy rule for a four-by-four magic square formed from the numbers 1–16 placed in four rows of four digits. The series of the numerals on each diagonal must be reversed. Again, it is not necessary to restrict oneself to the numbers 1–16. In fact, we can again build up the sequence of numbers from two arithmetical progressions, one denoting the order in the first row and the other the order in the first column. (See Fig. 7a.)

Adkins' rule also applies to a square of these numerals; the square of Figure 7a is converted to a magic square by reversing the diagonals. (See Fig. 7b.)

Practical applications

For adopting magic squares to classroom use it is desirable to have an easy method that enables us to construct a magic square for any given number. Most easily one may employ a diagonal for the construction of the auxiliary figures of either the odd-cell squares or the four-by-four squares. The elements of this diagonal are always part of an arithmetical progression.

In the odd-cell squares (see Fig. 6a) the increment of the horizontal diagonal (7) equals the sum of the increments of the border rows (2 and 5 respectively in the figure), while the increment of the vertical diagonal (3) equals the difference of the border row increments. The same applies to the oblique diagonals of the four-by-four squares (see Fig. 7a).

Furthermore, the sum of the first and last elements of either diagonal is an important quantity. For the odd-cell squares this sum equals twice the term in the middle, which in turn equals the magic sum divided by the rank of the square (see Fig. 6a: $[3+10+17+24+31+38+45]\div 7 = 24$). In the four-cell squares the sum of the first and the last elements of the diagonal equals half the total of the four elements. With the knowledge of this it is possible to construct any magic square with a given number as totals of rows, columns, or diagonals.

Columbus discovered America in the year 1492; let us make a four-by-four magic square with totals 1492. Starting with the diagonal of the auxiliary figure, the first and the last numbers of the diagonal must equal $1492 \div 2 = 746$. The last number of the diagonal equals the first one plus three increments, from which it follows that the difference of the num-

INCREMENT 3 \ INCREMENT 5				
	2	7	12	17
	5	10	15	20
	8	13	18	23
	11	16	21	26

Figure 7a

26	7	12	11
5	18	13	20
8	15	10	23
17	16	21	2

Figure 7b

bers must be divisible by three (if we want to avoid fractions). The numbers 100 and 646 suffice. The increment is then $(646 - 100) \div 3 = 182$. The entire diagonal is therefore 100, 282, 464, and 646. Further, we split this increment in two parts, e.g., 82 and 100, and use these parts to construct the first row and the first column. Then the rest is known. The example is demonstrated in Figures 8a and 8b.

INCREMENT 82

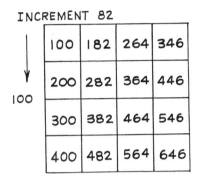

Figure 8a. Auxiliary square.

646	182	264	400
200	464	382	446
300	364	282	546
346	882	564	100

Figure 8b. Magic square after reversing diagonals.

After the teacher has made up the square the children check the results by performing the fifteen possible additions: 4 rows, 4 columns, 2 diagonals, 4 two-by-two squares in the corners, and 1 two-by-two square in the middle. By the choices of the first and last numbers of the diagonal and the increment, the teacher can vary the difficulty of the arithmetic calculation according to the progress of the class.

The year 1964 consists of 19 and 64. We shall make a magic square for 19 and 64. The number 19 is a prime number. Hence,

we must avail ourselves of fractions. In the auxiliary figure for the five-by-five square, one places the number $19 \div 5 = 3.8$ in the middle. (See Fig. 9a.)

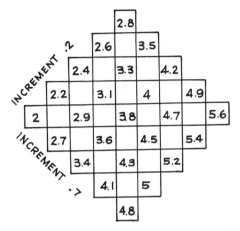

Figure 9a. Auxiliary figure for magic number 19.

The left-cell of the horizontal diagonal is less than 3.8, e.g., 2. The increment of the diagonal is then fixed, viz. $(3.8 - 2) \div 2 = 0.9$. The complete diagonal is now known. Then the increment 0.9 is split into two numbers, e.g., 0.2 and 0.7, which determines the border rows. We chose these numbers so that the entire square contains all different numbers. The resulting magic square is Figure 9b.

2.4	4.1	3.3	5	4.2
4.9	3.1	4.8	4	2.2
2.9	5.6	3.8	2	4.7
5.4	3.6	2.8	4.5	2.7
3.4	2.6	4.3	3.5	5.2

Figure 9b. The magic square of 19.

The magic square for 64 is made as the one for 1492. The diagonal is 4, 12, 20, and 28. The increment (8) we split in 7 and 1 (Figs. 10a and 10b).

If interested in the magic squares for 32 or 16, one can utilize the magic square of 64 by dividing all terms by 2 and 4 respectively. (See Figs. 10c and 10d.)

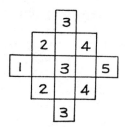

Figure 10a. *Auxiliary square for 64.*

4	11	18	25
5	12	19	26
6	13	20	27
7	14	21	28

Figure 10a. Auxiliary square for 64.

28	11	18	7
5	20	13	26
6	19	12	27
25	14	21	4

Figure 10b. Magic square after diagonal reverse.

14	$5\frac{1}{2}$	9	$3\frac{1}{2}$
$2\frac{1}{2}$	10	$6\frac{1}{2}$	13
3	$9\frac{1}{2}$	6	$13\frac{1}{2}$
$12\frac{1}{2}$	7	$10\frac{1}{2}$	2

Figure 10c. Magic square for 32.

7	$2\frac{3}{4}$	$4\frac{1}{2}$	$1\frac{3}{4}$
$1\frac{1}{4}$	5	$3\frac{1}{4}$	$6\frac{1}{2}$
$1\frac{1}{2}$	$4\frac{3}{4}$	3	$6\frac{3}{4}$
$6\frac{1}{4}$	$3\frac{1}{2}$	$5\frac{1}{4}$	1

Figure 10d. Magic square for 16.

It is not always possible to split the increment of the diagonal in such a manner that no two numbers occur twice, particularly in squares with a low magic sum as might be used in lower grades. Suppose a child in the third grade has his ninth birthday, for which occasion the teacher plans to make a magic square with totals equal to nine. The middle term in a three-by-three square is then $9 \div 3 = 3$; the increment of the diagonal is $3 - 1 = 2$. This is divided in 1 and 1 (if no fractions are to be used). (See Figs. 11a and 11b.)

Figure 11a. Auxiliary square for 9.

2	3	4
5	3	1
2	3	4

Figure 11b. Magic square for 9.

The magic square with totals 10 can be made with a four-by-four square. The diagonal is 1, 2, 3, and 4. The fraction $\frac{1}{2}$ is chosen for both increments. (See Figs. 12a and 12b.)

1	$1\frac{1}{2}$	2	$2\frac{1}{2}$
$1\frac{1}{2}$	2	$2\frac{1}{2}$	3
2	$2\frac{1}{2}$	3	$3\frac{1}{2}$
$2\frac{1}{2}$	3	$3\frac{1}{2}$	4

Figure 12a. Auxiliary square for 10.

4	$1\frac{1}{2}$	2	$2\frac{1}{2}$
$1\frac{1}{2}$	3	$2\frac{1}{2}$	3
2	$2\frac{1}{2}$	2	$3\frac{1}{2}$
$2\frac{1}{2}$	3	$3\frac{1}{2}$	1

Figure 12b. The magic square for 10.

The magic square for 11 is easily built on a three-by-three square. The middle number is

$$11 \div 3 = 3\frac{2}{3}.$$

Choose $\frac{1}{3}$ for the left cell in pyramid.

Magic squares for 12 and 13 are obtained from Figure 4b and Figure 3 respectively. The numbers in the cells are divided by five.

The magic square for 14 can be obtained by adding one to each cell of Figure 12b, by constructing a four-by-four square (diagonal 2, 3, 4, and 5; increments $\frac{1}{6}$ and $\frac{5}{6}$), or by constructing a seven-by-seven square (in the middle $14 \div 7 = 2$; left-cell $\frac{1}{2}$; increments $\frac{1}{20}$ and $\frac{9}{20}$).

These examples are suitable for all elementary school grades from the third grade up, according to the age of the pupils, providing decimal fractions can be used in seventh- and eighth-grade arithmetic.

The high school teacher can use negative numerals; for example, he can subtract 10 from each cell in Figure 1. He can also change the numerals to another numeration system.

Machines*

MARVIN KARLIN

Introduction

Drop some meat into a meat grinder, turn the handle, and out comes hamburger. Drop a number into our number machine, turn the handle, and out comes a number (Fig. 1).

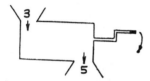

Figure 1

For kindergarten and first grade children the machine accepts objects (Fig. 2).

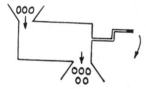

Figure 2

The machine is a versatile tool for (1) illuminating basic properties of numbers under certain operations, and (2) providing drill with motivation.

Inverse operations

Addition-subtraction

Introduce the machine and a table for numbers simultaneously. Give two ex-

* The original machine idea is attributed to Edward Begle, Stanford University. It is used in programmed algebra material of the University of Illinois Committee on School Mathematics (UICSM) and in Wirtz-Botel, *Math Workshop for Children* (Chicago: Encyclopaedia Britannica Press, Inc., 1961).

amples, then ask children to tell what comes out for numbers you put in.

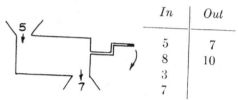

In	Out
5	7
8	10
3	
7	

Figure 3

After presenting quite a few examples like the one in Figure 3, ask the children to name the machine according to what it does. The machine above is an "add 2" machine.

One may utilize the machine in Figure 3 to introduce subtraction to children without new language or notation (Fig. 4). The

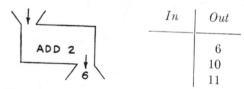

In	Out
	6
	10
	11

Figure 4

inverse relationship between addition and subtraction is clearly demonstrated: subtraction undoes addition. The machine in

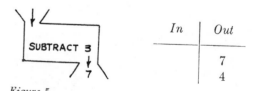

In	Out
	7
	4

Figure 5

Reprinted from *The Arithmetic Teacher* (May 1965), 327–334, by permission of the author and the publisher.

Figure 5 demonstrates that addition undoes subtraction.

Another technique for emphasizing inverse operations is to rotate the machine 180 degrees (Fig. 6). The numerals don't

In	Out
5	7
3	5
7	9

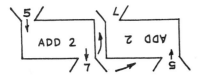

Figure 6

really need to be upside down, of course. Now, let's alter the direction of the answer so that "out" becomes "in," and vice

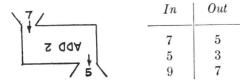

In	Out
7	5
5	3
9	7

Figure 7

versa (Fig. 7). What kind of machine do we now have? (Subtract 2.)

A third technique for exploring inverse operations is the use of several machines

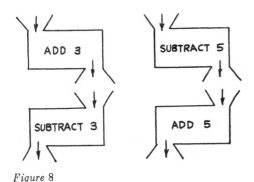

Figure 8

in combination (Fig. 8). Here a number emerges from the first machine and immediately drops into the second. Be certain to keep an In-Out table.

Multiplication and Division

Since multiplication and division are inverse operations, the procedures above can be applied to demonstrate this relationship.

Squaring and unsquaring

Have the children guess what emerges if seven enters the machine in Figure 9.

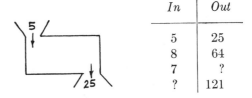

In	Out
5	25
8	64
7	?
?	121

Figure 9

Children may name this a "multiply by itself" machine. Fine. There's no rush to use the proper term, "square" or "squaring" machine.

Suggest that this machine puts out several perfect squares, such as 121 and 36. "What number entered in each case?" Lo and behold, your children will be extracting square roots.

By rotating the machine to get the inverse we obtain what children might call an "unsquaring" machine or, more properly, a "square root" machine. We wait far too long to introduce square root. Here we have a technique which permits early introduction and which is pedagogically sound, since it avoids new terminology and new notation.

Identity elements

Exhibit the machine and In-Out table in Figure 10. Children might call this a "do-nothing" machine. But suppose it must do something. It could "add 0,"

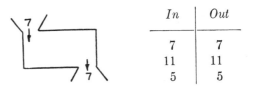

In	Out
7	7
11	11
5	5

Figure 10

399

"subtract 0," "multiply by 1," or "divide by 1."

Zero is the *identity element* under addition, i.e., adding zero to any number of arithmetic leaves that number identically the same: for any number of arithmetic $\square$, $\square + 0 = \square$. Subtracting zero is merely the inverse (undoing) of adding zero.

One is the *identity element* under multiplication: for any number of arithmetic $\square$, $\square \times 1 = \square$. Dividing by one is the inverse of multiplying by one.

Closure properties

Consider the machines in Figure 11.

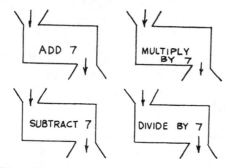

Figure 11

Suppose we choose a whole number and drop it into each machine. Is it possible that any of the machines would not be able to discharge a whole number? Yes, any whole number less than seven would frustrate the "Subtract 7" machine. We find that the set of whole numbers is not closed under subtraction, i.e., we don't always get a whole number as the difference between two given whole numbers. But the set of whole numbers is closed under addition, i.e., the sum of any two whole numbers is always a whole number. The set of whole numbers is also closed under multiplication, but *not* under division, since division by zero is meaningless. But if zero is excluded as a divisor, we shall always obtain a whole number as the quotient of two whole numbers.

Commutative properties

Consider the two machines and the table in Figure 12. Suppose we agree that

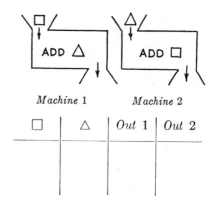

Figure 12

the number we represent in '$\square$' in machine 1 must also be represented in '$\square$' in machine 2, and the same for '$\triangle$.' For example, if we write '3' in '$\square$' and '5' in '$\triangle$,' we obtain the results shown in Figure 13.

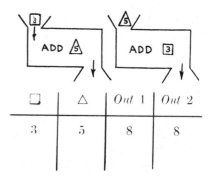

$\square$	$\triangle$	*Out* 1	*Out* 2
3	5	8	8

Figure 13

Now try other possibilities for '$\square$' and '$\triangle$.' Always we obtain the same number in Out 1 as in Out 2. This demonstrates the commutative property for addition: for any whole number $\square$, and for any whole number $\triangle$, $\square + \triangle = \triangle + \square$.

Now use the same format but change the word "add" in each machine to "subtract" (Fig. 14). Since $3 - 5$ is not a whole number, $3 - 5$ is not the same number as $5 - 3$. We thus discover that no commutative property for subtraction exists for the set of whole numbers.

Now try "multiply" and then "divide," as above. Children will find that there is a commutative property for multiplica-

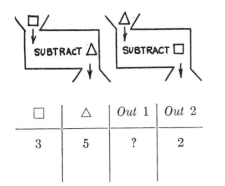

□	△	Out 1	Out 2
3	5	?	2

Figure 14

tion: for any whole number □, and for any whole number △, □×△ = △×□.

They also will find that there is no commutative property for division. For example, 3÷5 is not the same number as 5÷3.

Associative properties

Consider the machines in Figure 15.

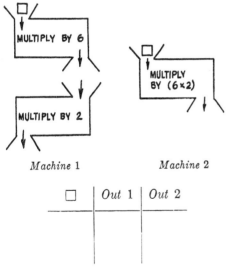

Machine 1   *Machine 2*

□	Out 1	Out 2

Figure 15

Again, we enter the same number in each machine. We find that we obtain the same output from each machine. If we change the two multipliers from 6 and 2 to 7 and 4, or ⅓ and 3, we find that the output is

the same if the same number enters each machine.

We have recognized the associative property for multiplication: for any number of arithmetic □, for any whole number △, and for any whole number ▽, (□×△)×▽=□×(△×▽). For example, if '□' is replaced by '4,' '△' by '6,' and '▽' by '2,' we obtain (4×6)×2 =4×(6×2).

Suppose we change the word "multiply" in the machines to "divide." If 12 enters machine 1, we obtain (12÷6)÷2, or 1. But in machine 2 we obtain 12÷(6÷2), or 4. Therefore, there is no associative property for division of numbers of arithmetic. (One false instance of a generalization [a counterexample] is sufficient to disprove that generalization.)

By trying addition and subtraction we would find that there is an associative property for addition, but none for subtraction. The statement of each property (the true as well as the false) can be obtained merely by changing the multiplication signs in the sentence with frames given above to all addition signs, and then to all subtraction signs.

Distributive property
for multiplication over addition

Consider the machine and table in Figure 16. Again the same number is

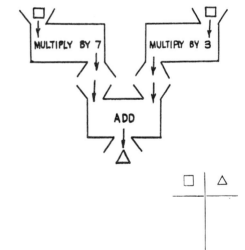

□	△

Figure 16

dropped into each of the top machines. The two numbers discharged from the upper machines are added together in the bottom machine. The following pattern— or one like it—will appear in the table:

□	△
4	40
5	50
9	90
12	120

Change 7 and 3 to 4 and 6, respectively, in the machines. The table remains the same. Change the machine numbers to 8 and 2. Again, the table is unchanged.

We have discovered the left distributive property for multiplication over addition: for any whole number □, for any whole number △, and for any whole number ▽, $\Box \times (\triangle + \triangledown) = (\Box \times \triangle) + (\Box \times \triangledown)$.

Therefore, if '4' replaces '□,' '7' replaces '△,' and '3' replaces ▽, we have $4 \times (7+3) - (4 \times 7) + (4+3)$. The machine performs the work at the right of =, but the pattern in the table advises that the arithmetic to the left of = is equivalent.

The property called the distributive property for multiplication over addition from which the above property is derived is stated as follows: for any whole number △, for any whole number ▽, and for any whole number □, $(\triangle + \triangledown) \times \Box = (\triangle \times \Box) + (\triangledown \times \Box)$.

Machine games

Simplifying complex machines

A table for the complex machine in Figure 17 might look like this:

In	Out
3	4
7	8
15	16
37	38

Children will see the emerging pattern as a table is developed, and their hands will shoot up as quickly as a number is

offered (either entering or leaving). After most children have seen the pattern, draw a single machine on the board and say, "I want this machine to have the same table as the one just completed. What shall we name it?"

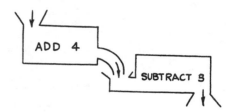

Figure 17

Now try a problem similar to the one in Figure 18 and discover what single machine satisfies.

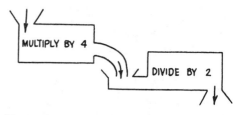

Figure 18

What single machine can replace the complex in Figure 19?

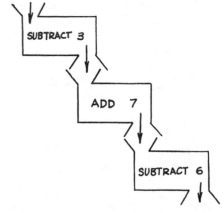

Figure 19

What single machine can replace the complex in Figure 20?

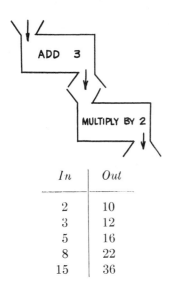

Figure 20

In	Out
2	10
3	12
5	16
8	22
15	36

No pattern is discernible from the table, and, indeed, no single machine can replace the given complex. The question thus arises, "When can a complex machine be replaced by a single machine and when not?" The answer, which children should discover for themselves, is that if, and only if, all machines involve only one operation and its inverse, then one machine utilizing one of those operations will suffice. Thus, if only addition and subtraction are involved or if only multiplication and division are involved, a single machine can replace the given complex.

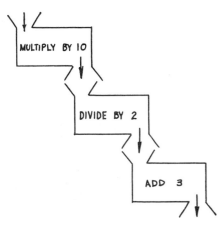

Figure 21

When can part of a complex machine be replaced? Consider the machines in Figure 21. The machine cannot be replaced by a single machine, but it can be simplified, as shown in Figure 22.

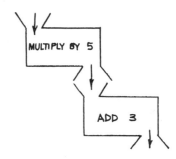

Figure 22

But suppose we change the order of the original machine (Fig. 23). For this machine no obvious simplification is possible.

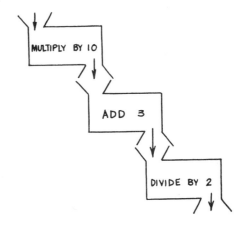

Figure 23

However, some bright youngster may recognize the possibility in Figure 24. Do

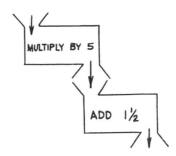

Figure 24

403

not point out this simplification if no child discovers it. Have the children invent machines for the class to simplify.

Name-the-blank machine

Try a name-the-blank machine (Fig. 25).

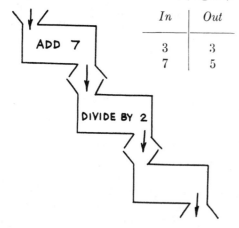

In	Out
3	3
7	5

Figure 25

Many variations on name-the-blank machines are possible, of course. Use the special machine in Figure 16, and omit the number in one of the machines. Try the same machine, but omit both numbers (Fig. 26). In this case the sum of the numbers in each of the top machines must be 13. An interesting sidelight on this problem is that if 10 is the number in the left machine and 3 in the right machine, then this machine demonstrates how we actually multiply a number represented

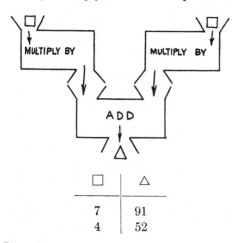

□	△
7	91
4	52

Figure 26

by a single digit and the number 13. For example,

$$\begin{array}{cc} 13 & (10+3) \\ \times 7 & \times 7 \\ \hline 91 & \overline{70+21=91} \end{array}$$

The '2' from the '21' is the carry.

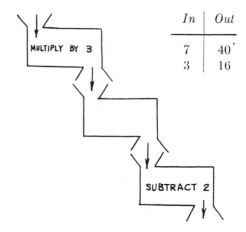

In	Out
7	40
3	16

Figure 27

Figure 27 shows a fairly difficult but intriguing problem. One determines that 21 entered the middle machine and 42 leaves the middle machine for the first entry in our table. For the second entry, 9 entered and 18 left the middle machine. A table for the middle machine makes the pattern clear:

In	Out
21	42
9	18

Yes, it's a "multiply by 2" machine.

Order of complex machines

If the "multiply by 3" machine preceded the "add 5" machine in Figure 28, would our table be the same? No! (In fact, the number emerging would be 10 less than each "out" number presently in the table.)

When can we change the order of the machines without altering the table? Children should discover that if only one operation and its inverse is involved, the order is immaterial.

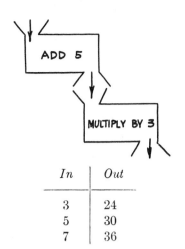

Figure 28

An unusual machine

See if the children can tell what emerges in Figure 29 for other entering numbers.

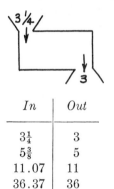

In	Out
$3\frac{1}{4}$	3
$5\frac{3}{8}$	5
11.07	11
36.37	36
14	14
14.7	14

Figure 29

The number line can be quite helpful. Have the children locate the number entering on the number line, then from there find the emerging number.

This machine could be called "the greatest whole number not greater than the number entering" machine. It is valuable for developing an ability to estimate.* For instance, note the following possibilities:

In	Out
$\frac{21}{4}$	
$\frac{16.05}{4}$	
$11.4-4.01$	
7×6.12	
$72\frac{1}{5}-10\frac{1}{2}$	
3.01	

Here we are not interested in the exact answer, only in the greatest whole number not greater than that answer.

Also note that this machine is *not* equivalent to a "rounding off to the nearest whole number" machine, for in our machine 3.7 emerges as 3, while it would be 4 if we were rounding off.

Final comments

An excellent concomitant with the use of machines is arithmetic with frames.† An "add 2" machine, for example, is equivalent to $\square + 2 = \triangle$, and the number entering goes in '$\square$,' while the number exiting goes in '$\triangle$.'

Machines are a great asset for illuminating basic properties and for encouraging discovery. My experience has been that children love to work with them, and accept much drill cheerfully when machines are used. In the words of W. S. Gilbert, "He who'd make his fellow creatures wise should always gild the philosophic pill."

* See David Page's *"Let's Do Something about Estimation,"* University of Illinois Arithmetic Project, Urbana, Illinois.
† See David Page's *"Arithmetic with Frames,"* University of Illinois Arithmetic Project, Urbana, Illinois, 1960.

MENTAL ARITHMETIC: SUGGESTIONS FOR INSTRUCTION

William E. Schall

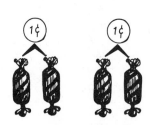

A six-year-old tries to find out how many pieces of two-for-a-penny candy he can get for a nickel.

An adult tries to total a grocery bill as he approaches the check-out counter.

Such experiences as these are common, and present many uses for mental arithmetic in finding quick solutions to everyday situations. Daily activities of children and adults often present a need for simple mental computations, making estimates, arriving at correct answers, and interpreting quantitative data, terms, and statements. However, when asked to perform some of the simplest operations, such as adding, subtracting, or multiplying simple numbers, or dividing by a power of ten, many students will resort to the use of pencil and paper. But the pencil-and-paper methods which comprise almost all arithmetic instruction are often inadequate for the real-life, store situations presented above.

It is important, then, not only that the content taught in the classroom be that which is most needed and most useful in life outside of school, but also that the methods by which it is learned and the way it is put into practice serve to make the daily activities of life easier to perform successfully.

Since written computation should seldom be necessary in dealing with simple quantitative situations like those illustrated, one is led to question whether arithmetic is being taught in a manner that promotes facility in the use of mental arithmetic outside the classroom. The amount of time spent on mental arithmetic is usually small compared with time spent on written arithmetic. In view of what is known about incidental learning, it seems doubtful that children will become adept at handling number situations mentally unless a special attempt is made to develop this type of skill. If pupils are given adequate opportunity to compute mentally, their familiarity with number combinations should gradually lead to automatic responses and provide a backlog of number information.

Reprinted with permission from the *New York State Mathematics Teachers' Journal* (October 1972), 175–181.

Mental arithmetic is also a good way to help children become independent of techniques learned by rigid memorization. It encourages the discovery of computational short cuts and can lead to deeper insights into the number system. Some examples of mental arithmetic lessons of approximately ten-minute duration follow:

I. "Follow-me" sequences, in booklet form, with spaces to be filled in by the pupil. For example, one day's lesson might be as follows:

 Hello, boys and girls. I'm Buffalo Bee, and I'm here to help you with arithmetic. FOLLOW-ME as we do some work with addition. Use your pencil only when you write the answers.

TUNE IN: HERE WE GO.

1. 2, ... add 1 more, ... plus 5 more, ... plus 6 more, ... your answer is □.

2. There are 3 sons and 2 daughters in the Jones family. How many children are there all together? □.

ISN'T THAT EASY?

NOW COMPLETE THE FOLLOWING SUMS FOR ME.

3. 2 + 3 = □   5. 5 + 1 = □   7. 4 + 2 = □
4. 4 + 0 = □   6. 7 + 4 = □   8. 7 + 3 = □

EXCELLENT!

NOW HELP ME FILL IN THE "NAMES–FOR–10" BOX.

	Names–for–10 Box
Example	7 + 3
9.	8 + __
10.	9 + __
11.	__ + 5
12.	1 + __

GOOD! WORK RAPIDLY BUT ACCURATELY.

CAN YOU RENAME THE NUMBERS?

 Buffalo Bee   14 = 1 ten + 4 ones

Examples   11 = 1 ten + 1 one

ARE YOU READY FOR SOME?

14.   48 = _____ tens + _____ ones.

13.   23 = _____ tens + _____ ones.

15.   201 = _____ hundreds + _____ tens + _____ ones.

USE RENAMING TO SOLVE THESE QUICKLY.

16. 21 + 14 = _____   17. 30 + 21 = _____

FIND THE SUMS ON THE ADDITION WHEEL.

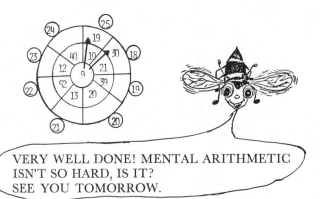

VERY WELL DONE! MENTAL ARITHMETIC ISN'T SO HARD, IS IT? SEE YOU TOMORROW.

The answers for each lesson might be given on the first page of the next lesson in the booklet. Examples of other activities in the short-answer booklet form, to be used according to the needs of a class could include the following:

A. FINISH THE "TWELVE BOX." Each product must be 12.

	The Twelve Box
Example	3 × 4
1.	___ × 3
2.	___ × 1
3.	___ × 6
4.	___ × 2
5.	___ × 12

B. Using estimation in division

SEE   THINK

1. $563 \div 81 \rightarrow 560 \div 80 =$
2. $223 \div 13 \rightarrow 220 \div 10 =$
3. $798 \div 41 \rightarrow \quad \div \quad =$
4. $403 \div 18 \rightarrow \quad \div \quad =$

C. Here comes the number train: CHARLIE ENGINEER says, "FOLLOW ME!"

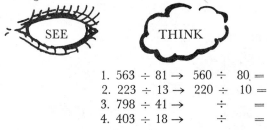

1. $7 - 3 + 12 \times 4 - 13 = ?$

2. $72 \div 8 + 3 \times 5 = ?$

D. Clock Arithmetic.

TUNE IN CLOSELY. We are going to work on a clock. Keep your eyes open, and use the clock to help you.

Examples: $6 + 3 =_{12} 9$
$6 + 7 =_{12} 1$
$6 + 3 + 5 =_{12} 2$

1. $3 + 4 =_{12}$ _____
2. $10 + 2 + 5 =_{12}$ _____
3. $5 + 4 + 2 =_{12}$ _____

4. $2 + 12 + 2 =_{12}$ _____
5. $10 + 9 =_{12}$ _____

E. Relationships

Buffalo Bee says, "Use the correct symbol: $=$ for equals, $>$ for is greater than, $<$ for is less than."

1. $35 - 15 \bigcirc 21$
2. $40 - 10 \bigcirc 30$
3. $14 - 0 \bigcirc 15$

4. $30 - 19 \bigcirc 11$
5. $26 - 17 \bigcirc 10$

II. "Follow-me" sequences, using audio tapes. Instead of a booklet with written instructions, similar activities might be put onto audio tapes, to be used either by individuals or by small groups. In the sample exercises which follow, each item to be put on the tape is in parentheses, and a pause is indicated by a row of dots. Each pupil should have an answer sheet on which to record his responses.

A. Word Problems:
   1. (Bob bought a watch for $24. He also paid $4 for a shirt and $5 for a sweater. How much money did he spend?) . . . . (Record your answer.) . . . . (Listen closely for the next question.)
   2. (A subway train left Central Station with 78 passengers aboard. At Brooklyn Station, 13 passengers got off and 25 got on.) . . . . (The train stopped next at Grant Street, where 45 passengers got off and no one got on.) . . . . (At the next stop, College Station, 105 passengers boarded the train.) . . . . (How many passengers were then on board?) . . . . (Record your answer.)

B. "Follow-me" Activities:
   1. (20 times 1) . . . . (multiply by 2) . . . . (add 4) . . . . (Record your result.)
   2. (10 times 10) . . . . (multiply by 4) . . . . (Record your result.)
   3. (14 times 6) . . . . (add 10) . . . (Record your result.)
   4. (14 times 0) . . . . (add 4) . . . . (subtract 4) . . . . (Record your result.)
   5. (16 times 1) . . . . (multiply by zero) . . . . (Record your result.)
   6. (1 times 0) . . . . (multiply by 16) . . . . (Record your result.)

C. Patterns:
   1. (What is my pattern? If you think you see it, tell the next five terms. 1/4; 1/2; . . . . 1; . . . . 2; . . . . 4; . . .)
   2. (Let's do another pattern. This time tell the next three terms. 5; 10; . . . . 15; . . . . 20; . . . .).

409

III. Oral sequences in mental arithmetic. During the school day opportunities frequently arise for arithmetic activities that are spontaneous, and that can be done without paper and pencil. It may be as the class waits for a special teacher to arrive, or when a few minutes are left before the lunch-break bell. It may even be in a dull time on a stormy day. Many varied practice exercises, word problems, and, "Follow-me" sequences can be done orally on such occasions. For example, you might make a practice of opening a class with an exercise like the following:

A. Follow-me sequences.
1. 8 plus 13, subtract 13, now multiply by 4, and record your answer.
2. 25 plus 65 .... subtract 65 .... now multiply by 5 .... record your answer.
3. 15 times 2 .... divide this by 10 .... add 5 .... multiply by zero .... record your result.

B. Verbal problems:
1. It was the championship football game for our team. They scored 6 points in the first quarter, 0 points in the second quarter, 14 points in the third, and 3 points in the fourth. What was our final score?
2. Susan was to be at school at 8:30 A.M. She didn't arrive at school until 9:12 A.M. How many minutes late was she?
3. If oranges cost 48¢ a dozen, how much will 18 oranges cost?

C. Verbal problems involving logic or approximate answers:
1. The Pirates beat the Mets 3–2. How many different players crossed homeplate and scored a run?
2. If I have 3 strawberries and give you 2, how many do you have?
3. Our sunflower is 5'8" tall. Is this taller than your brother?

D. Reading and using tables, graphs, and scales. Such activities in volve mental arithmetic and should not be overlooked. They might include:

1. Using tables, graphs, or scales found in census reports, encyclopedias, newspapers, and textbooks, predict the population trend for Brazil. Estimate Brazil's population for 1980, 2000, 2100.
2. Turn to a map of Australia in your Atlas. Find the scale of miles. Now estimate the length of the coastline of Australia (to the nearest 100 miles).
3. Using a graph from a local newspaper concerning the national budget, determine (a) the lagest expenditure and (b) the smallest expenditure; then estimate (c) the difference in dollars between the largest and smallest expenditures, and (d) the total in dollars of the two largest expenditures in the budget.

SUMMARY. Experience and research indicate that pupils will grow in ability to use mental arithmetic if they are given specific classroom experiences with it. Why not give it a try? Challenge your class with a follow-me exercise in your next meeting.

BIBLIOGRAPHY

Boulware, G. E. The emerging concept of mental arithmetic. Unpublished doctoral dissertation, Teachers College, Columbia University, 1950.
Flournoy, M. F. *A study of the effectiveness of an oral arithmetic program prepared for use at the intermediate grade level.* (Doctoral dissertation, State

University of Iowa) Ann Arbor, Michigan: University Microfilms, 1953, No. 523417.

Flournoy, F. Instruction in mental arithmetic. *The Elementary School Journal,* 1954, 55, 148–153.

Flournoy, F. Providing mental arithmetic experiences. *The Arithmetic Teacher,* 1959, 6, 133–139.

Hall, J. V. Solving arithmetic problems without pencil and paper. *Elementary School Journal,* 1947, 48, 212–217.

Hall, J. V. Mental arithmetic: misunderstood terms and meanings. *The Elementary School Journal,* 1954, 54, 349–353.

Kramer, K. *Mental computation, Teacher's Guide c.* Chicago: Science Research Associates, 1965.

Operations: *basic number operations.* Report No. 2, 1965, A series of informational pamphlets about elementary school mathematics, University of the State of New York, The State Education Department, Bureau of Elementary Curriculum Development, Albany, New York; p. 4.

Petty, O. *A study of non-pencil and paper methods of solving arithmetic word problems presented visually.* Unpublished doctoral dissertation, State University of Iowa, 1952.

Petty, O. Non-pencil and paper solution of problems, an experimental study. *The Arithmetic Teacher,* 1956, 12, 229–235.

Schall, W. E. A comparison of mental arithmetic modes of presentation in elementary school mathematics. Unpublished doctoral dissertation, The Pennsylvania State University, 1969.

Sisters of St. Joseph (Sister Josephine). *Mental Arithmetic* #3. Philadelphia: Catholic Students Press, 1961.

Spitzer, H. *The teaching of arithmetic.* Boston: Houghton-Mifflin, 1948; second edition 1961; third edition 1967.

Wolf, W. C. Non-written figuring: its role in the elementary school curriculum. *Educational Research Bulletin,* 1960, 39, 206–213.

SELECTED BIBLIOGRAPHY FOR CHAPTER 16
Remediation and Enrichment

Abeles, Francine and Zoll, Edward J. "Networks, Maps and Betti Numbers: An Eight-Year-Old's Thinking." *School Science and Mathematics* (May 1971), 369–72.

Adkins, Julia. "An Application of Modular Number Systems." *The Arithmetic Teacher* (December 1968), 713–14.

Arnold, William R. "Computation Made Interesting." *The Arithmetic Teacher* (May 1971), 347–50.

Ashlock, Robert B. "Planning Mathematics Instruction for Four- and Five-Year-Olds." *The Arithmetic Teacher* (May 1966), 397–400.

Balka, Don S. "Creative Ability in Mathematics." *The Arithmetic Teacher* (November 1974), 633–36.

Carroll, Emma C. "Logarithms for Ten-Year-Olds." *The Arithmetic Teacher* (March 1968), 273–75.

Deans, Edwina. "Independent Work in Arithmetic." *The Arithmetic Teacher* (February 1961), 77–80.

Ellison, Louise. "Mathematics for the Very Young." *Parents* (July 1969), 48, 77, 79–81.

Fitting, Marjorie A. Pickering. "SCUBA: Some Challenging Un-boring Arithmetic." *The Arithmetic Teacher* (April 1974), 294–97.

Gorts, Jeannie. "Magic Square Patterns." *The Arithmetic Teacher* (April 1969), 314–16.

Haines, Margaret. "Modular Arithmetic." *The Arithmetic Teacher* (March 1962), 127–29.

Heard, Ida Mae. "Developing Geometric Concepts in the Kindergarten." *The Arithmetic Teacher* (March 1969), 229–30.

Hiehle, Petronella. "Putting Frames to Work: An Enrichment Activity." *The Arithmetic Teacher* (November 1968), 649–51.

"Logic in the Construction of Magic Squares." *The Arithmetic Teacher* (November 1965), 560.

McCallon, Earl L. and Cowan, Paul J. "Enrich-ment With Exponents." *The Arithmetic Teacher* (January 1968), 70.

McClintic, Joan. "The Kindergarten Child Measures Up." *The Arithmetic Teacher* (January 1968), 26–29.

McCombs, Wayne E. "4 X 4 Magic Square for the New Year." *The Arithmetic Teacher* (January 1970), 79–80.

McKillip, William D. "Patterns—A Mathematics Unit for Three- and Four-Year-Olds." *The Arithmetic Teacher* (January 1970), 15–18.

Mastain, Richard and Nossoff, Bernice. "Mathematics in the Kindergarten." *The Arithmetic Teacher* (January 1966), 32.

Matthew, William. "Try Magic Squares." *Instructor* (January 1968), 98.

Oberlin, Lynn and Oberlin, Mary J. "Mathematics for Four-Year-Olds." *The Arithmetic Teacher* (January 1968), 10–12.

Parsons, Cynthia. "Unusual Arithmetic." *The Arithmetic Teacher* (February 1961), 69–74.

Ranucci, Ernest R. "Thumb-Tacktics." *The Arithmetic Teacher* (December 1969), 605, 630, 664.

Schell, Leo. "Horizontal Enrichment With Graphs." *The Arithmetic Teacher* (December 1967), 654–56.

Schlinsog, George W. "More About Mathematics in the Kindergarten." *The Arithmetic Teacher* (December 1968), 701–05.

Uprichard, A. Edward. "Focus on Research: The Effect of Sequence in the Acquisition of Three Set Relations; An Experiment with Pre-Schoolers." *The Arithmetic Teacher* (November 1970), 597–604.

Woods, Ruth. "Preschool Arithmetic is Important." *The Arithmetic Teacher* (January 1968), 7–9.

Yates, William E. "The Trachtenberg System as a Motivational Device." *The Arithmetic Teacher* (December 1966), 677–78.

Zahn, Karl. "Interest Getters." *The Arithmetic Teacher* (April 1968), 372–74.

CHAPTER 17

EVALUATION AND DIAGNOSIS

Two very important and interrelated elements of a program for teaching elementary school mathematics are *evaluation* and *diagnosis.* As teachers, we need to use effective evaluation, be it summative and/or formative,* to determine how well our program is succeeding. Diagnosis is a must if we are to find the points in the program where our pupils are having difficulty.

Evaluation and diagnosis are, therefore, integral parts of a continuous process of teaching and learning mathematics. The articles in this chapter were selected because of the practical suggestions or models they propose or demonstrate for use in evaluation and diagnosis of elementary mathematics instruction. Alan Riedesel in "The Theme in Arithmetic" offers a novel evaluation and diagnostic technique for the elementary level. Written compositions or themes can be very revealing in the way they reflect the level of thinking and performance of a pupil. This is a very practical suggestion that many teachers will wish to try in their classrooms.

Tommie West's article, "Diagnosing Pupil Errors: Looking for Patterns," should indeed be helpful to teachers. Children's papers provide vital information and the teacher who is aware of patterns of error is in a better position to provide the appropriate remediation.

Norbert Maertens and Clarence Schminke in "Teaching—For What?" indicate that children demonstrate their mathematics learning on four distinct levels. These are called (1) rote or associative learning, (2) concept learning, (3) principle learning, and (4) problem solving. Learning behavior at each level is assessable, and the information obtained is essential in determining how appropriate the instructional process has been and what the subsequent teaching sequences should be. Be sure and read this article for an excellent description of each learning level and ways to assess learning at each one.

Cecil Trueblood, in "A Model for Using Diagnosis in Individualizing Mathematics Instruction in the Elementary School Classroom," suggests that the present emphasis on the recognition of individual differences in learners has resulted in (1) a major change for the classroom teacher, (2) a revision in the teacher's conception of instruction, (3) the emergence of four types of individualized instruction, and (4) a need to revise the teacher's conception of diagnosis. He offers a model for conducting diagnosis which should prove helpful to teachers who are attempting to individualize mathematics instruction in the elementary classroom.

*Benjamin S. Bloom, "Some Theoretical Issues Relating to Educational Measurement," *The 68th Yearbook of The National Society for the Study of Education,* Part II (Chicago: University of Chicago Press, 1969), pp. 47–48.

413

The Theme in Arithmetic

Alan Riedesel

THE WRITTEN THEME is another means of improving arithmetic understanding. Writing a theme affords the pupil an opportunity to express in words his ideas and understanding of an arithmetic process on which he has been working. The ability to explain the steps of a procedure is a good indication that the pupil really understands the process.

During the tenth week of school the seventh grade class was completing a review of the study of multiplication of fractions. The group had used concrete materials, various techniques for solving examples and problems, and had arrived at what they considered the best method of working with fractions.

At the beginning of one period the teacher suggested that the class could clarify its thinking concerning the multiplication of fractions by using a written description of the method that they used.

This was the first time that the students had attempted to write short papers concerned with an aspect of arithmetic. Thus, a number of them had difficulty in beginning their themes. Such statements as "I think I know a lot about the multiplication of fractions but I haven't had to really think about it until now" were common. But, after a few minutes everyone was writing and, although the writing went slowly at the beginning of the period, it was gratifying to note that once the child did begin to write, ideas seemed to come to his mind and he was able to write rather rapidly. At the end of the period the papers were given to the teacher.

The next period the papers were returned to the group and ideas concerning them were exchanged. The teacher asked the group if they would be interested in reading some of the papers written by other members of the class. The class decided that they should be rewritten in ink so that they could be easily read as a part of a display on the bulletin board. The following are examples of the compositions:

(*Jay's paper*)

I think that the meaning of "multiplied by a fraction" is a short way of adding fractions.

The way that you multiply a fraction by a whole number is, first you change the whole number to an improper fraction. Then you multiply the numerator and the denominator by the denominator. If necessary change to a mixed number. For example:

$$3 \times \tfrac{1}{2} =$$

$$\tfrac{3}{1} \times \tfrac{1}{2} = \tfrac{3}{2} = 1\tfrac{1}{2}$$

The way you multiply a mixed number by a fraction is that you change the mixed number to an improper fraction. Then multiply the numerator by the numerator then multiply the denominator by the denominator. If necessary change the product to its lowest terms. For example:

$$2\tfrac{4}{6} \times \tfrac{1}{4} =$$

$$\tfrac{16}{6} \times \tfrac{1}{4} = \tfrac{4}{6} = \tfrac{2}{3}$$

The way you multiply a fraction by a fraction. You multiply the numerator by the numerator. Then multiply the denominator by the denominator and if it is necessary change to lowest terms. For example:

$$\tfrac{3}{4} \times \tfrac{2}{8} = \tfrac{6}{32} = \tfrac{3}{16}$$

MULTIPLYING FRACTIONS

(*Barbara's paper*)

1. To multiply $4 \times \tfrac{3}{4}$ you first have to change 4 into a fraction. That would make 4, $\tfrac{4}{1}$. Then you can cancel like this: $\tfrac{4}{1} \times \tfrac{3}{4}$. Four will go into the numerator one time. Four will go into the denominator one time. Then you multiply one times three, which is three and one times one which is one. One will go into three, three times, so the answer is three.

Reprinted from *The Arithmetic Teacher* (April 1959), 154–155, by permission of the author and the publisher.

2. To multiply $2\frac{1}{2} \times \frac{1}{2}$ you first have to make the $2\frac{1}{2}$ an improper fraction. You do this by multiplying the whole number by the denominator of the fraction and you get $\frac{5}{2}$. Your problem now looks like this $\frac{5}{2} \times \frac{1}{2}$. You then multiply the numerators 5 and 1, and you get 5. Then multiply the denominators 2 and 2, and you get 4. It is now an improper fraction so you divide 4 into 5 and get 1 and $\frac{1}{2}$. $1\frac{1}{2}$ is the answer.
3. To multiply $\frac{3}{4} \times \frac{5}{3}$ you first can cancel. You cancel like this: $\frac{3}{4} \times \frac{5}{3}$ the denominator three goes into itself one time and the numerator three goes into itself one time. The numerator 5 goes into itself one time and the denominator 5 goes into itself one time. You then multiply denominators 1 and 1 together and the numerators 1 and 1 together and you get $\frac{1}{4}$. Then divide 1 into 1 and you get the answer 1.

The above arithmetic themes are typical of the group's first attempt to use the theme. As the students continued their arithmetic work during the semester they were asked to write other explanations of mathematical processes. Following the completion of a unit of work on decimals the theme below was written.

What Are Decimals?

(Annie Lou's paper)

Decimals are based on ten and multiples of ten. The decimal point separates the whole number(s) from the decimal fraction.

We use decimals because it is easier to get an accurate measure with them, especially with very small measurements.

It is easier for me to use decimals, because there are not many steps to remember when I multiply, divide, add, and subtract them. It is easier for me to read them, too.

When you are adding decimals you have to remember to have the decimal points in the addends and the sum in a straight column. When you are subtracting decimals you have to remember to keep the decimal point in the minuend, subtrahend, and difference in a straight column. When you are dividing decimals you have to remember to move the decimal point in the divisor and the dividend to the right as many places as there are in the divisor. You have to remember to put the decimal point in the quotient, right, above the one in the dividend after it has been moved. When you multiply a decimal you have to remember to count how many figures there are to the right of the decimal point in the multiplicand and in the multiplier, then you count that many places off in the product.

We use decimals to show someone's batting average, to show how fast someone can swim, to show how high someone can jump, to show how long something is and to show how much something weighs.

Mr. Riedesel or who ever else reads this theme, I think you would be interested in knowing that the decimal system came into use in the seventeenth century.

Summary

In analyzing the above procedure, it can be noted that the following benefits can be derived from writing an arithmetic theme:

1. The teacher has a means of evaluating the individual student, because the themes show the variations in the level of understanding among the children.
2. A valuable tool is provided for use in diagnosing the difficulties that individual class members have with a particular arithmetical process.
3. The actual writing focuses attention on a particular skill and makes it necessary for the student to clarify his thinking on a particular arithmetic area.
4. Writing about mathematics entails the ability to develop accurate explanations and descriptions which will help the student in writing about other subjects, particularly in the scientific fields.
5. The children have a new approach to mathematics, a subject in which writing technique is not usually employed.
6. The students are helped to think about the *how* and *why* of a particular arithmetic process.
7. Class unity can be maintained at a high peak, because all of the students are able to participate in the same activity with a feeling of achievement.

While the arithmetic theme is not by any means a panacea, when properly used it can be a valuable aid in the development of a sound arithmetic program.

Editor's Note. The written composition or theme can be very revealing in the way it reflects the level of thinking and performance of a pupil. Teachers frequently assume that pupils have developed a higher level of understanding than is really true. A tape recording or a theme, as Mr. Riedesel suggests, is helpful. The most common procedure is merely the asking of questions and listening. Perhaps we teachers do too much talking and not enough listening.

Diagnosing pupil errors: looking for patterns

TOMMIE A. WEST

Diagnostic teaching is teaching that looks at the errors children make and subsequently structures the learning experiences so that the errors will be eliminated. There is hardly a skill in the teacher's repertoire that is more important than the ability to identify pupil errors and prescribe appropriate remedial procedures.

The most fruitful diagnosis is the diagnosis that looks for patterns of errors rather than random errors. Teaching is so much more effective if a class of errors can be eliminated by a single remedial procedure. An examination of the errors children make on their papers reveals generally two kinds of errors: one type we may call *careless* errors and the other *conceptual* errors.

The careless error is generally spotted as the error that is not consistently made. It occurs most likely because the child loses track of where he is in a procedure. Algorithms that require the child to momentarily store information while he performs another operation typically result in this kind of error.

It is useful to distinguish the careless error from the conceptual error because the remedial procedures may be quite different. The careless error will probably be caught if the child reworks the problem. If the child is given a procedure for checking his work or for estimating the reasonableness of his solutions, then careless errors may be reduced. The conceptual error is quite different. No matter how many times the child works the problem, he will probably make the same mistake.

We detect conceptual errors by seeing patterns of errors in a child's paper. The child may *consistently* make the error, which means that he has an incorrect understanding of the procedure; or he may make the error only sporadically, or only under certain circumstances, indicating that he has some grasp of the procedure but not complete understanding.

Diagnostic teaching requires that teachers (1) distinguish between conceptual and careless errors, (2) identify the precise nature of the careless errors, (3) infer the conceptual basis (cause) of the conceptual error, and finally, (4) prescribe appropriate remedial procedures.

Figures 1, 2, and 3 are samples of the work of three children. A diagnostic analysis of each can be made.

Diagnosis and prescription

MIKE'S PAPER

1. All Mike's errors are consistent ones, caused by a conceptual error; each one can be predicted. However, in D Mike also makes a careless error by reversing the digits in his answer.

2. When adding a two-digit number and a one-digit number, Mike first sums the ones column then adds to this sum the number in the tens column.

3. It is apparent that Mike does not understand the role of place-value in numerical representation. Neither is he disturbed by the impossibility of his answers.

4. The nature of his pattern of errors

Reprinted from *The Arithmetic Teacher* (November 1971), 467–469, by permission of the author and the publisher.

MIKE

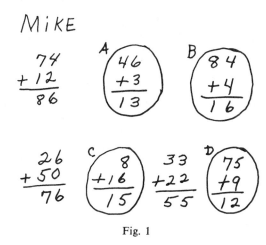

$$\begin{array}{r}74\\+12\\\hline 86\end{array}$$

A
$$\begin{array}{r}46\\+3\\\hline 13\end{array}$$

B
$$\begin{array}{r}84\\+4\\\hline 16\end{array}$$

$$\begin{array}{r}26\\+50\\\hline 76\end{array}$$

C
$$\begin{array}{r}8\\+16\\\hline 15\end{array}$$

$$\begin{array}{r}33\\+22\\\hline 55\end{array}$$

D
$$\begin{array}{r}75\\+9\\\hline 12\end{array}$$

Fig. 1

suggests that Mike needs much developmental work with concrete representations of numbers and with finding sums using concrete materials. Mike's work with symbols should be a *record* of what he finds with concrete materials. Base-ten blocks, bundles of sticks, place-value charts, or money (pennies and dimes) suggest themselves. Mike would also profit from instruction in estimating his answers.

Douglas's Paper

1. The error that Douglas makes is consistent; it is a conceptual error.

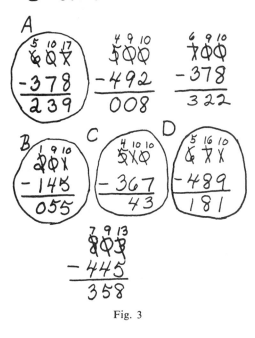

Douglas

$$\begin{array}{r}700\\-378\\\hline 322\end{array}$$

$$\begin{array}{r}607\\-378\\\hline 229\end{array}$$

A
$$\begin{array}{r}930\\-278\\\hline 622\end{array}$$

B
$$\begin{array}{r}510\\-367\\\hline 133\end{array}$$

C
$$\begin{array}{r}691\\-489\\\hline 112\end{array}$$

$$\begin{array}{r}803\\-445\\\hline 358\end{array}$$

Fig. 2

2. Notice that Douglas regroups hundreds to tens and *always* records the number as 9 tens, even when it is not appropriate. Notice also that sometimes it *is* appropriate.

3. Douglas does not understand the regrouping procedure in subtraction.

4. Having Douglas solve his problems with physical models of the place-value idea (such as base-ten blocks, bundles of sticks, place-value charts) will help him see the basis of the regrouping procedure in the subtraction algorithm. Giving him exercises in renaming will also help him, especially if accompanied by a physical model. Encouraging him to regroup 60 tens as 59 tens and 10 ones instead of double regrouping would probably keep him from making so many errors of this kind.

Susan

A
$$\begin{array}{r}607\\-378\\\hline 239\end{array}$$

$$\begin{array}{r}500\\-492\\\hline 008\end{array}$$

$$\begin{array}{r}700\\-378\\\hline 322\end{array}$$

B
$$\begin{array}{r}201\\-145\\\hline 055\end{array}$$

C
$$\begin{array}{r}500\\-367\\\hline 43\end{array}$$

D
$$\begin{array}{r}677\\-489\\\hline 181\end{array}$$

$$\begin{array}{r}803\\-445\\\hline 358\end{array}$$

Fig. 3

Susan's Paper

1. Problems *A* and *C* are probably careless errors. Problems *B* and *D* are examples

417

of a single conceptual error.

2. In problem *A*, Susan did not completely record the regrouping of 10 tens to 9 tens and 10 ones. This is obviously not a consistent error. Look at all the occasions when she did not make this error. She probably recorded the 17 when she first encountered the regrouping problem but in the process of regrouping hundreds as tens forgot that she had not in fact done the regrouping yet (by modifying both tens and ones).

Problem *C* is probably also a careless error. She simply forgot to finish the problem.

3. Problems *B* and *D* are interesting. Susan did not add 10 ones to 1 one in problems having a 1 in the ones digit. She does not make this mistake when any other numeral occurs in the ones digit.

4. The basis of her problem seems to be some misunderstanding of the regrouping procedure in subtraction problems, at least in certain cases. Base-ten blocks, bundles of sticks, place-value charts, or money (dimes and pennies) could well exemplify the regrouping procedure in the subtraction algorithm.

Teachers who develop the practice of looking at children's papers critically to detect such patterns of errors are in a better position to provide for remediation of learning difficulties. They will probably also develop better teaching procedures for *avoiding* the very appearance of such errors in children's work. Children's papers provide vital information to the teacher. Happy is the child whose teacher is a good diagnostician!

Teaching—for what?

NORBERT MAERTENS and
CLARENCE SCHMINKE

As classroom teachers and, concomitantly, students of the teaching of elementary school mathematics, we readily accept the idea that learning is accomplished on a variety of levels. Further, our teaching methods generally reflect this. When we want quick, automatic response to number facts, we use flash cards, timed tests, and the like; but when we want to develop understanding and the ability to solve problems, we use quite different procedures. In the latter instance we are likely to pace instruction more evenly, provide many concrete materials, encourage pupil inquiry through group discussion and give appropriate attention to the mathematics of our environment. Experience has taught us, however, that when we attempt to evaluate such instructional efforts, we seem much less effective than we assumed ourselves to be during the original teaching process. Our instruments are often haphazard and ill-conceived, and our interpretations of test results can be equally vague and sterile. The material that follows presents a hierarchical framework for evaluation—a framework that can help us systematically examine different levels of learning and subsequently determine appropriate evaluation activities more closely associated with the original purposes of the instruction.

Children may demonstrate their mathematics learning on four distinct levels. These levels are called (1) rote, or associative, learning, (2) concept learning, (3) principle learning, and (4) problem solving.[1] Learner behavior at each of these levels is assessable, and the information obtained is essential in determining how appropriate the instructional process has been as well as subsequent teaching sequences.

Rote, or associative, learning

Associative learning might best be described as memorization with little understanding. The learner is trained to make a particular response to a given stimulus, and understanding is not emphasized or required. For example, examine figure 1.

Example *A* illustrates a child's work with a division-of-fractions algorithm. If the child has been told to solve such exercises by inverting the divisor and multiplying, he is able to determine a quotient. As long as he remembers his teacher's instruction, he need not know why such a procedure is successful. Such a response is an associative response.

Example *B* illustrates a child's work

1. Robert M. Gagné, *The Conditions of Learning* (New York: Holt, Rinehart & Winston, 1965).

Reprinted from *The Arithmetic Teacher* (November 1971), 449–456, by permission of the authors and the publisher.

EXAMPLE A

$\frac{1}{4} \div \frac{1}{2} = $ _____

Child's Solution:

$\frac{1}{4} \times \frac{2}{1} = $ _____

EXAMPLE B

Thirty is what
percent of 180?

Child's Solution:

$$X = \frac{30}{180} \times 100$$

EXAMPLE C

Joan has three pennies,
and her mother gives her
three more. How many does
she have in all?

Child's Solution:

$3 + 3 = $ _____

Fig. 1

with percent. In the past, teachers spent many hours teaching the three types of percentage problems. The problem shown in *B* was known as a "case-three" problem. When children recognized a case-three problem they applied a case-three algorithm. This required little knowledge of percent, and most children were ignorant of why this procedure was used. Again, such a response is associative.

Example *C* illustrates a child's work with a story problem. In this situation the teacher has spent many hours coaching children to look for appropriate word clues. For example, in the problem shown, the child correctly recognized that the words *and* and *more* are generally associated with addition problems. Then the last sentence provided him with further proof that he must apply an addition algorithm. The words "how many . . . in all" are in contrast with words like *less* and *take away* or phrases such as *how many remaining,* all of which would be associated with the use of a subtraction algorithm. This type of re-

sponse, associating words and phrases with algorithms, is also called associative.

Assessment of associative learning

An in-depth knowledge of associative testing is essential, for such knowledge enables us to identify cases in which rote responses have been given where understanding is desired.

It is often prognosticated by teachers and parents that children who appear to have been quite successful in primary mathematics will be equally successful in the intermediate grades. However, as the content in mathematics becomes more complex, performance begins to falter. Failures begin to be more common than successes, until finally an adverse attitude toward mathematics is effected, thus compounding an already serious problem. Such a drop in performance is subsequently attributed to such things as lack of motivation, the fourth-grade slump, reading difficulties, personality conflicts, and so on. In some cases the child's performance may have been influenced by these factors, but such labeling does not adequately explain why the child's performance continues to drop. An equally plausible hypothesis might be that the child can no longer substitute an adequate memory for understanding. He is unable to memorize enough to be successful because the interference of similar content hinders new learning.

Many tests are constructed almost exclusively of items that test at the associative level or that allow associative response to be substituted for a demonstration of understanding. Examine the test items in figure 2.

Item *A* allows an associative response for at least two reasons: (1) if the teacher has used doll figures in his teaching of the concept *five,* a child may associate dolls with five or may infer that only doll figures are used to represent five. (2) The triangular configuration of the set, rather than the number of elements in the set, may have become associated with five. Errors of this nature become increasingly possible as the

EXAMPLE A

EXAMPLE B

8
×7

EXAMPLE C

Jane has 32 marbles to
give away. She decides to
divide them equally among
her four friends. How many
will each receive?

Fig. 2

number of set members is increased beyond
that which is readily perceivable by the
child. For example, many teachers use a
triangular shape to represent *ten* as in
figure 3.

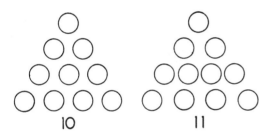

10   11

Fig. 3. Response bound to configuration

Because children cannot immediately
determine the number of objects by perception, they must count. This is fine. Unfortunately, if such a configuration is used
repeatedly, their concept of ten becomes
bound to the configuration, not the numerousness of the set.

Item *B* in figure 2 allows an associative
response because children may merely
memorize the correct response for the given
stimulus. Thus, the child who has had repeated practice with many exercises on a
sheet of paper may have simply developed

a series of associations with the stimuli. It
is very possible that the child could do this
without any knowledge of process.

Item *C* of figure 2 also generates an
associative response. The phrase "divide
them equally among her four friends"
may be a clue phrase for the child. If he has
learned to write the algorithm 4)32 as an
automatic response to the problem, he is
once again able to substitute memory for
understanding.

It is important to note that a child's
response to the test items in figure 2 need
not be exclusively associative. He may have
mentally or physically manipulated materials or constructed models to solve the
problems, and his understanding may be
complete and in-depth. However, from
such items the teacher has no way to find
this out. Contrast this with the responses
required to complete the items for the next
three levels successfully.

Concept learning

Concepts are mental structures that represent a class of experiences. They are
formed when a child generalizes from a
series of experiences having some element
of commonality. For example, observe the
schematic in figure 4. The circle on the left
represents the real world of the child. Represented here are those things that appeal
to the senses—things he can see, touch,
smell, hear, and taste. Thus, a child notes
that horns honk, limburger cheese has an
odor, and sticks are hard while feathers
are soft.

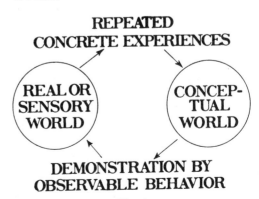

REPEATED
CONCRETE EXPERIENCES

REAL OR
SENSORY
WORLD

CONCEP-
TUAL
WORLD

DEMONSTRATION BY
OBSERVABLE BEHAVIOR

Fig. 4

The circle on the right in figure 4 represents a child's conceptual world. Represented here are a child's generalizations. For example, consider the concept *tree*. The child can easily determine whether or not an object is a tree. To him trees are large plants, or plants that will be large. He knows, for example, that a sidewalk is not a tree, and neither is a horse or a ball a tree. Yet he cannot precisely define tree. If he and several others are asked to describe what is meant by *tree,* each will very likely draw a picture that is unique. This is so because each child has his own personal concept of tree. His concept has been determined by his experiences.

Since concepts exist in the mind, they are abstract. However, Piaget found that young elementary school children generally understand only what they experience —what they can see, smell, touch, taste, or hear. Thus it is apparent that the five senses must be used if children are to gain a concept. This fact is represented in figure 4 by the label "Repeated Concrete Experiences."

Through repeated exposure to a variety of materials, children are able to attain a concept. For example, to illustrate the concept *three,* a teacher places a variety of objects on a table and asks a child to arrange them in sets according to shape (see fig. 5). When this is done, the teacher can then ask, "Is there anything that seems to be the same about all of these sets?" Among other things that seem "to be the same," children are likely to discover that there is a commonality in the numerousness of each of the sets. This can be verified by a series of one-to-one matchings. The teacher might then say that the number of each set is three. Other physical materials and activities would then be provided to extend the child's concept until the word *three* became bound to the cardinality of the sets. Following activities such as the above, and including other numbers, the teacher would most likely introduce the numeral and the written word for three.

Assessment of concept learning

Concepts are complex generalizations, and teachers are well advised to use a variety of questions to ascertain conceptual understanding. When used in combination, examples such as those suggested below illustrate how teachers might prepare a test to determine children's understanding of the concept *three.*

1. Circle those sets containing exactly three objects. (Use sets of unfamiliar objects, some having three elements, some more than three, some less than three.)

2. Go to the bookshelf and bring me three books.

3. Color three of the cans from the set shown (fig. 6).

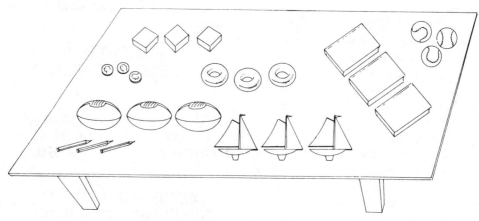

Fig. 5

422

Fig. 6

4. Draw a set of three objects. Do not use objects that have been used on this page.

5. Do both of these pictures contain a set of three objects? How do you know?

These examples merely illustrate the myriad of good test items that a teacher might choose to determine a child's understanding of the concept *three*. Note that correct responses on items such as the above require an understanding of the concept *three*. Rote association is not enough.

In question 1 the learner is asked to discriminate sets of three objects from sets containing other than three. Further, the opportunity to make an error is given by including some sets that are similar in physical characteristics but of different number.

In questions 2 and 3 the child is asked to select a subset from a total set. In question 2 he is asked to bring the objects to the teacher, whereas in question 3 he must color objects. The two items, although similar in the question they ask, allow for divergent responses. Question 2, for example, allows a child to get out of his seat and move around. Often this type of question is appealing to the child who is emotionally unresponsive to more sedentary types of questioning or who has a problem of motor coordination which makes coloring difficult.

In question 4 the child is asked to draw a set of three objects. You might vary this item by asking a child to arrange a set from a given quantity of physical objects or even by having him locate and describe a set of physical objects. By arranging or locating a set of unfamiliar objects representing *three*, the child demonstrates that his understanding is not bound to a particular physical arrangement or attribute.

Question 5 requires that a child describe his notions regarding cardinality. Often questions of this nature reveal rather serious problems of understanding which might otherwise go unnoticed.

Questions such as 2, 4, and 5 are rarely found in mathematics texts. However, improved evaluation activities such as these are necessary to complement the active-learning emphasis within contemporary programs. We must be alert to the child's environment as an aid to testing just as we are for teaching.

Principle learning

A principle is a mental structure relating two or more concepts. For example, consider the relationship involved in the operation of addition. Before one can understand and solve problems involving addition, he must first have a concept of each of the numbers, as well as concepts of the relationships of addition, and be able to relate the numbers to form a unique sum. Thus, before a child can understand and solve the addition sentence $8 + 5 = \square$, he must have attained concepts of eight and five as well as of plus and equal. Further, he must be able to relate these concepts to compute the sum, 13. Each child must be able to discriminate examples where other concepts are incorrectly used to complete the statement $8 + 5 = \square$. Thus they must know that this statement requires addition, not some other operation.

Testing for understanding of principles

Teachers must be alert not to confuse principle learning with associative learning. For example, the commutative property of multiplication is often shown in the algebraic form $a \times b = b \times a$. This is then illustrated by examples, such as $5 \times 4 = 4 \times 5$. Questions that ask children to "fill in a blank to complete the sentence 'The commutative property of multiplication is stated _____ '" simply call for substitution of the algebraic sentence $a \times b = b \times a$ or a particular instance, such as $5 \times 4 = 4 \times 5$. Such a response tells the teacher little more than that the child has a good memory. Limited to the results of such questions, even the most conscientious teacher could not plan an effective program. For example, such questions do not tell him whether he should provide more work on the commutative property of multiplication or progress to other topics. The question is essentially worthless for such decisions.

Examine the following questions designed to test a pupil's understanding of a principle. (Principle: In addition, the order of the addends does not affect the sum—the commutative property.)

1. Which property was used in the following example?

    $3 + (8 + 7) = (8 + 7) + 3$

    a) The closure property
    b) The commutative property
    c) The well-defined property
    d) The distributive property

2. Explain how the commutative property helps you in computing sums.

3. Solve the following exercise by reordering the numerals to make your work easier. Show your work and label each use of the commutative property.

    $23 + 16 + 54 + 17 = $ ____

4. Use the numerals 6 and 8 to illustrate what you mean by the commutative property of addition.

Each of the above questions requires a child to demonstrate his knowledge in a unique situation. In problem 1 the child is faced with more than two addends, yet the only action taken is to change the order. If he understands the commutative property as a change in order without change in sum, he should answer this item correctly. However, if he has merely memorized the algebraic sentences indicated earlier, he will probably miss this one. Item 2 provides still further information. A child may have a beginning awareness of what the commutative property means but may not understand how it is of any use. Questions such as this will help identify

this lack. Questions 3 and 4 require that children use numerals to illustrate their understanding of the commutative property. In question 3 the child must solve an addition exercise and use and label the commutative property, while in 4 the child shows his understanding by reordering a given set of numerals to illustrate the property. Thus, such items require a synthesis of concepts for successful performance.

Problem solving

The problem-solving level of learning is attained when a child is able to accommodate concepts and principles. He must modify his preexisting cognitive structure to the new learning and also modify the new learning in response to what he already knows. For example, consider the child who has developed an understanding of one-to-one and many-to-one correspondence in the primary grades. Assume that he is now faced with a problem. He is to buy six cans of corn. The listed price is two cans for 37¢, and the teacher has asked him to determine the price for six cans. He might use a process such as that shown in figure 7. He knows that two cans of corn cost 37¢ and reasons that he can show this by drawing a picture. His illustration is an example of many-to-one correspondence. Once this has been determined, the child simply adds the number of 37¢ increments to determine his total price. In this instance, understanding of many-to-one correspondence and the operation of addition is not enough. Until the child realizes that these two mathematical processes can be adapted to solve this problem, he remains at the principle level. It is only when he realizes the implications of concepts and principles for solving problems that he can move to the problem-solving level.

Testing for understanding at the problem-solving level

Evaluation of a child's ability to solve problems should not be confined to the

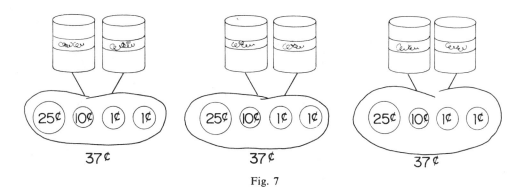

Fig. 7

more pedestrian story problems commonly found in problem-solving units or simply to pencil-and-paper tests labeled "Problem Solving." Instead, teachers should make use of the classroom and surrounding environment to supplement such items. Also, teachers need to provide a range of problem-solving experiences so they may readily identify areas needing attention. For example, to determine a child's ability to solve problems involving area, a teacher might suggest a problem like the following:

> Which of the two blacktopped surfaces has the greater area, the basketball court or the teachers' parking lot? Record your work on a paper.

If the child selects an appropriate instrument and correctly performs the measurements and computation required, you may assume that he has attained the problem-solving level of understanding. Suppose, however, that he has made an error. Examination of his work may reveal faulty concepts or principles. See figure 8, for example. The answer is correct: the basketball court is larger than the parking lot. However, examination of the child's work indicates that he did not compare area, but perimeter. Further examination is necessary to determine whether the error was due to a faulty concept of area or whether the child misread the question. Moreover, the teacher may obtain other information. By observing the work he can notice that the child selected an appropriate instrument and used it correctly. Further, he

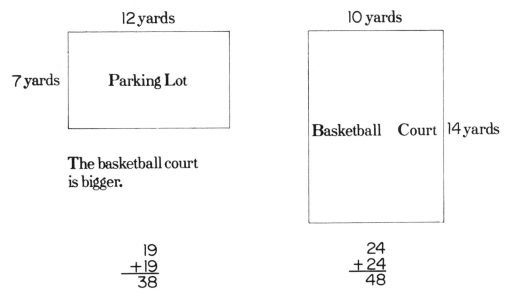

Fig. 8

425

might observe that the child has developed a beginning awareness of perimeter and how it is measured. Thus, it may be that the child's error was caused by a labeling problem rather than a more serious conceptual misunderstanding. Again, other problem situations or teacher questions are needed to complete the evaluation.

Although testing for problem-solving ability should not be exclusively a function of performance on story problems, neither should such problems be omitted. They do serve a useful purpose in that many real-life problems are on paper (tax computations, statistics, grocery ads, etc.). They are also efficient in that they save student time. Further, they may provide the teacher with essential information that is unavailable through concrete problems such as the above. Consider the following story problem:

> Nancy wants to buy a bicycle that sells for $45.99. She has saved $32.75 during the year. How much more does Nancy need to save? Nancy is eight years old and has an allowance of 75¢ a week.

If the child writes $32.75 + ____ = $45.99 and then rewrites the exercise as ____ = $45.99 − $32.75, he has abstracted the necessary information from the problem and demonstrated his understanding of the subtractive process necessary for its solution. He understands the process, and he can now compute the answer.

However, an incorrect solution or the inability to write a number sentence indicates a need for further analysis. Was the error one of computation? Did the child correctly interpret the problem? If not, was his error due to a reading problem, a prob-lem of understanding of the operation required, or perhaps the irrelevant information? Further examination and analysis are required to determine the child's misunderstanding. The teacher must also be aware that several correct answers to a list of story problems mean little or nothing unless the solutions suggest the child's understanding of the process and reflect his ability to compute.

In conclusion

Although not standardized, teacher-prepared tests have the advantage of being able to reflect a teacher's ongoing mathematics program. They need not be formal, nor must their results indicate any quantifiable score. Rather, such testing can provide a teacher with the continuous data he needs to evaluate the success of his teaching—even on a day-to-day basis if necessary.

Teacher-prepared tests should examine all levels of learning so that one can note whether an observed failure may be due to a child's faulty use of number facts (associative learning), his misunderstanding of the quantities involved with given numerals (concept learning), his inability to synthesize his understanding of numeration and joining to form an understanding of the operation of addition (principle learning), or finally, his inability to adapt these processes to the solution of problems (problem solving). Each level is assessable, and each provides the teacher with information he must have before he can adequately evaluate his program. He can test what he teaches.

A model for using diagnosis in individualizing mathematics instruction in the elementary school classroom

CECIL R. TRUEBLOOD

A revolution in mathematics instruction has begun and will probably continue through this decade. The revolution consists of a major change in the elementary school teacher's role in the educative process. The teacher of elementary school mathematics is modifying his traditional role as "director" and "lesson planner" for classroom-sized groups and is assuming the role of "instructional programmer" for individual learners. In assuming this role, the elementary mathematics teacher's concern has shifted to exercising more control over the instructional environment by arranging scope, sequence, content, feedback, evaluation, and materials appropriate for individual learners.

The acceptance of this new role in the educative process has brought about a series of other changes. These include a change in the elementary mathematics teacher's conception of instruction and the development of various types of individualized instruction. The purpose of this paper is to (1) present an overview of the two changes just mentioned, and (2) justify the need to modify and present a model for modifying the way in which elementary teachers conceptualize and use diagnosis to individualize mathematics instruction in their classrooms. The film *Using Diagnosis in the Mathematics Classroom*[1] illustrates how the suggested model of diagnosis could be used to individualize mathematics instruction in an elementary school classroom.

Elementary teachers' changing conception of instruction

Because of the teachers' role shift, there has been a corresponding change in the way in which elementary mathematics teachers are being encouraged to think about instruction. Traditionally, mathematics educators, school administrators, and mathematics supervisors have encouraged teachers to seek out and try new materials, techniques, and procedures. In response to this encouragement, elementary teachers have attempted to improve their teaching performance by using new materials and techniques with individuals, small groups, and the class. Hence, the elementary teachers' primary interest naturally became

1. This is one of a set of five films developed at The Pennsylvania State University for the Interpretive Study of Research and Development in Elementary School Mathematics, a project funded by the U.S. Office of Education, 1970.

Reprinted from *The Arithmetic Teacher* (November 1971), 505–511, by permission of the author and the publisher.

focused on the ways and means of changing their teaching performance.

As a result of this interest, the key question asked by most elementary teachers became "What new instructional procedures should I adopt to improve my teaching performance?" rather than "What should or can my students learn?" These two questions are obviously related, but under the means, or teaching-process, orientation just described the direct connection was obscured. Thus it is not surprising that most elementary teachers' conception of instruction became and has remained, until recently, means, or teaching-process, oriented.

Leading teacher educators, such as Bloom (1968), Glaser (1968), and Lindval (1961), have questioned the adequacy of this means-referenced conception of instruction. They point out that a means-referenced orientation is inadequate for most instructional decision making. In its place, they have suggested that teachers begin to ask a different, more fundamental question—a question based on a goal-referenced instructional model. This model is illustrated in figure 1.

In general, the goal-referenced instructional model asks the teacher to give first priority to the goals and objectives of instruction. This implies that the question the teacher should ask first is "What do I want my learners to learn?" rather than "What materials, techniques, and strategies should I use?" The essential point is that the goal-referenced model stresses the decision making the teacher engages in before and after instruction. This stress is accomplished by focusing the teacher's attention first on the role that objectives should play in planning and evaluating instruction rather than just by encouraging the teacher to use a particular set of materials without considering the outcomes it might produce.

**Emerging types
of individualized instruction**

The use of the goal-referenced instructional model raises two basic issues for

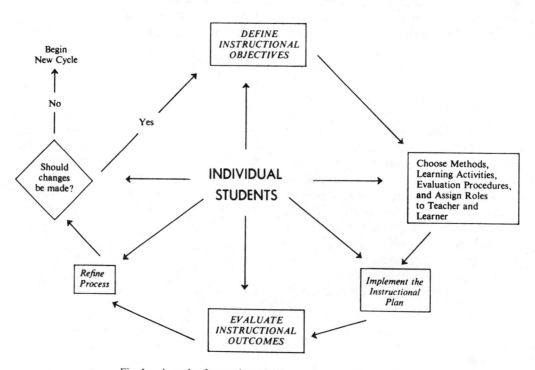

Fig. 1. A goal-referenced model for systematized instruction

428

teachers who wish to individualize instruction: Who should determine the instructional objectives, and who should select the means to attain them? In reality, there are situations where the teacher is the primary determiner of both what and how students should be taught, and then, of course, there are situations where the student is the primary determiner. Experienced teachers know that such issues are not clear-cut, and this certainly is true for individualized instructional settings. Figure 2 presents a matrix identifying four popular types of individualized instruction and specifies the source from which statements of objectives and means for achieving those objectives could originate. The writer's purpose in identifying and describing these four types of individualized instruction is not to narrow the concept of individualized instruction but rather to broaden it. He especially does not want to give the impression that a teacher should use one type of individualized instruction to the exclusion of the others. In fact, the essential point is that the selective use of each of the four types deserves serious consideration in order to meet the individual needs of students.

Source of Instructional Means	Source of Instructional objectives	
	Teacher	Learner
Teacher	A. Individually prescribed instruction	B. Personalized instruction
Learner	C. Self-directed instruction	D. Independent study

Fig. 2. A matrix showing the source from which objectives and instructional means originate for four types of individualized instruction

INDIVIDUALLY PRESCRIBED INSTRUCTION (IPI)

Individually prescribed instruction does not refer only to the Individually Prescribed Instruction Project of the Learning Research and Development Center at the University of Pittsburgh. The basic point of view represented by type *A* assumes the needs of each learner to be unique. It also maintains that the teacher, as a trained professional, should identify the learner's individual needs and prescribe appropriate instruction designed to meet those needs. Type *A*, therefore, is characterized by the following:

1. Carefully developed pretests designed to diagnose the specific needs of the learner
2. Clearly specified behavioral objectives that are formulated by the teacher, based on the results of diagnosis
3. Carefully constructed methods and materials developed to enable the learner to reach each specified behavior
4. Individually determined pacing, which is a characteristic of all types of individualized instruction

PERSONALIZED INSTRUCTION (PI)

The basic strategy represented by type *B* was developed to provide the learner with opportunities to practice decision making so that he will see and feel the results of his decision. This type of individualization is characterized by the following:

1. The learner choosing and defining personal learning objectives; a learning contract is usually employed
2. The teacher selecting the resources and instructional procedures to be used by the student in reaching his objectives
3. The learner and teacher selecting prescribed evaluation criteria to be used to determine when the learner has fulfilled his contract

SELF-DIRECTED INSTRUCTION (SDI)

The basic assumption of those teachers employing type *C* is that each individual's strengths, learning strategies, and learning style are unique. From this assumption it follows logically that any preconceived sequence or system does the learner an injustice by denying him freedom to fully develop his individuality by prescribing learning activities that he should be learning to prescribe for himself. One of the reasons learners go to school is to learn how to be self-directed, self-actualizing people. Hence, the type *C* approach is characterized by the following:

1. A well-developed testing program that is used by the teacher to identify the learner's needs

2. A set of clearly stated objectives that facilitate the teacher's attempt to communicate these instructional objectives to the learner

3. A well-developed learning laboratory with a wide variety of materials that can be placed at the disposal of the learner, since he is the one who chooses the instructional activities to achieve specified outcomes

INDEPENDENT STUDY (IS)

The basic philosophy underlying type *D* asserts that the end result of all education is to develop learners who can clearly identify their own objectives, which are consistent with their value system, and then choose appropriate means for achieving these objectives. Therefore, type *D* is characterized by the following activities of the learners:

1. Determining the objectives they desire to pursue

2. Selecting their own materials from a well-developed learning resource center

3. Evaluating their progress in consultation with the teacher

Need for diagnosis

When one compares the four types of individualized instruction just described with the goal-referenced model in figure 1, another important question arises: How crucial is diagnosis for teachers wishing to make effective use of a goal-referenced instructional model within the context suggested by the four types of individualized instruction?

As indicated in the introduction, the primary purpose of the individualized instruction is to change the typical group-paced classroom situation so that each student receives an *appropriate* set of instructional activities. This means an attempt should be made to increase motivation by identifying the learner's interests and providing learning experiences compatible with those interests and with the achievement of specified learning outcomes. When individualizing instruction, therefore, the teacher has a continuing need for information about each learner. The operation involved in obtaining essential data and background information about each learner and the process of analyzing that data and information is called diagnosis. Without the various types of diagnostic information, it is possible that teachers could waste instructional time teaching a student what he already knows or, as often occurs, frustrate the learner by attempting to develop behaviors and skills that he does not have the prerequisites or interests to successfully develop.

Modifying the teacher's conception of diagnosis

Current teacher-preparation programs and many mathematics-methods texts attempt to help teachers cope with readiness and sequencing of instruction by presenting a conception of diagnosis that is means oriented. That is, they focus primarily on the different types of diagnostic techniques and procedures that are available, how to administer and construct the various kinds of tests, and how to interpret the data obtained from standardized and teacher-made tests. All of these are important skills. This particular emphasis, however, leads the teacher to believe that the first and most important question to be considered before beginning an instructional unit is "What diagnostic techniques, instruments, or procedures should I use before and during my instructional unit?"

How adequate is this means-oriented conception of diagnosis?

There are at least two ways to test the adequacy of the means-oriented conception of diagnosis. The first is to determine whether the research in elementary school mathematics supports a means orientation.

The second is to examine the requirements of the context within which the diagnostic information is to be utilized in order to see whether the means of obtaining diagnostic information should be the teacher's first or primary consideration.

Consider first the research on individualizing instruction in elementary school mathematics. A comprehensive survey of this research literature has been summarized by Suydam and Weaver (1970). How does diagnosis aid in individualizing instruction? Suydam and Weaver conclude that the research dealing with diagnosis shows that teachers should ascertain the *specific* errors that a pupil is making, determine *specifically* how he works, and give *specific* remedial help designed to decrease his errors. This research finding is clearly describing the type of individualized instruction referred to in figure 2 as type *A*, or Individually Prescribed Instruction.

Consider next the requirements of the goal-referenced instructional model and its use in a IPI setting. Figure 1 and the characteristics of Individually Prescribed Instruction indicate that the first and most important requirement for using this instructional model in an IPI context would be the formulation of precise instructional objectives for each individual learner. This means that diagnosis should identify the pupil behaviors that should be used to formulate or select precise behavioral objectives for individual students.

A comparison of the research findings on individualizing mathematics instruction with the requirements of the goal-referenced instructional model as used in an IPI setting reveals a consistent pattern. The comparison seems to argue in favor of a conception of diagnosis that gives primary emphasis to the use the teacher makes of the results of diagnosis rather than just to his being concerned about the diagnostic means per se. That is, the comparison suggests that instead of asking a question concerning the means of diagnosis, the teacher should ask a question more directly related to the IPI setting—"What specific behaviors do my learners possess or lack at the beginning of an instructional unit?"

Such a goal-referenced conception of diagnosis and its relationship to the goal-referenced instructional model as used in an IPI context is presented in figure 3.

Essentially, the procedural model presented in figure 3 indicates that the teacher should begin Individually Prescribed Instruction after diagnosing the learner's readiness or status in steps that lead to precisely defined instructional objectives. Designing an instructional plan suitable for each learner is not easy; it requires additional, careful planning by the teacher. Next, the model suggests that the teacher should continually monitor the learner's progress as he negotiates the instructional plan. The final steps of the model show that the accuracy and usefulness of the initial diagnosis should be evaluated in terms of how well it facilitated the teacher's choice of effective instructional activities.

To use the goal-referenced model of diagnosis in the context of the other three types of individualized instruction presented in figure 2 would require some modification. First, depending on the type of individualized instruction involved, the student would be asked to assume some responsibility for defining learning outcomes, selecting instructional activities, and evaluating his own performance. Obviously, the amount and extent of this type of participation would depend on several factors, such as age, grade level, and the student's past experience with the various types of individualized instruction. Second, the learner would be required to do some self-diagnosis; therefore, the type of instructional objectives would differ in form and substance from the type of teacher-developed outcomes used in the IPI setting. Third, since the student would be asked to do self-evaluation, the evaluation procedures and the use made of the results would differ. For example, some self-reporting techniques might be developed by the teacher and used by the student to report his progress for an independent-study unit.

431

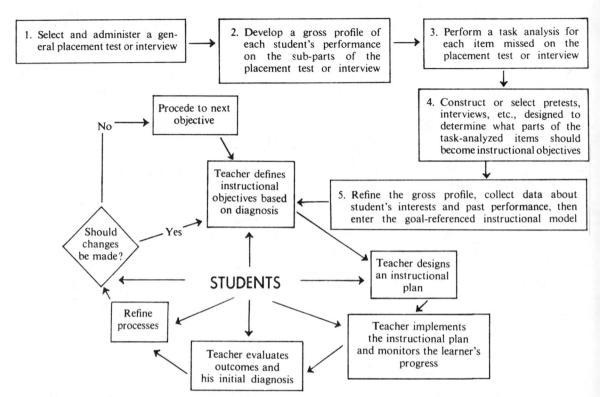

Fig. 3. A model for using goal-referenced diagnosis in the elementary school classroom to individualize instruction

From this brief discussion, one can see that use of the goal-referenced model of diagnosis in the other three types of individualized instruction raises some interesting topics for further discussion, as well as some exciting possibilities for future curriculum development and research. It is hoped that the example given for the IPI context illustrates how diagnosis can be combined with the goal-referenced instructional model to individualize mathematics instruction in the elementary school classroom.

diagnosis from a means-referenced to a goal-referenced orientation. Finally, the writer presented a model for conducting diagnosis which might facilitate a teacher's attempt to use diagnosis for individualizing mathematics instruction in the elementary school classroom.

Summary

This paper suggests that the current emphasis on the recognition of individual differences in learners has resulted in (1) a major change for the teacher, (2) a revision in the teacher's conception of instruction, (3) the emergence of four types of individualized instruction, and (4) a need to revise the teacher's conception of

References

Bloom, Benjamin S. "Learning for Mastery." *Evaluation Comment* (published by the Center for the Study of Evaluation of Instructional Programs, UCLA) 1 (May 1968).

Glaser, R. *Adapting the Elementary School Curriculum to Individual Performance.* Proceedings of the 1967 Invitational Conference on Testing Problems. Princeton: Educational Testing Service, 1968.

Lindval, E. M. *Testing and Evaluation: An Introduction.* New York: Harcourt, Brace & World, 1961.

Suydam, Marilyn N., and J. Fred Weaver. "Individualizing Instruction." In *Using Research: A Key to Elementary School Mathematics.* University Park: Pennsylvania State University, 1970.

SELECTED BIBLIOGRAPHY FOR CHAPTER 17
Evaluation and Diagnosis

Ashlock, Robert B. "A Test of Understandings for the Primary Grades." *The Arithmetic Teacher* (May 1968), 438–41.

Brownell, William. "The Evaluation of Learning Under Dissimilar Systems of Instruction." *The Arithmetic Teacher* (April 1966), 267.

Epstein, Marion. "Testing in Mathematics? Why? What? How?" *The Arithmetic Teacher* (April 1968), 311–19.

Fellows, Martha M. "A Mathematics Attitudinal Device." *The Arithmetic Teacher* (March 1973), 222–23.

Gray, Roland. "An Approach to Evaluating Arithmetic Understandings." *The Arithmetic Teacher* (March 1966), 187.

Leeseberg, Norbert H. "Evaluation Scale for a Teaching Aid in Modern Mathematics." *The Arithmetic Teacher* (December 1971), 592–94.

Maertens, Norbert. "An Analysis of the Effects of Arithmetic Homework Upon the Arithmetic Achievement of 3rd Grade Pupils." *The Arithmetic Teacher* (May 1969), 383–89.

Peterson, John C. "Four Organizational Patterns for Assigning Mathematics Homework." *School Science and Mathematics* (October 1971), 592–96.

Rea, Robert and Reys, Robert. "Mathematics Competencies of Entering Kindergarteners." *The Arithmetic Teacher* (January 1970), 65–74.

Richard, June V. "Interviews to Assess Number Knowledge." *The Arithmetic Teacher* (May 1971), 322–26.

Smith, Robert F. "Diagnosis of Pupil Performance on Place-Value Tasks." *The Arithmetic Teacher* (May 1973), 403–08.

Snipes, Walter. "Mobility on Arithmetic Achievement." *The Arithmetic Teacher* (January 1966), 43.

Weaver, J. Fred. "Some Factors Associated with Pupils' Performance Levels on Simple Open Addition and Subtraction Sentences." *The Arithmetic Teacher* (November 1971), 513–19.

PART VIII

THE WRAP-UP

CHAPTER 18

ELEMENTARY SCHOOL MATHEMATICS TODAY AND TOMORROW

During the last two decades, a number of significant changes have taken place in teaching methods, materials, and media used in elementary school mathematics. These may be summed up briefly as follows:

(1) The introduction of modern programs in mathematics, with an emphasis on discovery modes of learning.

(2) Attempts to design courses for individual children, instead of having the entire class do the same thing at the same time.

(3) Programs arranged in hierarchies of learning suggested by Gagné, Bloom, Piaget, and others.

(4) The use of new media and materials.

Many of the changes have resulted in a changing role for the classroom teacher. The direction of the change has been from that of an authority figure who gives out knowledge ("the teacher knows and shows, the kid sits and gits"), to that of a course designer, guide, facilitator, and diagnostician.

The articles in this chapter are offered as a thought-providing synthesis for this book of readings. Read Jo Stephens' article, "Where Are We Now?," and seriously consider some of the issues that she raises. Hopefully, we will all ask ourselves from time to time, "Where are we now?".

Ruric Wheeler, in his article "The Role of the Teacher," claims that we still have substantial illiteracy, widespread incompetence, and incomprehension in all areas of the curriculum. Even their best students, teachers say, are in many cases unskillful in performance and superficial in understanding. The dropouts indicate a contempt for our forms and methods of education at least as great as that directed at any other facet of our social organization. What then is the role of the teacher? Be sure to read this article for some interesting observations.

As a final wrap-up, you will certainly want to read Morris Kline's "The Proper Direction for Curriculum Reform." Hopefully, we can all help bring to fruition Kline's final statement: "When we reach the stage where 50 or 60 percent of the high school graduates can truthfully say that they liked the mathematics they took, then we shall have attained some measure of success."

Good luck and good teaching of mathematics.

Where are we now?

JO STEPHENS

'STANDARD IN THE THREE R'S WAS HIGHER IN 1914.'

So read the headline in a national daily in 1969, heralding the publication of an article by Sir Cyril Burt. The editor commented:

'Sir Cyril's observations at least provide a little handy ammunition for those awful reactionary parents who – through personal observation on a day-to-day basis – have come to the conclusion that their children are not learning to handle the basic tools of civilisation properly.'

The blame for this *appalling* state of affairs is often given to the *progressive* primary schools. All this *new maths.*, *creative writing*, *project work*, *free play* and (worst of all) the *complete lack of discipline* in modern schools must be the reason for the decline in standards!

Is there truth behind the emotional journalism?

Sir Cyril Burt, now in his eighties, was appointed to the London County Council as a psychologist before the First World War. One of his first tasks was to produce a series of attainment tests covering reading, comprehension, spelling, mechanical arithmetic and problem arithmetic (his terms, not mine). His writings tell us that he was asked to do this because the School Board for London was worried about the decline in standard of the three R's due to the introduction of progressive methods in primary schools (what was that about a pendulum?).

Results of the same tests have been recorded by various investigators since then. If the same tests are not used then, clearly, comparisons cannot be made. Based on the 1914 level taken as 100% the 1969 data[1] gives

Reading accuracy	95.4%
Comprehension	99.3%
Spelling	91.1%
Mechanical arithmetic ..	92.5%
Problem arithmetic	96.3%

'Appalling results,' said a newspaper report but, more cautiously, Burt himself was reported as saying, '. . . appreciably lower than they were 55 years ago', and later in the argument, '. . . the trend has shown, not an improvement but, if anything, a decline'. I wondered if his *if anything* threw doubt on the interpretation by the Press.

The figures *do* show a decline in standard on those tests. Whether an individual teacher calls it appalling or just fractional will depend upon a whole lot of other things. It is all too easy to make modern primary school method the scapegoat. I want to hazard a few other guesses as to the reason for it.

For example, one might interpret the effect of such methods in quite another way. Figures show that achievement on these tests reached a very low point (compared with 1914) at the end of World War 2 and Burt comments on the surprise of the subsequent recovery. This recovery took place whilst the criticised methods were gaining ground in many primary schools. Could it be that they contributed to this improvement? We do not know.

Clearly the figures alone are not enough and, worse, they encourage us to make up stories, depending on our own beliefs, to fit them. No interpretation makes sense until we know something of the aims of teachers then and now and of the methods used to try to achieve those aims.

Let us assume that Burt's tests reflect the work actually being done in the schools of the time. (Is this reasonable? Could we say the same of today?) A section of arithmetic includes:

(For age 11 years)
$$2\tfrac{7}{15} + 1\tfrac{20}{21} + \tfrac{16}{35} - \tfrac{12}{105}.$$

(For age 12 years)
Simplify $\dfrac{4236.4 \times .008}{1.0591}$

and

Find the sum of 1.7 of 5 lbs. + 3.75 of 1 lb. 4 oz.

(For age 14 years)
The area of a square is 1,722.25 square feet; find the length of the side in yards, feet and inches.

Burt's writings include compositions about school. Maybe they give a glimpse of the pattern of work. A 10 year old says: 'The lessons are sums, dictation, spelling and drawing. We do arithmetic

This article is reproduced from *Mathematics Teaching* (Summer 1971), 2–5, the Journal of the Association of Teachers of Mathematics, by permission of the author and the editor.

every morning, history on Tuesday and drawing Friday afternoons'. Other similar passages contribute to a picture of the school day and indicate something of weightings given to various aspects of the curriculum.

The overall impression I get from the children's writing is of teaching directed strongly towards the techniques of arithmetic and the elements of English grammar. (It would be interesting to read similar reports from today's pupils especially if, in the same group, some had been shown these earlier writings and some had not. Any offers?) It may be that these goals were right for the times but compare them with the wide range of skills, knowledge and experience which we would encourage today. Many of our children read, comprehend and experiment with materials of a scientific and technological nature never known in 1914. The physical development and activity of young children is now generally more advanced. Should we assume that if our 11 year old is less adept than his grandfather at the computational techniques with vulgar fractions then he is 'not learning to handle the basic tools of civilisation properly'? Is it not possible that he could be better educated, in modern terms, than his 1914 counterpart? I hope so. No doubt under the same circumstances our pupils would do at least as well as their grandparents did *but do we want even to put it to the test?*

The plain fact is that objectives have changed in many respects. Time spent on activities now regarded as inappropriate at certain stages is more than used up on other studies. The new studies are thought to be more relevant at *this* end of the century. It would scarcely be a matter for pride if educational objectives had not been subject to review in 55 years. It is clear that we have failed to interpret these changes to the layman. Perhaps we have only ourselves to blame if journalists write, seriously, such rubbish.

Look again at the arithmetic question on square roots for 14 year olds. The six digits lead me to think that the standard arithmetical technique is expected for I doubt if 7 figure logs were in use or that an iterative method would have been common. Did you ever learn that technique? It used to be needed for that expansion of metals question in A-level Physics and I've got vague memories of marking off the digits in two's from the right (or was it left?). I wonder how often it is taught to 14 year olds now? I suppose I would be able to re-learn the method if I had to and that I might remember it if there was enough motivation . . . after all I still remember the logs of 2, 3 and π and it is several years since I last used them. The point is that, even under the specific title of square roots, a test appropriate for one objective *may not* be appropriate for another.

The question of changing aims is a central one as the curriculum evolves but, in the emphasis on *what* and *how*, an understanding of the precise

nature of changes may be tacitly assumed and not explicitly stated. Hence we may be disillusioned about what we hoped might be an improvement in our work simply because we have not fully considered appropriate changes in evaluation. If a change is made either in method or content (or both) then it seems to me to be important to ascertain whether this involves a different goal and hence new criteria of success. Otherwise we may not recognise success when it occurs or may wonder – as do some parents, it appears – why the old goals are not reached. ('All this work on shape is all very well but how does it help their arithmetic?')

Consider, as a further example, work on measures in primary schools. The aim used to be to develop the ability to perform, rapidly, multi-unit calculations. Success was rightly judged by testing this ability and, at that level, a knowledge of the reality which the calculations could represent was unimportant. Today it is accepted that complicated multi-unit calculations seldom occur in practical situations and consequently there is less emphasis on calculation and a good deal of emphasis on practical measurement in preparation for a real understanding of measures as used in science, commerce or industry.

In measuring activities we try to prepare children for reality and not just for a pencil and paper test at some stage. I have been careful not to imply that calculations are superfluous but I still have an uneasy feeling that some teachers are using measuring activities simply to sugar the pill of the inevitable calculations which still, in such cases, remain the ultimate goal because that's what 'they' put in the tests. I suggest, therefore, that an assessment which covers *only* calculating ability is inappropriate and a new assessment which reflects both the practical aspects and realistic calculations might better be substituted.

Indeed, despite the widespread changes, practical work is still unknown in a few classes. Only two years ago I observed 8 year olds who were all engaged in calculations in yds. ft. and . . ssch . . you know what! I talked to a boy (who was doing well) and he could not show me the approximate length of any of the units although he could reel off the relationships between them. Later I asked the teacher if they did any practical measuring. 'No, there isn't time to waste. You see these sums will be in the 11+.' I asked if she would be changing any of her methods when the 11+ was abolished, as was expected in that area. 'No, we shall not lower our standards just because there's no examination.' There was a sense in which I envied the clarity of her position; no qualms, no personal responsibility, no need to wonder.

In the days before CSE I did my share of cramming with borderline O-level students. Even now we all know that there are more cases of teaching being made to fit examinations than examinations

made to fit teaching for the whole problem of ensuring that assessments are compatible with aims is complex; the more so when outside interests are involved. I think it is fair to say that secondary teachers have done more *themselves* to try to make their external examinations (CSE in particular) fit modern methods of teaching than have primary teachers (where 11+ examinations exist). Most primary teachers, on the other hand could be said to have a broader overall view of educational aims and can, I believe, rightfully expect to be able to obtain more guidance about the objectives in particular subject areas from subject specialists. I have the feeling that where we (e.g., as CSE panel members, as examiners, as advisers) find ourselves responsible for preparing or choosing testing material used by others then we could make a more positive contribution.

Unfortunately it is all too easy to find oneself a party to 'we must put . . . in because *they* always teach it' and 'if only *they* would stop putting . . . in then we could be doing something more valuable instead'. I can't imagine that the situation will improve until far more teachers are prepared to take greater personal responsibility for the methods of assessment of their own pupils, even within a public examination system. Great strides have been made in secondary schools with both CSE and O-level. Many ATM members and others have been involved in re-thinking the examinations they use but even now many of the new examinations reflect only an exchange of mathematical topics, and, it is rare to find proposals for public examinations which encompass new emphases in mathematical activity[2].

Changes in content have proved relatively easy to cope with. They may be additions, may be exchanges, and they arise, broadly speaking, from new discovery and invention, shifts in relative importance and sometimes just plain fashion. Once the idea of letting developments in knowledge, or changes in use, influence school studies then it is simply a case of up-dating one's knowledge of subject matter. Unfortunately, there is a good deal of resistance amongst adults to learning more mathematics.

Changes in school mathematics are often labelled 'new mathematics' or 'modern mathematics'. These terms have a variety of meanings in different contexts and do not communicate accurately un-

less used in an agreed way in a group. In the secondary field they often indicate a change in content – scarcely modern as some aspects are over 200 years old – although we are only just getting round to the point where they are introduced to most school children. Content changes do occur in primary schools and their 'new mathematics' may sometimes mean such topics as graphs and shape in the infants school and multi-base arithmetic in the juniors. Some of these changes are 'new' to primary schools in the sense that they have been handed down from secondary schools (this happens at all levels of course). Others such as logic, sets and relations reflect content changes and basic re-thinking at all levels.

Popular articles which attempt to explain content change in mathematics teaching to parents often confuse two ideas. They sometimes give the impression that new content is intended to get the same results, but better. Occasionally this *is* the purpose (e.g., multi-base arithmetic is intended to promote better understanding of the number system and computational methods) but sometimes the purpose is for the sake of the new mathematical topic itself (e.g., topology, logic). In our conversations with parents I think we must make these two ideas explicit. If we do not then parents may reject the unfamiliar work as irrelevant and the effect on the child's future learning may be significant.

In all this period of change my own hope is that the most significant developments for the future are not those of content, nor new ways along old paths, but those which affect the ways in which we work with children and the ways in which children are involved in their own work. When a primary teacher talks about 'new mathematics' he is often referring to this sort of change from teacher-dominated mass instruction to a more personal approach. Of course this is an over-simplification and I have been careful not to use the words 'individualised learning' as this too is rapidly becoming a multi-meaning term.

It is the changes of approach and levels of pupil involvement which have been the hardest for me. The worst aspect is the insecurity of not knowing whether you are in fact improving your pupils' mathematical education. Subsequently, in advocating such changes, it is only comparatively recently (unfortunately) that I have made a conscious and serious attempt to develop specific discussion about aims and compatible methods of evaluation.

The story of the teacher whose class was working a calculation concerning measures may serve again as an example. To persuade her to provide measuring activities and then to leave her unsupported and using the old criteria of success would almost inevitably result in disenchantment and in what she would describe as a lowering in standards.

Even when we try to be careful to sort out all these goals and consequences in our own minds we are not always very good at communicating them to those who depend upon such knowledge in order to operate well themselves. For example, if one of the aims of an infant school is to let each child work at his own pace as far as possible then, when the transfer to junior school arrives, not all the children will have worked successfully at the same topics to the same depth. Elementary, but what will happen if the next teacher likes to keep her class working together! If the new teacher is to have a chance of success then, having accepted the diversity, she needs to know a good deal about the overall development used and the rough position of any pupil in that development. It *is* possible for the receiving teacher to undertake some sort of check-up but it must necessarily be compatible in content, language and method with previous experiences. Ideally, of course, all the necessary information is passed from teacher to teacher and the responsibility of both in this exchange is, for the children's sake, equally great.

Well, where *are* we now? The traditional picture of the authoritarian teacher has largely disappeared and all too often the layman has a mistaken view of an anti-disciplinarian in his place. As a profession we must do a better PR job. Destructive criticism from within the profession is given disproportionate coverage by the mass media. Critical colleagues might do more good, and certainly less harm, in making their contribution to discussion in teachers' centres and professional associations.

What is emerging is a differently based authority. For all that the unconvinced and the uninformed say about 'discovery methods' we do, or should, know and should be able to explain our aims. At any time we must have an overall pattern of development and philosophy for education. Of course we adapt and develop our earliest ideas as we read, teach, talk and add to our experience but without a good theory we are just child-minders. (Where did I read 'There is nothing more practical than a good theory'?) When confronted with an idea we may try it and maybe accept or reject it for the future. Sometimes we may try something even more radical than usual in order to make judgments from first-hand experience (you know when you do this because your critics call it jumping on every bandwagon!).

Occasionally whole new fields of mathematics are suggested as suitable for a particular age group, or new unifying ideas or mathematical methods are developed. For the teacher these are often hard to accept as they usually call for greater faith in the end product, great need for adjustment in teaching and greater need for good public relations.

A current example is the rising interest in logic and algebraic structures especially at primary level. It may be that here we shall see the innovations in the coming decade. It will be a pity if this development fails to find its place in the whole fabric of mathematical education simply because the pioneers are blind to anything else. It will help if lecturers and writers make an effort to help teachers to develop a full understanding of the aims of the work, ways of evaluating it and its relationship with other aspects of education. My present feeling is that there may be too much emphasis on logic activities for the classroom (many called 'games' – see MT No. 49 page 57) and not enough on the philosophy behind their introduction. The dilemma for any lecturer is that plenty of classroom material shows him to be a real feet-on-the-ground type, and yet if he skimps the theory he knows that he is just proliferating a collection of pastimes.

While I'm at it I may as well thump one more tub! There is, I think rightly, tremendous enthusiasm for practical mathematics, for giving pupils the means to use materials and develop important ideas through personal experience. Unfortunately the wide acceptance of this approach *can* give those who feel their own mathematical ability to be limited the opportunity to become just providers of packages and organisers of classrooms. They may really believe that if the materials and instructions are there then the children will learn mathematics. They are misled by that neat-sounding and now oh-so-familiar ancient Chinese proverb, 'I do and I understand'. Shall we change it to, '*I do and I am more likely to be able to understand*'? The role of the teacher in discussing ideas and results, in imparting facts and asking questions, is as important as ever it was. Sometimes the communication is between a teacher and one child, sometimes a small group, sometimes a normal class-size group, sometimes more. If we neglect this person-to-person communication then I think we probably deny our pupils more than we offer them.

References
1 BURT, C. 1969 "Recent studies of abilities and attainments", Journal of the Association of Educational Psychologists, II pp 4f.
2 See however 'Examinations and Assessment' (ATM pamphlet), the ATM Sixth Form Bulletins and the proposals of the ATM A-level and O-level groups.

Acknowledgements
The illustration showing work at the water tray is reproduced from the *Capacity* workbook of *Willbrook Discovery Mathematics*, by permission of the Educational Supply Association Ltd.
The other drawings are by Joan Dean, General Adviser for Primary Education for Berkshire LEA and are reproduced from the Berkshire Working Party Report *Mathematics in the Primary School*.

The Role of the Teacher

DAVID WHEELER

If we know that ineffective teaching of mathematics is not due to the difficulty of the subject matter, and if we know that changing the classroom environment by grouping the children, promoting individual work, introducing a variety of materials, and liberating the structure of the timetable, does not contain within itself the possibility of acting directly on the awarenesses of children, and if we then do not re-examine in the most fundamental way how as teachers we should act, we are guilty of a total failure in seriousness for we have stopped our progress towards a better education for children just short of the point at which we can make a contribution to it.

At this point in time it is not at all necessary to say yet again that the teacher can only begin to be effective if he respects each child and works for each child's autonomy. We have made considerable adjustments in our beliefs and attitudes about children and about teaching. We know, for example, that education is not just concerned with filling children with knowledge or with slotting them into a role in society. We believe that children can take some responsibility for their own learning, and that we do not have to programme them as if they were machines. We no longer believe that they are entirely formed and fixed by what happened to them before they came to school or what happens to them while they are in it. Although policies and administration may lumber a long way behind, at least in the classroom the child may make his personal contribution and find his contribution valued.

But if we measure this progress, positive as it is, against the distance still to go, aren't we forced to say that it is almost negligible? We still have substantial illiteracy, widespread incompetence and incomprehension in all areas of a curriculum which, God knows, is not particularly demanding. The best students we have are said by their teachers to be in many cases unskilful in performance and superficial in understanding. The drop-outs indicate a contempt for our forms and methods of education at least as great as that which they direct at any other facet of our social organisation. Whatever may be found to stand in our defence, it certainly cannot be said that we have yet broken through. Judging by results rather than intentions – and what else is there to go by? – SMP and the Leicestershire primary schools, say, move us an inch when we need to go a mile.

I do not wish to deny the inch of progress. Other things being equal, SMP is undeniably a better programme of mathematics than the syllabuses it replaces; the children in Leicestershire schools are having a better chance to develop their powers than their predecessors had. But in spite of this improvement, the effects of these reforms bear very little relation in magnitude to their aims and reputations.* Gratitude may make us acknowledge the inch; honesty compels us to admit the remaining mile.

We could say, of course, that reforms in any part of education have never been more than marginal improvements in the situation at the time. Looking back, we may see everywhere only sluggish and reluctant change. If we look outside education, though, we must wonder if this slow crawl is inevitable. The technology which could put men on the moon was developed in less than fifteen years. Is there any *a priori* reason why illiteracy cannot be eliminated at least as quickly as tuberculosis has been?

I believe we know that we can tinker all we like with programmes and classroom environments without even approaching the radical reform in effectiveness that is necessary, because what we are occupying ourselves with is not at the heart of the dynamic process of educating. We have only to acknowledge to ourselves that we are educating for a future, and are not able to know what requires to be known in it, to understand that the

* I have deliberately chosen to mention two reforms which are viewed with considerable admiration by educators outside this country. There is not meant to be any implication that they are the only ones worth considering.

This article is reproduced from *Mathematics Teaching* (Spring 1970), 23–29, the Journal of the Association of Teachers of Mathematics, by permission of the author and the editor.

detailed content of the curriculum cannot possibly be of the first importance. (The present is no different from the past in this respect but the pace of change is so much greater that the impossibility of prediction has become more obvious.) We have only to perceive that our minds can continue to function normally in all but the most austere and oppressive physical conditions to know at once that since our minds are not the slaves of our physical environment they will not be miraculously transformed by alterations in it. If at some point in our lives we have experienced the frustration of not being able to do what it was necessary to do because we had earlier failed to equip ourselves with a particular skill, we know at firsthand that freedom to choose our activities does not necessarily develop self-reliance and autonomy.

I also believe that we know that the barrier to rapid progress in education has little to do with the need to re-define our aims. We are ineffective not because we don't know what it is we want. If I say that the aim of a school education is to assist every child to become operational in the basic skills current in his particular society, and beyond that to acquire access to his own powers so that he can function in that society as an autonomous person, I cannot think that I will find anyone to disagree. I shall be accused of being vague and imprecise, of course, but to that I shall say that I am only being concise. There is no difficulty in expanding the definition into a working brief if we wish, but this seems hardly essential if we already understand what it means. The attempts of the philosophers and the 'behavioural objective' psychologists to get us to define our aims in shopping-list form would never have assailed us if we had got near to achieving our real aims. They are only urging us to settle for so much less in the hope that we will be able to manage at least that.

Our problems lie in the inadequacy of the means we adopt to reach our aims. I have indicated the view that we spend too much of our reforming zeal in working on aspects which, although valuable aids to the kind of education we want to give, cannot be the primary medium for the dynamic process of educating children. Suppose the disparity between aims and achievement can be traced entirely to the actions of the teacher. It is here, after all, that the process that we hope will culminate in children learning begins. There is no inference to be drawn that our intentions are not of the best; the eagerness with which proposals for reform are embraced, at some cost to ourselves, is evidence to the contrary. But we are, it seems to me, very muddled and confused in what we are doing, and unsure of what our functions in the classroom are. And before I hear any 'yesses' let me urge that what I am discussing is not verified by anyone's humility but only by a study of the facts. I am reminded, in this connec-

tion, of those extraordinary research results which suggest that differences in teacher personality have the largest detectable correlations with differences in competence. Such results, I feel, would be unlikely to hold for plumbers, artists or astronauts (although they might hold for politicians and television interviewers). We can read the results as a vindication of the significance of the teacher's personality for the high art of teaching, but an equally plausible interpretation is that the overall level of competence happens to be extremely low. The best of a bunch of incompetent plumbers are no doubt the ones with nice personalities. However, that is merely a joke and an indication. I use it to introduce the notion that we tend to regard teaching as an art but we could make it into a science.

To say, as we often do, that each teacher can only teach in his own way, or that good teachers are born not made, or that a teacher's personality is the most important part of his equipment, is to turn one's back on the possibility that teaching can be scientific. Perhaps these beliefs, or others like them, are so deeply rooted in us that they account for our neglect of the science of teaching – for indeed, if they are true, there can be no way of improving teaching except through altering the secondary characteristics of the teaching situation. But if we have experienced the fact that we can use ourselves in different ways, and that these ways are under our control, to some extent at least, then we can begin to entertain the possibility of making our teaching scientific and consider what that will require of us.

The objective of science is to get to know the truth. It is concerned to remove ignorance, illusions and preconceived ideas, in order to reach reality. The method of science is observation. Sometimes when the phenomena that the scientist wishes to study are beyond his power to affect them, or when his intervention would destroy the things themselves, he can only be alert and wait upon the occurrence and repetition of the events. In other cases he devises techniques to step up the observational yield: by inventing instruments, for example, or by setting up experiments in which the phenomena occur as and when he wants them to. But the principal technique of the scientist is to act upon the situation in which a phenomenon occurs in order to change it, and to observe the consequent changes in the phenomenon. It is this technique which permits the scientific study of how children learn. It does not require that the phenomena being studied can be controlled or isolated from the situations in which they happen. The role of the scientist here is not to stand back in order to watch, but to intervene and continue to watch.

If the scientist is described as an impartial observer it can mislead us because it suggests that he is detached from the situation and lacks concern for the outcome of his observations. But the scientist is very concerned – concerned to know the truth – and he is not detached from the situation but immersed in it. The impartiality of the scientist is an attitude to his own past that will allow him to be entirely sensitive to the present which he is observing. Being human he has expectations, hopes, preconceptions, theories; but if he is acting as a scientist, he is able to abandon any or all of them at once without regrets if his observations tell him that they are wrong.

The casting of hypotheses or theories to account for the behaviour he observes is a second-order activity which he undertakes for his own benefit and to enable him to communicate his discoveries to others. It is a means by which he can 'make sense' of his observations, by relating them to other observations he has made or to what is already known, and it serves to draw his attention to inconsistencies, or to gaps, or to further questions, which in turn send him back to making more observations. Prediction is only a relatively infrequent by-product of this general activity and not, as is often said, central to the scientific process.

When we have understood that science is only a particular mode of using oneself in order to reach reality, many of the problems which we may suppose present when we try to make the study of human behaviour scientific are seen to be illusory. Because we first meet science as a study of aspects of the physical world, where it seems to be concerned with the discovery of laws or principles governing the behaviour of all objects of a particular type in particular circumstances, we may find it difficult to conceive how any study of human behaviour can be scientific because we know that a human being can always choose to behave as a counter-example to any law or principle we have formulated. But science shouldn't be confused with collections of 'laws', which are only objectifications of the agreement of many people that the result of some scientific observation is valid for many situations. If this external view of science is taken we can find ourselves, if we are not careful, forced into the indefensible position of holding that the truth of an observation depends on the number of people who accept it, or the number of times it can be repeated. (Probabilistic models of degrees of belief fall into this absurdity as soon as they blur the distinction between believing and knowing.) We are not less sure that the earth is round because there are some people still alive who believe it is flat. Crick and Watson's breakthrough was not contingent upon anyone else duplicating what they had done.

The generality of scientific truth does not hinge on agreement or replication but on the fact that the scientist can say, if he has obeyed the rules of his discipline, that his perceptions owe nothing to skills or qualities peculiar to himself and that therefore it was and still is open to anyone else to discover the truth he has found. With this understanding, we can see that what is scientific is the method of study, and that the method may, in principle, be applied to situations of any kind to yield results of infinite generality.

With those preliminaries about the nature of science in mind, we can begin to consider some of the ways in which teaching can be scientific.

Let us start by turning on their heads two interlinked aspects of common classroom practice. The first things that a teacher usually does when faced with a new class of children is to establish himself in relationship with them. (Teacher-trainers have one sort of language for this: experienced teachers another!) This may or may not be a one-sided relationship (the class may be 'it' or 'them' or 'us'), but in either case the teacher consciously uses himself as a focus of attention for the children, frequently exploiting those of his skills or personal characteristics which he knows have a strong effect on others. Having established this relationship as a framework, he turns his attention to the tasks for which the class has met. Because he is at some level aware that he has stressed his individuality, and that this has coloured the activities in which he and the children have engaged, he distrusts his judgments about the individual children, and bolsters or replaces these with information gathered informally from chats with other teachers or formally from tests and examinations. In fact, for any important decisions concerning the scholastic progress of his pupils, he prefers the discriminating to be done by someone else.

The teacher who approaches his job scientifically will start with the tasks to be done and will consider how the attention of the children can be focused on them. He will consciously withdraw as much of himself as possible so that he will not be an interference to the activity he wants to promote. In not drawing the attention of the children to himself, but to the tasks on hand, he can be an impartial observer of their actions as they tackle them. Because he knows now what the children are able to achieve by working at tasks themselves, his judgment of their capabilities and attainments is more reliable than anyone else's, and he will reject external assessments if they conflict with his own.

No doubt both of these descriptions have been caricatured in order to contrast them, but the purpose here is to suggest that our conventional wisdom about how the teacher should work may be entirely in error. The real picture isn't so simple, but this is because we tolerate contradictory elements in our picture of what a teacher is. For instance, we are quite capable of saying that the teacher should begin by establishing working relationships with his class *and* that since it is they who are to learn it is they who must do the work. These are by implication contradictory because if the teacher takes the initiative in establishing relationships before there are any tasks, the children will know that the tasks do not have first priority; they are being thoroughly logical in subsequently working on the relationships instead of on the tasks – which can lead, for example, to the kind of classroom behaviour John Holt has described for us.

I have so far only given a taste, though, of what is implied by making teaching scientific, and to give time for the first wry flavour to dissipate, let me offer some reasons why teaching should change in this direction:

☐ no other reforms have yet altered teaching effectiveness substantially enough

☐ science has achieved radical changes in a short time in other fields

☐ the scientific method affords teachers the only technique available in our culture which can show up incorrect preconceptions and bogus folklore, and subordinate fantasy to reality

☐ science sharply distinguishes knowing from believing and can make teaching independent of persuasion, propaganda and personality (the cult of)

☐ scientific method is a tool which can be handed on to make its users independent and autonomous, recognizing truth as their only authority.

That teaching should aim to assist children to achieve independence and autonomy is a principle carrying a good deal of weight at this time, but we have confused it with issues of freedom of choice which are not central. Children don't have to be given freedom; they already possess it. No-one can teach anything to anyone against his will. The deplorable consequences of some of our teaching – the absurd behaviour designed to avoid engagement with the tasks and values of the classroom – can be read as signs that the children *are* essentially free and struggling madly to preserve their freedom. Their autonomy is expressed through their acts of rejection. So we'd better consider what kind of autonomy is worth having, rather than brandish the unqualified word as a magic cure-all.

Autonomy is intimately linked with competence. The six month old baby is no less an individual, free to be himself, than the six year old, and the six year old no less free to be himself than the sixteen year old. But the baby is much less able to act alone than the others; unaided, he literally cannot stand on his feet. The six year old is less autonomous than the sixteen year old since there are so many things he cannot yet do for himself that the older child can. It is unscientific to ignore the types and degrees of competence through which children pass as they move towards maturity and to forget that autonomy is relative, related to the competences that have been acquired. To fail to assist children to acquire the greatest possible range of competences is therefore to force

them to forfeit some of the autonomy which is potentially theirs. How, then, can we *dare* to equate giving children autonomy to giving them freedom?

But autonomy is also independence and self-reliance, which are matters concerning the will and the feelings as much as the extent of skill and know-how. If we observe children scientifically, we see that these aspects are there from birth (or even before), and that we will not have to teach them to achieve independence or self-reliance, but only to do what we can to stop ourselves taking them away.

A third ingredient which everyone needs for autonomy and independent functioning is a 'sense of autonomy', that is: an awareness of what skills are at one's disposal, that one can use them to be more self-reliant, and also one's self-reliance to acquire further skills. Without this self-awareness and consciousness of one's own powers, autonomy will not be realized in action.

Conventional teaching only meets the first of these three requirements; 'liberal progressive' teaching only the first two (or, more frequently, only the second, since in preserving children's independence it is inclined to reduce to a minimum the acquisition of competence). The three requirements together can only be met by scientific teaching, which is to say: teaching done by scientists to produce scientists. 'The learner as scientist' will not be a bad catch-phrase to adopt when we embark on this radical renewal of our teaching. It will demonstrate that we want the emphasis put on the threefold mastery, self-reliance and self-awareness of the child, and that we believe that the proper technique for teachers will be to confront the child with problems and challenges which he must tackle in the spirit of the scientist, knowing what he knows and recognizing the truth when he sees it.
(Easier said than done, but better said than not said.)

We must take it for granted here that teachers also have mastery, self-reliance and self-awareness in the same areas of functioning, and we leave this for teachers of teachers to consider. We will also assume that it is possible (and necessary) to build up a repertory of games, problems, and challenges, that each teacher can select from and add to, and which are of a sort to elicit from the children spontaneous activity which they can work on to yield an awareness of their own powers. Since ATM members have already made contributions to this repertory it is only necessary that this activity should continue, with perhaps a much greater awareness on our own part of the characteristics that go to make good inventions for this kind of work in the classroom (i.e. we ought to become more scientific in our work too).

I feel I can now return to the point, which may have given difficulty earlier, that the teacher must withdraw as much of himself as possible in the teaching situation, and perhaps it has become clearer what this implies. He must use every means he can find to focus the attention of the children on the problem, and this means that he *must* efface himself from their attention. On the other hand, the children will be at the centre of *his* attention because he must study them to know how to help them keep to their task. (It may not be apparent at first just how different this description is from the usual behaviour of teacher and children; we are adept at mistaking our intentions for our deeds.)

If we watch this teacher at work we see that:

☐ he teaches the whole class or a group of the whole class much of the time

☐ he sets the situation, giving essential information, but beyond that tells the children nothing

☐ he obtains as much feedback from the children as possible, by observing, asking questions, and asking for particular actions

☐ he works with this feedback immediately

☐ he never collects answers from the whole group to check that they all have the correct one, nor does he collect all the different answers they may have obtained without working on their differences

☐ except on rare occasions he does not indicate whether a response is right or not, though he often asks the children which it is

☐ he accepts errors as important feedback telling him more than correct responses, and by directing the children's attention back to the problem he urges them to use what they know to correct themselves

☐ he does not praise or blame particular responses or particular children, though he may exhort or reassure by expressing his faith in the children's capabilities

☐ his dialogues with the children contain phrases like:

Look at the problem
Look at what you have done
Listen to what you are saying
Don't guess: tell me what you know
You know you can do this: try it again
Is it right?
How do you know that?
Are you sure?

☐ he respects a child's right to opt out, and if the feedback from the children assures him he is trying to make them go too fast or too far he abandons his attack at once

☐ if he sees that a particular child in a group who is not immediately succeeding can be challenged so that he conquers his hesitations, he gives all his attention to that child at once

☐ if he sees that a particular child in a group is working on the problem but not contributing to the group, he leaves the child alone until he wishes to take part

☐ he appears to an uninvolved observer to be impersonal in his approach, giving no favours to anyone and taking success for granted, but showing no disappointment if it does not arrive.

There are enough controversial points here to indicate that this kind of teaching is *not* what most of us practice, yet it is only a matter of detailed and unprejudiced observation (scientific observation) to assure ourselves that this teacher's method works, and works better than the normal, 'friendly' way. Watching him, we are conscious that *everything* he does directs the children to attend to the problem and to their own actions in tackling it. Naturally this way of teaching requires great skill, sensibility and integrity, and to be successful with it requires a serious and disciplined apprenticeship. But don't let us be confused by its difficulty into saying that it cannot be the right method; it is open to everyone, in his classroom, to prove that it is.

It isn't easy to describe the way in which this teacher works; I wonder what picture you have made of it? I hope that you see that it is quite different from the teacher's way in conventional classrooms on the one hand, and from the teacher's way in 'liberal progressive' classrooms on the other. I can imagine that some of you, particularly teachers in primary schools, will either find yourselves lacking sympathy with the description, or will reject it because you could not teach in that way in your situation. It seems to me to be one of the grossest educational errors of recent times that we have managed – for the best of reasons, of course – to construct situations in the classroom which make it difficult or impossible for the teacher to have much choice of ways to act. The irony of this is that reorganisation has often been introduced in order to liberate the children from unnecessary restrictions but has ended up by shackling the teacher. The 'open' progressive classroom where children work alone or in pairs 'at their own pace', perhaps on topics of their own choice, for all of the time, has exchanged one form of restriction for a worse. It restricts the children, who must work from pre-planned and packaged material, and the teacher, who cannot teach more than two children at any one time (if we allow that teaching involves more than answering the occasional question and marking the occasional piece of work). Contrary to belief, the progressive classroom *delays* the arrival of autonomy in the child because the teacher is never sufficiently in control of the situation to be able to throw him to the edge of his resources at the moment when it is necessary.

I suppose I understand that we have moved in this direction because we have become much more aware of the dreadful things that teachers have done to children in the past. I am only questioning whether the best solution is to take the pressure off, or to assume that teachers *can* learn other ways in which to act. If we don't choose the latter, we have in effect thrown in the sponge and denied ourselves that mile of progress again.

We must not minimise how much teachers will have to change if they are to act with maximum efficiency. There is still a horrifying amount of insidious folklore in the air which has *never* been put to scientific test; there are any number of liars, charlatans and fools making reputations and money out of the education business; there are, too, all the constraints – financial policy, administrative machinery, public opinion, promotion prospects, buildings, equipment, racial tensions, party politics – that impinge. Only pressure groups and political action can deal with the last category, but there is nothing, except our preconceptions, that gets in the way of our being clearsighted and courageous in what we do in our classrooms. At various points in the past science has shown us how to remove ignorance, prejudice and irrational fears, and it could come to the rescue again if we *want* to know the truth.

Of course the siren voice of reasonableness can mimic the accent of the scientist. Take as an example the model of teaching a complex skill by breaking it down into a sequence of components which must be successively mastered, and then compare it with how we learned to play an instrument, or to play golf, or to swim, or to ride a bicycle, or to walk (if we remember). The model sounds so rational and logical that it is quite an effort to face the fact that it doesn't correspond *at all* to the manner in which we learned any of the skills we possess. In all the cases mentioned, and in many others, the learner has to cope with a complex situation (not necessarily the total possible complexity: in learning to drive he will begin on quiet roads and at a slow speed; but he must nevertheless use all the controls from the start). In the early stages he makes very many errors and progress is so slow as to be almost undetectable; he has to face and deal with recurrent failure. Eventually, if he persists, the moment suddenly arrives when a partial coordination is achieved – the first staggering step, the first ball hit in approximately the intended direction – and from now on progress is assured (even though there will be setbacks, the ground can be reclaimed) until the next major challenge has to be faced. If this is what happens in learning a skill – and we know it is – how can we be so stupid as to be seduced by improper suggestions? The consequences of the truth here are that when we want to teach a child a skill, we must offer enough of the complexity to work on from the beginning; that we must expect and accept very many mistakes in the early stages, conscious that work is continuing even though it is hard to see the evidence; that we do not help by isolating difficulties, but rather by continually drawing the children's attention to aspects of the complexity that they may have overlooked; that polishing and perfecting cannot start until coordination of the strands has been achieved – that, for the learner, analysis follows synthesis.

But don't let's go just by the reasonableness of this story either, or we will likely be in the same boat as before. *Is it true? How do you know that? Are you sure?*

It is a further question to ask how much we are concerned with teaching skills and how much with other things. For the moment it is sufficient to say that if mathematics is seen as an activity, it must require a fair amount of sheer operational competence and know-how to get it going. It is a whole series of further questions to ask how we shall know what all the challenges are that have to be met so that a child can act as a mathematician, and it is not my purpose to give any answers here. If the rest of the article has made any sense, you will know that the answers will be obtained, if we don't know them already, from a scientific study of mathematical activity.

I cannot end without saying something to those who will feel that I have been concerned solely with the didactic function of the school and that I have omitted almost all the other aspects of school education, the ones which they happen to value most. There is only space to make a few remarks which I cannot hope will persuade, but which may clarify my position for anyone who is interested.

Children do not choose whether to go to school or not. It might be better if they did, but that is not a realistic hope in most countries of the world at present. In the enormous stretch of their lives that they spend, compulsorily, in school, they are entitled to an adequate return for their efforts just as when they are at work they are entitled to an adequate wage. Putting it another way, teachers have a responsibility not to waste children's time. If it is in fact possible by developing the science of teaching to teach all that we regard as essential for each child to know more efficiently than we do at present, and in much less time, then he is entitled to the benefits of this new productivity. One of the benefits is that there will be time for him to learn more than he can now, and another, that he will have more time in school for cultural and social activities, for the exercise of free choice and the development of initiative. Certainly nothing need be lost from all that any school can now provide for its children.

But I can imagine that some of you will agree that this is a possibility without going so far as to want it to happen. You would say, I think, that this is giving school a different character from the one you wish it to have. All right; you must work for the school you want. I will just add that for me the school is most likely to make its best contribution when it is sharply differentiated from other social organisations. It shouldn't be an extension of the home if the home happens to be 'good', or a therapeutic community if the home happens to be 'bad'. (A small number of schools may have to be therapeutic communities for those children who are most seriously deprived and disturbed.) In England most parents and teachers *want* schools to be an extension of or substitute for the home. They want children to have a middle class education because they are simply afraid of what may happen to children if they leave school without that particular protection. This attitude over-rates middle class values* whilst simultaneously under-rating children. It is an understandable attitude as long as most of our teaching is so ineffective, because it seems better for children to have a partial substitute for education than to have no education at all. But if parents and teachers could see somewhere, in a few classrooms at first, that professionally skilful teaching could offer each child access to and awareness of his own powers, they would be so amazed at what all children could achieve that they would see that protection was no longer necessary, and we could put away placebos and palliatives and get on with the real thing.

No one article can possibly do justice to all the facets of the role of the teacher. This has concentrated on the teacher's functionings in the classroom, ignoring everything else that might be said to belong to his responsibilities, and moreover has dealt exclusively with ways in which his work in the classroom can become scientific. Teaching is indeed an art too. But a thousand years (or more) spent in developing it as an art have taken us almost nowhere, good teaching being as rare as it has always been, and bad teaching very little better. It can be left for another article, or another writer, to consider how the art of teaching can be developed from the solid base of a science of teaching.†

* Value systems, like politics, are about priorities. Western middle class values rank ambition above the will to learn and success above self-knowledge. To this extent they are really on the side of schools as they are and against schools as they might be.

† I am aware that at several points in this article I have gone beyond what can legitimately be said in my own capacity as a scientist, and to this extent I have betrayed the principal cause for which I attempt to argue. But because some, at least, of what I have written is scientifically verifiable, and because the exercise of disentangling this from the rest that is not may be a valuable exercise for readers as well as for me, I have allowed myself to publish this inadequate and unrevised attempt to confront a central educational issue.

THE PROPER DIRECTION FOR CURRICULUM REFORM

Morris Kline

Introduction. It is at least fifteen years since the modern mathematics movement appeared and about ten years since modern mathematics curricula began to be taught somewhat extensively. My experiences with students who come to college today as compared with those who came ten years ago is that they do not like mathematics any more than their predecessors used to—which means that they dislike it intensely—and do not understand it any better. In fact they know less technique. They do know that $3 + 4 = 4 + 3$ by the commutative law, but they do not know that $3 + 4 = 7$. The theory of sets has not proved to be the elixir of life for mathematical pedagogy. And the sterile, desiccated axiomatic mathematics has not promoted understanding. It seems to me that a review of what we have been doing is called for, and I should like to attempt one.

The Traditional Curriculum. For several generations we have been teaching what I shall call the traditional curriculum. In grades seven and eight we have taught informal and simple algebra and geometry. We then move on to algebra proper, Euclidean geometry, more algebra, trigonometry, and, usually in the fourth high-school year, solid geometry and advanced algebra.

What is right and wrong about this curriculum? As far as subject matter is concerned it is basically right. A few topics have become outmoded, notably the logarithmic solution of triangles in trigonometry, and Horner's method of finding irrational roots of equations, commonly taught in advanced algebra. There is some question whether solid geometry warrants a full semester at this stage of a young person's education. I personally believe that it does but would not quarrel with anyone who disagrees.

One defect of the traditional curriculum is the striking difference between the method of presenting algebra and that used for geometry. Algebra is presented as a series of processes with little rhyme or reason, whereas geometry is presented deductively. Surely if both are part of mathematics the presentations should have the same character. Moreover, because algebra is presented as a series of processes, the students are obliged to learn them by rote. This type of learning is not only boring but readily forgotten.

The gravest defect, to my mind, is the lack of motivation. Why should young people take an interest in any of this mathematics? Mathematics proper is admittedly abstract. The child can perhaps see the need to learn arithmetic. He will use a modicum of it later in life. But he can see no point to learning algebra and geometry. Why should he learn addition of fractions, solution of equations, exponents, factoring, and other processes? The situation is not much better in geometry. It is true that students can see what geometry is about. The figures tell this story. But the question of why one should study this material is not answered. One can readily understand what the history of China is about, but not be convinced that one

Reprinted with permission from the *New York State Mathematics Teachers' Journal* (April 1970), 56–65.

should be obliged to learn it. Why is it important to know that the opposite angles of a parallelogram are equal or that the altitudes of a triangle meet in a point?

Clearly one cannot defend algebra, geometry, and trigonometry on the grounds that they will be useful later in life. The educated layman does not have occasion to use this knowledge unless he becomes a scientist, engineer, or mathematician. But those who do cannot be more than a small percentage of the high-school population. Moreover, even if all the students were to use mathematics later in life, this usage cannot be the motivation. Young people cannot be asked to take seriously material that will be useful only many years later. This motivation is often described as offering "pie in the sky." When, some years ago, the schools did introduce interest and discount in the seventh and eighth grades in an effort to motivate students, they found that twelve- and thirteen-year-olds did not take to the material, and the experiment has been conceded a failure.

Much of the mathematics taught is often defended as training the mind. There may or may not be such training, but if there is it can be achieved with subject matter that is far more agreeable and understandable. One can teach the commonly used forms of reasoning by resorting to simple social or legal problems whose relevance to life is far more apparent to the students.

Another commonly advanced justification for teaching mathematics at the high-school level is the beauty of the subject. But we know that these subjects were not selected because they are beautiful. They were selected because they are necessary for further work in mathematics. There is no beauty in adding fractions, in the quadratic formula, or in the law of sines. Moreover, novitiates are not likely to find beauty in a subject they are still striving to master, any more than one who is learning French grammar can appreciate the beauty of French literature.

To sum up, then, the defects in the traditional curriculum are some outmoded topics, the rote learning, and the stark lack of motivation.

The Modern Mathematics Curriculum. As we all know, a change in curriculum has been urged upon us, the change to what is known as modern mathematics or the new mathematics, and many schools have adopted it.

What is modern mathematics? Surprisingly, perhaps, this question is not easy to answer. Dozens of groups have been working on modern mathematics, some having started well back in the 1950s, and any number of texts have been published that represent themselves as modern; but one cannot find any single document that tells us what modern mathematics is, why it is superior to the traditional curriculum, or why students will be able to absorb it. These curriculum groups have made many claims for the superiority of their materials over the traditional subject matter, claims that students will now be able to absorb mathematics and understand it rather than having to learn by rote, and claims that their material meets the modern needs of students. But there has been no substantiation of these claims. The word "modern" was undoubtedly deliberately chosen to imply that the traditional curriculum is outdated, and to connote that what is modern must necessarily be better than what is not. Just how modern modern mathematics is we shall shortly see. At the moment we shall pursue the question of what it offers.

Perhaps the modernists think that the material speaks for itself. Well, since this is our only recourse we shall resort to it. There are many versions, but the principal features are common to all.

The outstanding feature of the modern mathematics curriculum is the approach it makes to the standard high-school material. The new ap-

449

proach is entirely deductive. In particular, the number system is built up deductively by starting with the natural numbers, then introducing the counting numbers, the integers, the rational numbers, and finally, by fudging, the irrational numbers. In all this work definitions, axioms, and proofs are the essence.

The second feature of the new curricula is that the deductive structures are to be rigorous: the level of rigor is to be that which professional mathematicians regard as satisfactory. We are no longer to be content with the unconscious assumptions made in the past, but are to spell out in detail every assumption and its consequences. We must assume or prove, then, that a line divides a plane into two parts, that every line segment has one and only one midpoint, and that the result of subtracting 4 from 6 is unique.

The third feature of the modern mathematics curricula is that mathematics is built up as an isolated, self-sufficient, pure body of knowledge. By chance, presumably, the deductive structures fit some physical phenomena, and so they could be applied to real problems. However, even this seemingly fortuitous value is not utilized. Mathematics is self-justifying. It exists in and for itself. In the seventh- and eighth-grade material, some of the modern mathematics curricula make minor concessions to applications. Having eased their consciences, the curriculum-makers feel justified in subsequently ignoring all applications.

According to the modern curricula, mathematics is also self-generating; that is, the axioms, concepts, and theorems come from purely mathematical sources, insights, and needs. Negative numbers, irrationals, complex numbers, functions, matrices, and the many concepts and theorems of geometry are all arrived at by mathematical considerations. For example, negative numbers are introduced to solve equations one cannot solve without them. Further, in generating mathematics the leading principle is to extend to new concepts the axioms that hold for the original ones. Accordingly, the associative, commutative, and distributive axioms are extended to all types of numbers, matrices, and so on. (There is a slight[!] difficulty in extending the commutative principle of multiplication to matrices, but no doubt this point will be cleared up in some revision.)

Students are asked to learn abstract concepts in the expectation that, if they learn these, the concrete realizations will be automatically understood. Thus, if a student learns the general definition of function, he will presumably understand the specific functions he will have to deal with; and if he learns what a field is, he will know all about the rational numbers and other mathematical objects.

The sixth principle that I see incorporated in the modern mathematics texts is the reliance upon terminology. Every concept is defined carefully (except, of course, for the few necessarily undefined terms). Thus angle, triangle, polygon, numeral, equation, sentence, phrase, compound, open and closed sentences, binary operation, closure, uniqueness of inverse, the nonnull set which contains the empty set, line, line segment, point pair, distance, length, and several hundred other concepts are defined formally. The definitions must satisfy all the requirements of strict logical definition.

The seventh principle is that symbols are always better than words. Hence one finds masses of parentheses, braces, brackets, vertical bars, quantifier symbols, and many others.

The eighth feature of the new curricula is new subject matter. Set theory, inequalities, bases of number systems, group and field, Boolean algebra, and matrices now take the place not only of those topics that I would agree are outmoded but of many standard topics in the traditional curriculum.

450

Criticism of the Modern Mathematics Curriculum. I could devote this entire article to a criticism of the modern mathematics principles and subject matter. I shall confine myself to a few criticisms and give references to others. The basic principle that mathematics can be understood through a logical presentation is to my mind entirely fallacious. Mathematics is understood intuitively and checked by deductive arguments. By "intuition" I mean physical understanding, sensory impressions, experience, inductive arguments, probable inference, and arguments by analogy. The whole history of mathematics shows that even the great mathematicians gained their understanding intuitively and in many cases never bothered to give deductive proofs.[1] In any case the proofs came afterwards, often centuries after full understanding and utilization. For example, the logical development of the number system was not attained until the 1870s. How did Newton, Gauss, Euler, Laplace, and hundreds of others manage to work with the number system?

Many mathematicians have expressed themselves on the role of logic. Descartes said, "I found that, as for Logic, its syllogisms and the majority of its other precepts are useful rather in the communication of what we already know or . . . in speaking without judgment about things of which one is ignorant." Jacques Hadamard remarked that logic merely sanctions the conquests of the intuition, and Herman Weyl described logic as "the hygiene which the mathematician practices to keep his ideas healthy and strong." Pascal said, "Reason is the slow and tortuous method by which those who do not understand the truth discover it." One cannot reject completely the use of deductive proof. It has a place, but the most important thing about deductive proof is to keep it in its place. An exclusively deductive approach to mathematics is not the way to get to understand the subject.[2]

The second criticism of the modern mathematics movement for which I shall take time is that it does not at all tackle the problem of motivation. I have already pointed out that one principle of the modern mathematics curriculum is that mathematics is built up as an isolated, self-sufficient, pure body of knowledge. Mathematics exists in and for itself. Unfortunately, mathematics proper is abstract, not a natural human interest. The need for motivation and meaning for each topic we take up is to my mind the gravest deficiency of the traditional curriculum. This defect is not only not remedied in the modern mathematics curriculum but is even graver because the latter removes mathematics further from reality by treating the subject as self-sufficient and abstract.

By introducing new topics as generated by purely mathematical needs, the modern curriculum foregoes an excellent opportunity to motivate. Negative numbers may indeed be introduced to solve equations such as $x + 5 = 0$, but this motivation is hardly a forceful one to the student. Why should anyone wish to solve such equations? What is accomplished by solving them except to add to the burden of learning more mathematics?

Modern mathematics is a presentation from the point of view of the mathematician, the shallow mathematician who can see only the petty puerile deductive details of set theory, of the real number system and Euclidean geometry, of minor and useless distinctions such as that between number and numeral, of trivial symbolism. The modern mathematics people offer a sophisticated, abstract, rigorous mathematics that erodes

[1] See my article, "The Pedagogical Implications of History," this *Journal*, Oct. 1957, pp. 3–8, Jan. 1958, pp. 18–22, April 1958, pp. 13–18.

[2] For a more extensive substantiation of this point see my "Logic Versus Pedagogy", *American Mathematical Monthly*, Vol. 77, March 1970.

the substance; they present uninspiring and pointless abstractions; they have isolated the subject from all other bodies of knowledge; and they offer dogmatic presentations of final versions of branches of mathematics. The formalism of this curriculum, and for that matter of the traditional algebra, can lead only to a decline of vitality and to authoritarianism in mathematics.

The Proper Direction for Reform. The first consideration in the design of a curriculum is to meet the problem of motivation, a problem not solved by either the traditional or the modern mathematics curriculum. Mathematics proper seems meaningless to young boys and girls, and they constantly ask, Why should I have to learn this material?

The natural motivation is the study of real, largely physical problems. This is also the historical and currently valid reason for the importance of mathematics. Practically all of the major branches of mathematics arose in response to such problems, and certainly on the elementary levels this motivation is genuine. Would it meet the interests of high-school students? They live in the real world and, like all human beings, either have some curiosity about real phenomena or can be far more readily aroused to take an interest in them than in abstract mathematics. Hence there is an excellent prospect that the genuine motivation will also be the one that interests high-school students. Some limited experience has shown that this is indeed the case, but whether or not it is, the value of mathematics, the primary reason for its existence, is to help man understand and utilize natural phenomena, and we should feel obliged to present this value.

The use of real problems, especially of physical ones, serves not only to motivate mathematics but to give meaning to it. Negative numbers are not just inverses under addition to positive integers but are the number of degrees below zero on a thermometer. Functions are not only sets of ordered pairs but also relationships between real things, such as the height and time of flight of a ball thrown up into the air, the distance of a planet from the sun at various times of the year, or the population of a country over some period of years. The ellipse is not just a peculiar locus but the path of a planet or a comet. Mathematical concepts arose from such physical situations or phenomena, and their meanings were physical for those who created mathematics in the first place. To rob the concepts of their meaning is to throw away the fruit and leave the husks.

In determining content, then, the first criterion is motivation. There is another vital consideration: the role of elementary and high-school education. The primary role is to open up the various academic worlds to novitiates. Hence the choice of subject matter should be made accordingly. There should be no thought of training future mathematicians or scientists. In fact very few of these people will be professional users of mathematics; most will be laymen. We should therefore choose material that will be of significance to future laymen. Toward this end the role of mathematics in helping man to understand and take advantage of the workings of nature is certainly one of the great features of mathematics that should be inculcated.

However, this is not enough. The elementary and secondary schools should be offering a liberal education and the mathematics courses should contribute their share. Hence, in addition to the motivation recommended, the significance of mathematics or of primarily mathematical accomplishments should also be presented. Mathematical work has determined our view of nature; it has given us power over nature; and it has helped to fashion our leading philosophical doctrines. Mathematics has enabled painters to paint realistically. It has given us not only an understanding

452

of musical sounds but the analysis of such sounds that is indispensable in the design of the telephone, the phonograph, the radio, and other sound-recording and -reproducing instruments. The question "what is truth" cannot be discussed without involving the role that mathematics has played in convincing man that he can or cannot find the truth. Much of our literature is permeated with mathematical ideas and mathematical accomplishments. Indeed, it is often impossible to understand our writers and our poets unless one knows what ideas and accomplishments of mathematics they are treating or reacting to.

What we should be fashioning and teaching are the relationships of mathematics to these other human interests; in other words, a broad cultural course in which mathematics is the core. Some of these relationships can serve as motivation, others can be applications, and still others can supply interesting reading and discussion material that would vary and enliven the content of our mathematics courses. Does such material exist and can it be used at the high school level? I know that it does and I have tried to present some of it in my *Mathematics in Western Culture, Mathematics and the Physical World* and *Mathematics: A Cultural Approach*. Fortunately, mathematics has roots in painting, music, architecture, and philosophy, as well as in physical science, and the achievements of mathematics bear on literature, the social sciences, biology, geology, and even religion. In fact, if our subject did not have intimate relationships with these major and vital branches of our culture it would not warrant the regard in which it is now held nor its important place in our curriculum.

I would like to point out that by catering to the general student we do not in fact ignore the future professional. A small percentage will be physicists, chemists, engineers, social scientists, technicians, statisticians, actuaries, and so on. Obviously it is desirable that these students get, as early as possible, some knowledge of how mathematics can help them in their future work. In fact, if they are already inclined toward one of these careers and if we show them how mathematics is useful in it, they will take an interest in the subject because of their proposed careers. Even if students do not already incline toward a particular career, we are obliged to open up the world to them and to show them what the various professions involve. One vital way of doing this is to show them how mathematics is utilized in other fields.

Though I have urged a choice of subject matter that would fulfill the goals of motivation and cultural enrichment, it would of course be desirable, in view of the sequential nature of mathematics, to incorporate the subjects that are normally taught at the various levels, so that not only would the elementary and secondary school programs be intelligible but also those students who will pursue the subject further at the college level would not be delayed by the omission of necessary subject matter. Fortunately it is possible to teach the standard material of the traditional curriculum with the proper motivation and treatment of its significance. Except for a few points, this is the material that has been found useful, and this is why it has been taught in preference to many other possible topics. However, one should not hesitate to depart from some of it if a richer and more vital course can be fashioned. Any topics that may have to be omitted and that are necessary to the further pursuit of mathematics can be incorporated at the stage where the courses address students who are definitely committed to mathematics. It will be necessary to reorder the topics, notably in algebra, to motivate the subject properly.

As to technical presentation, I would recommend several principles.

The first is that mathematics must be developed not deductively but constructively. We must build up the concepts, techniques, and theorems from the simplest cases to the slightly more involved, and then to the still more involved. The correct presentation, the best one pedagogically, treats mathematics not as a finished structure but as something that has grown and continues to grow in parallel with changes in other aspects of human life.

Only after thoroughly understanding what we have studied do we consider a deductive formulation. In fact, the very idea of a deductive structure must be learned, and this should therefore be introduced gradually. In no case do we start with a deductive approach, even after students have come to know what that means. The deductive organization is the final step. The constructive approach includes having the students do the building, the guessing, the conjecturing, and the fashioning of proofs. This approach insures understanding and teaches independent, productive thinking.

In building mathematics constructively the genetic principle is enormously helpful. This principle says that the historical order is usually the right order and that the difficulties which mathematicians themselves have experienced are just the difficulties our students will experience. Let me illustrate this last assertion. If it took mathematicians 1,000 years, as indeed it did, from the time that first-class mathematics appeared, to arrive at the concept of negative numbers, and if it took another 1,000 years, as it did, for mathematicians to accept negative numbers, you may be sure that the students are going to have trouble with negative numbers. Hence we must be prepared for, and help them overcome, the difficulties. Extending the distributive law to negative numbers will be of no help at all in understanding negative numbers.

Instead of presenting mathematics as rigorously as possible, present it as intuitively as possible. Accept and use without mention any facts that are so obvious that students do not recognize they are using them. Students will not lose sleep worrying about whether a line divides the plane into two parts. Prove only what the students think requires proof. The ability to appreciate rigor is a function of the age of the student and not of the age of mathematics. As Professor Max M. Schiffer of Stanford University has put it, "In teaching never put logical carts before heuristic horses."

To remedy the disparate treatments of algebra and geometry a modicum of proof should enter into the development of algebra. However, it is not the elementary and obvious steps that should be supported by deductive arguments, but those that are not intuitively evident.

Mathematics is not an isolated self-sufficient body of knowledge. It exists primarily to help man understand and master the physical world and, to a slight extent, the economic and social worlds. Mathematics serves ends and purposes. It does something. We must constantly show what mathematics accomplishes in domains outside of mathematics. We can try to inculcate interest in mathematics proper and the enjoyment of mathematics, but these must be byproducts of the larger goal of showing what mathematics accomplishes.

To introduce mathematical concepts, operations, and properties we should rely as far as possible on real processes or real phenomena. For example, our method of adding fractions was adopted because the sum represents what results physically when fractional entities are combined. Moreover, the fact that this operation is commutative and associative does not derive from some inner mathematical consideration but from the physical truth of these properties.

In place of abstractions we must, as far as possible, present concrete material. It does not matter if a student cannot give a general definition of function. It suffices if he knows that $y = x$, $y = 2x$ and $y = x^2$ are functions and if he learns how to work with such functions. After enough experience with functions the student will be able to make his own definition. And if after further experience the definition has to be modified, this is not calamitous. Our insight into all knowledge grows with experience. Again, it does not matter whether a student can define a polygon as long as he can recognize and work with it.

Introduce as few terms as possible. Use common words, preferably those already familiar to students. Keep the new terminology to a minimum.

Use as few symbols as possible. Symbols scare students. Moreover, the meaning of a symbol must be remembered and so is often more of a burden than a help. The gain in brevity may not compensate for the disadvantages.

My recommendations place an obligation on the teachers. They will have to learn how mathematics serves science and man if they are to communicate this value to the student. I have the decided impression that most of us who became mathematics teachers did so for personal reasons. But our education did not show us how mathematics is important. Hence, while we are assured that it is, we are not convinced by our knowledge of the actual details. Moreover, if we are to teach the subject so as to reach the student we must get to see mathematics from the standpoint and attitude of nonmathematicians. Only then will we succeed in presenting it so that young people can appreciate it.

We like to think today that we have been more successful than previously in teaching mathematics because more high school students are being subjected to the calculus. Our hothouse breeding of mathematics students has produced this crop of purported devotees of mathematics. But I would propose another criterion as to whether we have been successful in putting mathematics across. When we reach the stage where 50 or 60 per cent of the high school graduates can truthfully say that they liked the mathematics they took then we shall have attained some measure of success.

SELECTED BIBLIOGRAPHY FOR CHAPTER 18
Elementary School Mathematics Today and Tomorrow

Ballew, Hunter. "Starting the New School Year in Mathematics." *The Arithmetic Teacher* (October 1972), 427-30.

Clark, John R. "Elementary School Mathematics in the 1970s." *The Arithmetic Teacher* (October 1971), 385.

Davis, Robert. "New Math: Success/Failure." *Instructor* (February 1974), 53.

——. "Report from the States." *Mathematics Teaching* (Spring 1970), 6-12.

Frye, Helen. "Mathematics Throughout the Curriculum." *The Arithmetic Teacher* (December 1969), 647-50.

Glennon, Vincent J. ". . . And Now Synthesis: A Theoretical Model for Mathematics Education." *The Arithmetic Teacher* (February 1965), 134-41.

——. "Too-Heavy Impact of Mathematicians." *Instructor* (February 1974), 45.

Hlavaty, Julius. "A Message to Teachers of Elementary Mathematics." *The Arithmetic Teacher* (May 1968), 397-99.

"A Look at Mathematics Education Today." *The Arithmetic Teacher* (October 1973), 503-10.

Nichols, Eugene D. "Is Individualization the Answer?" *New York State Mathematics Teachers' Journal* (April 1972), 66-77.

Reynolds, P. "Have We Gone Too Far?" *New York State Mathematics Teachers' Journal* (April 1970), 99-100.

Riedesel, C. Alan. "Recent Research Contributions to Elementary School Mathematics." *The Arithmetic Teacher* (November 1970), 245-52.

Sandel, Daniel. "Teach So Your Goals Are Showing!" *The Arithmetic Teacher* (April 1968), 320-23.

"Status Report: Mathematics Curriculum-Development Projects Today." *The Arithmetic Teacher* (May 1972), 391-94.

Suydam, Marilyn. "Continuing the Mathematics Revolution." *American Education Journal* (January/February 1970), 26.

Todd, Paul and Stoughton, George. "Laps for Progress." *New York State Mathematics Teachers' Journal* (April 1972), 98-101.

Vaughan, B. W. "Network Analysis: Applied to the Design of a Primary School Mathematics Syllabus." *Mathematics Teaching* (Spring 1971), 13-16.

Willoughby, Stephen S. "Mathematics Education in New York State." *New York State Mathematics Teachers' Journal* (January 1972), 3-11.

——. "Mathematics Education in New York State." *New York State Mathematics Teachers' Journal* (April 1972), 57-66.

Wilson, Malcolm G. "We All Teach Mathematics." *The Arithmetic Teacher* (February 1969), 86-87.

APPENDIX

Suggested Bibliography of Materials for Elementary School Mathematics

This appendix is meant to serve as a guide to free materials, including addresses of publishers and distributors. It also mentions other booklets and publications listing free or inexpensive materials in mathematics for the elementary classroom. Teachers often find that they can obtain many worthwhile materials at very little cost.

The materials in this appendix are available free from the addresses listed. Additional sources of free or inexpensive materials may be found in the addresses of publishers and distributors listed or in some of the suggested books (not all free) on such materials.

A very short list of children's books emphasizing mathematical concepts is provided. Primary teachers should be able to extend this list and make very suitable use of these materials in the classroom.

One final note—local businesses may be the source of many very good teaching aids and materials. A hardware store, lumber company, bank, etc., may provide many free or inexpensive items. A local survey could be very worthwhile.

Company	Publication or Materials

Banking and Economics

The American Bankers Association
Banking Education Committee
90 Park Avenue
New York, NY 10016

How Banks Help. A 49-page booklet containing a series of human interest stories about banks, banking and people. Useful for consumer or business mathematics. *Using Bank Services.* A 40-page illustrated booklet on the services of banks and how to use them. It discusses safeguarding money and valuables, transferring funds (checking accounts, certified checks, bank drafts, etc.), making loans, providing trust services, and other services of banks. It includes sketches of checks, deposit slips, withdrawal slips, and promissory notes. *Vinny and Billy – The Boys with a Piggy Bank.* A colorful 18-page booklet which tells about two boys who earn money to buy their father a birthday gift, and open a savings account with their extra money. Classroom quantities may be available through a local bank.

Better Business Bureau of Metropolitan
New York, Inc.
Educational Division
220 Church Street
New York, NY 10013

Facts You Should Know About Savings. A 16-page pamphlet that discusses the need for savings, various types of savings institutions, and types of investments. Useful for units in consumer mathematics.

Credit Union National Association, Inc.
(CUNA)
1617 Sherman Avenue
Box 431
Madison, WI 53701

A Brief History of the Credit Union Movement; Consumer Facts leaflet; *Credit Union National Association, Inc. Yearbook 1974; Credit Unions – What They Are, How They Operate, How to Join, How to Start One; Everybody's Money; A Teacher's Guide to Credit Unions. Using Credit Wisely.* A 36-page booklet which discusses the history and uses of credit. Various types of credit and true interest rates are examined.

Department of National Revenue, Taxation,
c/o 873 Heron Road
Cartier Square "A," Room 4409
Ottawa, Ontario, Canada KLA OL8

Teaching Taxes. Teacher's kit (Canada only)

Dun and Bradstreet, Inc.
Public Relations & Advertising,
99 Church Street
New York, NY 10007

Cost of Doing Business – Corporations; Cost of Doing Business – Proprietorships & Partnerships; The Dun & Bradstreet Report; The Dun & Bradstreet Reference Book; The Failure Record (analysis of business failures & their causes); *Growth of the Credit Function; How to Build Profits by Controlling Costs; How to Control Accounts Receivable for Greater Profits; Key Business Ratios in 125 Lines; Pitfalls in Managing a Small Business*

Federal Reserve Bank of Atlanta
Research Department

Counterfeit?; Fundamental Facts About U.S. Money. A 16-page pamphlet which

Federal Reserve Station,
Atlanta, GA 30303

Federal Reserve Bank of New York
Public Information Department
Federal Reserve P. O. Station
New York, NY 10045

Household Finance Corporation
Money Management Institute
Prudential Plaza
Chicago, IL 60601

Institute of Life Insurance
277 Park Avenue
New York, NY 10017

Internal Revenue Service
District Director
161 Washington Avenue
Albany, NY 12210

Manufacturers Hanover Trust
Advertising and Public Relations
350 Park Avenue
New York, NY 10022

Merrill, Lynch, Pierce, Fenner & Smith, Inc.
70 Pine Street
New York, NY 10005

Publications Services,
Division of Administrative Services
Board of Governors of the Federal
Reserve System
Washington, D.C. 20551
(includes listings for all Federal Reserve
Banks that distribute free materials)

U.S. Secret Service
P. O. Box 1087
Syracuse, NY 13201

discusses the history of bills and coins, and tells about the design and circulation of U.S. money.
Keeping Our Money Healthy. A 16-page illustrated booklet on the purposes of the Federal Reserve System. Classroom quantities are available. *Money: Master or Servant.* A colorful 45-page illustrated booklet that discusses the role of money, banking and the Federal Reserve System in our economy. Classroom quantities are available.
Money Management Program.

Making the Most of Your Money. A booklet, planned for adult education, but applicable to high school consumer mathematics courses. Includes budgeting, shopping, credit, buying a car, and payroll deductions. Classroom quantities are available.

The Teaching Taxes Program. A set of materials for teaching about income taxes and how to fill out income tax forms. "Understanding Taxes" publication 21 is about the ordinary 1040 form and publication 22 is about farm taxes. Classroom quantities and teachers' manuals are available.

How Our Bank Helps Our City. A 20-page illustrated booklet that explains in simple language what banks do.

How to Read a Financial Report. A 23-page booklet on reading and understanding financial reports.

Federal Reserve Bulletin; Free Publications of the Federal Reserve Banks (FR 346) (catalog)

Counterfeiting and Forgery. A 4-page leaflet on the detection of counterfeit money and forgery. Contains a partial picture of a genuine dollar bill and a counterfeit dollar bill.

Careers and Career Applications

American Mathematical Society
P. O. Box 6249
Providence, RI 02940

American Petroleum Institute
1271 Avenue of the Americas
New York, NY 10020

Choosing a University; Scholarships and Fellowships; Finding Employment in the Mathematical Sciences.
Mathematics in the Petroleum Industry. A 23-page booklet containing units on statistics, geometry, game theory, slide

Also:

New York State Petroleum Council
142 State Street
Albany, NY 12207

American Statistical Association
Room 703
810-18th Street, N.W.
Washington, D.C. 20006

Bell Telephone Laboratories
Public Relations and Publication
Division
Mountain Avenue
Murray Hill, NJ 07974

Bureau of Labor Statistics
U.S. Department of Labor
341 Ninth Avenue
New York, NY 10001

Chrysler Corporation
P. O. Box 1687
Detroit, Michigan 48231

Department of Commerce
1200 18th Street, N.W.
Washington, D.C. 20036

Department of Commerce of New York
State
112 State Street
Albany, NY 12207

U.S. Department of Labor
Room 110
1371 Peachtree Street, N.E.
Atlanta, GA 30309

U.S. Department of Labor
Bureau of Labor Statistics
Occupational Outlook Service
Washington, D.C. 20212

U.S. Department of Labor
Women's Bureau
Washington, D.C. 20210

General Electric Company
Educational Relations Publications
One River Road
Schenectady, NY 12305

General Electric Company
Public Relations Operations, WID 2
Fairfield, CT 06431

rule, and linear programming. Good graphs
and diagrams.

Careers in Statistics. A 23-page illustrated
booklet that discusses various careers in the
field of statistics. (One to 25 copies only.)

*On the Nature of Mathematical Research
in Industry.* A 20-page booklet by H. O.
Pollak which discusses how mathematics
is used in finding the shortest distance
connecting a series of points, setting up cir-
cuits in computers and in research in
general. Contains diagrams.

Employment Outlook—Mathematicians,
Bulletin 1650–37. A 10-page booklet on
employment opportunities for mathematicians, statisticians, and actuaries.

Math Problems from Industry. A 55-page
booklet which illustrates problems that
are met in the work of Chrysler Corporation.
Most of the problems are involved with
finding the dimensions, areas, or volumes
of some complicated figures. Available in
classroom quantities.

*Careers in Science at the National Bureau of
Standards.* A 30-page illustrated booklet
that discusses opportunities in mathematics,
physics, chemistry, and electronics in the
National Bureau of Standards.

Why Should Girls Study Mathematics? A
2-page leaflet by Nura D. Turner discussing
opportunities for women in the field of
mathematics.

Math and Your Career. A leaflet that
explains the importance of studying mathematics for various occupations. A list of
bulletins that describe the occupations in
more depth is included.

Math and Your Career (single to 35 only)

*Careers for Women in the 70's; Why Not
Be a Technical Writer? Why Not Be A
Mathematician?* A 5-page leaflet that
discusses the opportunities for women
in the field of mathematics.

You and Tomorrow. A colorfully illustrated
14-page booklet on scientific and mathematical careers of the future.

Quincy Looks Into His Future

General Motors Corporation
  Public Relations Staff
  Room 1-101
  General Motors Building
  Detroit, MI 48202

Mathematics At Work In General Motors.
A set of twelve 4-page illustrated leaflets.
Each contains a unit on topics such as
rates of speed, parabolas, gear ratios, the
conics, other number bases, computers,
logic and measurement. They show the
applications of the mathematics to industrial
use. They are punched for a three-ring notebook. Available only for the teacher.

Institute of Life Insurance
  277 Park Avenue
  New York, NY 10017

A 17-page illustrated booklet on careers in the
insurance field. Included are actuarial,
accounting, electronics, investment analysis
and research.
Career Packets

The Mathematical Association of America
  1225 Connecticut Avenue, NW
  Washington, D.C. 20036

Society of Actuaries
  208 South LaSalle Street
  Chicago, IL 60604

So You're Good at Math

Society for Industrial and Applied
  Mathematics
  33 South 17th Street
  Philadelphia, PA 19103

*Bibliography on Careers in Mathematics
and Related Fields*

Van Valkenburgh, Nooger and Neville, Inc.
  15 Maiden Lane
  New York, NY 10038

*Directory of Achievement Tests for Occupa-
tional Education*

Computers and Calculators

American Petroleum Institute
  1271 Avenue of the Americas
  New York, NY 10020

Yes, No . . . One, Zero. A 27-page booklet
that discusses the language of the computer—
the base-two system and how the computer
works. Some of the history, present uses,
and possible future uses of the computer
are also mentioned.

Association for Computing Machinery
  1133 Avenue of the Americas
  New York, NY 10036

Computopics. A bibliography of pamphlets
and films about careers in computer science.
For teachers.

U.S. Atomic Energy Commission
  P. O. Box 62
  Oak Ridge, TN 37830

Computers. A 56-page illustrated booklet
by William R. Corliss on the history and
development of the computer from the
abacus to the present day. Contains a good
section on what makes a computer work.
Utilizes diagrams and problems effectively.

Bell Telephone Laboratories
  (contact your local business office)

Cardiac. Cardiac is an acronym for CARD-
board Illustrative Aid to Computation. A
cardboard model illustrates the operation of
a computer. One cardboard model is supplied
for each student. An instruction manual
and an overhead projectual of the CARDIAC
are also available.

Consolidated Edison Co. of New York
  4 Irving Place
  New York, NY 10003

Easy Arithmetic. A 3-page reprint from
Consolidated Edison Company's Employee
Magazine. The article discusses base-2 arith-
metic and its application to the computer.

General Electric Company
  Educational Publications

You and the Computer. Many calculator
and computer companies have teaching

One River Road
Schenectady, NY 12305

Monroe Calculator Company
550 Central Avenue
Orange, NJ 07051

Wang Laboratories, Inc.
c/o SWAP
836 North Street
Tewksbury, MA 01876

Decimals, Fractions and Conversion Tables
Contact your local Air Force Recruiting
Station or:
3501st U.S.A.F. Recruiting group (DA)
Westover Air Force Base
MA

Automatic Electric
Northlake, IL 60164

Brown & Sharpe Mfg. Co.
Industrial Products Division
Precision Park
North Kingstown, RI 02852

Charles Bruning Company
1800 W. Central Road
Mount Prospect, IL 60056
Illinois Tool Works, Inc.
2501 North Keeler Avenue
Chicago, IL 60639

K.O. Lee Company
1110 First Avenue South East
Aberdeen, SD 57401

materials available to users of their machines.
A letter on school stationery to any of
these companies should result in information
about their teaching materials.
Arithmetic Minus Mystery=Understanding.
A 13-page illustrated booklet on teaching
the properties of arithmetic with a calculator.
Addition, subtraction, multiplication and
division are discussed. *Can Children Really
Enjoy Arithmetic.* A 15-page booklet that
describes a calculator program run by
Memorial School in Cedar Grove, New
Jersey, at the fifth-grade level.
Past reference issues of *Programmer* magazine;
reprints of application articles from *Programmer* magazine.

An 8½″ by 11″ sheet of heavy durable
paper, punched for a ring binder. Logarithms,
square roots, trigonometric functions and
simple formulas are on one side. The periodic
table of the elements for chemistry and a
list of formulas for physics are on the other
side.
Conversion Factors. A 24-page booklet
on conversions used by practicing engineers
involving area, angles, force, length, linear
velocity, power, temperature, volume and
weight. *Tables and Formulae.* A 32-page
booklet of tables and formulas used by
engineers. Included are linear, square, cubic,
liquid, dry, circular, and nautical measures.
Also inch/millimeter conversions, temperature conversions, squares, cubes and roots,
and various powers of numbers and functions
of degrees.
"Decimal Equivalents of Common Fractions."
Wall chart WC-1 is a yellow and white chart
20″ by 27″. Decimal equivalents are included for fractions from 1/64 to 1. "Decimal Equivalents and Tap Drill Sizes." A
2″ by 5¼″ pocket chart of decimal equivalents for fractions from 1/64 to 1.
"Decimal Equivalents." A 6″ by 8½″ chart
showing decimal and fractional equivalents
from 1/64 to 1.
Trigonometry Tables and Involute Functions.
A 74-page booklet containing the six trig.
functions for each angle in one-minute
intervals. Involute tables and decimal
equivalents are also included.
"Millimeter Conversion Chart." An 8″ by
12″ card listing millimeter/inch equivalents
and fraction/decimal equivalents.

Monroe Calculator Company
    550 Central Avenue
    Orange, NJ 07051
Contact your local Navy Recruiting Station
    or: Navy Recruiting Station
    U.S. Post Office and Court House
    Albany, NY 12201
    Attention: TAO

Ohaus Scale Corporation
    29 Hanover Road
    Florham Park, NJ 07932

Reynolds Metals Company
    Public Relations Department
    Richmond, VA 23261

South Bend Lathe
    400 West Sample Street
    South Bend, IN 46623

L. S. Starrett Company
    Athol, MA 01331

"Decimal Equivalents." An 8½″ by 11″ cardboard sheet that lists decimal equivalents for fractions from 1/16 to 1.

"Tables and Formulas" (RAD 68914). An 8½″ by 11″ sheet of heavy, durable paper, punched for a ring binder. Trigonometric functions and four-place logarithms are on one side and a list of formulas and conversion factors on the other.

"Useful Data." An 8½″ by 11″ sheet, punched for three-ring notebooks, which lists various types of linear, liquid, volume and weight measures, and conversions.

Factors and Formulas. A 75-page booklet with logarithms, trigonometric values, and a very wide variety of conversion tables and formulas.

"Decimal Equivalents." Chart No. CE 777 is a blue and white wall chart, 19″ by 13″. It lists fraction/decimal equivalents from 1/64 to 1.

"Decimal Equivalent Chart." A 23″ by 37″ wall chart that lists fractional and decimal equivalents. Also lists tap, drill, and pipe thread sizes.

Maps

National Industrial Conference
    Board, Inc.
    845 Third Avenue
    New York, NY 10022

Silva, Inc.
    LaPorte, IN 46350

Road Maps of Industry. A series of 8½″ by 11″ charts that are printed semi-monthly. A source of statistics and graphs about wages, prices, population, employment, retail sales, etc. for bulletin board display and classroom use. (Will be mailed only to school or professional address.)

Elementary Map and Compass Instruction. A 14-page booklet that describes a possible program for map and compass instruction. An outdoor training area is described and some math lab techniques could be used.

Mathematical Concepts
    algebra

Barron's Educational Series, Inc.
    113 Crossways Park Drive,
    Woodbury, NY 11797
    (all requests must be sent on school stationery)
Key Curriculum Project
    Box 2304, Station A
    Berkeley, CA 94702

College Algebra (high school edition, complementary texts of all publications available); *Regents Examinations*—Algebra, Geometry, Trigonometry (complementary texts of all publications available).
Clave Para El Algebra, Cuadernos No. 1–4; *Key to Algebra,* Booklets 1–4

    arithmetic

John D. Caddy
    Box 251
    Canoga Park, CA 91305

The 390 Basic Arithmetic Facts

463

Easy Learn
11008 Easy Street
St. Ann, MI 63074

McGraw-Hill Book Company
Webster Division
Manchester Road
Manchester, MO 63011

geometry

Bonded Scale and Machine Company
2176 South Third Street
Columbus, OH 43207

Scott, Foresman and Company
99 Bauer Drive
Oakland, NJ 07436

numbers and numeration

Allyn and Bacon, Inc.
Rockleigh, NJ 07647

The Duodecimal Society of America
20 Carlton Place
Staten Island, NY 10304

Charles E. Merrill Publishing Co.
1300 Alum Creek Drive
Columbus, OH 43216

sets, probability, and statistics

Automobile Manufacturers Assoc.
320 New Center Building
Detroit, MI 48202

Institute of Life Insurance
Educational Division
277 Park Avenue
New York, NY 10017

Insurance Information Institute
Director of Educational Division
110 William Street
New York, NY 10038

Scott, Foresman and Company
99 Bauer Drive
Oakland, NJ 07436

10-piece addition tray; 10-piece subtraction
tray; 20-piece subtraction tray

*Novel Ways to Reintroduce Addition and
Multiplication* (X 19). An 8-page leaflet
that discusses "partial sums" and "lattice
multiplication."

*Original Investigation Or How To Attack
Exercise in Geometry.* A 63-page illustrated
booklet by Elisha S. Loomis that gives
suggestions and presents a systematic
procedure for solving problems in geometry.
Geometry in the World Around You. A 22"
by 34" poster with pictures of everyday
objects that may be used to illustrate various
geometric figures. Good for bulletin board
material.

Ancient Systems of Numeration N-B-642.
A 17" by 22" sheet on ancient numeration
systems and clock arithmetic.
An Excursion in Numbers. A 14-page booklet
by F. Emerson Andrews about a base-12
number system and its value for our times.
Zero Is A Hero. A 12-page booklet by
Frederick Robinson that tells the uses of zero
in story form.

*Facts for Study. Automobile Facts and
Figures. Motor Truck Facts.* Three booklets
of 30 to 60 pages each that contain many
graphs and tables of statistics for use in
mathematics classes.
Sets, Probability and Statistics. A 36-page
illustrated booklet that discusses sets, sample
spaces, probability, statistics, and their uses
in life insurance. It contains over 75 problems
and experiments. Classroom quantities
are available.
Chances Are. . . . A 32-page booklet on
probability that uses programmed instruction to explain the law of large numbers and
how it applies to insurance rates and tables.
Classroom quantities are available.
A.I.M. Punchout No. A 1264. A punchout
that, when used with a pencil, forms a
"spinner-top." The device takes the place
of a die when used for probability experiments. It is one of the devices used in the
"Activities in Mathematics" program.

miscellaneous

Aero Educational Products Ltd.
P. O. Box 71
St. Charles, IL 60174
(Evaluation kits for each Lab are
available to supervisory or administra-
tive personnel.)

Math Lab Sample and Information Kit (math
practice slate).

Educulture Inc.
1220 Fifth Street
Santa Monica, CA 90401

Basic Applied Mathematics (sample kit of
audio-cassettes, response manuals and workbooks, etc.); *Mini-Courses in Math: Algebra
& Mini-Courses in Math; Trigonometry*
(sample kits of audio-cassettes & workbooks,
etc.); *Think Metric, USA
Quickstrips* (review sheets).

Eye Gate House
146-01 Archer Avenue
Jamaica, NY 11435

Field Enterprises Educational Corporation
Educational Services, Sta. 8
˒   Merchandise Mart Plaza
Chicago, IL 60654

New Math - SA 2286. A 17-page reprint
from the World Book Encyclopedia. It is
well-illustrated and discusses "Sets" and "Numeration Systems" in a readable fashion.

D.C. Heath and Company
125 Spring Street
Lexington, MA 02173

*New Directions in Elementary School
Mathematics.*

Litton Industries
Beverly Hills, CA

Problematical Recreations. A 52-page booklet
of challenging methematical problems. The
problems are not easy but the answers are
given.

Ohaus Scale Corporation
29 Hanover Road
Florham Park, NJ 07932

An Introduction to Mathematical Concepts.

Scott, Foresman and Company
1900 East Lake Avenue
Glenview, IL 60025

*Problem Solving Strategies for Elementary
Mathematics; How to Read and Construct
a Graph;* posters for middle grade classrooms;
posters for primary classrooms.

Victor Comptometer Corporation
Victor Educational Services
3900 North Rockwell Street
Chicago, IL 60618

The Victor Refresher (No. 561-124-2). A
15-page booklet, punched for a three-ring
binder, that lists basic mathematical terms and
formulas. The rules are mechanical and are
illustrated with calculator tape pictures.

Measurement

Contact your local Air Force
  Recruiting Station or:
    3501st U.S.A.F. Recruiting Group (DA)
    Westover Air Force Base
    MA

A plastic six-inch ruler is available in limited
quantities.

American Gas Association, Inc.
  Educational Services
  1515 Wilson Boulevard
  Arlington, VA 22209

How Your Gas Meter Works Kit. A unit
showing how a gas meter works and how to
read one. There are 36 student work sheets
included.

Arco Travel Service
  600 Fifth Avenue
  New York, NY 10020

"Transcontinental Mileage Chart." A chart
showing mileages and driving times between
cities.

Bendix Automation & Measurement
  Division

Roundness Measurement Kit. A kit containing die-cut parts that are used to illustrate

721 Springfield Street
P. O. Box 1127
Dayton, OH 45401
Brown & Sharpe Mfg. Co.
  Industrial Products Division
  Precision Park
  North Kingstown, RI 02852

The Cooper Group
  P. O. Box 728
  Apex, NC 27502
U.S. Department of Commerce
  National Bureau of Standards
  Washington, D.C. 20234

Field Enterprises Educational Corp.
  Merchandise Mart Plaza
  Chicago, IL 60654
Ford Motor Company
  The American Road
  Dearborn, MI 48121

General Motors Corporation
  Public Relations Staff
  Room 1-101
  General Motors Building
  Detroit, MI 48202
Lufkin Rule Company
  Box 728
  Apex, NC 27502

Sealtest Consumer Service
  605 Third Avenue
  New York, NY 10016

Standard Milling Company
  1009 Central Street
  Kansas City, MO 64105

L.S. Starrett Company
  Athol, MA 01331

diametric measurement, V-block measurement, and radial measurement. Useful for technical or vocational math.
Principles and Practice of Precision Measurement. A series of lesson plans and suggestions for teaching precision measurement. The material is oriented toward industrial measurement and would be well suited for use in technical or vocational math courses.
The Amazing Story of Measurement.

A Look At the National Bureau of Standards— 1970. A 26-page booklet that discusses the activities of the National Bureau of Standards. There are chapters on measurement and standards with a section on the metric study.
Units and Systems of Weights and Measures. Letter Circular LC 1035. A 32-page booklet on the origin, history, development, and present status of the systems of weights and measures. The booklet is quite thorough and lists a wealth of information about all types of measurements.
Metric System/Weights and Measures.

A History of Measurement. A colorful 36″ by 48″ chart which explains the history of such measurements as the cubit, foot, inch, fathom, yard, and furlong. Very good for bulletin boards. Multiple copies available.
Precision—A Measure of Progress. An interesting 64-page booklet on measurement from the time of Noah's Ark to the present and a glimpse into the future. It is well illustrated.
The Amazing Story of Measurement. A colorful 24-page booklet in comic book format which discusses the history of measurement. The need for standardizing measuring devices and improving precision is emphasized. Classroom quantities are available.
"Kitchen Measures." An 8½″ by 11″ sheet, punched for a three-ring binder, that lists the common equivalents used in measurement in cooking.
"Height-Weight Chart." A chart that can be hung on the wall for measuring the height of students. It gives the average weight for boys and girls of various heights.
The Tools and Rules for Precision Measuring. An 80-page illustrated booklet that discusses precision measurement in industry. The use

Tempil Corporation Hamilton Boulevard South Plainfield, NJ 07080	of rules, tapes, calipers, dividers, microm- eters, gauges, etc., is explained. Useful for shop or vocational mathematics. *How Temperatures Are Measured.* A 7-page booklet by G.M. Wolten on temperature measurement by various techniques. Ther- mometers, pyrometers, changes in color and chemical signals are discussed. For shop or vocational mathematics.

The Metric System

Acme Ruler Company Ltd. Foster Street, P. O. Box 239, Mount Forest, Ontario, Canada NOG 2LO	Meter sticks; metric classroom desk rulers.
American National Metric Council 1625 Massachusetts Avenue, NW Washington, D.C. 20036	*American National Metric Council Brochure;* *Metric Reporter*
Headquarters, U.S. Army Recruiting Command, USAREC (USARCASP-E) Fort Sheridan, IL 60037 (contact your local Army recruiter for copies; these are available in the U.S. only)	"The Modernized Metric System," RPI-911 (poster).
LaPine Scientific Company 375 Chestnut Street Norwood, NJ 07648	Metric pocket rule.
Metric Supply International 1906 Main Street Cedar Falls, IO 50613	Metric educational bulletins.
National Bureau of Standards Metric Information Office Washington, D.C. 20234	Metric packet; "All You Will Need to Know About Metric" (poster).
National Council of Teachers of Mathe- matics 1906 Association Drive Reston, VA 22091	*NCTM Metrication Update and Guide to* *Suppliers of Metric Materials; Sources of* *Materials Available on Careers in Mathematics.*
Polymetric Services, Inc. & AMJ Publishing Company 18314 Oxnard Street, Box 696 Tarzana, CA 91356	*American Metric Journal.*
Sears, Roebuck and Company Consumer Information Service D/703, Sears Tower Chicago, IL 60684	*An Educator's Guide to Teaching Metrica-* *tion; Facts and Figures about Revolving* *Credit.*
Schloat Productions 150 White Plains Road Tarrytown, NY 10591	*How Do You Evaluate a Metric Program?* *Measuring: A Metric Approach* (catalog).
Spectrum Educational Supplies, Ltd. 8 Denison Street Markham, Ontario Canada L3R 2P2	*Recommendations for Equipment for* *Teaching the Metric System of Measurement.*
Willow House Publishers 121 N. San Joaquin Street Stockton, CA 95202	*Everyday Metrics; Measuring Metric* (Book 1); *Measuring Metric* (Book 2); *Metric Funda-* *mentals; Metric Workshop for Teachers;*

Science and Mathematics

U.S. Department of the Interior
  Geological Survey
  Reston, VA 22092

Selected packet of geological teaching aids;
teacher's packet of geological materials.

General Motors Corporation
  Public Relations Staff
  Room 1-101
  General Motors Building
  Detroit, MI 48202

Optics and Wheels. A 32-page booklet that
tells how scientific principles of optics and
light have been applied to the automobile
headlight. Could be used when studying
conic sections, reflection angles, etc.

Moody Institute of Science
  Educational Film Division
  12000 E. Washington Blvd.
  Whittier, CA 90606

Mathematics Teacher Reprints—"The
Mathematics of the Honeycomb" and "Of
Bees and Mathematicians" (supply is
limited); utilization guide for "Mathematics
of the Honeycomb."

The Slide Rule

Charvoz-Carsen Corporation
  5 Daniel Road
  Fairfield, NJ 07006

*Aristo Slide Rule Bulletin Number 1; Aristo
Slide Rule Bulletin Number 2.* Two 6-page
leaflets on the slide rule. Number 1 discusses the construction of the slide rule and
examines the various scales. Number 2 gives
a history of the slide rule and goes into some
applications. Some of the problems may be
somewhat advanced for high school students.

Pickett Industries
  436 E. Gutierrez Street
  P. O. Box 1515
  Santa Barbara, CA 93102

A Teaching Guide for Slide Rule Instruction.
A 48-page illustrated booklet on teaching
the use of the slide rule. A thorough instruction booklet.

Time and Time-Telling

Bulova Watch Company, Inc.
  630 Fifth Avenue
  New York, NY 10020

The Pursuit of Accuracy. A 15-page mimeo-
graphed booklet on the history of time
from the ancient sun dial to the modern
tuning-fork movements in watches. *Time-
keeping in the Space Age.* A 5-page mimeo-
graphed leaflet on the various time zones
around the world. Both of the above booklets
are available in classroom quantities.

Colorco
  P. O. B. 1546
  Garland, TX 75040

"Colorclock"; "Colorcolor," "Colornumer-
als," and "Colortables" (posters).

Eastman Kodak Company
  343 State Street
  Rochester, NY 14650

*George Eastman and the Thirteen-Period
Calendar.* A 5-page excerpt from Carl
Ackerman's biography, *George Eastman.*
The value and business uses of a 13-month
year are discussed.

Florida Citrus Commission
  Youth and School Service
  P. O. Box 148
  Lakeland, FL 33802

*Four Seasons 4–S. Monthly Calendar MC.
The Orange Clock OC.* Three spirit duplica-
tor masters that will run up to 500 copies. "4-S"
can be used for teaching about the four
seasons. "MC" is a monthly calendar to be
filled in each day by the students. "OC" is

a clock face in the shape of an orange cross section. The students' copies can be made with movable clock hands.

World Time Chart. An 8½″ by 11″ chart that shows time zones in the U.S. and lists time differences between other countries and the U.S.

"300 Year Calendar" (poster).

Manufacturers Hanover Trust
  Advertising and Public Relations
  350 Park Avenue
  New York, NY 10022

Merrill Analysis Inc.
  Box 228
  Chappaqua, NY 10514
  (must enclose a self-addressed,
  stamped envelope with each request.)

Pan American Airways
  Educational Services
  Pan Am Building
  New York, NY 10017

Time Selector. A disk, which, when set at local time gives times for a number of cities around the world.

The World Calendar
  P. O. Box 224
  Lenox Hill Station
  New York, NY 10021

The World Calendar. A 12-page booklet advocating a "world calendar" that would be the same every year. It gives some history of our calendar and presents reasons for changing to the "world calendar."

Miscellaneous

Cuisenaire Company of America, Inc.
  12 Church Street
  New Rochelle, NY 10805

The Cuisenaire Reporter.

Fawcett Publications, Inc.
  Fawcett Place
  Greenwich, CT 06830

Teachers desk and review copies (complementary tests of all publications available); *College Catalog; Educational Catalog.*

W. H. Freeman and Company
  660 Market Street
  San Francisco, CA 94104

Jacob's Newsletter.

Miller-Brody Productions, Inc.
  342 Madison Avenue
  New York, NY 10017

Bookmark (describing the MATH AUDIO RESPONSE SYSTEM).

Mu Alpha Theta
  University of Oklahoma
  601 Elm Avenue, Room 423
  Norman, OK 73069

Mathematical Booklist for High School Libraries; Mathematical Projects, Exhibits and Reports.

Pathescope Educational Films Inc.
  71 Weyman Avenue
  New Rochelle, NY 10802

"No Four Letter Words, Please" (poster).

Wff'n Proof Learning Games Assoc.
  1490 South Blvd.
  Ann Arbor, MI 48104

The Wff'n Proof Newsletter.

OTHER SOURCES OF FREE MATERIAL LISTS FOR EDUCATORS

Aids to Education, General Motors Corporation, Public Relations Staff, General Motors Building, Detroit, MI 48202, FREE.

Catalog of Free Teaching Materials, Catalog of Free Teaching Materials, P. O. Box 1075, Ventura, CA 93001, $2.68.

Education Guide to Free Science Materials, Educators Progress Service, Randolph, WI 53956, $8.25.

Education PL 31, Superintendent of Documents, Government Printing Office, Washington, D.C. 20402, FREE.

Educators Grade Guide to Free Teaching Aids, 1975, Patricia H. Suttles, M.B.A., and Raymond H. Suttles, B.S., eds., Educators Progress Service, Inc., Randolph, WI 53956, $23.00.

Educators Guide to Free Films, 1975, John C. Diffor and Mary F. Horkheimer, Educators Progress Service, Randolph, WI 53956, $12.75.

Educators Guide to Free Filmstrips, 1975, John C. Diffor and Mary F. Horkheimer, Educators Progress Service, Randolph, WI 53956, $9.25.

Elementary Guide to Free Curriculum Materials, Educators Progress Service, Randolph, WI 53956, $8.75.

Free and Inexpensive Learning Materials, 17th ed., 1974, Office of Educational Services, George Peabody College for Teachers, Nashville, TN 37203, $3.50. (An 18th edition will be available in early 1976.)

Free and Inexpensive Materials, National Council of Teachers of Mathematics, 1201 Sixteenth Street, N.W., Washington, D.C. 20036, FREE.

Free and Inexpensive Materials for Teaching Family Finance, Council for Family Financial Education, Twin Towers, Silver Spring, MD 20910, $1.00.

Free Learning Materials for Classroom Use, Extension Service, State College of Iowa, Cedar Falls, IO, $1.00.

Scientific Tests, Standards, Mathematics, Physics PL 64, Superintendent of Documents, Government Printing Office, Washington, D.C. 20402, FREE.

Selected Free Materials for Classroom Teachers, 5th ed., 1975, Ruth St. Aubrey, Fearon Publishers, 6 Davis Drive, Belmont, CA 94002, $2.50.

Sources of Free and Inexpensive Teaching Aids, Bruce Miller Publications, Box 369, Riverside, CA 92502, $.60.

Teachers' Resource Reference, New York State Petroleum Council, 142 State Street, Albany, NY 12207, FREE.

Teaching Aids, School and College Relations, New York Stock Exchange, 11 Wall Street, New York, NY 10005, FREE.

Where to Get and How to Use Free and Inexpensive Teaching Aids, Atherton Press, 70 Fifth Avenue, New York, NY 10011, $2.00.

NEWSLETTERS AND MISCELLANEOUS TEACHERS' BOOKLETS

Aids for Mathematics Education—Space-Oriented Mathematics for the Early Elementary Grades, Circular No. 741. A 50-page illustrated booklet on space-oriented mathematics. It contains a series of lessons ranging from helping Col. Glenn with the countdown to finding weights and distances in space. The pages are perforated and may be removed and duplicated.
    Office of Education
    Publications Distribution Unit
    Room SB–5026
    U.S. Department of Health, Education and Welfare
    Washington, D.C. 20202

Education Today—Mathematics. Four-page bulletins published for the information of teachers. They contain diagrams, graphs, and suggestions for enriching elementary school mathematics. Included are: "The Meaning of Sets—#46," "Statistics in the Elementary Grades—#47," and "Mathematics in the Kindergarten?—#48."
    Charles E. Merrill Publishing Company
    1300 Alum Creek Drive
    Columbus, OH 43216

Elementary School Notes—Mathematics Issue. A 4-page newsletter published at infrequent
   intervals. The 1970 issue was entitled, "Modern Math—Making It Relevant."
> Ginn and Company
> 125 Second Avenue
> Waltham, MA 02154

Every Teacher Is A Health Teacher. Two 4-page leaflets with sections that show how health can be
   used in the mathematics curriculum. Losing and gaining weight and counting calories are
   mentioned.
> Health and Welfare Division
> Metropolitan Life Insurance Company
> One Madison Avenue
> New York, NY 10010

Math Lab Tips. A newsletter that presents articles on mathematics laboratories, open education,
   activities for the mathematics classroom, teaching lessons, activity cards, etc.
> Olivetti Education Center
> 155 White Plains Road
> Tarrytown, NY 10591

Mathematical Projects, Exhibits, Reports. A 6-page listing of mathematics projects and a bibliography useful for developing individual interests. A list of *Selected Mathematical Books* is also
   available.
> Mu Alpha Theta
> National High School & Junior College Math Club
> The University of Oklahoma
> 1000 Asp Avenue, Room 215
> Norman, OK 73069

Men of Modern Mathematics. A 24" by 144" poster that presents a mathematics time line from
   1000 AD to 1900 AD. It presents a wealth of historical information in an interesting and
   colorful way.
> International Business Machines
> 590 Madison Avenue
> New York, NY 10022

NASA Facts. A publication of the National Aeronautics and Space Administration distributed
   at irregular intervals. Such issues as "Space Navigation (NF-37)" and "Orbits and Revolutions
   (S-7)" are useful for mathematics classes.
> Publications Distribution
> FAD-1
> National Aeronautics and Space Administration
> Washington, D.C. 20546

Photograph: Vital Tool in Reaching The Moon. A series of 12 beautiful photographs on 9" by
   12" sheets. Included are pictures of the moon, the Gemini spacecraft, and the earth. Useful
   for a bulletin board display concerning mathematics in space exploration.
> Eastman Kodak Company
> 343 State Street
> Rochester, NY 14650

*Ranger Rithmetic for First and Second Grade Teachers. Ranger Rithmetic for Third Grade
   Teachers. Ranger Rithmetic for Fourth Grade Teachers. Ranger Rithmetic for Fifth Grade
   Teachers. Ranger Rithmetic for Sixth Grade Teachers. Ranger Rithmetic for Seventh
   Grade Teachers.* A series of 12-page booklets that contain problems about forestry and con-
   servation for use in mathematics classes. A good aid for environmental education.
> United States Department of Agriculture
> Forestry Service
> Washington, D.C. 20250

The Resourceful Teachers—Mathematics. A 4-page newsletter published at infrequent intervals.
   No. 1 is entitled "A Sensible Approach to Functions in the Elementary Grades" and defines
   function and gives a number of illustrations. No. 2 is entitled "Evaluating Mathematics Programs for Grades 1–8" and contains a checklist that can be used in picking an elementary
   mathematics textbook.

Silver Burdett Company
Morristown, NJ 07960

Teaching and Learning in Elementary Mathematics. A newsletter covering various topics in elementary mathematics.

Harcourt, Brace and World, Inc.
757 Third Avenue
New York, NY 10017

Teaching Trends. An occasional newsletter that discusses some current aspect of education. The 1971 mathematics issue is entitled "Mathematics and the Computer."

Scott, Foresman and Company
99 Bauer Drive
Oakland, NJ 07436

Tests and Measurement Kit. A collection of booklets that are not specifically mathematics but would help a teacher learn more about evaluation. The set includes "Making the Classroom Test," "Short-cut Statistics for Teacher Made Tests," and "Multiple-Choice Questions: A Close Look."

Educational Testing Service
Princeton, NJ 08540

Timely Topics in Modern Mathematics. An occasional newsletter on enrichment topics. Problems, suggestions, and games are given. No. 1 is "Fun with Base Two" and No. 2 is "An Unsophisticated Approach to Computation with Logarithms."

Houghton-Mifflin Company
53 West 43rd Street
New York, NY 10036

BOOKS

There are many books written for children in the primary grades that emphasize mathematical concepts. Following is a very selective and abbreviated list of some of the books that are generally found in elementary school libraries or in the children's section of public libraries.*

1. Ambler, Gifford C. *Ten Little Foxhounds.* New York: Grosset & Dunlap, Inc., 1958. (number order)
2. Barr, Jene, *Big Wheels – Little Wheels.* Chicago: Albert Whitman & Co., 1955. (size, shape)
3. Clark, Ann Nolan. *Third Monkey.* New York: The Viking Press, Inc., 1956. (ordinals 1st-3rd)
4. DeRegniers, Beatrice S. *Cats, Cats, Cats, Cats.* New York: Pantheon Books, Inc., 1958. (number)
5. Jean, Priscilla. *Pattie Round and Wally Square.* Astor-Honor, Inc., 1965. (patterns, shapes)
6. Kohn, Bernice. *Everything Has a Shape and Everything Has a Size.* Englewood Cliffs, N.J.: Prentice-Hall, Inc., 1964. (pattern, shape, size)
7. Nic Leodhas, Sorche. *All in the Morning Early.* New York: Holt, Rinehart & Winston, Inc., 1963. (number, counting, order)
8. Scott, Louise, and J. J. Thompson, *Rhymes for Fingers and Flannel-Boards.* New York: McGraw-Hill Book Company, 1960. (rhymes for number)
9. Sefridge, Oliver G. *Fingers Come In Fives.* Boston: Houghton Mifflin Co., 1966. (counting)
10. Seuss, *Dr. McElligot's Pool.* New York: Random House, Inc., 1947. (number, comparison, shape, size, length, height)

*See the Teacher's Edition of the Beginners' Book of *Modern Mathematics Through Discovery*, Silver Burdett Company.

11. True, Louise. *Number Men*. Chicago: Childrens Press, Inc., 1948. (numbers 1–10; writing numbers 1–10)
12. Ungerer, Tomi. *Snail, Where Are You?* New York: Harper & Row, Publishers, 1962. (pattern, shape)
13. Wing, H. R. *Ten Pennies for Candy*. New York: Holt, Rinehart & Winston, Inc., 1963. (money)
14. Ziner, Feenie, and P. Galdone. *Counting Carnival*. New York: Coward-McCann, Inc., 1962. (counting)
15. Zolotow, Charlotte. *One Step, Two*. New York: Lothrop, Lee & Shepard Co., Inc., 1955. (number comparison, size, shape)

SUPPLIERS OF CLASSROOM MATERIALS

Academic Industries, Inc.
1754 Walton Avenue
Bronx, NY 10453
Addison-Wesley Pub. Co.
Reading, MA 01867
Aero Educational Services
St. Charles, IL 60174
Concept Company
P. O. Box 273
Belmont, MA 02178
Continental Press
Elizabethtown, PA 17022
Cooper Bros. Co.
24 Kinkel Street
Westbury, NY
Creative Playthings
Princeton, NJ 08540
Cuisenaire Company of
America
9 Elm Avenue
Mount Vernon, NY
10550
Edmund Scientific Co.
Barrington, NJ 08007
Education Service, Inc.
P. O. Box 219
Stevensville, MI 49127
Encyclopaedia Britannica
Press
425 North Michigan
Chicago, IL 60611
ETA School Materials Division
159 East Kinzie Street
Chicago, IL 60610
Frederick Post Co.
Chicago, IL 60690
Garrard Publishing Co.
Champaign, IL
Ginn and Co.

72 Fifth Avenue
New York, NY 10011
J. L. Hammet Co.
2392 Vaux Hall Road
Union, NJ 07083
Harcourt, Brace & World
757 Third Avenue
New York, NY 10017
Harper and Row, Publishers
Elmsford, NY
Harr Wagner
609 Mission Street
San Francisco, CA
Hayes School Publishing Co.
201 Rebecca Avenue
Wilkinsburg, PA 15221
Herder and Herder, Inc.
232 Madison Avenue
New York, NY 10017
Holt, Rinehart and Winston,
Inc.
383 Madison Avenue
New York, NY 10017
Houghton Mifflin Co.
53 West 43rd Street
New York, NY 10036
IMOUT Arithmetic Drill
Games
706 Williamson Bldg.
Cleveland, OH 44114
Imperial Products
Kankakee, IL 60901
Instructo Corp.
Paoli, PA 19301
Laidlaw Brothers
Thatcher and Madison
River Forest, IL 60305
Geyer Instructional Aids Co.
1229 Maxine Drive
Fort Wayne, IN 46807

Math-Master Labs
Division of Gamco Industries, Inc.
Box 310
Big Spring, TX 79720
Math Media Division
H and M Associates
P. O. Box 1107
Danbury, CT 06810
Minnesota Mining and Manufacturing Co. (3M)
Visual Products Division
2501 Hudson Road
St. Paul, MN 55101
N.C.T.M. (National Council
of Teachers of Mathematics)
1201 16th St., N.W.
Washington, D.C. 20036
F. A. Owen Pub. Co.
Instructor Park
Danville, NY 14437
Pickett, Inc.
Pickett Square
Santa Barbara, CA 93102
Scholastic Book Services
904 Sylvan
Englewood Cliffs, NJ
07632
Science Research Associates
(SRA)
259 E. Erie Street
Chicago, IL 60611
Scott Foresman and Co.
Oakland, NJ 07436
Selective Educational Equipment (SEE)
3 Bridge Street
Newton, MA 02195

Sigma Enterprises, Inc.
  Box 15485
  Denver, CO 80215
Systems for Education, Inc.
  612 N. Michigan Avenue
  Chicago, IL 60611
TUF
  P. O. Box 173
  Rowayton, CT 06853
Madison Project
  8356 Big Bend Blvd.
  St. Louis, MO 63119
A. C. Vroman, Pub.
  367 S. Pasadena Avenue
  Pasadena, CA

J. Weston Walch, Pub.
  Box 1075
  Portland, ME 04104
Walter Teaching Programs
  and Teaching Aids
  720 Fifth Avenue
  New York, NY 10019
Webster Division, McGraw-
  Hill
  Manchester Road
  Manchester, MO 63011
or: 330 West 42nd St.
  New York, NY 10036
Wff'n Proof
  P. O. Box 71

New Haven, CT 06501
John Wiley & Sons, Inc.
  One Wiley Drive
  Somerset, NJ 08873
World Wide Games
  Delaware, OH 43015
Xerox Education Division
  Learning Materials, Inc.
  600 Madison Avenue
  New York, NY 10022
or: P. O. Box 601
  Rochester, NY 14602
Yoder Instruments
  East Palestine, OH 44413

NAME INDEX *

Adams, John Quincy, 201
Adkins, Bryce E., 392, 394
Ainsworth, Nathan, 325
Albert, Irwin, 158
Ando, Masue, 91
Archimedes, 292
Armstrong, Charles, 267
Armstrong, Jenny R., 371
Austwick, G., 272

Ball, Joyce, 159
Ball, Linda, 161
Barton, A., 23
Baumgartner, Margery, 241
Beaton, Mary, 26
Beberman, Max, 30
Beeson, Lenon, 271
Beougher, Elton E., 332
Biggs, Edith E., 31, 33, 288
Blackwell, F. Frank, 31
Blatt, Mary M., 208
Bloom, Benjamin S., 413, 428
Blough, Glenn O., 256
Bohan, Harry, 138
Bowles, D. Richard, 200
Bradwell, Julia, 271
Brown, V. K., 186
Brownell, William A., 138
Bruner, Jerome, 76, 371
Buccos, Barbara J., 302
Burt, Bruce C., 164
Burt, Sir Cyril, 437

Cacha, Frances B., 98
Callahan, John J., 379
Calvo, Robert, 268
Cantor, 45
Cappon, John, Sr., 392
Carter, Ron, 271
Cathcart, W. G., 297
Chabe, Alexander, 145
Clarke, R., 272
Cohen, Louis S., 121

*Names listed in each chapter
bibliography not included.

Cooper, Frayda F., 346
Costello, S., 271
Cotter, Stanley, 116
Crawford, Ronald W., 199

D'Augustine, Charles H., 312
Davis, Robert, 17
Deans, Edwina, 144
Descartes, Rene, 273
Deutsch, Martin, 365
Dienes, Zoltan, 27, 42, 293
DiSpigno, Joseph, 106
Dixon, J. A., 272
Duncan, Ernest R., 131

Einstein, 4
Eiss, Albert, 187
Engel, A., 272
Euclid, 4, 27

Fernhoff, Robert, 359
Fibonacci, Leonardo, 325
Fielker, David S., 270

Gagne, Robert M., 419
Gauss, Karl Frederick, 3
Gibbs, Glenadine E., 138
Ginther, John L., 302
Glaser, R., 428
Godsave, Bruce F., 273
Goulding, E. M., 42
Green, Roberta, 376

Hamilton, E. W., 80
Hammer, R., 23
Hawins, R. W., 271
Heard, Ida Mae, 153, 276
Hebb, Donald O., 366
Herod, Joan, 94
Hervey, Margaret, 69
Hewitt, Frances, 392, 394
Hilbert, David, 27
Higgins, Jon L., 243
Hives, B. M., 42, 44

Hogan, James R., 316
Hugget, Albert J., 256

Ikeda, Hitoshi, 91

Jacobson, Ruth S., 379
Jefferson, Thomas, 200
Johnson, Donovan A., 138
Johnson, Harry C., 142

Kael, P., 23
Kaplan, Jerome D., 367
Karlin, Marvin, 398
Kieren, T. E., 178
King, Irv, 75
Kline, Morris, 448
Knapp, Clifford E., 330
Knight, Carlton W., II, 198
Knowles, Evelyn, 42
Koegler, Ronald, 365
Kohl, H. R., 23
Kolesnick, Teodore, 144
Krich, Percy, 142
Kurlik, Stephen, 278

LaGanke, Lucile, 65
Laplace, 3
Leibnitz, 13
Lemaitre, 261
Lemmon, Robert A., 328
Liedtke, W., 178, 297
Lincoln, Abraham, 201
Lindval, E. M., 428
Litwiller, B., 69
Luth, Lois M., 113

McIntosh, Alistair, 87, 89, 121
McMeen, George, 190
Maertens, Norbert, 419
Matthews, Geoffrey, 295
Mauthe, Albert H., 215
May, Lola J., 104, 281, 303, 305, 307, 385

475

Mercier, Raymond, 271
Milne, Esther, 248, 283
Montessori, Maria, 365
Moser, James M., 48
Myers, Donald E., 72

Nelson, L. D., 297

Oates, F. H. C., 271
Omejc, Eve, 128
Overton, B., 271

Page, David, 405
Papendick, A., 272
Papy, Frederique, 261
Parker, Robert, 250
Parten, M. B., 367
Pascal, Blaise, 161
Paschal, Billy J., 365
Payne, Joseph, 172
Peters, Charlotte, 154
Peterson, Wayne, 59
Phillips, Jo McKeeby, 349
Piaget, 28, 76
Pierce, Benjamin, 3
Pierson, R. C., 155
Polya, George, 20

Prielipp, Robert W., 97

Quast, W. G., 388

Rahmlow, Harold F., 63
Ranucci, Ernest R., 167, 189
Ray, William J., 357
Reid, Constance, 45
Richardson, Elwyn, 31
Riedesel, Alan, 210, 414
Riemann, 4
Roberge, James J., 52, 229
Robins, John A., 199
Rockcastle, Verne N., 342
Rode, Joann, 136
Ruchlis, Hy, 319
Russell, Bertrand, 4

Schaeffer, Anne W., 215
Schall, William E., 256, 316, 339, 406
Schmidt, Harold, 371
Schminke, Clarence, 419
Schrage, Merry, 95
Schweitzer, James P., 198
Scott, J., 24
Sherzer, Laurence, 111

Shipley, Sara, 254
Silvey, Ida Mae, 81
Sims, Jacqueline, 223
Smart, F. O., 272
Smith, C. Winston, Jr., 78
Stephens, Jo, 437
Strangman, Kathryn B., 194
Sueltz, Ben A., 3
Sullivan, John J., 182
Suydam, Marilyn N., 431
Swart, William L., 184
Swartz, Evelyn, 351

Tomashow, Beatrice E., 353
Trivett, John, 239
Trueblood, Cecil R., 227, 427

Van Engen, Henry, 75, 138

Walter, Marion, 272
Washington, George, 200
Weaver, Fred, 431
West, Tommie A., 416
Wheeler, David, 441
Whitehead, A. N., 288
Whittaker, Dora, 31
Williams, H. C., 271
Wilson, 227

SUBJECT INDEX